Estimators for Uncertain Dynamic Systems

Mathematics and Its Applications

Volume 458

Estimators for Uncertain Dynamic Systems

by

A.I. Matasov

Faculty of Mechanics and Mathematics,
Department of Applied Mechanics and Control,
M.V. Lomonosov Moscow State University,
Moscow, Russia

KLUWER ACADEMIC PUBLISHERS
DORDRECHT / BOSTON / LONDON

A C.I.P. Catalogue record for this book is available from the Library of Congress

ISBN 0-7923-5278-5

Published by Kluwer Academic Publishers,
P.O. Box 17, 3300 AA Dordrecht, The Netherlands.

Sold and distributed in North, Central and South America
by Kluwer Academic Publishers,
101 Philip Drive, Norwell, MA 02061, U.S.A.

In all other countries, sold and distributed
by Kluwer Academic Publishers,
P.O. Box 322, 3300 AH Dordrecht, The Netherlands.

Printed on acid-free paper

Printed in the Netherlands.

Table of Contents

Preface

When solving the control and design problems in aerospace and naval engineering, energetics, economics, biology, etc., we need to know the state of investigated dynamic processes. The presence of inherent uncertainties in the description of these processes and of noises in measurement devices leads to the necessity to construct the estimators for corresponding dynamic systems. The estimators recover the required information about system state from measurement data. An attempt to solve the estimation problems in an optimal way results in the formulation of different variational problems. The type and complexity of these variational problems depend on the process model, the model of uncertainties, and the estimation performance criterion. A solution of variational problem determines an optimal estimator.

Howerever, there exist at least two reasons why we use nonoptimal estimators. The first reason is that the numerical algorithms for solving the corresponding variational problems can be very difficult for numerical implementation. For example, the dimension of these algorithms can be very high. Sometimes the solving algorithms are unknown. The second reason is that, in applied problems, the algorithms that are optimal for a one noise model must operate under another, more complex assumptions about noises. Thus the simplified estimation algorithms should be used. In this case, it is important to evaluate the level of nonoptimality for these nonoptimal algorithms. Since the original variational problems can be very difficult, we must obtain the estimate of nonoptimality *without solving the original variational problem.* If the loss in optimality is not great, for example, about 10-20 percent, the use of the simplified algorithm is justified.

A one more circumstance that is of great importance in applied problems should be emphasized. The system models often are not quite accurate. The models of noises are inaccurate as a rule. The matter is as follows. On the one hand, the mathematicians are not the experts in a specific phenomena; so, they can not have an exact idea about noises. On the other hand, the engineers, which are the experts, have quite informal idea about the noises. Thus there is a gap between the actual noises and the models of the noises. Therefore a designer, when he makes his decision on the acceptance or rejection of an estimation algorithm, mainly relies upon the orders of the quantities that describe the system performance than on the exact characteristics of inaccurate models. The designer is interested in the question: whether the loss in optimality is critical? From this point of view, if the loss in optimality is 100 or even 200 percent, the simplified algorithm may be also accepted.

Our approach consists of the following steps: (a) derivation of a simplified algorithm (or choice of an already known algorithm) which can be easily implemented; (b) computation of a guaranteed level of nonoptimality for the chosen algorithm with the help of simple numerical means (in particular, without solving the difficult original problem); (c) use of the simplified algorithm with the known guaranteed level of nonoptimality instead of the optimal one.

It is well known that the use of the methods of functional analysis in automatic control is highly effective. The mathematical ground for our approach is the duality theory of infinite-dimensional convex extremal problems. This monography is likely to be the first book devoted to the construction of the guaranteed levels of nonoptimality for simplified estimation and control algorithms in dynamic systems. Most of the results of the book are original. Some parts of the book were presented in special courses given by the author at the Faculty of Mechanics and Mathematics, M.V. Lomonosov Moscow State University for graduated and post-graduated students.

In this monography we consider the linear dynamic systems. A considerable part of the book is devoted to guaranteed estimation problems. In particular, these problems arise when the system perturbations and measurement noises are unknown but bounded deterministic functions. In this case, the estimation error depends on the samples of perturbations and noises. Therefore, for an estimation performance, it is natural to take a maximum of estimation error over all admissible samples. This maximum is called the guaranteed estimation error. So, the optimal guaranteed estimation problem is to minimize the guaranteed estimation error. Clearly, this is a minimax problem.

Chapter 1 is of an introductory character. In this chapter, we consider some typical guaranteed estimation problems. Necessary and sufficient conditions for the optimality of estimators for deterministic systems are obtained. With the help of these conditions the exact solutions of some simple guaranteed estimation problems are constructed. Linear programming methods for numerical solving a general class of guaranteed estimation problems are breifly discussed.

The main idea of our approach is expounded for the simplest static systems, i.e., for systems with constant state vector. The optimal guaranteed *a priori* estimation problem for static linear systems with unknown but bounded measurement noises is considered. It reduces to a nonsmooth variational problem. We propose to change the nonsmooth problem by an appropriate linear-quadratic one. It is equivalent to the use of the least-square method for the estimation. With the help of the duality theory for convex variational problems the bound for the nonoptimality level of the least-squares method in the optimal guaranteed problem is obtained. Simple analytical examples verify the effectiveness of our approach. A new formalization of the guaranteed estimation problem with outliers in measurements is also presented.

In Chapter 2 the optimal deterministic *a priori* guaranteed estimation problem for linear dynamic systems is considered. The basic facts from the optimization theory and convex analysis are collected in a separate section. The optimal guaranteed estimation problem reduces to a quite difficult nonsmooth variational problem. The necessary and sufficient conditions for the optimality of estimator for dynamic systems are obtained. The optimality of linear estimators is established.

In Chapter 3 we use the Kalman filter for an approximate solving the guaranteed estimation problem. The Kalman filter is easy for numerical implementation. Using the duality theory, we obtain the bound for the nonoptimality level of the Kalman filter. It is important that this bound can be calculated basing on the Kalman filter characteristics only, *not solving the initial complicated problem.* Nontrivial numerical examples are investigated in detail.

In Chapter 4 the previous results are extended to the stochastic optimal guaranteed *a priori* estimation problem where, along with unknown bounded perturbations, the white noises with uncertain intensities are also present. The corresponding variational problem is very difficult and our estimate of nonoptimality seems to be the only effective tool to justify the use of the Kalman filter. The classical stochastic mean-square filtering problem is also studied. The sensitivity analysis of the Kalman filter to the uncertainty in the intensities of white noise covariances is performed.

The construction of the optimal mean-square estimate for the state of linear dynamic system requires the accurate values of noise covariance intensities for its implementation. In applied problems an uncertainty in these intensities is always present. So, a designed Kalman filter with designed intensities is used. The estimation quality of the designed filter can not be evaluated because of very complicated dependence of this quality on true intensity parameters (especially for the case of time-varying intensities). An upper bound for the sensitivity of the designed filter is constructed. This bound has a considerably simpler dependence on the true covariance intensities. Moreover, it can be calculated basing on the designed filter characteristics only.

Chapter 5 is devoted to estimation and control problems for linear dynamic systems with aftereffect. These problems are much more complicated than the estimation problems for systems without delay. So, the simplification of optimal algorithms and the construction of the levels of nonoptimality are still more desirable. The optimal guaranteed filtering problems for dynamic systems with delay (both deterministic and stochastic) are considered. In accordance with our approach, we change the corresponding nonsmooth variational problems by an appropriate linear-quadratic one. This linear-quadratic problem is equivalent to a corresponding stochastic mean-square filtering problem (as for systems without delay). The linear-quadratic problem has a constructive numerical solving algorithm. The results of Chapters 2 and 3 are generalized to dynamic systems with delay. In this case, the proofs require substantional modifications.

The stochastic mean-square filtering problem for dynamic systems with delay is also investigated. In this problem the disturbances are white noises (as for the classical filtering problems). Though the optimal estimator in the stochastic mean-square filtering problem can be constructed, the implementation of the optimal estimator requires the solving of the system of functional partial differential equations. Therefore, it is highly desirable to simplify the optimal algorithm, having excluded the block of solving the system of functional partial differential equations. These simplified algorithms for solving the filtering problem are obtained. With the help of the duality theory the estimates of the nonoptimality levels for the simplified algorithms are constructed. These estimates can be calculated *without solving the initial optimal filtering problem.* Moreover, a new state-space representation of the optimal filter is obtained. It follows from this representation that the optimal filter for systems with delay is an immediate extension of the classical Kalman filter in its form as well.

The control problems for systems with lumped and distributed delay with the quadratic cost functional and with L_1-cost functional are studied. They are equivalent to the corresponding filtering problems. The levels of nonoptimality for simplified control laws are also constructed.

Thus, we obtain reliable and relatively simple estimation and control algorithms, that provide the "key" solutions to the posed problems with the evaluation of the influence of neglected factors. The main feature of our approach is that the estimates for the levels of nonoptimality for the simplified algorithms can be calculated with considerably lesser efforts than the exact solutions of the corresponding initial problems.

We hope that our book will be of interest to researchers and engineers who deal with estimation and control systems. Besides, the mathematicians who are the experts in the optimization theory and the theory of functional differential equations will find new applications of these theories to automatic control. A student with good knowledge of calculas would find the book accesible, but the material of the monography would also be appropriate for two-semester graduate and post-graduate courses.

The results presented in this book were initiated by the works by Profs. M.L. Lidov and N.A. Parusnikov. Our treatment for systems with aftereffect was pioneered by Prof. V.B. Kolmanovskii.

The author is deeply indebted to Profs. V.V. Alexandrov, E.A. Devyanin, A.M. Formalskii, V.B. Kolmanovskii, L.M. Krol, M.L. Lidov, L.A. Mironovskii, V.M. Morozov, N.A. Parusnikov, Ya.N. Roytenberg, who taught him. The author is also grateful to his colleagues Profs. B.Ts. Bakhshiyan, L.Yu. Belousov, A.A. Golovan, V.N. Soloviov for discussions and friendly help.

This book was written under invaluable support by Profs. V.B. Kolmanovskii and N.A. Parusnikov. The work was partially supported by the Russian Foundation for Basic Research (Grant no. 96-01-00056).

Chapter 1

Guaranteed Parameter Estimation

Classical estimation problems are based on the assumption that the probabilistic distributions of disturbances are known or, at least, there are known such statistical characteristics as mathematical expectations and covariance matrices. If the system is described by ordinary differential equations and the disturbances are white Gaussian noises, then the optimal estimate (i.e., the estimate with a minimal variance of estimation error) is yielded by the well-known Kalman filter [84], [41], [79], [143], [47]. However, in many applications the stochastic description is not complete. For example, for many unique measurement systems (which are used, for example, for the ensuring of cosmic experiments) there is no sufficient amount of experimental data. Therefore, it is impossible to determine the statistical characteristics with required accuracy and reliability. Moreover, sometimes even the fact of statistical stability of experimental data is questionable. In the latter case, the stochastic description is impossible in principle. Therefore, along with the statistical description, a researcher has to use a set-membership description of uncertainty.

In 60's there appeared nonclassical estimation methods that are called guaranteed (minimax, robust, stable). Extensive literature is devoted to these methods: see, e.g., [130], [113], [189], [215], [117], [190], [123], [72], [43].

Below, an elementary introduction to the guaranteed estimation is given. Some typical guaranteed estimation problems are discussed. A new idea for the construction of estimators is presented for simplest guaranteed estimation problems. Other chapters are devoted to the use of this idea in more complex deterministic and stochastic dynamic systems, in particular, in systems with aftereffect.

1. Simplest Guaranteed Estimation Problems

In this book we mainly follow a trend in guaranteed estimation initiated by M.L. Lidov [130]. A survey of this trend is presented in [138] and [137]. To expose the idea of the guaranteed estimation, in this chapter we discuss the simplest problem statements. We start with the problem that is called "box scheme" [130].

1.1. *"Box Scheme"*

Let us consider a simplest guaranteed estimation problem. Assume that the equations for measurements have the form

$$z_k = H_k' q + \varrho_k, \qquad k = 1, \ldots, N, \tag{1.1}$$

where z_k are measurements, $q \in \mathbf{R}^n$ is an unknown vector parameter, $H_k \in \mathbf{R}^n$ are given vectors, ϱ_k are measurement errors; for typical cases, $n \ll N$. A prime denotes the transposition sign.

The problem is to estimate a parameter

$$l_* = a'q, \tag{1.2}$$

where $a \in \mathbf{R}^n$ is a given vector. For example, if $a = (1, 0, \ldots, 0)'$ then $l_* = q_1$ and it is required to estimate the first component of q.

Suppose that there is a lack of experimental data for statistical description of ϱ_k. So, we assume that the measurement error vector $\varrho = (\varrho_1, \ldots, \varrho_N)'$ belongs to a parallelepiped

$$\mathcal{P} = \left\{ \varrho \in \mathbf{R}^N \;\middle|\; |\varrho_k| \le \sigma_k, \quad k = 1, \ldots, N \right\}, \tag{1.3}$$

where σ_k, $\quad k = 1, \ldots, N$ are given nonnegative numbers.

Consider a linear estimator

$$\hat{l} = \sum_{i=1}^{N} \Phi_k z_k, \tag{1.4}$$

where Φ_k are estimator weight coefficients that should be determined. Since the estimator (1.4) is completely defined by its weight coefficients $\{\Phi_k\}_1^N$, we refer to the vector $\Phi = (\Phi_1, \ldots, \Phi_N)'$ also as an estimator.

The estimation error $\hat{l} - l_*$ depends on the estimator Φ and the realization of measurement noise ϱ. In accordance with the guaranteed approach, the estimation quality is defined by a guaranteed estimation error

$$d(\Phi) = \max_{q \in \mathbf{R}^n,\ \varrho \in \mathcal{P}} |\hat{l} - l_*|. \tag{1.5}$$

The optimal *a priori* guaranteed estimation problem is to find an estimator Φ^0 such that

$$d(\Phi^0) \le d(\Phi), \qquad \forall\, \Phi \in \mathbf{R}^N. \tag{1.6}$$

So, the problem (1.1)–(1.6) is a minimax problem

$$I_0 = \min_{\Phi \in \mathbf{R}^N} \max_{q \in \mathbf{R}^n,\, \varrho \in \mathcal{P}} |\hat{l} - l_*|. \tag{1.7}$$

REMARK 1.1. Consider the estimation problem in a linear dynamic system

$$\frac{d}{dt}x(t) = A(t)x(t), \qquad z(t_k) = \bar{H}'(t)x(t_k) + \varrho(t_k), \quad k = 1, \ldots, N,$$

where $l_* = a'x(T)$. Using a fundamental matrix $\theta(t)$, we represent the measurements in the form

$$z(t_k) = H'(t_k)q + \varrho(t_k), \qquad H(t_k) = \left(\theta^{-1}(T)\right)' \theta'(t_k)\tilde{H}(t_k), \quad q = x(T).$$

For example, if A is constant, then

$$\theta(t) = \exp(At), \qquad H(t_k) = \exp(A'(t_k - T))\tilde{H}(t_k).$$

Thus if disturbances do not act on the dynamic sytem, then the estimation problem reduces to the static problem (1.1).

REMARK 1.2. The problem (1.5), (1.6) is called *a priori* problem because the maximum in (1.5) is taken over all possible realizations of measurements $z = (z_1, \ldots, z_N)'$. Another, *a posteriori* guaranteed estimation problem could be also considered [190], [123], [43]. In this problem statement, z is fixed, and the uncertainty in q and ϱ must be restricted by constraints (1.1) with the unique fixed measurement vector $z = (z_1, \ldots, z_N)'$. In our book, only *a priori* guaranteed estimation problem will be considered (except for Section 9).

Let us reduce the guaranteed estimation problem to a variational problem. It follows from (1.1), (1.2), (1.4) that

$$\hat{l} - l_* = \sum_{i=1}^{N} \Phi_i z_i - a'q = \sum_{i=1}^{N} \Phi_i \left(H_i'q + \varrho_i\right) - a'q \tag{1.8}$$

$$= \sum_{i=1}^{N} q'\left(H_i\Phi_i - a\right) + \sum_{i=1}^{N} \Phi_i \varrho_i.$$

Therefore,

$$d(\Phi) = \begin{cases} \sum_{i=1}^{N} \sigma_i |\Phi_i| & \text{if } \sum_{i=1}^{N} H_i \Phi_i = a \\ \infty & \text{otherwise.} \end{cases} \tag{1.9}$$

Thus the optimal estimator $\Phi = (\Phi_1, \ldots, \Phi_N)'$ is a solution of the following variational problem

$$\sum_{i=1}^{N} \sigma_i |\Phi_i| \to \min_{\Phi} \tag{1.10}$$

subject to

$$\sum_{i=1}^{N} H_i \Phi_i = a. \tag{1.11}$$

The constraint (1.11) is called *the unbiasness condition.* If the measurements are exact, i.e., $\varrho = 0$, and (1.11) holds, then it follows from (1.8) that $\hat{l} = l_*$ for any q.

It will be shown in Section 3 that the problem (1.10), (1.11) is equivalent to a linear programming problem with m constraints of equality type (see, e.g., [46]). So, the solution of the guaranteed estimation problem can be computed by well-known numerical procedures. This fact is very important for applications.

Let us describe a stochastic version of the "box scheme". Assume that, instead of deterministic restrictions (1.3), the measurement errors are random variables with zero means and bounded nonnegative variances but with unknown covariances:

$$\mathbf{E}\varrho_i = 0, \quad \mathbf{E}\varrho_i^2 \le \sigma_i^2, \quad |\mathbf{E}\varrho_i \varrho_j| \le (\mathbf{E}\varrho_i^2)^{\frac{1}{2}} (\mathbf{E}\varrho_j^2)^{\frac{1}{2}}, \quad i, j = 1, \ldots, N. \tag{1.12}$$

Denote by $\mathcal{P}_S$ the set of all random vectors ϱ satisfying (1.12). The unknown vector q is still deterministic.

Consider also linear estimates (1.4) and introduce the guaranteed value of estimation error variance

$$D(\Phi) = \sup_{q \in \mathbf{R}^n,\ \varrho \in \mathcal{P}_S} \mathbf{E}(\hat{l} - l_*)^2. \tag{1.13}$$

Obviously, given estimator Φ,

$$\mathbf{E}(\hat{l} - l_*)^2 = \Big(q'\big(\sum_{i=1}^{N} H_i \Phi_i - a\big)\Big)^2 + \mathbf{E}\Big(\sum_{i,j=1}^{N} \varrho_i \varrho_j \Phi_i \Phi_j\Big)$$

$$= \Big(q'\big(\sum_{i=1}^{N} H_i \Phi_i - a\big)\Big)^2 + \sum_{i,j=1}^{N} c_{ij} \Phi_i \Phi_j, \quad c_{ij} = \mathbf{E}\varrho_i \varrho_j$$

(note that $\mathcal{C} = \{c_{ij}\}_1^N \ge 0$, i.e., $\mathcal{C}$ is a positive semidefinite matrix [144]).

Recall that a coefficient of correlation k_{ij} is defined by the equality $c_{ij} = \sigma_i \sigma_j k_{ij}$ [144].

Then it is easy to prove that

$$\sup_{\varrho \in \mathcal{P}_S} \sum_{i,j=1}^{N} c_{ij} \Phi_i \Phi_j = \max_{-1 \leq k_{ij} \leq 1,\ \mathcal{K}=\{k_{ij}\} \geq 0} \sum_{i,j=1}^{N} \sigma_i \sigma_j k_{ij} \Phi_i \Phi_j = \left(\sum_{i=1}^{N} \sigma_i |\Phi_i| \right)^2$$

and this maximum is attained at

$$k_{ij}^0 = \operatorname{sign}(\Phi_i \Phi_j) = \operatorname{sign} \Phi_i \operatorname{sign} \Phi_j, \qquad \operatorname{sign} x = \begin{cases} +1 & \text{if } x > 0 \\ 0 & \text{if } x = 0 \\ -1 & \text{if } x < 0. \end{cases} \tag{1.14}$$

Note that if α is a zero-mean random variable with unit variance, then the stochastic process $\varrho_i = \sigma_i \operatorname{sign}(\Phi_i)\, \alpha$ has the covariance matrix $\{c_{ij}^0\}_1^N = \{\sigma_i \sigma_j k_{ij}^0\}_1^N$, where k_{ij}^0 is defined by (1.14). Therefore k_{ij}^0 are realizable.

Since k_{ij}^0 define the worst correlations, the problem under consideration is called the "worst correlation" problem [138]. Similar to (1.9), the optimal estimator for the "worst correlation" problem is also defined by the variational problem (1.10), (1.11).

1.2. *Polyhedral Constraints*

Let us discuss a useful generalization of the "box scheme" [141]. Consider measurements (1.1) in a vector form:

$$z = H'q + \varrho, \qquad H = (H_1, \ldots, H_N), \tag{1.15}$$

$z = (z_1, \ldots, z_N)'$, $\varrho = (\varrho_1, \ldots, \varrho_N)'$. It is required to estimate the quantity (1.2). Assume that instead of component-wise restriction (1.3) the measurement errors are bounded by a polyhedral:

$$\varrho \in \mathcal{G} = \left\{ \varrho \in \mathbf{R}^N \,\middle|\, -\beta \leq C\varrho \leq \beta \right\}, \tag{1.16}$$

where C is a given $s \times N$ matrix, $s \geq N$, $\beta \in \mathbf{R}^s$ is a given vector with nonnegative elements; the notation $x \leq y$ for $x, y \in \mathbf{R}^s$ means that $x_i \leq y_i$, $i = 1, \ldots, s$. To be definite, assume also that $\operatorname{rank} H = n$ and $\operatorname{rank} C = N$.

Consider the class of linear estimates (1.4):

$$\hat{l} = \Phi' z.$$

and the optimal guaranteed estimation problem (cf. (1.7)):

$$I_0 = \min_{\Phi \in \mathbf{R}^N} \max_{q \in \mathbf{R}^n,\ \varrho \in \mathcal{G}} |\hat{l} - l_*|. \tag{1.17}$$

Problem (1.17) is a generalizaiton of problem (1.7) with $s=N$, $\beta=(\sigma_1,\ldots,\sigma_N)'$, $C=E_N$, where E_N is identity matrix.

Let us point to some important special cases of (1.17).

Variation of Bounded Noise

Suppose that the noise samples are not only bounded but also have bounded increments:

$$|\varrho_i| \le \sigma_i, \quad i=1,\ldots,N, \qquad |\varrho_{j+1}-\varrho_j| \le \nu_j, \quad j=1,\ldots,N-1. \tag{1.18}$$

This means that the noise samples cannot vary too fast. Clearly, constraints (1.18) are equivalent to (1.16) with the matrix ($s=2N-1$)

$$C=\begin{pmatrix} E_N \\ D_{N-1\times N}\end{pmatrix}, \qquad D_{N-1\times N}=\begin{pmatrix} -1 & 1 & 0 & \ldots & 0 \\ & & \ldots & & \\ 0 & \ldots & 0 & -1 & 1 \end{pmatrix},$$

and $\beta=(\sigma_1,\ldots,\sigma_N,\nu_1,\ldots,\nu_{N-1})'$. Obviously, the boundness of higher differences can be also considered.

Averaging of Noise

In some cases, the noise samples can possess an averaging:

$$|\varrho_i| \le \sigma_i, \quad i=1,\ldots,N, \qquad \left|\frac{1}{N}\sum_{i=1}^{N}\varrho_i\right| \le m_\epsilon \quad \left(m_\epsilon < \frac{1}{N}\sum_{i=1}^{N}\sigma_i\right).$$

This feature is described by the row $\left(N^{-1},\ldots,N^{-1}\right)$ in C.

Degenerate Noise

Suppose that

$$\varrho = B\zeta, \tag{1.19}$$

where B is an $N\times k$ matrix, $\zeta=(\zeta_1,\ldots,\zeta_k)'$, $|\zeta_j|\le 1$, $j=1,\ldots,k$. Supose also that $N\ge k$ and $\operatorname{rank} B=k$ [18]. Let us show that under these assumptions (1.19) reduces to (1.16).

Let us complement B with $N-k$ columns up to a nonsingular $N\times N$ matrix B_e. Let us also complement ζ up to a vector $\zeta_e=(\zeta',0'_{N-k})'$, where 0_{N-k} is an $(N-k)\times 1$ zero vector. Then,

$$B_e^{-1}\varrho=\zeta_e, \qquad \zeta_e=(\zeta_{e1},\ldots,\zeta_{eN})', \tag{1.20}$$

$$|\zeta_{ej}| \leq 1, \quad j = 1, \ldots, k, \qquad \zeta_{ej} = 0, \quad j = k+1, \ldots, N.$$

Thus, (1.20) are equivalent to (1.16) with $s = N$,

$$C = B_e^{-1}, \qquad \beta_j = 1, \quad j = 1, \ldots, k, \qquad \beta_j = 0, \quad j = k+1, \ldots, N.$$

Multiplying (1.15) by C (and assuming that C is of full rank), we get the equivalent equation:

$$\tilde{z} = \tilde{H}'q + \tilde{\varrho}, \qquad \tilde{z} = Cz, \quad \tilde{H} = HC', \quad \tilde{\varrho} = C\varrho, \tag{1.21}$$

where

$$-\beta \leq \tilde{\varrho} \leq \beta. \tag{1.22}$$

At first sight, it is obvious that the problem (1.21), (1.22) is equivalent to the initial problem (1.1), (1.3). So, the optimal estimator Φ^0 is determined by the formula

$$\Phi^0 = C'\tilde{\Phi}^0, \tag{1.23}$$

where $\tilde{\Phi}^0$ satisfies the following variational problem

$$\min_{\tilde{\Phi} \in \mathbf{R}^s} \beta'|\tilde{\Phi}|, \qquad HC'\tilde{\Phi} = a, \qquad |\tilde{\Phi}| \triangleq \left(|\tilde{\Phi}_1|, \ldots, |\tilde{\Phi}_s|\right)'. \tag{1.24}$$

But this is not so trivial by the following reason. Though inequalities (1.22) are valid, the set

$$C\varrho, \qquad \varrho \in (1.16),$$

generally, does not exhaust the s-dimensional parallelepiped

$$\{\xi \in \mathbf{R}^s \mid -\beta \leq \xi \leq \beta\}.$$

Nevertheless, the equivalence does take place.

LEMMA 1.1. *The optimal estimator Φ^0 is determined by (1.23) and (1.24).*

PROOF. Similar to (1.8),

$$\hat{l} - l = (H\Phi - a)'q + \Phi'\varrho. \tag{1.25}$$

We see from (1.25) (cf. (1.9)) that $|\hat{l} - l|$ is bounded if and only if

$$H\Phi = a. \tag{1.26}$$

Then,

$$\hat{l} - l = \Phi'\varrho. \tag{1.27}$$

Let us denote the set of estimators satisfying the unbiasness condition (1.26) by $\mathcal{E}$. Obviously, $\mathcal{G} \neq \emptyset$, as $\varrho = 0 \in \mathcal{G}$. The set $\mathcal{G}$ is centrally symmetric: if $\varrho \in \mathcal{G}$, then $-\varrho \in \mathcal{G}$. Therefore the problem (1.17), (1.26), (1.27) is equivalent to the problem

$$I_0 = \min_{\Phi \in \mathcal{E}} \max_{\varrho \in \mathcal{G}} \Phi' \varrho. \tag{1.28}$$

ASSERTION 1.1. *The following equality holds*:

$$\max_{\varrho \in \mathcal{G}} \Phi' \varrho = \min_{\tilde{\Phi} \in \mathcal{Q}} \beta' |\tilde{\Phi}|, \qquad \mathcal{Q} = \left\{ \tilde{\Phi} \,\middle|\, C' \tilde{\Phi} = \Phi \right\}. \tag{1.29}$$

Assertion 1.1 is a special case of the so-called *conjugate duality relation*. This assertion is discussed below in Section 5 (Remark 5.1).

It follows from (1.16), (1.26), (1.28), and (1.29) that

$$I_0 = \min_{\Phi, \tilde{\Phi}} \beta' |\tilde{\Phi}|, \quad \text{under constraints} \quad H\Phi = a, \quad C'\tilde{\Phi} = \Phi.$$

Obviously, these constraints are equivalent to the constraints $HC'\tilde{\Phi} = a$. □

We see from this Section that the noise models in the guaranteed estimation problems posed above are quite flexible. They cover wide and practically important classes of disturbances. Moreover, we shall later show that the corresponding variational problems reduce to linear programming problem with m equality constraints, which can be successfully solved by efficient numerical methods.

2. Continuous Measurements Case

Let us perform a continuous version of the "box scheme" [113], [117], [28]. Assume the measurements are continuous:

$$z(t) = H'(t)q + \varrho(t), \quad t \in [0, T], \quad z(t), \varrho(t) \in \mathbf{R}^1, \quad H(t), q \in \mathbf{R}^n, \tag{2.1}$$

where $z(t)$ are the measurements, $H(t)$ are given vectors, q is an unknown vector parameter, $\varrho(t)$ is a measurement noise. The components of $H(t)$, $\varrho(t)$ are assumed to be piecewise continuous and

$$\varrho \in \mathcal{P}_C, \qquad \mathcal{P}_C = \{\varrho(t) \mid |\varrho(t)| \leq \sigma(t)\}, \tag{2.2}$$

where $\sigma(t)$ is a given nonnegative piecewise continuous function.

As before, it is required to estimate the quantity (1.2) with the help of linear functionals

$$\hat{l} = \int_0^T \Phi_{pc}(t)\,dt + \sum_{s=1}^{M} \Phi_{(s)} z(t_s), \qquad t_s \in [0, T], \tag{2.3}$$

where $\Phi_{pc}(t)$ is a piecewise continuous function, $\Phi_{(s)}$ are numbers. Using the notion of the Dirac delta-function, formula (2.3) can be represented in the compact form

$$\hat{l} = \int_0^T \Phi(t) z(t)\,dt,$$

$$\Phi(t) = \Phi_{pc}(t) + \sum_{s=1}^{M} \Phi_{(s)} \delta(t - t_s), \qquad \int_0^T \delta(t - t_s) z(t)\,ds \triangleq z(t_s). \tag{2.4}$$

By $\mathcal{F}$ we denote the set of estimators $\Phi(t)$ defined by (2.4). Obviously, it forms a linear space. Equiping $\mathcal{F}$ with the norm

$$||\Phi||_{\mathcal{F}} = \int_0^T |\Phi_{pc}(t)|\,dt + \sum_{s=1}^{M} |\Phi_{(s)}|,$$

we can treat $\mathcal{F}$ as a normed space.

REMARK 2.1. The unique aim of introducing the delta-functions is to reduce a notation. Another, traditional rigorous way to represent (2.3) is to use the notion of the integral with respect to a signed Borel measure:

$$\hat{l} = \int_0^T z(t)\,d\mu_\Phi,$$

where μ_Φ is a measure such that, for any Borel set $C \subset [0, T]$,

$$\mu_\Phi(C) = \int_C \Phi_{pc}(t)\,dt + \sum_{s=1}^{M} \Phi_{(s)} \chi_C(t_s), \quad \chi_C(t_s) = \begin{cases} 1, & t_s \in C, \\ 0, & t_s \notin C. \end{cases} \tag{2.5}$$

Then $||\Phi||_{\mathcal{F}} = \operatorname{Var} \mu_\Phi$, where $\operatorname{Var} \mu_\Phi$ is the total variation of the measure μ_Φ. Since the rigorous form (2.5) does not introduce more clarity in our consideration, we prefer to use the simple "engineering" form (2.4).

The guaranteed estimation error is defined by the expression

$$d(\Phi) = \sup_{q \in \mathbf{R}^n,\, \varrho \in \mathcal{P}_C} |\hat{l} - l_*| = \sup_{q \in \mathbf{R}^n,\, \varrho \in \mathcal{P}_C} \left| \int_0^T \Phi(t) z(t)\,dt - a'q \right|$$

$$= \sup_{q \in \mathbf{R}^n,\, \varrho \in \mathcal{P}_C} \left| q' \left(\int_0^T H(t)\Phi(t)\,dt - a \right) + \int_0^T \Phi(t)\varrho(t)\,dt \right|.$$

Let us show that

$$\sup_{\varrho \in \mathcal{P}_C} \int_0^T \Phi(t)\varrho(t)\,dt = \int_0^T \sigma(t)|\Phi(t)|\,dt. \tag{2.6}$$

Clearly, for $\Phi \in \mathcal{F}$,

$$\left| \int_0^T \Phi(t)\varrho(t)\,dt \right| \leq \int_0^T \sigma(t)|\Phi(t)|\,dt = \int_0^T \Phi(t)\varrho_{\max}(t)\,dt,$$

where

$$\int_0^T \sigma(t)|\Phi(t)|\,dt \triangleq \int_0^T \sigma(t)|\Phi_{pc}(t)|\,dt + \sum_{s=1}^{M} \sigma(t_s)|\Phi(t_s)|,$$

$$\varrho_{\max}(t) = \sigma(t)\operatorname{sign}\Phi(t), \qquad \operatorname{sign}\Phi(t) \triangleq \begin{cases} \operatorname{sign}\Phi_{(s)} & \text{for } t = t_s \\ \operatorname{sign}\Phi_{pc}(t) & \text{for } t \neq t_s. \end{cases}$$

Generally, $\varrho_{\max}(t)$ is not piecewise continuous. But it can be approximated by a piecewise continuous function $\varrho_a(t)$ satisfying three conditions: (a) $|\varrho_a(t)| \leq \sigma(t)$, $t \in [0,T]$; (b) $\varrho_a(t_s) = \varrho_{\max}(t_s)$, $s = 1,\ldots,M$; (c) $\varrho_a(t) = \varrho_{\max}(t)$ at all points of $[0,T]$ except for a set of arbitrarily small Lebesgue measure.

Therefore,

$$d(\Phi) = \begin{cases} \displaystyle\int_0^T \sigma(t)|\Phi(t)|\,dt & \text{if } \displaystyle\int_0^T H(t)\Phi(t)\,dt = a \\ \infty & \text{otherwise.} \end{cases} \tag{2.7}$$

Thus the optimal estimator that minimizes the guaranteed estimation error is a solution of the following continuous variational problem

$$\int_0^T \sigma(t)|\Phi(t)|\,dt \to \inf_{\Phi \in \mathcal{F}}, \tag{2.8}$$

$$\int_0^T H(t)\Phi(t)\,dt = a. \tag{2.9}$$

The problem (2.8), (2.9) is called the *moment problem* [173], [117], [118], [183].

Certainly, a continuous analog to the "worst correlation problem" can be constructed as well. It is easy to prove that this extension also leads to the problem (2.8), (2.9) [28].

Remark 1.1 is valid for continuous measurements as well.

3. Linear Programming

The canonical linear programming problem has the form [46]

$$\sum_{i=1}^{K} c_i x_i \to \inf_x, \qquad x = (x_1, \ldots, x_K)' \in \mathbf{R}^K. \tag{3.1}$$

$$\sum_{i=1}^{K} A_i x_i = a, \qquad x_i \geq 0, \quad i = 1, \ldots, K, \tag{3.2}$$

where c_i, $i = 1, \ldots, K$, are given numbers, $A_i \in \mathbf{R}^n$, $i = 1, \ldots, K$, are specified vectors.

Let us show how the variational problem (1.10), (1.11) can be reduced to the canonical linear programming problem.

THEOREM 3.1. 1^0. *An estimator* $\Phi^0 = (\Phi_1^0, \ldots, \Phi_N^0)'$ *is a solution of the problem* (1.10), (1.11) *if and only if* $\Phi^0 = u^0 - v^0$, *where* $\{u^0, v^0\}$, $u^0 = (u_1^0, \ldots, u_N^0)'$, $v^0 = (v_1^0, \ldots, v_N^0)'$ *is a solution of the linear programming problem*

$$\sum_{i=1}^{N} \sigma_i (u_i + v_i) \to \inf_{u,v}, \tag{3.3}$$

$$\sum_{i=1}^{N} H_i (u_i - v_i) = a, \qquad u_i, v_i \geq 0, \quad i = 1, \ldots, N. \tag{3.4}$$

2^0. *Let the pair* $\{u^0, v^0\}$ *be a solution of the linear programming problem* (3.3), (3.4); *then*

$$\sigma_i u_i^0 v_i^0 = 0, \qquad i = 1, \ldots, N.$$

Moreover, there exists at least one solution of (3.3), (3.4) *such that* $u_i^0 v_i^0 = 0$, $i = 1, \ldots, N$.

PROOF. Let Φ^0 be a solution of (1.10), (1.11) and

$$\sum_{i=1}^{N} \sigma_i |\Phi_i^0| = I_0,$$

$$\sum_{i=1}^{N} H_i \Phi_i^0 = a.$$

We put

$$u_i^0 = \frac{1}{2}\left(|\Phi_i^0| + \Phi_i^0\right), \qquad v_i^0 = \frac{1}{2}\left(|\Phi_i^0| - \Phi_i^0\right).$$

Obviously, for $i = 1, \ldots, N$,

$$u_i^0 \geq 0, \quad v_i^0 \geq 0, \qquad u_i^0 - v_i^0 = \Phi_i^0, \quad u_i^0 + v_i^0 = |\Phi_i^0|, \qquad u_i^0 v_i^0 = 0.$$

The pair $\{u^0, v^0\}$ is feasible for (3.3), (3.4), i.e.,

$$\sum_{i=1}^{N} H_i\left(u_i^0 - v_i^0\right) = a$$

and

$$\sum_{i=1}^{N} \sigma_i\left(u_i^0 + v_i^0\right) = I_0.$$

Assume that $\{u^0, v^0\}$ is not a solution of (3.3), (3.4). Then there exists a feasible $\{u, v\}$ such that

$$\sum_{i=1}^{N} \sigma_i\left(u_i + v_i\right) = I < I_0.$$

Further, $\Phi = u - v$ is feasible for (1.10), (1.11) and, since $|u_i - v_i| \leq (u_i + v_i)$ for $u_i, v_i \geq 0$,

$$\sum_{i=1}^{N} \sigma_i|\Phi_i| = \sum_{i=1}^{N} \sigma_i|u_i - v_i| \leq \sum_{i=1}^{N} \sigma_i\left(u_i + v_i\right) = I < I_0.$$

The last relation contradicts the optimality of Φ^0. Consequently $\{u^0, v^0\}$ is a solution for (3.3), (3.4).

Conversely, let $\{u^0, v^0\}$ be a solution for (3.3), (3.4). Consider $\Phi^0 = u^0 - v^0$. Obviously, Φ^0 is a feasible element for (1.10), (1.11) and

$$\sum_{i=1}^{N} \sigma_i|\Phi_i^0| = \sum_{i=1}^{N} \sigma_i|u_i^0 - v_i^0| \leq \sum_{i=1}^{N} \sigma_i\left(u_i^0 + v_i^0\right).$$

Assume that Φ^0 is not optimal for (1.10), (1.11). Hence there exists a feasible estimator Φ such that

$$\sum_{i=1}^{N} \sigma_i|\Phi_i| < \sum_{i=1}^{N} \sigma_i|\Phi_i^0| \leq \sum_{i=1}^{N} \sigma_i\left(u_i^0 + v_i^0\right).$$

Then the pair $\{u, v\}$ with $u_i = 1/2\left(|\Phi_i| + \Phi_i\right) \geq 0$, $v_i = 1/2\left(|\Phi_i| - \Phi_i\right) \geq 0$ is a feasible pair for (3.3), (3.4), and

$$\sum_{i=1}^{N} \sigma_i |\Phi_i| = \sum_{i=1}^{N} \sigma_i \left(u_i + v_i\right) < \sum_{i=1}^{N} \sigma_i \left(u_i^0 + v_i^0\right).$$

The last inequality contradicts the optimality of $\{u^0, v^0\}$.

Let us prove the second part of the theorem. If $\{u^0, v^0\}$ is a solution, then the pair $\{u, v\}$ with

$$u_i = u_i^0 - \min(u_i^0, v_i^0) \geq 0, \quad v_i = v_i^0 - \min(u_i^0, v_i^0) \geq 0, \qquad i = 1, \ldots, N,$$

is also a feasible pair, $u_i v_i = 0$, $i = 1, \ldots, N$, and

$$\sum_{i=1}^{N} \sigma_i \left(u_i^0 + v_i^0\right) \leq \sum_{i=1}^{N} \sigma_i \left(u_i + v_i\right) = \sum_{i=1}^{N} \sigma_i \left(u_i^0 + v_i^0\right) - 2\sum_{i=1}^{N} \sigma_i \min(u_i^0, v_i^0)$$

$$\leq \sum_{i=1}^{N} \sigma_i \left(u_i^0 + v_i^0\right).$$

The last formula implies that $\{u, v\}$ is also optimal and $\sigma_i \min(u_i^0, v_i^0) = 0$, i.e., $\sigma_i u_i^0 v_i^0 = 0$. □

It follows from Theorem 3.1 that the guaranteed estimation problems posed above reduces to the canonical linear programming problem (3.1), (3.2) with

$$K = 2N, \qquad c_i = c_{i+N} = \sigma_i, \qquad x_i = u_i, \quad x_{i+N} = v_i,$$

$$A_i = -A_{i+N} = H_i, \qquad i = 1, \ldots, N.$$

This problem has $2N$ variables and n equality constraints.

Highly efficient numerical algorithms are developed for solving the linear programming problem (in particular, the simlex method and its modifications [46]). Moreover, the linear programming problem possesses a remarkable qualitative feature: there exists a solution $x = (x_1, \ldots, x_K)'$ with at most n nonzero entries, where n is the number of constraints. This fact generates an important property of the guaranteed estimation problem: an optimal guaranteed estimator has at most n nonzero entries, where n is the dimension of unknown vector q. This means that n measurements only are necessary for the optimal *a priori* guaranteed estimation; the other measurements are useless or even worsen the estimation accuracy. We now turn to rigorous statements.

A conic hull of vectors $A_1, \ldots, A_K \in \mathbf{R}^n$ is the set of vectors A such that

$$A = \sum_{k=1}^{K} \lambda_k A_k, \quad \lambda_k \geq 0, \quad k = 1, \ldots, K. \tag{3.5}$$

Let us first formulate an auxiliary theorem, which is a version of the Carathéodory theorem [185], [5].

THEOREM 3.2. *Any vector A of a conic hull can be represented as an element of the conic hull of at most s ($s \leq n$) linearly independent vectors from* $\{A_1, \dots, A_K\}$.

PROOF. Suppose a vector A belongs to a conic hull and have the representation (3.5). In the definition of the conic hull $\lambda_k \geq 0$, but it is clear that we can assume that $\lambda_k > 0$, $k = 1, \dots, K$. If the vectors $A_1, \dots, A_K$ are linearly dependent, then there exist μ_k, $k = 1, \dots, K$, not all zero, such that

$$\sum_{k=1}^{K} \mu_k A_k = 0$$

and without loss of generality at least one of μ_k is negative.

Let us choose

$$\alpha = \min_k \left\{ \frac{-\lambda_k}{\mu_k} \;\middle|\; \mu_k < 0 \right\}; \quad \text{obviously, } \alpha > 0.$$

then

$$A = \sum_{k=1}^{K} \lambda_k A_k + \alpha \sum_{k=1}^{K} \mu_k A_k = \sum_{k=1}^{K} (\lambda_k + \alpha \mu_k) A_k = \sum_{k=1}^{K} \tilde{\lambda}_k A_k,$$

$$\tilde{\lambda}_k = \lambda_k + \alpha \mu_k, \qquad k = 1, \dots, K.$$

In accordance with the choice of α, all $\tilde{\lambda}_k \geq 0$ and at least one of them is equal to zero. Deleting A_k with the zero $\tilde{\lambda}_k$, we get a representation of (3.5) type with at most $K - 1$ vectors. This procedure can be repeated until only linearly independent vectors will remain in the conic hull. □

Consider the linear programming problem (3.1), (3.2). Assume that the set of x satisfying (3.2) is not empty.

THEOREM 3.3. *Suppose the objective function* (3.1) *is bounded from below for all feasable x; then there exists at least one solution of* (3.1), (3.2) $x^0 = (x_1^0, \dots, x_K^0)'$ *with at most n non-zero entries.*

PROOF. Let $x = (x_1, \dots, x_K)' \in \mathbf{R}^K$ be a feasible vector, i.e., a vector satisfying (3.2). Consider the following system of equalities

$$\sum_{k=1}^{K} c_k x_k = L \tag{3.6}$$

$$\sum_{k=1}^{K} A_k x_k = a, \qquad x_k \geq 0, \quad k = 1, \dots, K.$$

It follows from (3.6) that the vector $(L, a')' \in \mathbf{R}^{n+1}$ belongs to the conic hull of augmented vectors $(c_k, A_k')' \in \mathbf{R}^{n+1}$. Then, by Theorem 3.2, there exist $s \leq n+1$ linearly independent vectors from $\mathbf{R}^{n+1}$ (without loss of generality these vectors are the first s ones) such that

$$\begin{aligned}\sum_{k=1}^{s} c_k \tilde{x}_k &= L \\ \sum_{k=1}^{s} A_k \tilde{x}_k &= a, \qquad \tilde{x}_k > 0, \quad k = 1, \dots, s.\end{aligned} \tag{3.7}$$

First assume $s \leq n$. Then it follows from (3.7) that there exists a vector $\tilde{x} = (\tilde{x}_1, \dots, \tilde{x}_K)'$ with at most n nonzero entries, which yields the same value to the objective function as the initial one.

Let now $s = n+1$. The system (3.7) can be rewritten in the form

$$\sum_{k=1}^{n+1} \begin{pmatrix} c_k \\ A_k \end{pmatrix} \tilde{x}_k + \begin{pmatrix} 1 \\ 0 \end{pmatrix} \tilde{x}_{n+2} = \begin{pmatrix} L \\ a \end{pmatrix}, \tag{3.8}$$

$$\tilde{x}_k > 0, \quad k = 1, \dots, n+1, \quad \tilde{x}_{n+2} = 0.$$

Since the number of vectors in system (3.8) is equal to $n+2$, there exists μ_k, $k = 1, \dots, n+2$, such that

$$\sum_{k=1}^{n+1} \mu_k \begin{pmatrix} c_k \\ A_k \end{pmatrix} + \mu_{n+2} \begin{pmatrix} 1 \\ 0 \end{pmatrix} = 0. \tag{3.9}$$

By Theorem 3.2, the first $n+1$ vectors in (3.9) are linearly independent. Hence $\mu_{n+2} \neq 0$ and we can put $\mu_{n+2} = 1$. It follows from (3.8) and (3.9) that for any $\alpha > 0$

$$\sum_{k=1}^{n+1} \begin{pmatrix} c_k \\ A_k \end{pmatrix} (\tilde{x}_k + \alpha\mu_k) + \begin{pmatrix} 1 \\ 0 \end{pmatrix} \alpha = \begin{pmatrix} L \\ a \end{pmatrix}. \tag{3.10}$$

If $\mu_k \geq 0$, $k = 1, \dots, n+1$, then we have from (3.10) that

$$\begin{aligned}\sum_{k=1}^{n+1} c_k \bar{x}_k &= L - \alpha, \qquad \forall \alpha > 0, \\ \sum_{k=1}^{n+1} A_k \bar{x}_k &= a, \qquad \bar{x}_k = \tilde{x}_k + \alpha\mu_k > 0, \quad k = 1, \dots, n+1.\end{aligned}$$

Consequently the objective function can be arbitrarily small that contradicts the condition of Theorem 3.3. Then there exists at least one $\mu_i < 0$, and by the choice of α we can make at least one of $\bar{x}_k$ equal to zero. Without loss of generality $\bar{x}_k \geq 0$, $k = 1, \ldots, n$ and $\bar{x}_{n+1} = 0$. Therefore (3.10) can be represented in the form

$$\sum_{k=1}^{n} c_k \bar{x}_k = L - \alpha, \qquad \alpha > 0,$$
$$\sum_{k=1}^{n} A_k \bar{x}_k = a, \qquad \bar{x}_k \geq 0, \quad k = 1, \ldots, n.$$

Thus, for any initial vector x, we can find a feasible vector with at most n nonzero entries that yields the same or lesser value to the objective function.

Let us now prove that the solution of (3.1), (3.2) exists. Assume the optimal objective function value is equal to $L^0 > -\infty$. Then there exists a sequence of vectors $x^{(p)}$ with at most n nonzero entries such that $\sum_{k=1}^{K} c_k x_k^{(p)} \to L^0$, $p \to \infty$. Since K is finite and the total number of all possible sets of indices for nonzero entries in $x^{(p)}$ is finite, we can select a subsequence $x_k^{(\bar{p})}$ such that the nonzero entries have common indices. In other words, after renumeration we have a sequence of the following systems of equalities:

$$\sum_{k=1}^{n} c_k x_k^{(\bar{p})} = L^{(\bar{p})} \to L^0, \quad \bar{p} \to \infty, \tag{3.11}$$
$$\sum_{k=1}^{n} A_k x_k^{(\bar{p})} = a, \qquad x_k^{(\bar{p})} \geq 0, \quad k = 1, \ldots, n. \tag{3.12}$$

Further we consider two alternative versions:

(a) $\det\{A_1, \ldots, A_n\} \neq 0$; (b) $\det\{A_1, \ldots, A_n\} = 0$.

In case (a), the system (3.12) is solvable and has the unique solution $x^{(\bar{p})} = (x_1^{(\bar{p})}, \ldots, x_n^{(\bar{p})}, 0'_{K-n})' = x^0, \forall \bar{p}$. In case (b), there exists a nonsingular matrix C such that

$$CA_k = \begin{pmatrix} \bar{A}_k \\ 0_{n-v} \end{pmatrix}, \quad Ca = \begin{pmatrix} \bar{a} \\ 0_{n-v} \end{pmatrix}, \qquad \bar{A}_k, \bar{a} \in \mathbf{R}^v, \quad 0_{n-v} \in \mathbf{R}^{n-v},$$

$$\operatorname{rank}\{\bar{A}_1, \ldots, \bar{A}_n\} = v,$$

and we can proceed to a subspace of a lesser dimension $v < n$. Then the sequence of the systems of equalities (3.11), (3.12) reduces to the form

$$\sum_{k=1}^{n} c_k x_k^{(\bar{p})} = L^{(\bar{p})} \to L^0, \quad \bar{p} \to \infty$$
$$\sum_{k=1}^{n} \bar{A}_k x_k^{(\bar{p})} = \bar{a}, \qquad x_k^{(\bar{p})} \geq 0, \quad k = 1, \ldots, n, \qquad \bar{A}_k, \bar{a} \in \mathbf{R}^v.$$

Arguing as above, we see that there exists a subsequence of vectors $\tilde{x}^{(\tilde{p})} = (\tilde{x}_1^{(\tilde{p})}, \ldots, \tilde{x}_v^{(\tilde{p})}, 0'_{K-v})'$ with at most $v < n$ nonzero entries such that

$$\begin{aligned} \sum_{k=1}^{v} c_k \tilde{x}_k^{(\tilde{p})} &= L^{(\tilde{p})} \to L^0, \quad \tilde{p} \to \infty \\ \sum_{k=1}^{v} \bar{A}_k \tilde{x}_k^{(\tilde{p})} &= \bar{a}, \qquad \tilde{x}_k^{(\tilde{p})} \geq 0, \quad k = 1, \ldots, v. \end{aligned} \tag{3.13}$$

If $\det\{\bar{A}_1, \ldots, \bar{A}_v\} \neq 0$, then the solution of (3.13) $\tilde{x}^{(\tilde{p})}$ is the solution of the linear programming problem for every $\tilde{p}$. If $\det\{\bar{A}_1, \ldots, \bar{A}_v\} = 0$, then the procedure of reduction should be repeated with a $\tilde{v} < v$. □

Using Theorem 3.1 we get the following

COROLLARY 3.1. *At least one optimal estimator Φ^0 has at most n nonzero entries.*

Let us consider the moment problem (2.8), (2.9). If $\sigma(t) \geq \sigma^0 > 0$, $t \in [0, T]$, then, using the trivial substitution

$$\tilde{\Phi}(t) = \sigma(t)\Phi(t), \qquad \tilde{H}(t) = \sigma^{-1}(t)H(t), \tag{3.14}$$

we can represent the variational problem in the form (the tilde is omitted)

$$\int_0^T |\Phi(t)|\, dt \to \inf_{\Phi \in \mathcal{F}}, \tag{3.15}$$

$$\int_0^T H(t)\Phi(t)\, dt = a. \tag{3.16}$$

Assume that (3.16) is resolvable, i.e., the feasable set is not empty.

THEOREM 3.4. *Suppose $H(t) \in \mathbf{R}^n$ is continuous on $[0, T]$; then there exists at least one solution of* (3.15), (3.16) *Φ^0 such that*

$$\Phi^0(t) = \sum_{s=1}^{n} \Phi^0_{(s)} \delta(t - t_s). \tag{3.17}$$

PROOF. Let the optimal value for the moment problem (3.15), (3.16) be equal to I_0. Consider a minimizing sequence

$$\Phi^{(k)}(t) = \Phi^{(k)}_{pc}(t) + \sum_{s=1}^{M_k} \Phi^{(k)}_{(s)} \delta(t - t_s^{(k)}), \qquad k \to \infty,$$

$$\int_0^T |\Phi^{(k)}(t)|\,dt = I_0 + \epsilon_k, \qquad \int_0^T H(t)\Phi^{(k)}(t)\,dt = a, \tag{3.18}$$

$$\epsilon_k \to 0, \quad k \to \infty.$$

For arbitrary k, we approximate the integrals in (3.18) for $\Phi_{pc}^{(k)}(t)$ by Riemann sums. Then there exist impulse functions

$$x^{(k)}(t) = \sum_{i=1}^{N_k} x_i^{(k)}\delta(t - \bar{t}_i^{(k)}), \qquad k \to \infty,$$

such that

$$I_0 \le \int_0^T |\Phi^{(k)}(t)|\,dt = \int_0^T |x^{(k)}(t)|\,dt + \eta_k^{(1)} = \sum_{i=1}^{N_k} |x_i^{(k)}| + \eta_k^{(1)} = I_0 + \epsilon_k,$$

$$a = \int_0^T H(t)\Phi^{(k)}(t)\,dt = \int_0^T H(t)x^{(k)}(t)\,dt + \eta_k^{(2)} = \sum_{i=1}^{N_k} H(\bar{t}_i^{(k)})x_i^{(k)} + \eta_k^{(2)},$$

$$\bar{t}_i^{(k)} \in [0,T] \quad \text{and} \quad \epsilon_k,\ \eta_k^{(1)},\ \|\eta_k^{(2)}\| \to 0 \ \text{ as } \ k \to \infty.$$

In accordance with Corollary 3.1 there exists a sequence

$$y^{(k)}(t) = \sum_{s=1}^{n} y_s^{(k)}\delta(t - \tau_s^{(k)})$$

with at most n nonzero components such that

$$0 \le \sum_{s=1}^{n} |y_s^{(k)}| \le \sum_{i=1}^{N_k} |x_i^{(k)}| = I_0 + \epsilon_k - \eta_k^{(1)},$$

$$\sum_{s=1}^{n} H(\tau_s^{(k)})y_s^{(k)} = \sum_{i=1}^{N_k} H(\bar{t}_i^{(k)})x_i^{(k)} = a - \eta_k^{(2)},$$

$$\{\tau_s^{(k)}\}_{s=1}^{s=n} \subset \{\bar{t}_i^{(k)}\}_{i=1}^{i=N_k}, \qquad k = 1, 2, \ldots.$$

Since $\tau_s^{(k)} \in [0,T]$ and $y_s^{(k)}$ are bounded for $k = 1, 2, \ldots$, there exist subsequences $\tau_s^{(\bar{k})}$ and $y_s^{(\bar{k})}$ such that

$$\tau_s^{(\bar{k})} \to t_s^0 \in [0,T], \qquad y_s^{(\bar{k})} \to \Phi_{(s)}^0, \qquad s = 1, \ldots, n, \qquad \bar{k} \to \infty.$$

Construct the function

$$\Phi^0(t) = \sum_{s=1}^{n} \Phi_{(s)}^0 \delta(t - t_s^0).$$

Let us show that $\Phi^0(t)$ is a solution for the moment problem (3.15), (3.16). In fact, taking into account that $H(t)$ is continuous, we have

$$\int_0^T H(t)\Phi^0(t)\,dt = \sum_{s=1}^n H(t_s^0)\Phi^0_{(s)} = \lim_{\bar{k}\to\infty} \sum_{s=1}^n H(\tau_s^{(\bar{k})})y_s^{(\bar{k})}$$

$$= \lim_{\bar{k}\to\infty} \left(a - \eta_{\bar{k}}^{(2)}\right) = a$$

and

$$I_0 \le \int_0^T |\Phi^0(t)|\,dt = \sum_{s=1}^n |\Phi^0_{(s)}| = \lim_{\bar{k}\to\infty} \sum_{s=1}^n |y_s^{(\bar{k})}|$$

$$\le \lim_{\bar{k}\to\infty} = \left(I_0 + \epsilon_{\bar{k}} - \eta_{\bar{k}}^{(1)}\right) = I_0.$$

□

4. Necessary and Sufficient Conditions for Optimality

Necessary and sufficient conditions for the optimality of an estimator for the problem (2.8), (2.9) define the structure of the solution. In some cases, they yield the solution in explicit form. Moreover, basing on these conditions, other variational problems can be obtained to find the optimal estimator. Let us put $\sigma(t) = 1$ (cf. (3.14)) and consider the problem of moments in the simplest form (3.15), (3.16). If an estimator satisfies the unbiasness condition (3.16), then we say that the estimator is unbiased.

THEOREM 4.1. 1^0. *Let an estimator*

$$\Phi^0(t) = \Phi^0_{pc}(t) + \sum_{s=1}^M \Phi^0_{(s)}\delta(t - t_s)$$

be a solution of the problem (3.15), (3.16); *then there exists a vector* $\beta \in \mathbf{R}^n$ *such that*

(a) $|H'(t)\beta| \le 1, \qquad t \in [0,T];$ (4.1)

(b) *On the set* $\mathcal{T}^1 = \left\{t \,\middle|\, |H'(t)\beta| < 1\right\}, \qquad \Phi^0_{pc}(t) = 0;$ (4.2)

On the set $\mathcal{T}^2 = \left\{t \,\middle|\, |H'(t)\beta| = 1\right\},$

either $\Phi^0_{pc}(t) = 0$ *or* $\operatorname{sign}\Phi^0_{pc}(t) = \operatorname{sign}\big(H'(t)\beta\big)$

(possibly except for a finite number of discontinuity points of $H(t)$ *and* $\Phi^0_{pc}(t)$);

(c) *If* $t_s \in \mathcal{T}^1$, *then* $\Phi^0_{(s)} = 0$; (4.3)

If $t_s \in \mathcal{T}^2$, *then*

$$\text{either} \quad \Phi^0_{(s)} = 0 \quad \text{or} \quad \operatorname{sign} \Phi^0_{(s)} = \operatorname{sign}\big(H'(t_s)\beta\big).$$

2^0. *Suppose an unbiased estimator* $\Phi^0 \in \mathcal{F}$ *and a vector* $\beta \in \mathbf{R}^n$ *satisfy conditions* (4.1) – (4.3); *then* Φ^0 *is the solution of the problem* (3.15), (3.16).

REMARK 4.1. The set $\mathcal{T}^2$ consists of the points of a maximum for $H'(t)\beta$. In generic case, $\mathcal{T}^2$ contains only finite number of points. Therefore in "almost all cases" the piecewise component $\Phi^0_{pc}(t)$ is equal to zero and the optimal estimator is described by impulse functions. This is the reason why we introduce delta-functions into consideration.

PROOF. Consider the problem (3.15), (3.16). The functional (3.15)

$$I: \mathcal{F} \to \mathbf{R}^1, \qquad I(\Phi) = \int_0^T |\Phi(t)|\, dt$$

is convex (and continuous) on $\mathcal{F}$:

$$I(\alpha\Phi_1 + (1-\alpha)\Phi_2) \le \alpha I(\Phi_1) + (1-\alpha) I(\Phi_2),$$

$$0 \le \alpha \le 1, \qquad \Phi_1, \Phi_2 \in \mathcal{F}.$$

The constraint (3.16) is represented in the form

$$\Lambda(\Phi) = a, \qquad \Lambda\colon\ \mathcal{F} \to \mathbf{R}^n, \qquad \Lambda(\Phi) = \int_0^T H(t)\Phi(t)\, dt,$$

where $\Lambda(\Phi)$ is a linear (and continuous) operator. Let us apply the infinite-dimensional Lagrange theorem [75] (see also Chapter 2 for rigorous statement) to our convex variational problem

$$I_0 = \inf_{\Phi \in \mathcal{F}} I(\Phi), \qquad \Lambda(\Phi) = a. \tag{4.4}$$

By this theorem, if Φ^0 is a solution of (4.4), then there exist a nonnegative number $\beta_0 \in \mathbf{R}^1$ and a vector $\beta \in \mathbf{R}^n$ (not both zero) such that

$$\beta_0 I(\Phi^0) + \beta'\left(a - \Lambda(\Phi^0)\right) \le \beta_0 I(\Phi) + \beta'\left(a - \Lambda(\Phi)\right), \qquad \forall\, \Phi \in \mathcal{F}. \tag{4.5}$$

Let us show that we can set $\beta_0 = 1$ in the latter inequality. In fact, suppose $\beta_0 = 0$. Then it follows from (4.5) that

$$-\,\beta'\Lambda(\Phi^0) \le -\,\beta'\Lambda(\Phi), \qquad \beta \in \mathbf{R}^n, \quad \beta \ne 0, \qquad \forall\, \Phi \in \mathcal{F}$$

or, equivalently,

$$-\int_0^T \beta' H(t)\Phi^0(t)\,dt \leq -\int_0^T \beta' H(t)\Phi(t)\,dt, \qquad \forall\,\Phi \in \mathcal{F}. \tag{4.6}$$

First let us study the case when the linear hull of $H(t)$ coincides with $\mathbf{R}^n$, i.e., the unbiasness condition is not degenerate. Let $\{H(\tau_1),\dots,H(\tau_n)\}$ be a basis in $\mathbf{R}^n$. Introduce also a vector $\lambda = (\lambda_1,\dots,\lambda_n)'$ such that

$$\sum_{i=1}^n H(\tau_i)\lambda_i = \beta.$$

Let us choose $\Phi(t) = c\sum_{i=1}^n \lambda_i\delta(t-\tau_i)$, $c \in \mathbf{R}^1$, $c > 0$. Then we have from (4.6) that for any $c > 0$

$$-\int_0^T \beta' H(t)\Phi^0(t)\,dt \leq -c\beta'\int_0^T H(t)\sum_{i=1}^n \lambda_i\delta(t-\tau_i)\,dt = -c\|\beta\|^2,$$

$\|\beta\| = (\beta'\beta)^{1/2}$. Obviously, the last assertion is incorrect. Therefore, $\beta_0 > 0$ and we can put $\beta_0 = 1$.

Assume that the linear hull of $\{H(t)\}$, $t \in [0,T]$, does not coincide with $\mathbf{R}^n$. Then there exists a nonsingular $n \times n$ matrix C such that

$$C'C = I_n, \quad CH(t) = \begin{pmatrix} \bar{H}(t) \\ 0_{n-r} \end{pmatrix}, \quad \bar{H}(t), \bar{a} \in \mathbf{R}^r, \tag{4.7}$$

and the linear hull of $\{\bar{H}(t)\}$, $t \in [0,T]$, coincides with $\mathbf{R}^r$. The decomposition (4.7) means that the unbiasness condition is equivalent to the relation

$$C\int_0^T H(t)\Phi(t)\,dt = \begin{pmatrix} \int_0^T \bar{H}(t)\Phi(t)\,dt \\ 0_{n-r} \end{pmatrix} = Ca = \begin{pmatrix} \bar{a} \\ 0_{n-r} \end{pmatrix}, \quad \bar{a} \in \mathbf{R}^r.$$

(Since Φ^0 is a solution, the unbiasness condition (3.16) is resolvable.)

Then the variational problem takes the form

$$\inf_{\Phi\in\mathcal{F}} I(\Phi), \qquad \int_0^T \bar{H}(t)\Phi(t)\,dt = \bar{a}, \qquad \bar{H}(t) \in \mathbf{R}^r.$$

Note that this problem has a nondegenerate unbiasness condition.

Arguing as above, we see that if Φ^0 is a solution, then

$$I(\Phi^0) + \bar{\beta}'\left(\bar{a} - \int_0^T \bar{H}(t)\Phi^0(t)\,dt\right) \tag{4.8}$$

$$\leq I(\Phi) + \bar{\beta}'\left(\bar{a} - \int_0^T \bar{H}(t)\Phi(t)\,dt\right), \quad \forall\,\Phi \in \mathcal{F}, \quad \bar{\beta} \in \mathbf{R}^r.$$

Returning to the initial dimension of the unbiasness condition, we get from (4.8) that for any $\tilde{\beta} \in \mathbf{R}^{n-r}$

$$I(\Phi^0) + \begin{pmatrix} \bar{\beta} \\ \tilde{\beta} \end{pmatrix}' \left[\begin{pmatrix} \bar{a} \\ 0_{n-r} \end{pmatrix} - \int_0^T \begin{pmatrix} \bar{H}(t) \\ 0_{n-r} \end{pmatrix} \Phi^0(t)\, dt \right]$$

$$\leq I(\Phi) + \begin{pmatrix} \bar{\beta} \\ \tilde{\beta} \end{pmatrix}' \left[\begin{pmatrix} \bar{a} \\ 0_{n-r} \end{pmatrix} - \int_0^T \begin{pmatrix} \bar{H}(t) \\ 0_{n-r} \end{pmatrix} \Phi(t)\, dt \right],$$

or,

$$I(\Phi^0) + \left(C' \begin{pmatrix} \bar{\beta} \\ \tilde{\beta} \end{pmatrix} \right)' \left[C' \begin{pmatrix} \bar{a} \\ 0_{n-r} \end{pmatrix} - \int_0^T C' \begin{pmatrix} \bar{H}(t) \\ 0_{n-r} \end{pmatrix} \Phi^0(t)\, dt \right] \tag{4.9}$$

$$\leq I(\Phi) + \left(C' \begin{pmatrix} \bar{\beta} \\ \tilde{\beta} \end{pmatrix} \right)' \left[C' \begin{pmatrix} \bar{a} \\ 0_{n-r} \end{pmatrix} - \int_0^T C' \begin{pmatrix} \bar{H}(t) \\ 0_{n-r} \end{pmatrix} \Phi(t)\, dt \right].$$

Putting

$$\beta = C' \begin{pmatrix} \bar{\beta} \\ \tilde{\beta} \end{pmatrix},$$

we get (4.5) from (4.7), (4.9) with $\beta_0 = 1$:

$$I(\Phi^0) + \beta' \left(a - \int_0^T H(t)\Phi^0(t)\, dt \right) \tag{4.10}$$

$$\leq I(\Phi) + \beta' \left(a - \int_0^T H(t)\Phi(t)\, dt \right), \qquad \forall \Phi \in \mathcal{F}.$$

It follows from the Lagrange theorem [75] (see Chapter 2) that (4.10) is also a sufficient condition for Φ^0 to be optimal. Note that the sufficiency follows directly from (4.10). In fact, for any Φ satisfying the unbiasness condition (3.16), the expression within parentheses in (4.10) is equal to zero. Thus inequality (4.10) is the necessary and sufficient condition for Φ^0 to be optimal.

Let us show that (4.10) is equivalent to the conditions of Theorem 4.1. Inequality (4.10) is represented in the form

$$\int_0^T \left(|\Phi^0(t)| - \beta' H(t)\Phi^0(t) \right) dt \leq \int_0^T \left(|\Phi(t)| - \beta' H(t)\Phi(t) \right) dt. \tag{4.11}$$

Inequality (4.11) implies (4.1). In fact, suppose that for some t^0

$$|\beta' H(t^0)| > 1.$$

Taking

$$\Phi(t) = c \operatorname{sign}\left(\beta' H(t^0)\right) \delta(t - t^0), \qquad c > 0,$$

we have from (4.11) that

$$\int_0^T \left(|\Phi^0(t)| - \beta' H(t)\Phi^0(t)\right) dt \le c\left(1 - |\beta' H(t^0)|\right), \qquad \forall c > 0.$$

This contradiction concludes the proof of (4.1).

By virtue of (4.1), it follows from (4.11) that

$$\int_0^T \left(|\Phi^0(t)| - \beta' H(t)\Phi^0(t)\right) dt = 0, \tag{4.12}$$

or,

$$\begin{cases} \displaystyle\int_0^T \left(|\Phi^0_{pc}(t)| - \beta' H(t)\Phi^0_{pc}(t)\right) dt = 0, \\ \displaystyle\sum_{s=1}^{M} \left(|\Phi^0_{(s)}| - \beta' H(t_s)\Phi^0_s\right) = 0. \end{cases} \tag{4.13}$$

Since (cf.(4.1)),

$$|\Phi^0_{pc}(t)| - \beta' H(t)\Phi^0_{pc}(t) \ge 0, \qquad |\Phi^0_{(s)}| - \beta' H(t_s)\Phi^0_{(s)} \ge 0,$$

formulas (4.13) imply that for all continuity points of $H(t)$ and $\Phi^0_{pc}(t)$

$$|\Phi^0_{pc}(t)| - \beta' H(t)\Phi^0_{pc}(t) = 0, \qquad t \in [0, T], \tag{4.14}$$

and

$$|\Phi^0_{(s)}| - \beta' H(t_s)\Phi^0_{(s)} = 0, \qquad s = 1, \ldots, M. \tag{4.15}$$

Relations (4.2), (4.3) follow directly from (4.14), (4.15). Thus the first assertion of Theorem 4.1 is proved.

Let us prove the second assertion of Theorem 4.1. Obviously, if (4.2), (4.3) hold, then (4.14) – (4.15), and, consequently, (4.13) are valid. Therefore, (4.11) follows from (4.1), and (4.10) holds. □

Let us consider simple examples.

EXAMPLE 4.1. Suppose that the measurements have the form

$$z(t) = q + \varrho(t), \quad t \in [0, T], \qquad |\varrho(t)| \le 1, \qquad q, \varrho(t) \in \mathbf{R}^1.$$

It is required to estimate q. Thus, $n = 1$, $H(t) = 1$, $a = 1$ and the problem of moments has the form

$$\int_0^T |\Phi(t)|\,dt \to \inf, \qquad \int_0^T \Phi(t)\,dt = 1. \tag{4.16}$$

Theorem 3.4 claims that an optimal estimator exists and at least one of optimal estimators is $\Phi^*(t) = \alpha^*\,\delta(t - t^*)$. Using Theorem 4.1, we see that there exists a scalar β, $|\beta| \le 1$ such that if $|\beta| < 1$, then $\Phi^0(t) \equiv 0$ and the unbiasness condition does not hold. Hence, $|\beta| = 1$ and we have $\mathcal{T}^1 = \emptyset$, $\mathcal{T}^2 = [0, T]$. Conditions (4.2), (4.3) imply that if $\Phi_{pc}(t) \neq 0$ ($\Phi_{(s)} \neq 0$), then $\operatorname{sign}\Phi_{pc}(t) = \operatorname{sign}\beta$ ($\operatorname{sign}\Phi_{(s)} = \operatorname{sign}\beta$). It follows from (4.16) that $\operatorname{sign}\beta = \beta = 1$. Thus the optimal estimator has the form

$$\Phi^0(t) = \Phi^0_{pc}(t) + \sum_{s=1}^{M} \Phi^0_{(s)}\delta(t - t_s),$$

$$\Phi^0_{pc}(t) \ge 0, \quad \Phi^0_{(s)} \ge 0, \qquad \int_0^T \Phi^0_{pc}(t)\,dt + \sum_{s=1}^{M} \Phi^0_{(s)} = 1. \tag{4.17}$$

The second assertion of Theorem 4.1 confirms that every estimator satisfying (4.17) is optimal. In particular, $\Phi(t) = \delta(t - t^*)$ is an optimal estimator for any $t^* \in [0, T]$.

Note that our conclusion has a simple explanation. In fact, we have from (4.16) that

$$\int_0^T |\Phi(t)|\,dt \ge \int_0^T \Phi(t)\,dt = 1.$$

Example 4.2. Consider a material point moving along a straight line with a constant velocity. The equation of motion has the form:

$$s(t) = v_0 t + s_0, \qquad t \in [0, T],$$

where s_0 is an initial location of the point, v_0 is a velocity of the point. It is required to estimate v_0 by the measurement of $s(t)$ corrupted by bounded noise $|\varrho(t)| \le 1$. In other words, $n = 2$, $q = (v_0, s_0)'$, $H(t) = (t, 1)'$, $a = (1, 0)'$. The corresponding problem of moments takes the form

$$\int_0^T |\Phi(t)|\,dt \to \inf, \qquad \int_0^T t\Phi(t)\,dt = 1, \quad \int_0^T \Phi(t)\,dt = 0. \tag{4.18}$$

Due to Theorem 3.4, an optimal estimator exists and is concentrated at two instants. But this theorem does not say what are these instants and whether this impulse estimator is unique. Let us use Theorem 4.1. By virtue of (4.1), there exists a vector $\beta = (\beta_1, \beta_2)'$ such that

$$|\beta_1 t + \beta_2| \leq 1, \qquad t \in [0, T].$$

Furthermore, if $\beta_1 \neq 0$, then $|\beta_1 t + \beta_2| < 1$ except for two instants: 0 and T. Therefore, by (4.2), $\Phi^0_{pc}(t) = 0$ and $\Phi^0(t) = \Phi^0_{(1)}\delta(t) + \Phi^0_{(2)}\delta(t - T)$, $\Phi^0_{(1)}, \Phi^0_{(2)} \in \mathbf{R}^1$. It follows from (4.18) that

$$T\Phi^0_{(2)} = 1, \qquad \Phi^0_{(1)} + \Phi^0_{(2)} = 0;$$

consequently,

$$\Phi^0_{(1)} = -\frac{1}{T}, \qquad \Phi^0_{(2)} = \frac{1}{T}.$$

Obviously the last estimator satisfies the sufficient conditions of Theorem 4.1 with $\beta_1 = 2/T$, $\beta_2 = -1$.

Let us now show that $\beta_1 \neq 0$. Suppose that $\beta_1 = 0$; then $|\beta_2| \leq 1$. If $|\beta_2| < 1$, then $\Phi^0(t) = 0$ and the unbiasness condition does not hold. Hence, $|\beta_2| = 1$, $\mathcal{T}^1 = \emptyset$ and $\mathcal{T}^2 = [0, T]$. Therefore, by virtue of (4.2), (4.3), $\Phi^0_{pc}(t) \geq 0$, $\Phi^0_{(s)} \geq 0$. This contradicts the unbiasness condition in (4.18). Thus our optimal estimator is unique.

4.1. *Discrete Measurements*

Since we admit piecewise continuous matrices $H(t)$, the case of discrete measurements

$$z(t_i) = H_i x(t_i) + \varrho(t_i), \qquad i = 0, \ldots, N, \tag{4.19}$$

is a special case of the problem under consideration. In fact, let us put

$$H(t) = \begin{cases} H_i & \text{if } t = t_i \\ 0 & \text{otherwise.} \end{cases} \tag{4.20}$$

Then the values of $\Phi(t)$ at $t \neq t_i$ do not influence on the integral in the unbiasness condition (3.16) but on the integral in (3.15). Therefore, $\Phi(t) = 0$, $t \neq t_i$. This means that the optimal estimator has the impulse component only and

$$\hat{l} = \int_0^T \Phi(t) z(t)\, dt = \sum_{i=0}^{N} \Phi(t_i) z(t_i) = \sum_{i=0}^{N} \Phi_i z(t_i).$$

Then the necessary and sufficient conditions for the optimality of estimator take the form

THEOREM 4.2. 1^0. *Let an estimator* $\Phi^0 = (\Phi_1^0, \ldots, \Phi_N^0)'$ *be a solution of the problem*

$$\sum_{i=0}^{N} |\Phi_i| \to \inf_{\Phi} \tag{4.21}$$

under constraint

$$\sum_{i=0}^{N} H_i \Phi_i = a; \tag{4.22}$$

then there exists a vector $\beta \in \mathbf{R}^n$ *such that*

(a1) $|H_i'\beta| \leq 1, \qquad i = 1, \ldots, N;$ (4.23)

(b1) *On the set* $\mathcal{T}^1 = \left\{ t_i \;\middle|\; |H_i'\beta| < 1 \right\}, \qquad \Phi_i^0 = 0;$

On the set $\mathcal{T}^2 = \left\{ t_i \;\middle|\; |H_i'\beta| = 1 \right\},$ (4.24)

either $\Phi_i^0 = 0$ *or* $\operatorname{sign} \Phi_i^0 = \operatorname{sign}(H_i'\beta)$.

2^0. *Suppose an unbiased estimator* Φ^0 *and a vector* $\beta \in \mathbf{R}^n$ *satisfy conditions* (4.23), (4.24); *then* Φ^0 *is a solution of the problem* (4.21), (4.22).

Let us modify Theorem 4.2 for a more general problem (1.10), (1.11), when $\sigma_i \geq 0$.

THEOREM 4.3. 1^0. *Let an estimator* $\Phi^0 = (\Phi_1^0, \ldots, \Phi_N^0)'$ *be a solution of the problem* (1.10), (1.11); *then there exists a vector* $\beta \in \mathbf{R}^n$ *such that*

($\mathrm{a1}_\sigma$) $|H_i'\beta| \leq \sigma_i, \qquad i = 1, \ldots, N;$ (4.25)

($\mathrm{b1}_\sigma$) *On the set* $\mathcal{T}_\sigma^1 = \left\{ t_i \;\middle|\; |H_i'\beta| < \sigma_i \right\}, \qquad \Phi_i^0 = 0;$

On the set $\mathcal{T}_\sigma^2 = \left\{ t_i \;\middle|\; |H_i'\beta| = \sigma_i, \quad \sigma_i \neq 0 \right\},$ (4.26)

either $\Phi_i^0 = 0$ *or* $\operatorname{sign} \Phi_i^0 = \operatorname{sign}(H_i'\beta)$.

2^0. *Suppose an unbiased estimator* Φ^0 *and a vector* $\beta \in \mathbf{R}^n$ *satisfy conditions* (4.25), (4.26); *then* Φ^0 *is a solution of the problem* (1.10), (1.11).

The proof of Theorem 4.3 is analogous to that of Theorem 4.1.

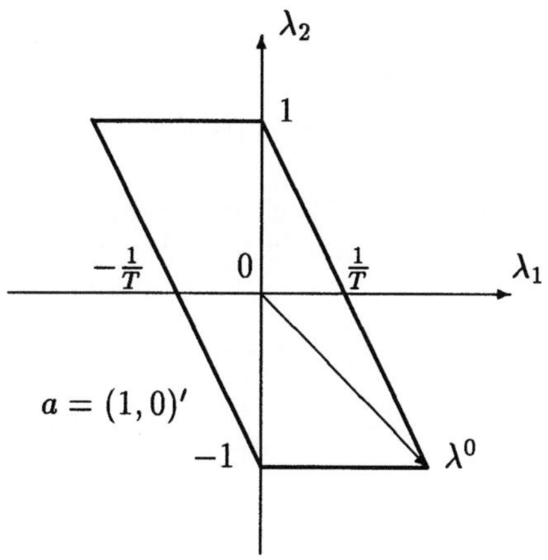

Fig. 1. Dual problem

5. Dual Problem and Chebyshev Approximation

Theorem 4.1 implies a very important duality relation. To present this relation let us introduce *a dual problem*:

$$I^0 = \sup_{\lambda \in \mathbf{R}^n} a'\lambda \tag{5.1}$$

under constraint

$$|H'(t)\lambda| \leq 1, \qquad t \in [0, T]. \tag{5.2}$$

The dual problem has an obvious geometric interpretation: on a convex set (5.2), it is required to find a vector that has a maximal projection on a given direction a. The set (5.2) and the optimal vector λ^0 for Example 4.2 are shown at Figure 1.

In this section, it is assumed that $H(t)$ is continuous. Then, by Theorem 3.4, the solution of the primal problem (3.15), (3.16) (the optimal estimator) Φ^0 exists.

LEMMA 5.1 (Duality relation). *Let Φ^0 be a solution of* (3.15), (3.16). *Then a solution of* (5.1), (5.2) *exists, and the values of primal and dual problems are equal:*

$$I_0 = I^0. \tag{5.3}$$

PROOF. For any $\Phi \in \mathcal{F}$ satisfying the unbiasness condition (3.16) and for any $\lambda \in \mathbf{R}^n$ satisfying the restriction (5.2)

$$a'\lambda = \int_0^T \Phi(t)H'(t)\lambda\, dt \leq \int_0^T |\Phi(t)||H'(t)\lambda|\, dt \leq \int_0^T |\Phi(t)|\, dt.$$

Maximizing the left-hand side of this inequality in λ under constraint (5.2) and minimizing the right-hand side in Φ under constraint (3.16), we get

$$I^0 = \sup a'\lambda \leq \inf \int_0^T |\Phi(t)|\, dt = I_0. \tag{5.4}$$

If Φ^0 is an optimal solution, then, due to Theorem 4.1, there exists a Lagrange multiplier β such that

$$a'\beta = \int_0^T \Phi^0(t)H'(t)\beta\, dt = \int_{T^2} \Phi^0(t)H'(t)\beta\, dt \tag{5.5}$$

$$= \int_{T^2} |\Phi^0(t)|\, dt = \int_0^T |\Phi^0(t)|\, dt.$$

Therefore (5.3) holds and β is a solution of the dual problem (5.1), (5.2).

□

The relation between the primal problem (3.15), (3.16) and the dual problem (5.1), (5.2) is a general result of the optimization theory. There is a rule for constructing the dual problem [75], [51] (see also Chapter 2). Consider the Lagrange function

$$\mathcal{L}(\Phi, \lambda) = I(\Phi) + \lambda' \left(a - \int_0^T H(t)\Phi(t)\, dt \right).$$

Define a problem in minimum

$$\inf_{\Phi \in \mathcal{F}} \left(\sup_{\lambda \in \mathbf{R}^n} \mathcal{L}(\Phi, \lambda) \right) \tag{5.6}$$

and a problem in maximum

$$\sup_{\lambda \in \mathbf{R}^n} \left(\inf_{\Phi \in \mathcal{F}} \mathcal{L}(\Phi, \lambda) \right). \tag{5.7}$$

It can be easily verified that (5.6) is equivalent to (3.15), (3.16) and (5.7) is equivalent to (5.1), (5.2). In fact,

$$\sup_{\lambda \in \mathbf{R}^n} \left\{ \int_0^T |\Phi(t)|\, dt + \lambda' \left(a - \int_0^T H(t)\Phi(t)\, dt \right) \right\}$$

$$= \begin{cases} \int_0^T |\Phi(t)|\, dt & \text{if } \int_0^T H(t)\Phi(t)\, dt = a \\ \infty & \text{otherwise,} \end{cases}$$

and

$$\inf_{\Phi\in\mathcal{F}} \left\{ \int_0^T |\Phi(t)|\, dt + \lambda' \left(a - \int_0^T H(t)\Phi(t)\, dt \right) \right\}$$

$$= a'\lambda + \inf_{\Phi\in\mathcal{F}} \int_0^T \Big(|\Phi(t)| - \lambda' H(t)\Phi(t) \Big)\, dt$$

$$= \begin{cases} a'\lambda & \text{if } |\lambda' H(t)| \le 1,\ t \in [0,T], \\ -\infty & \text{otherwise.} \end{cases}$$

Thus the duality relation is a minimax relation [75], [51] (see also Chapter 2):

$$\inf_{\Phi\in\mathcal{F}} \sup_{\lambda\in\mathbf{R}^n} \mathcal{L}(\Phi,\lambda) = \sup_{\lambda\in\mathbf{R}^n} \inf_{\Phi\in\mathcal{F}} \mathcal{L}(\Phi,\lambda). \tag{5.8}$$

General formulas (5.6) – (5.8) allows us to construct corresponding dual problems and to obtain dual relations for much more complex variational problems. We shall systematically use these relations in our book.

REMARK 5.1. Consider the discrete problem (1.10) (1.11) ($\operatorname{rank} H = n$). Using Theorem 4.3, a discrete version of Lemma 5.1 can be easily proved:

$$\min_{\Phi\in(1.11)} \sum_{i=1}^{N} \sigma_i |\Phi_i| = \max_{\lambda\in(4.25)} a'\lambda.$$

Obviously, the equality (5.5) is valid for any solution λ^0 of the dual problem:

$$a'\lambda^0 = \int_0^T \Phi^0(t) H'(t)\lambda^0\, dt = \int_0^T |\Phi^0(t)|\, dt.$$

Hence,

$$\text{if} \quad |H'(t)\lambda^0| < 1, \quad \text{then} \quad \Phi^0(t) = 0 \tag{5.9}$$

(possibly except for a finite number of discontinuity points of $\Phi^0_{pc}(t)$ [1]). Therefore the estimator is nonzero only if

$$|H'(t)\lambda^0| = 1. \tag{5.10}$$

[1] Recall that in this section $H(t)$ is continuous.

Usually, the equality (5.10) is attained only at finite number of points $\{t_1, \ldots, t_d\}$. According to (5.9) they define the instants when the measurements should be made. If their number is less than or equal to the dimension of the state space and the corresponding vectors form a nonsingular matrix, i.e.,

$$d \leq n, \qquad \text{rank}\,(H(t_1), \ldots, H(t_d)) = d, \tag{5.11}$$

then the unbiasness condition (3.16) determines uniquely the estimator components by the system of algebraic equations:

$$\sum_{s=1}^{d} H(t_i)\Phi^0_{(s)} = a. \tag{5.12}$$

Conditions (5.11) usually hold. In the case $d > n$, condition (5.10) considerably constricts the set of possible measurement instants. As a rule, the optimal measurement instants can be easily selected from $\{t_1, \ldots, t_d\}$.

Thus the problem of finding the optimal estimator can be reduced to the dual problem. Let us show that the dual problem is equivalent to a certain functional approximation problem. First assume that

$$a = (1, \ldots, 0)'. \tag{5.13}$$

Obviously, the general case $a = (a_1, \ldots, a_n)'$, where $a_s \neq 0$ for some s, reduces to (5.13) by the following change of variables:

$$\tilde{H}(t) = CH(t), \qquad \tilde{a} = (1, \ldots, 0)' = Ca, \tag{5.14}$$

$$C = \left(\begin{array}{c:c:c} 0_{1\times(s-1)} & a_s^{-1} & 0_{1\times(n-s)} \\ \hdashline & -a_1 & \\ a_s I_{s-1} & \vdots & 0_{(s-1)\times(n-s)} \\ & -a_{s-1} & \\ \hdashline & -a_{s+1} & \\ 0_{(n-s)\times(s-1)} & \vdots & a_s I_{n-s} \\ & -a_n & \end{array}\right),$$

where I_k and $0_{p\times q}$ are identity matrix and zero matrix of dimension $k \times k$ and $p \times q$, respectively.

A polynomial $H_1(t) + \sum_{j=2}^{n} y_j H_j(t)$ is called a generalized polynomial. With $H_j(t) = t^{n-j}$, $j = 1, \ldots, n$, it reduces to an ordinary polynomial. Introduce a linear space of continuous functions on $[0, T]$. The norm $\|f\|_C = \max_{[0,T]} |f(t)|$ in this space is called the Chebychev norm. This normed space is denoted by $C[0,T]$. The generalized Chebyshev polynomial is a generalized polynomial with a minimal norm, i.e., it yields a solution to the minimax problem:

$$L = \min_{y_j,\, j=2,\dots,n} \max_{t\in[0,T]} \left| H_1(t) + \sum_{j=2}^{n} y_j H_j(t) \right| \tag{5.15}$$

$$= \min_{y_j,\, j=2,\dots,n} \left\| H_1(t) + \sum_{j=2}^{n} y_j H_j(t) \right\|_C .$$

The problem (5.15) is the problem of the best approximation of a function $H_1(t) \in C[0,T]$ by a linear finite-dimensional manifold.

Suppose $\lambda^0 = (\lambda_1^0, \dots, \lambda_n^0)'$ is a solution of the dual problem. Since $H(t)$ is uniformly bounded, $\lambda = 0$ is an interior point of set (5.2); therefore, $\lambda_1^0 \neq 0$. Clearly, $\lambda_1^0 > 0$. Moreover, $\max_{t\in[0,T]} |\lambda^{0\prime} H(t)| = 1$; otherwise there is a vector collinear to λ^0 that yields a greater value to the objective function $a'\lambda$.

Let us prove that the coefficients $y_j^0 = \lambda_j^0 / \left(\lambda_1^0\right)^{-1}$, $j = 2, \dots, n$, resolve the approximation problem (5.15) and define the generalized Chebyshev polynomial.

The proof is by *reductio ad absurdum*. Let a solution of (5.15) is attained at y_j, $j = 2, \dots, n$. This means that

$$L = \max_{t\in[0,T]} \left| H_1(t) + \sum_{j=2}^{n} y_j H_j(t) \right| < \max_{t\in[0,T]} \left| H_1(t) + \sum_{j=2}^{n} y_j^0 H_j(t) \right| \tag{5.16}$$

$$= \left(\lambda_1^0\right)^{-1} \max_{t\in[0,T]} \left| \lambda_1^0 H_1(t) + \sum_{j=2}^{n} \lambda_j^0 H_j(t) \right| = \left(\lambda_1^0\right)^{-1} .$$

The number $L > 0$, otherwise for any $\alpha \in \mathbf{R}^1$, $\alpha > 0$,

$$\alpha H_1(t) + \sum_{j=2}^{n} \alpha\, y_j H_j(t) \equiv 0$$

and, consequently, the optimal value of the dual problem is equal to $+\infty$. It follows from (5.16) that

$$\left| L^{-1} H_1(t) + \sum_{j=2}^{n} L^{-1} y_j H_j(t) \right| \leq 1, \qquad L^{-1} > \lambda_1^0 .$$

This condraticts the assumption that λ^0 is a solution of the dual problem.

Conversely, let y_j^0, $j = 2, \dots, n$, be a solution of (5.15) and $L > 0$. Then the vector

$$\lambda^0 = \left(L^{-1}, L^{-1} y_2^0, \dots, L^{-1} y_n^0\right)' \tag{5.17}$$

satisfies (5.2). Suppose a vector $\lambda = (\lambda_1, \dots, \lambda_n)'$ satisfying (5.2) yeilds a greater value to the objective function, i.e., $\lambda_1 > L^{-1} > 0$. Consider the set $y_j = \lambda_j(\lambda_1)^{-1}$, $j = 2, \dots, n$. Then

$$\max_{t\in[0,T]} \left|H_1(t) + \sum_{j=2}^{n} y_j H_j(t)\right| = (\lambda_1)^{-1} \max_{t\in[0,T]} \left|\lambda' H(t)\right| \le (\lambda_1)^{-1} < L.$$

This contradicts the optimality of y_j^0.

Thus, (5.17) defines the direct relation between the dual problem and the generalized Chebyshev polynomial.

Clearly, the instants where $|\lambda^{0\prime} H(t)| = 1$ coincide with the instants where the modulus of the corresponding generalized Chebyshev polynomial attains its maximal values. Thus we obtained a constructive algorithm for finding an optimal estimator. First the corresponding generalized Chebychev polynomial should be found. It determines the points where (5.10) holds. Then the optimal estimator is calculated from (5.11), (5.12).

5.1. *Sufficient Conditions for the Optimality of Generalized Polynomial*

Let us formulate a classical result for the generalized Chebyshev polynomial (see, e.g., [129]).

DEFINITION 5.1. The system $\{H_2(t), \dots, H_n(t)\}$, $H_j(t) \in C[0,T]$ $(j = 2, \dots, n)$ satisfies the Haar condition (is said to be the Chebyshev system) if the equation

$$\sum_{j=2}^{n} \delta y_j H_j(t) = 0 \tag{5.18}$$

has at most $n-2$ roots on the segment $[0,T]$ for any δy_j, $\sum_{j=2}^{n} \delta y_j^2 \neq 0$.

THEOREM 5.1. *Suppose the functions $H_j(t)$, $j = 2, \dots, n$, satisfy the Haar condition; then a polynomial $S^0(t) = H_1(t) + \sum_{j=2}^{n} y_j^0 H_j(t)$ is the generalized Chebyshev polynomial if and only if there exist points $\tau_0 < \dots < \tau_{n-1}$ from $[0,T]$ such that*

$$S^0(\tau_i) = -S^0(\tau_{i+1}), \qquad i = 0, \dots, n-2,$$

$$|S^0(\tau_i)| = \max_{[0,T]} |S^0(t)|, \qquad i = 0, \dots, n-1.$$

This theorem is called the theorem of *Chebyshev alternance*. It can be used for constructive finding the generalized Chebyshev polynomial for various specific problems.

Note that the Haar condition guarantees the uniqueness of the generalized Chebyshev polynomial [129].

Sometimes the following sufficient conditions can be successfully used. Let the functions $H_j(t) \in \mathbf{R}^1$ be differentiable on $[0, T]$.

LEMMA 5.2. *Suppose a polynomial*

$$S^0(t) = H_1(t) + \sum_{j=2}^{n} y_j^0 H_j(t), \qquad t \in [0, T],$$

satisfies the conditions

$$S^0(t_0) = -S^0(t_1) = \ldots (-1)^s S^0(t_s) = \ldots = (-1)^{n-1} S^0(t_{n-1}), \tag{5.19}$$

$$0 = t_0 < t_1 < \ldots < t_{n-2} < t_{n-1} = T,$$

$$\left.\frac{dS^0(t)}{dt}\right|_{t=t_i} = 0, \qquad i = 1, \ldots, n-2, \tag{5.20}$$

$$\frac{dS^0(t)}{dt} \neq 0 \quad \textit{for} \quad t \neq t_i, \qquad i = 0, \ldots, n-1,$$

and the equation

$$\sum_{j=2}^{n} \delta y_j H_j(t) = 0 \tag{5.21}$$

for any δy_j, $\sum_{j=2}^{n} \delta y_j^2 \neq 0$, has at most $n-2$ roots in the open interval $(0, T)$; then $S^0(t)$ is the generalized Chebyshev polynomial:

$$\min_{y_j} \max_{t \in [0,T]} |S(t)| = \max_{t \in [0,T]} |S^0(t)|.$$

PROOF. Clearly, under conditions (5.19), (5.20) the points

$$\{0, t_1, \ldots, t_{n-2}, T\}$$

are the unique maximum and minimum points of the function $S^0(t)$ on $[0, T]$. Assume that $\max_{[0,T]} |S(t)| < \max_{[0,T]} |S^0(t)|$ for a certain function $S(t)$. This means that the values of $S(t)$ belong to a strip $[-\xi, \xi]$ and $\xi < \max_{[0,t]} |S^0(t)|$. Then there exist instants

$$0 < t_1' < t_1'' < t_1 < t_2' < t_2'' < \ldots < t_{n-2}' < t_{n-2}'' < t_{n-2} < t_{n-1}' < t_{n-1}'' < T$$

such that

$$S^0(t_i') = -S^0(t_i''), \qquad |S^0(t_i')| = |S^0(t_i'')| = \xi, \qquad i = 1, \ldots, n-1.$$

Therefore the graph of $S(t)$ must intersect the curve $S^0(t)$ at least $n-1$ times in the open interval $(0, T)$. This contradicts condition (5.21).

□

REMARK 5.2. Though Lemma 5.2 provides only sufficient conditions and does not guarantee the uniqueness, the condition (5.21) is less restrictive than the Haar condition.

EXAMPLE 5.1. Consider the problem from Example 4.2. The corresponding polynomial has the form

$$S(t) = H_1(t) + yH_2(t) = t + y, \qquad t \in [0, T].$$

Let us show that the polynomial

$$S^0(t) = t - \frac{T}{2}$$

is optimal.

Indeed,

$$S^0(0) = -S^0(T) = -\frac{T}{2}, \qquad \frac{dS^0(t)}{dt} \equiv 1,$$

and the equation

$$\delta y\, H_2(t) = \delta y = 0, \qquad \delta y \neq 0,$$

has no roots in $(0, T)$. Therefore, by virtue of Lemma 5.2, $S^0(t)$ is optimal.

Generally, the optimal measurement instants are different for different vectors a, which define the estimated parameter. Nevertheless, the optimal measurement instants that were found for the special case $a = (1, \dots, 0)'$ can coincide with the optimal instants for a given vector $a \neq (1, \dots, 0)'$. Let us discuss this question.

Consider the case $a = (a_1, \dots, a_n)'$ with $a_s \neq 0$ for a certain $s \geq 2$. By the change of variables (5.14), the initial problem reduces to the problem with the goal vector $\tilde{a} = (1, \dots, 0)'$ and the measurement matrix

$$\tilde{H}(t) = \begin{pmatrix} a_s^{-1} H_s(t) \\ a_s H_1(t) - a_1 H_s(t) \\ \vdots \\ a_s H_{s-1}(t) - a_{s-1} H_s(t) \\ a_s H_{s+1}(t) - a_{s+1} H_s(t) \\ \vdots \\ a_s H_n(t) - a_n H_s(t) \end{pmatrix}.$$

Then the corresponding problem for the finding of the generalized Chebyshev polynomial takes the form

$$\max_{t \in [0,T]} |\tilde{H}_1(t) + \sum_{j=2}^{n} v_j \tilde{H}_j(t)| \to \min_{v_j}.$$

COROLLARY 5.1. *Suppose the conditions of Lemma 5.2 hold, the number* $a_1 + \sum_{j=2}^{n} a_j y_j^0 \neq 0$, *and the equation*

$$\sum_{j=2}^{n} \delta v_j \tilde{H}_j(t) = 0 \tag{5.22}$$

with not all zero δv_j *has at most* $n-2$ *roots in the segment* $(0, T)$; *then among generalized polynomials*

$$V(t) = \tilde{H}_1(t) + \sum_{j=2}^{n} v_j \tilde{H}_j(t)$$

the polynomial

$$V^0(t) = V(t)\big|_{v_j = v_j^0,\ j=2,\dots,n}$$

with

$$v_2^0 = a_s^{-1}\left[a_1 + \textstyle\sum_{i=2}^{n} a_i y_i^0\right]^{-1}, \tag{5.23}$$

$$v_j^0 = \begin{cases} y_{j-1}^0 a_s^{-1}\left[a_1 + \sum_{i=2}^{n} a_i y_i^0\right]^{-1}, & j = 3, \dots, s, \\ y_j^0 a_s^{-1}\left[a_1 + \sum_{i=2}^{n} a_i y_i^0\right]^{-1}, & j = s+1, \dots, n, \end{cases}$$

has the least deviation from zero,

$$V^0(t) = \left[a_1 + \sum_{i=2}^{n} a_i y_i^0\right]^{-1} S^0(t), \tag{5.24}$$

and

$$\max_{t \in [0,T]} |V^0(t)| = \left|a_1 + \sum_{i=2}^{n} a_i y_i^0\right|^{-1} \max_{t \in [0,T]} |S^0(t)|.$$

PROOF. By the choice of $\{v_j\}_2^n = \{v_j^0\}_2^n$ let us attempt to satisfy the identity

$$V^0(t) = \alpha S^0(t), \qquad t \in [0, T], \qquad \text{for an } \alpha > 0$$

or, equivalently,

$$H'(t) C' \begin{pmatrix} 1 \\ v^0 \end{pmatrix} = \alpha H'(t) \begin{pmatrix} 1 \\ y^0 \end{pmatrix}, \qquad t \in [0, T],$$

where $v^0 = (v_2^0, \ldots, v_n^0)'$, $y^0 = (y_2^0, \ldots, y_n^0)'$. Obviously, the system of equations with respect to α and v^0

$$C'\begin{pmatrix} 1 \\ v^0 \end{pmatrix} = \alpha \begin{pmatrix} 1 \\ y^0 \end{pmatrix}$$

has the solution such that $\alpha = \left[a_1 + \sum_{i=2}^n a_i y_i^0\right]^{-1}$ and v^0 is determined by (5.23). Thus the identity (5.24) holds. Therefore $V^0(t)$ satisfies the conditions of Lemma 5.2 with $H(t)$ replaced by $\tilde{H}(t)$. In fact, (5.19), (5.20) follow from (5.24); condition (5.21) is equivalent to (5.22). It follows from (5.24) that $V^0(t)$ and $S^0(t)$ have common extremal points. Thus the optimal instants are taken from the same set $\{0, t_1, \ldots, t_{n-2}, T\}$. □

EXAMPLE 5.2. Consider a two-dimensional system

$$\dot{x}(t) = Ax(t), \quad x(t) = (x_1(t), x_2(t))', \quad t \in [0, T],$$

$$A = \begin{pmatrix} a_{11} & a_{12} \\ a_{21} & a_{22} \end{pmatrix},$$

$$z(t) = x_1(t) + \varrho(t), \qquad |\varrho(t)| \le 1.$$

To be definite, assume that the solution of this system is oscillating, i.e., the discriminant of the characteristic equation is negative. Then the fundamental matrix of the system is determined by the formula

$$e^{At} = e^{\mu t}\begin{pmatrix} \cos\omega t + \left(\dfrac{a_{11}-a_{22}}{2}\right)\dfrac{\sin\omega t}{\omega} & a_{12}\dfrac{\sin\omega t}{\omega} \\ a_{21}\dfrac{\sin\omega t}{\omega} & \cos\omega t + \left(\dfrac{a_{22}-a_{11}}{2}\right)\dfrac{\sin\omega t}{\omega} \end{pmatrix},$$

where

$$\mu = \frac{a_{11}+a_{22}}{2}, \qquad 2a_{11}a_{22} - a_{11}^2 - a_{22}^2 - 4a_{12}a_{21} > 0, \tag{5.25}$$

$$\omega = \frac{1}{2}\sqrt{2a_{11}a_{22} - a_{11}^2 - a_{22}^2 - 4a_{12}a_{21}} > 0.$$

It follows from (5.25) that $a_{12}a_{21} < 0$.

So, the measurements can be represented in the static form

$$z(t) = H'(t)q + \varrho(t), \qquad q = (x_1(T), x_2(T))',$$

where

$$H(t) = e^{-A'(T-t)}e_1, \qquad e_1 = (1,0)',$$

or,

$$H(t) = (H_1(t), H_2(t))',$$

$$H_1(t) = \mathrm{e}^{-\mu(T-t)} \left\{ \cos\omega(T-t) + \left(\frac{a_{22}-a_{11}}{2}\right) \frac{\sin\omega(T-t)}{\omega} \right\},$$

$$H_2(t) = -\,\mathrm{e}^{-\mu(T-t)} \left\{ \frac{\sin\omega(T-t)}{\omega} a_{12} \right\}.$$

The generalized polynomial for the estimation of $q_1 = x_1(T)$ has the form

$$S(t) = H_1(t) + yH_2(t). \tag{5.26}$$

The conditions of Lemma 5.2 reduce to the restrictions:

(a) $S(0) = -S(T) = -1$, or, equivalently,

$$y = y^0 = \frac{\mathrm{e}^{\mu T} + \cos\omega T + \left(\frac{a_{22}-a_{11}}{2}\right)\frac{\sin\omega T}{\omega}}{a_{12}\frac{\sin\omega T}{\omega}};$$

(b) $\frac{dS(t)}{dt} \neq 0, \quad t \in (0,T),$ or, equivalently, the equation

$$a\cos\omega(T-t) + b\sin\omega(T-t) = 0,$$

where

$$a = \frac{\omega\left(\mathrm{e}^{\mu T} + \cos\omega T\right) + \mu\sin\omega T}{\sin\omega T}, \qquad b = \frac{\omega\sin\omega T - \mu\left(\mathrm{e}^{\mu T} + \cos\omega T\right)}{\sin\omega T},$$

has no solution in $(0,T)$;

(c) the equation

$$\delta y\,\mathrm{e}^{-\mu(T-t)}\frac{\sin\omega(T-t)}{\omega}a_{12} = 0,$$

has no root in the open interval $(0,T)$ for $\delta y \neq 0$.

If we evaluate $x_2(T)$, i.e., if $a = (0,1)'$, then we can use Corollary 5.1 with $s = 2$. The conditions of Corollary 5.1 take the form:

(d) $y^0 \neq 0$;

(e) the equation

$$\delta y\, H_1(t) \equiv \delta y\,\mathrm{e}^{-\mu(T-t)} \left\{ \cos\omega(T-t) + \left(\frac{a_{22}-a_{11}}{2}\right) \frac{\sin\omega(T-t)}{\omega} \right\} = 0$$

has no root on $(0, T)$ for $\delta y \neq 0$.

Clearly, if

$$T \leq \frac{\pi}{2\omega}, \qquad a_{22} - a_{11} \geq 0, \tag{5.27}$$

$$\omega \left(e^{\mu T} + \cos \omega T\right) + \mu \sin \omega T > 0, \qquad \omega \sin \omega T - \mu \left(e^{\mu T} + \cos \omega T\right) > 0,$$

then conditions (a)–(e) hold.

Thus, by virtue of Corollary 5.1, the optimal instants for estimating $x_1(T)$ and $x_2(T)$ are common.

Note that $H_2(t)$ does not satisfy the Haar condition as $H_2(T) = 0$ but the restriction (c) holds.

Since $S^0(0) = -S^0(T) = \max_{t \in [0,T]} |S^0(t)|$, then 0 and T are the optimal instants for measurements. Therefore the optimal estimator has the form

$$\Phi^0(t) = \Phi^0_{(1)}\delta(t) + \Phi^0_{(2)}\delta(t - T),$$

where $\Phi^0_{(1)}$ and $\Phi^0_{(2)}$ are determined from the unbiasness condition

$$H(0)\Phi^0_{(1)} + H(T)\Phi^0_{(2)} = a.$$

So, for $a = (1, 0)'$,

$$\Phi^0_{(1)} = 0, \qquad \Phi^0_{(2)} = 1,$$

and for $a = (0, 1)'$,

$$\Phi^0_{(1)} = e^{\mu T} \frac{\omega}{a_{12} \sin \omega T}, \qquad \Phi^0_{(2)} = \frac{\omega \cos \omega T}{a_{12} \sin \omega T} + \frac{a_{22} - a_{11}}{2a_{12}}.$$

Thus conditions (5.27) describe a class of two-dimensional systems with common measurement instants: $t = 0$ and $t = T$ (for $x_1(T)$ only the last measurement is needed).

6. Combined Model for Measurement Noise

Let us consider the case of continuous measurements:

$$z(t) = H'(t)q + \varrho(t), \quad t \in [0, T], \quad z(t), \varrho(t) \in \mathbf{R}^1, \quad H(t), q \in \mathbf{R}^n, \tag{6.1}$$

where $z(t)$ are the measurements, $H(t)$ are given vectors, q is an unknown vector parameter, $\varrho(t)$ is a measurement noise. In contrast to Section 2 it is assumed that, along with uncertain bounded disturbance, white noise is also present [132]:

$$\varrho(t) = \varrho^{(1)}(t) + \varrho^{(2)}(t), \tag{6.2}$$

where $\varrho^{(1)}(t)$ is a deterministic uncertain bounded disturbance

$$|\varrho^{(1)}(t)| \leq \sigma(t), \qquad t \in [0,T], \tag{6.3}$$

and $\varrho^{(2)}(t)$ is a white Gaussian noise:

$$\mathbf{E}\varrho^{(2)}(t) = 0, \qquad \mathbf{E}\varrho^{(2)}(t)\varrho^{(2)}(s) = r^{\varrho}(t)\delta(t-s), \qquad r^{\varrho}(t) \leq r(t). \tag{6.4}$$

The components of $H(t)$, $\varrho^{(1)}(t)$, $\sigma(t)$, $r^{\varrho}(t)$, and $r(t)$ are assumed to be piecewise continuous; $\sigma(t) \geq 0$, $r^{\varrho}(t) \geq 0$, $r(t) > r^0 > 0$.

It is convenient to represent our noise model in the form of a white Gaussian noise with uncertain first and second moment:

$$|\mathbf{E}\varrho(t)| \leq \sigma(t), \qquad \mathbf{E}\big(\varrho(t) - \mathbf{E}\varrho(t)\big)\big(\varrho(s) - \mathbf{E}\varrho(s)\big) = r^{\varrho}(t)\delta(t-s), \tag{6.5}$$

$$r^{\varrho}(t) \leq r(t), \qquad r(t) > r^0 > 0.$$

Then

$$\varrho^{(1)}(t) = \mathbf{E}\varrho(t), \qquad \varrho^{(2)}(t) = \varrho(t) - \varrho^{(1)}(t).$$

Denote by $\mathcal{P}_{\varrho}$ the set of all stochastic processes $\varrho(t)$ described by (6.5).

It is required to estimate a quantity $l_* = a'q$ (cf. (1.2)) with the help of linear functionals

$$\hat{l} = \int_0^T \Phi(t)z(t)\,dt, \qquad \Phi \in L_2, \tag{6.6}$$

where L_2 is the space of square-integrable functions.

A guaranteed value for the second moment of the estimation error is defined by the expression (cf. (1.13))

$$D(\Phi) = \sup_{q\in\mathbf{R}^n,\,\varrho\in\mathcal{P}_{\varrho}} \mathbf{E}\big(\hat{l} - l_*\big)^2. \tag{6.7}$$

Obviously,

$$\mathbf{E}\big(\hat{l} - l_*\big)^2 = \left(q'\left(\int_0^T H(t)\Phi(t)\,dt - a\right)\right)^2 \tag{6.8}$$

$$+\,\mathbf{E}\int_0^T\int_0^T \varrho(t)\varrho(s)\Phi(t)\Phi(s)\,dtds = \left(q'\left(\int_0^T H(t)\Phi(t)\,dt - a\right)\right)^2$$

$$+\left(\int_0^T \Phi(t)\varrho^{(1)}(t)\,dt\right)^2 + \int_0^T r^{\varrho}(t)\Phi^2(t)\,dt.$$

Using (6.5) and arguing as in Section 2, it is easy to prove that

$$D(\Phi) = \begin{cases} \mathcal{J}(\Phi) & \text{if } \int_0^T H(t)\Phi(t)\,dt = a \\ \infty & \text{otherwise,} \end{cases}$$

where

$$\mathcal{J}(\Phi) = \left(\int_0^T \sigma(t)|\Phi(t)|\,dt\right)^2 + \int_0^T r(t)\Phi^2(t)\,dt. \tag{6.9}$$

Thus the optimal guaranteed estimation problem reduces to the following variational prolem

$$\inf_{\Phi\in L_2} \mathcal{J}(\Phi) \tag{6.10}$$

under constraint

$$\int_0^T H(t)\Phi(t)\,dt = a. \tag{6.11}$$

Let us obtain necessary and sufficient conditions for the optimality of estimator.

LEMMA 6.1 [132]. *An estimator* $\Phi(t)$ *is a solution of* (6.10), (6.11) *if and only if there exist a vector* $\lambda \in \mathbf{R}^n$ *and a bounded measurable function* $e(t)$, $|e(t)| \leq 1$, *such that for almost every* $t \in [0,T]$ *the following condition holds*

$$\int_0^T \sigma(t)|\Phi(t)|\,dt\ \sigma(t)\,\text{sign}^*\Phi(t) + r(t)\Phi(t) - \frac{\lambda'}{2}H(t) = 0, \tag{6.12}$$

$$\text{sign}^*\Phi(t) = \begin{cases} +1 & \text{for } t \text{ such that } \Phi(t) > 0 \\ e(t),\ |e(t)| \leq 1, & \text{for } t \text{ such that } \Phi(t) = 0 \\ -1 & \text{for } t \text{ such that } \Phi(t) < 0. \end{cases}$$

REMARK 6.1. Note that (6.12) is analogous to the classical condition $d\mathcal{L}/d\Phi = 0$, where $\mathcal{L}$ is a Lagrange function and $d|x|/dx \sim \text{sign}x$. Condition (6.12) can be also obtained from the necessary and sufficient condition of a minimum of a convex function:

$$\partial\mathcal{L} \ni 0,$$

where $\partial\mathcal{L}$ is a subdifferential of $\mathcal{L}$ at the point Φ. Nevertheless, we do not use this technique in this Section; here we prefer elementary methods.

PROOF. First let us prove the necessity. Let us apply the Lagrange theorem to the nonsmooth variational problem (6.10), (6.11) [75], [51] (see, also,

Chapter 2 for rigorous general statement). If Φ is optimal, then there exist a number $\lambda_0 \geq 0$ and a vector $\lambda \in \mathbf{R}^n$, not both zero, such that

$$\lambda_0 \mathcal{J}(\Phi) + \lambda'\Big(a - \int_0^T H(t)\Phi(t)\,dt\Big) \tag{6.13}$$

$$\leq \lambda_0 \mathcal{J}(\Phi + \delta\Phi) + \lambda'\Big(a - \int_0^T H(t)\big(\Phi(t) + \delta\Phi\big)\,dt\Big), \qquad \forall\,\delta\Phi \in L_2.$$

Let us show that we can put $\lambda_0 = 1$. Further, we shall suppose for simplicity that

$$W = \int_0^T H(t)H'(t)\,dt \quad \text{is nonsingular.} \tag{6.14}$$

Condition (6.14) is the well-known observability condition (see Remark 6.2 below).

Assume that $\lambda_0 = 0$; consequently, $\lambda \neq 0$. Then (6.13) takes the form

$$\lambda' \int_0^T H(t)\delta\Phi\,dt \leq 0 \qquad \forall\,\delta\Phi \in L_2.$$

With $\delta\Phi(t) = H'(t)W^{-1}\lambda$ the latter inequality turns into the inequality $\lambda'\lambda \leq 0$, which is incorrect for nonzero λ. So, $\lambda_0 > 0$ and we can put $\lambda_0 = 1$.

Thus, there exists a vector λ such that the Lagrange function

$$\mathcal{L}(\Phi) = \mathcal{J}(\Phi) + \lambda'\Big(a - \int_0^T H(t)\Phi(t)\,dt\Big)$$

attains a minimum at the optimal estimator Φ:

$$\mathcal{L}(\Phi) \leq \mathcal{L}(\Phi + \delta\Phi), \qquad \forall\,\delta\Phi \in L_2.$$

The variation of $\mathcal{L}$ has the form

$$\mathcal{L}(\Phi + \delta\Phi) - \mathcal{L}(\Phi) = \Big(\int_0^T \sigma(t)|\Phi(t) + \delta\Phi(t)|\,dt\Big)^2 - \Big(\int_0^T \sigma(t)|\Phi(t)|\,dt\Big)^2$$

$$+ \int_0^T r(t)\,(\Phi(t) + \delta\Phi(t))^2\,(t) - \int_0^T r(t)\Phi^2(t)\,dt - \int_0^T \lambda' H(t)\delta\Phi(t)\,dt$$

$$= \Big(\int_0^T \sigma(t)|\Phi(t) + \delta\Phi(t)|\,dt - \int_0^T \sigma(t)|\Phi(t)|\,dt\Big)$$

$$\times \Big(2\int_0^T \sigma(t)|\Phi(t)|\,dt + \int_0^T \sigma(t)|\Phi(t) + \delta\Phi(t)|\,dt - \int_0^T \sigma(t)|\Phi(t)|\,dt\Big)$$

$$+2\int_0^T r(t)\Phi(t)\delta\Phi(t)\,dt + \int_0^T r(t)\left(\delta\Phi(t)\right)^2 dt - \int_0^T \lambda' H(t)\delta\Phi(t)\,dt$$

$$= 2\int_0^T \sigma(t)|\Phi(t)|\,dt\left\{\int_0^T \sigma(t)\Big[|\Phi(t)+\delta\Phi(t)| - |\Phi(t)| - \operatorname{sign}^*\Phi(t)\,\delta\Phi(t)\Big]\,dt\right\}$$

$$+\,2\int_0^T \sigma(t)|\Phi(t)|\,dt\int_0^T \sigma(t)\operatorname{sign}^*\Phi(t)\,\delta\Phi(t)\,dt + 2\int_0^T r(t)\Phi(t)\,\delta\Phi(t)\,dt$$

$$+\int_0^T r(t)\left(\delta\Phi(t)\right)^2 dt + \left(\int_0^T \sigma(t)\Big[|\Phi(t)+\delta\Phi(t)| - |\Phi(t)|\Big]\,dt\right)^2$$

$$-\int_0^T \lambda' H(t)\,\delta\Phi(t)\,dt.$$

Hence,

$$\delta\mathcal{L}(\Phi) = \mathcal{L}(\Phi+\delta\Phi) - \mathcal{L}(\Phi) \tag{6.15}$$

$$= \int_0^T \left\{2\int_0^T \sigma(t)|\Phi(t)|\,dt\;\sigma(t)\operatorname{sign}^*\Phi(t) + 2r(t)\Phi(t) - \lambda' H(t)\right\}\delta\Phi(t)\,dt$$

$$+\,2\int_0^T \sigma(t)|\Phi(t)|\,dt$$

$$\times\int_0^T \sigma(t)\Big[\,|\Phi(t)+\delta\Phi(t)| - |\Phi(t)| - \operatorname{sign}^*\Phi(t)\,\delta\Phi(t)\,\Big]\,dt$$

$$+\int_0^T r(t)\left(\delta\Phi(t)\right)^2 dt + \left(\int_0^T \sigma(t)\Big[|\Phi(t)+\delta\Phi(t)| - |\Phi(t)|\Big]\,dt\right)^2.$$

Assume that condition (6.12) is violated, i.e., for any function $e(t)$, $|e(t)| \le 1$, there exists a set S of nonzero measure such that

$$F(t) = 2\int_0^T \sigma(t)|\Phi(t)|\,dt\;\sigma(t)\operatorname{sign}^*\Phi(t) + 2r(t)\Phi(t) - \lambda' H(t) \neq 0, \quad t \in S.$$

Let us decompose the set S into three subsets S_-, S_0, and S_+, on which the optimal function $\Phi(t)$ takes negative, zero, and positive values, respectively. Obviously, S_- and S_+ do not depend on $e(t)$. Let us investigate three cases: (a) $\mu(S_-) > 0$; (b) $\mu(S_0) > 0$; (c) $\mu(S_+) > 0$, where $\mu(\cdot)$ is the Lebesgue measure. First we consider case (a). Let us select a set $S_-^\eta \subset S_-$ such that $\Phi(t) < -\eta$, $\eta > 0$, on S_-^η. The set S_-^η is of full measure, as

$$\{t\,|\,\Phi(t) < 0\} = \bigcup_{m=1}^{\infty}\left\{t\,\middle|\,\Phi(t) < -\frac{1}{m}\right\}$$

and at least one set $\left\{t \,\middle|\, \Phi(t) < -\frac{1}{m}\right\}$ must have a nonzero measure.

On S_-^η, for $|\delta\Phi(t)| < \eta$,

$$F(t) = -2\sigma(t)\int_0^T \sigma(t)|\Phi(t)|\,dt + 2r(t)\Phi(t) - \lambda' H(t) \neq 0$$

and

$$|\Phi(t) + \delta\Phi(t)| - |\Phi(t)| - \text{sign}^*\Phi(t)\,\delta\Phi(t) = 0.$$

Let us choose

$$\delta\Phi(t) = \begin{cases} -\epsilon\,\text{sign}F(t), & t \in S_-^\eta, \quad \epsilon > 0 \\ 0, & t \notin S_-^\eta. \end{cases}$$

Then, since

$$\int_0^T r(t)\,(\delta\Phi(t))^2\,dt + \left(\int_0^T \sigma(t)\Big[|\Phi(t) + \delta\Phi(t)| - |\Phi(t)|\Big]\,dt\right)^2$$

$$\leq \int_0^T r(t)\,(\delta\Phi(t))^2\,dt + \left(\int_0^T \sigma(t)|\delta\Phi(t)|\,dt\right)^2 \leq C\int_0^T (\delta\Phi(t))^2\,dt,$$

we have

$$\delta\mathcal{L}(\Phi) \leq -\epsilon\int_{S_-^\eta} |F(t)|\,dt + \epsilon^2 C_1,$$

where C, C_1 are some positive constants.

Hence, $\delta\mathcal{L}(\Phi) < 0$ for sufficiently small ϵ. This contradicts the optimality of Φ.

Case (c) can be studied similarily to case (a).

Consider case (b). Assume that $\mu(S_-) = \mu(S_+) = 0$, $\mu(S_0) > 0$.

Note that there exists a set $S_0^0 \subset S_0$, $\mu(S_0^0) > 0$ such that

$$2\int_0^T \sigma(t)|\Phi(t)|\,dt\;\sigma(t) < \left|2r(t)\Phi(t) - \lambda' H(t)\right|, \qquad t \in S_0^0. \tag{6.16}$$

In fact, if

$$2\int_0^T \sigma(t)|\Phi(t)|\,dt\;\sigma(t) \geq \left|2r(t)\Phi(t) - \lambda' H(t)\right| \quad \text{a.e. on } S_0, \tag{6.17}$$

then either

$$2\int_0^T \sigma(t)|\Phi(t)|\,dt\;\sigma(t) = \left|2r(t)\Phi(t) - \lambda' H(t)\right|, \tag{6.18}$$

or,

$$2\int_0^T \sigma(t)|\Phi(t)|\,dt\;\sigma(t) > \big|2r(t)\Phi(t) - \lambda' H(t)\big|. \tag{6.19}$$

Equality (6.18) implies

$$2\int_0^T \sigma(t)|\Phi(t)|\,dt\;\sigma(t)e(t) = -\big(2r(t)\Phi(t) - \lambda' H(t)\big), \tag{6.20}$$

where

$$e(t) = -\operatorname{sign}\big(2r(t)\Phi(t) - \lambda' H(t)\big), \qquad |e(t)| \le 1.$$

Equality (6.19) implies (6.20) with

$$\int_0^T \sigma(t)|\Phi(t)|\,dt\,\sigma(t) > 0, \qquad e(t) = \frac{-\big(2r(t)\Phi(t) - \lambda' H(t)\big)}{2\int_0^T \sigma(t)|\Phi(t)|\,dt\,\sigma(t)}, \qquad |e(t)| < 1.$$

Formula (6.20) contradicts the initial assumption that (6.12) does not hold and (6.16) is valid.

Let us choose

$$\delta\Phi(t) = \begin{cases} -\epsilon\operatorname{sign}\big(2r(t)\Phi(t) - \lambda' H(t)\big), & t \in S_0^0, \quad \epsilon > 0, \\ 0, & t \notin S_0^0. \end{cases}$$

Then

$$\delta\mathcal{J}(\Phi) = \int_{S_0^0}\Big\{2\int_0^T \sigma(t)|\Phi(t)|\,dt\,\sigma(t)e(t) + 2r(t)\Phi(t) - \lambda' H(t)\Big\}\delta\Phi(t)\,dt$$

$$+\,2\int_0^T \sigma(t)|\Phi(t)|\,dt \int_{S_0^0} \sigma(t)\Big[\,|\delta\Phi(t)| - e(t)\,\delta\Phi(t)\Big]\,dt$$

$$+\int_{S_0^0} r(t)\,(\delta\Phi(t))^2\,dt + \Big(\int_{S_0^0}\sigma(t)|\delta\Phi(t)|\,dt\Big)^2.$$

Let us take $e(t) = \operatorname{sign}\delta\Phi(t)$; then $|\delta\Phi(t)| - e(t)\delta\Phi(t) = 0$ and

$$\delta\mathcal{J}(\Phi) \le -\epsilon\int_{S_0^0}\Big\{-2\int_0^T \sigma(t)|\Phi(t)|\,dt\,\sigma(t) + \big|2r(t)\Phi(t) - \lambda' H(t)\big|\Big\}\,dt$$

$$+\,C_2\epsilon^2, \qquad C_2 = \text{const} > 0. \tag{6.21}$$

It follows from (6.16) and (6.21) that $\delta\mathcal{J}(\Phi)$ is negative if ϵ is sufficiently small. This contradicts the initial assumption that Φ is optimal.

Let us now prove the sufficiency. Obviously,

$$|\Phi(t)+\delta\Phi(t)|-|\Phi(t)|\geq \operatorname{sign}^{*}\Phi(t)\cdot\delta\Phi(t), \qquad t\in[0,T].$$

Then from (6.15) it follows that (6.12) is a sufficient condition for Φ to be optimal.

REMARK 6.2. If W is singular, then the dimenision of restriction (6.11) can be reduced. Suppose that $W\lambda=0$ or, equivalently, $\lambda' W\lambda=0$. This means that

$$\int_0^T (\lambda' H(t))^2\,dt = 0 \quad \text{or} \quad \lambda' H(t)=0 \quad \text{a.e. on } [0,T]. \tag{6.22}$$

If the dimension of subspace defined by (6.22) is equal to $n-m$, then there exists an orthogonal matrix C such that

$$CH(t)=\begin{pmatrix}\bar H(t)\\ 0_{n-m}\end{pmatrix}, \qquad Ca=\begin{pmatrix}\bar a\\ 0_{n-m}\end{pmatrix}, \qquad \bar a,\bar H(t)\in\mathbf{R}^m, \tag{6.23}$$

$$\int_0^T (\bar\lambda' \bar H(t))^2\,dt>0 \qquad \forall\bar\lambda\in\mathbf{R}^m$$

or, equivalently,

$$\int_0^T \bar H(t)\bar H'(t)\,dt \qquad \text{is nonsingular.}$$

Then constraint (6.11) turns into the following reduced constraint

$$\int_0^T \bar H(t)\Phi(t)\,dt=\bar a. \tag{6.24}$$

Using (6.23) and (6.24), the singular case can be studied similar to that of Section 4. Thus, we can get that $\lambda_0>0$ in the singular case as well.

□.

It follows from Lemma 6.1 that if the optimal estimator exists, then it is determined by the formula

$$\Phi(t)=\begin{cases}\Phi^+(t)= -\dfrac{\lambda' H(t)+2\sigma(t)\zeta}{2r(t)} & \text{for } t\in t^+;\\[2ex] \Phi^-(t)= -\dfrac{\lambda' H(t)-2\sigma(t)\zeta}{2r(t)} & \text{for } t\in t^-;\\[2ex] 0 & \text{for } t\notin t^+\cup t^-.\end{cases} \tag{6.25}$$

Here

$$\zeta = \int_0^T \sigma(t)|\Phi(t)|\,dt, \tag{6.26}$$

$$t^+ = \Big\{ t \in [0,T] \;\Big|\; \lambda' H(t) + 2\sigma(t)\zeta < 0 \Big\}, \tag{6.27}$$

$$t^- = \Big\{ t \in [0,T] \;\Big|\; \lambda' H(t) - 2\sigma(t)\zeta > 0 \Big\}.$$

Conversely, if Φ is defined by (6.25) – (6.27), then Φ satisfies (6.12) and, hence, is optimal.

Formula (6.25) contains $n+1$ unknown parameters: ζ and λ. These parameters can be determined from $n+1$ equations (6.11) and (6.26). In general case, this is a quite complex nonlinear problem that can be solved by various numerical methods. Formula (6.25) makes clear the structure of the optimal estimator. In particular, we see that the optimal estimator can vanish on a set of nonzero measure.

In some cases, Lemma 6.1 allows us to solve the problem (6.10), (6.11) completely.

EXAMPLE 6.1 [132]. Consider the measurements of the form (cf. (6.1))

$$z(t) = q_1 \cos t + q_2 \sin t + \varrho(t), \qquad t \in [0, 2\pi],$$

where $\varrho(t)$ satisfies (6.2) – (6.4) with constant parameters σ and r. In our notation, $n=2$, $m=1$, $H(t) = (\cos t, \sin t)'$. Let us represent the Lagrange multiplier in (6.12) in the form

$$\frac{1}{2}\lambda = \sigma\zeta\xi(\cos\kappa, \sin\kappa)', \tag{6.28}$$

where ζ, κ, and ξ are yet unknown parameters. It follows from (6.25) that $\Phi \not\equiv 0$ if $|\xi\cos(t-\kappa)| > 1$. So, the problem has a nontrivial solution under $\xi > 1$; this means that we can put

$$\xi = 1/\cos z, \qquad 0 < z < \pi/2. \tag{6.29}$$

With this notation,

$$\Phi(t) = \begin{cases} \Phi^+(t) = -\dfrac{\sigma\zeta}{r}\left[\dfrac{\cos(t-\kappa)}{\cos z} + 1\right] & \text{for } t \in t^+; \\[2ex] \Phi^-(t) = -\dfrac{\sigma\zeta}{r}\left[\dfrac{\cos(t-\kappa)}{\cos z} - 1\right] & \text{for } t \in t^-; \\[2ex] 0 & \text{for } t \notin t^+ \cup t^-. \end{cases} \tag{6.30}$$

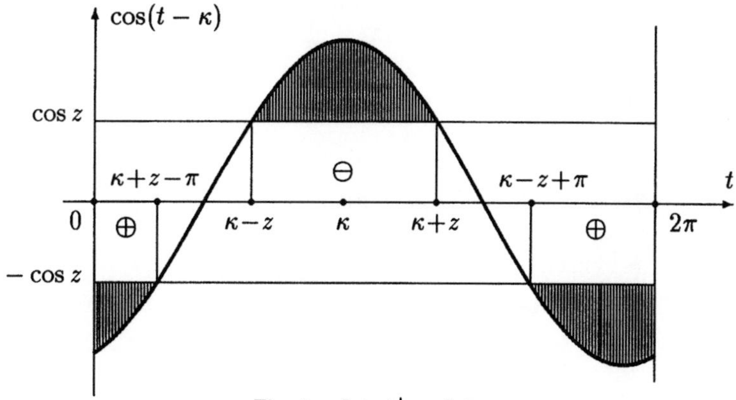

Fig. 2. Sets t^+ and t^-

To determine ζ, κ, and z we must substitute (6.30) into (6.26) and (6.11):

$$\zeta = \int_{t^+} \Phi^+(t)\, dt - \int_{t^-} \Phi^-(t)\, dt, \tag{6.31}$$

$$a = \int_{t^+} H(t)\Phi^+(t)\, dt + \int_{t^-} H(t)\Phi^-(t)\, dt. \tag{6.32}$$

Generally, the sets

$$t^+ = \Big\{ t \in [0, 2\pi] \;\Big|\; \cos(t-\kappa) + \cos z < 0 \Big\},$$

$$t^- = \Big\{ t \in [0, 2\pi] \;\Big|\; \cos(t-\kappa) - \cos z > 0 \Big\},$$

defined by (6.27), are quite complicated (see Figure 2). We shall use a trick to compute quadratures (6.31) and (6.32). Let us represent (6.31) and (6.32) in the form

$$\zeta = \int_0^{2\pi} \max\left(\Phi^+(t), 0\right) dt - \int_0^{2\pi} \min\left(\Phi^-(t), 0\right) dt, \tag{6.33}$$

$$a = \int_0^{2\pi} H(t) \max\left(\Phi^+(t), 0\right) dt + \int_0^{2\pi} H(t) \min\left(\Phi^-(t), 0\right) dt. \tag{6.34}$$

The integrand functions in (6.33) and (6.34) are 2π-periodic. Consequently,

$$\zeta = \int_{-\kappa}^{2\pi-\kappa} \max\left(\Phi^+(t+\kappa), 0\right) dt - \int_{-\kappa}^{2\pi-\kappa} \min\left(\Phi^-(t+\kappa), 0\right) dt$$

$$= \int_{-\pi/2}^{3\pi/2} \max(\Psi^+(t), 0)\, dt - \int_{-\pi/2}^{3\pi/2} \min(\Psi^-(t), 0)\, dt,$$

$$a = \int_{-\kappa}^{2\pi-\kappa} H(t+\kappa) \max(\Phi^+(t+\kappa), 0)\, dt$$

$$+ \int_{-\kappa}^{2\pi-\kappa} H(t+\kappa) \min(\Phi^-(t+\kappa), 0)\, dt$$

$$= \int_{-\pi/2}^{3\pi/2} H(t+\kappa) \max(\Psi^+(t), 0)\, dt + \int_{-\pi/2}^{3\pi/2} H(t+\kappa) \min(\Psi^-(t), 0)\, dt,$$

where

$$\Psi^+(t) = -\frac{\sigma\zeta}{r}\left[\frac{\cos t}{\cos z} + 1\right], \qquad \Psi^-(t) = -\frac{\sigma\zeta}{r}\left[\frac{\cos t}{\cos z} - 1\right].$$

Since $\Psi^+(t) > 0$ with $-z + \pi < t < z + \pi$ and $\Psi^-(t) < 0$ with $-z < t < z$, we have that (6.33) and (6.34) turn into the equalities

$$\zeta = \int_{-z+\pi}^{z+\pi} \Psi^+(t)\, dt - \int_{-z}^{z} \Psi^-(t)\, dt,$$

$$a = \int_{-z+\pi}^{z+\pi} H(t+\kappa)\Psi^+(t)\, dt + \int_{-z}^{z} H(t+\kappa)\Psi^-(t)\, dt$$

or, computing the quadratures,

$$\frac{r}{\sigma^2} = 4\,(\tan z - z)\,, \tag{6.35}$$

$$a = -\frac{\zeta\sigma}{r}\left(\frac{2z - \sin 2z}{\cos z}\right)\begin{pmatrix} \cos\kappa \\ \sin\kappa \end{pmatrix}, \tag{6.36}$$

respectively.

Define more exactly the form of the parameter being estimated. Let us set $l_* = a'q$, $a = (\cos\eta, \sin\eta)'$. In particular, with $\eta = 0$ we have $l_* = q_1$, and with $\eta = \pi/2$ we have $l_* = q_2$. It follows from (6.36) that

$$\kappa = \eta + \pi, \qquad \zeta = \frac{4\sigma\,(\sin z - z\cos z)}{2z - \sin 2z}. \tag{6.37}$$

Finally, ζ, κ, and z (ξ) are determined by (6.37) and (6.35) ((6.29)); obviously, (6.35) has a unique soluiton. Then the optimal guaranteed variance (6.9) can be easily computed:

$$D(\Phi) = \frac{4\sigma^2\,(\tan z - z)}{2z - \sin 2z}.$$

Let us study some limiting cases:
(a) assume that uncertain deterministic errors dominate: $r \to 0$; then $z \to 0$. If z is small, then Φ is nonzero only on small intervals t^+ and t^-. The total length of the intervals is equal to $4z$. If $r \to 0$, then the intervals contract into two points. It can be easily shown that

$$D(\Phi) \to \sigma^2, \qquad z \approx \left(\frac{3r}{4\sigma^2}\right)^{1/3}.$$

(b) assume that the measurement noise is almost white: $\sigma \to 0$. In this case, $z \to \pi/2$ and almost all points of $[0, 2\pi]$ are substantial for estimation; $D(\Phi) \to r/\pi$.

Consider also the case when both components of the noise are important:

$$\frac{r}{\sigma^2} = 4 - \pi \approx 1, \quad z = \pi/4, \qquad D(\Phi) = 2\sigma^2 \frac{4-\pi}{\pi - 2} = \frac{2r}{\pi - 2} \approx 2\sigma^2 \approx 2r.$$

EXAMPLE 6.2. Let us study another problem [132], in which Lemma 6.1 gives a complete solution. Assume that the optimal estimator has a definite sign:

$$\operatorname{sign} \Phi(t) = \delta, \qquad \delta = \pm 1, \qquad t \in [0, T].$$

It follows from (6.25) that

$$\Phi(t) = -\frac{1}{2r(t)}\Big[\lambda' H(t) + 2\sigma(t)\,\delta\,\zeta\Big], \qquad t \in [0, T]. \tag{6.38}$$

Then, from (6.26), (6.11),

$$\delta\zeta = \frac{1}{2} b'\lambda, \qquad \frac{\lambda}{2} = A^{-1} a, \tag{6.39}$$

where

$$b = -\int_0^T \frac{\sigma(t)}{r(t)} H(t)\,dt \left(1 + \int_0^T \frac{\sigma^2(t)}{r(t)}\,dt\right)^{-1},$$

$$A = \int_0^T \frac{1}{r(t)}\Big[H(t)H'(t) + \sigma(t)H(t)b'\Big]\,dt.$$

Let us now show that if

$$\int_0^T \frac{1}{r(t)} H(t)H'(t)\,dt \quad \text{is nonsingular,} \tag{6.40}$$

then A is nonsingular (obviously, (6.40) is equivalent to (6.14)). If (6.40) is not valid, then we can pass to an observable subspace.

The matrix A can be represented in the form

$$A = \int_0^T \frac{1}{r(t)} H(t) H'(t)\, dt - \frac{\displaystyle\int_0^T \frac{\sigma(t)}{r(t)} H(t)\, dt \int_0^T \frac{\sigma(t)}{r(t)} H'(t)\, dt}{\displaystyle 1 + \int_0^T \frac{\sigma^2(t)}{r(t)}\, dt}. \tag{6.41}$$

Consequently, A is symmetric. Further, for any nonzero $\alpha \in \mathbf{R}^n$, we have from (6.41) that

$$\alpha' A \alpha = \int_0^T \frac{1}{r(t)} f^2(t)\, dt - \frac{\displaystyle\left(\int_0^T \frac{\sigma(t)}{r^{1/2}(t)} \cdot \frac{f(t)}{r^{1/2}(t)}\, dt\right)^2}{\displaystyle 1 + \int_0^T \frac{\sigma^2(t)}{r(t)}\, dt},$$

where $f(t) = \alpha' H(t)$. Using the Cauchy-Schwarz-Bunyakovsky inequality and (6.40), we get

$$\alpha' A \alpha \geq \int_0^T \frac{1}{r(t)} f^2(t)\, dt - \frac{\displaystyle\int_0^T \frac{\sigma^2(t)}{r(t)}\, dt \cdot \int_0^T \frac{f^2(t)}{r(t)}\, dt}{\displaystyle 1 + \int_0^T \frac{\sigma^2(t)}{r(t)}\, dt}$$

$$= \int_0^T \frac{1}{r(t)} f^2(t)\, dt > 0.$$

So, under (6.40) the matrix A is nonsingular.

Finally, from (6.38) and (6.39),

$$\Phi(t) = \frac{1}{r(t)} \left[H'(t) + \sigma(t) b' \right] A^{-1} a. \tag{6.42}$$

In line with our initial assumption the following condition must hold

$$\Phi(t)\, \delta > 0, \qquad t \in [0, T]. \tag{6.43}$$

Thus, to obtain the optimal estimator we have to compute (6.42) and to check (6.43). If (6.43) holds, then, due to Lemma 6.1, Φ is optimal. If (6.43) is not valid, then our initial assumption is false.

7. Least-Squares Method in Guaranteed Parameter Estimation

The finding of optimal estimator can present a challenge. Sometimes, an algorithm that is optimal for one model of noises must operate under another assumptions about noises. It is not unfrequent in applications that a mathematical description of real plants and measurement devices is not quite accurate. Therefore the design decision is often based on the values of orders of system performance rather than on exact characteristics of inaccurate model.

So, in many cases, instead of searching for an optimal estimator, it is more preferable to use the existing simpler estimation algorithms. In such an approach, however, we must evaluate *without finding the optimal algorithm* the increment in estimation error due to the rejection of optimal filter. In this section, we shall obtain such an estimate for the simplest guaranteed estimation problem studied in Section 2. An advantage of our method lies in the fact that our method can be generalized to dynamic systems, in particular, to dynamic systems with aftereffect. Furthermore, this approach can be applied not only to guaranteed estimation, but to other important estimation problems, in which we have to use simplified algorithms, e.g., to the Kalman filtering problem in systems with uncertain noise intensities, to the filtering problem for weakly connected systems, etc. The main content of the book is devoted to the detailed application of our method to the problems mentioned.

Consider the guaranteed estimation problem (2.1)–(2.4), (2.7). For simplicity, we assume that $\sigma(t) \equiv 1$. It was shown in Section 2 that this estimation problem reduces to the convex variational problem (3.15), (3.16):

$$I_0 = \inf_{\Phi \in \mathcal{F}} \int_0^T |\Phi(t)|\, dt \quad \text{under constraint} \quad \int_0^T H(t)\Phi(t)\, dt = a. \tag{7.1}$$

Instead of the optimal solution Φ^0 of nonsmooth problem (7.1), let us use a simple estimator φ that corresponds to the least-squares method; the estimator φ is called LSM-estimator.

Recall that a classical least-squares method (LSM) for estimating $l_* = a'q$ by measurements (2.1)

$$z(t) = H'(t)q + \varrho(t), \qquad t \in [0, t],$$

is defined by the following variational problem

$$\hat{l} = a'\hat{q}, \qquad \hat{q} = \arg\min_{q \in \mathbf{R}^n} \int_0^T \Big(z(t) - H'(t)q\Big)^2 dt, \tag{7.2}$$

It can be easily shown that the solution of (7.2) has the form

$$\hat{l} = \int_0^T \varphi(t)z(t)\, dt, \quad \varphi(t) = a'W^{-1}H(t), \quad W = \int_0^T H(t)H'(t)\, dt. \tag{7.3}$$

We assume for simplicity that $H(t) \in \mathbf{R}^n$ is continuous and W is nonsingular (cf. (6.14) and Remark 6.2). Note that the heuristic estimation method (7.2) does not use any assumption about noise. Let us give another interpretation for φ.

LEMMA 7.1. 1^0. *LSM-estimator φ is a solution of the linear-quadratic problem*

$$J_0^2 = \inf_{\Phi \in L_2} \int_0^T \Phi^2(t)\, dt, \qquad \int_0^T H(t)\Phi(t)\, dt = a. \tag{7.4}$$

2^0. *The optimal value of the problem is determined by the formula*

$$J_0^2 = a'W^{-1}a. \tag{7.5}$$

PROOF. It follows from the Lagrange theorem [75], [51] (see also Chapter 2) that if an estimator Φ is optimal, then Φ minimizes the Lagrange function

$$\mathcal{L}(\Phi) = \lambda_0 \int_0^T \Phi^2(t)\, dt + \lambda' \left(a - \int_0^T H(t)\Phi(t)\, dt \right), \tag{7.6}$$

where $\lambda_0 \geq 0$, $\lambda_0^2 + \lambda'\lambda \neq 0$. (Since W is nonsingular, $\lambda_0 = 1$ as well as in Section 6.) Moreover, if Φ minimizes (7.6) (with $\lambda_0 = 1$), then Φ is optimal.

Differentiating (7.6) with respect to Φ, we have

$$\Phi(t) = \frac{1}{2}\lambda' H(t). \tag{7.7}$$

Substituting (7.7) into the constraint in (7.4), we get

$$\frac{1}{2}\lambda = W^{-1}a. \tag{7.8}$$

Then (7.4) and (7.5) folow from (7.7) and (7.8). □

REMARK 7.1. If $\varrho(t)$ is a white noise process:

$$\mathbf{E}\varrho(t) = 0, \qquad \mathbf{E}\varrho(t)\varrho(s) = r\delta(t-s), \quad r > 0,$$

then it can be easily shown that the estimation problem

$$\inf_{\Phi \in L_2} \sup_{q \in \mathbf{R}^n} \mathbf{E} \left(\hat{l} - l_* \right)^2$$

reduces to (7.4) and, in this case, LSM-estimator is an optimal linear estimator.

Thus the use of φ instead of Φ^0 means that the problem (7.1) is replaced by the following simple problem

$$J_0 = \inf_{\Phi \in L_2} \left(\int_0^T \Phi^2(t)\, dt \right)^{\frac{1}{2}}, \qquad \int_0^T H(t)\Phi(t)\, dt = a. \tag{7.9}$$

The power $\frac{1}{2}$ emphasizes that we replaced one norm $\int_0^T |\Phi(t)|\,dt$ by another one $\left(\int_0^T \Phi^2(t)\,dt\right)^{1/2}$.

We define the level of nonoptimality for the LSM-estimator by the ratio

$$\Delta = \frac{\sup_{q\in\mathbf{R}^n,\,|\varrho(t)|\leq 1} |\hat{l}_\varphi - l_*|}{\sup_{q\in\mathbf{R}^n,\,|\varrho(t)|\leq 1} |\hat{l}_0 - l_*|}, \tag{7.10}$$

where $l_\varphi = \int_0^T \varphi(t)z(t)\,dt$ and $l_0 = \int_0^T \Phi^0(t)z(t)\,dt$ are the LSM-estimate and the optimal guaranteed estimate, respectively. The estimators φ and Φ^0 satify the unbiasness condition; consequently, it follows from (2.7) that

$$\Delta = \frac{I(\varphi)}{I(\Phi^0)} = \frac{I(\varphi)}{I_0}, \qquad I(\Phi) = \int_0^T |\Phi(t)|\,dt. \tag{7.11}$$

Since we prefer not to solve (7.1), the denominator in (7.11) is unknown and, therefore, Δ is also unknown. Our aim is to construct an upper bound for Δ, which can be computed *without solving the nonsmooth problem* (7.1). If this upper bound is not too large, then the use of the simple LSM-estimator will be justified.

THEOREM 7.1. *Let* φ *be the LSM-estimator* (7.3). *Then*

$$1 \leq \Delta \leq \Delta^0, \qquad \Delta^0 = \frac{\displaystyle\int_0^T |\varphi(t)|\,dt \cdot \max_{[0,T]} |\varphi(t)|}{\displaystyle\int_0^T \varphi^2(t)\,dt} = \frac{||\varphi||_1 \cdot ||\varphi||_\infty}{||\varphi||_2^2}, \tag{7.12}$$

where $||\cdot||_i$ *is the norm in* L_i, i=1,2,∞.

Theorem 7.1 shows that we must know the LSM-estimator only to compute Δ^0. Formula (7.12) is a central result of our book. In subsequent chapters Theorem 7.1 will be generalized in various directions.

PROOF OF THEOREM 7.1. The dual problem for the primal problem (7.1) has the form (see (5.1), (5.2) from Section 5, where for technical reason we assumed that $H(t)$ is continuous)

$$I^0 = \sup_{\lambda\in\mathbf{R}^n} a'\lambda \quad \text{under constraint} \quad \max_{[0,T]} |H'(t)\lambda| \leq 1. \tag{7.13}$$

Moreover, due to Lemma 5.1,

$$I_0 = I^0. \tag{7.14}$$

Arguing as in Section 5, we can see that (7.9) is equivalent to (5.6), where

$$\mathcal{L}(\Phi,\lambda) = \left(\int_0^T \Phi^2(t)\,dt\right)^{\frac{1}{2}} + \lambda'\left(a - \int_0^T H(t)\Phi(t)\,dt\right). \tag{7.15}$$

Let us construct the problem dual to (7.9) by the rule (5.7).

LEMMA 7.2. 1^0. *The dual problem for the primal problem* (7.9) *has the form*

$$J^0 = \sup_{\lambda\in\mathbf{R}^n} a'\lambda \quad \text{under constraint} \quad (\lambda' W\lambda)^{\frac{1}{2}} - 1 \le 0; \tag{7.16}$$

2^0. *The solution of* (7.16) *has the form*

$$\lambda^0 = \left(a'W^{-1}a\right)^{-\frac{1}{2}} W^{-1}a; \tag{7.17}$$

3^0. *The duality relation holds*:

$$J_0 = J^0 = \left(a'W^{-1}a\right)^{\frac{1}{2}}. \tag{7.18}$$

PROOF OF LEMMA 7.2. Denoting by $\langle\Phi,\Psi\rangle$ the inner product in L_2, let us represent the Lagrange function (7.15) in the form

$$\mathcal{L}(\Phi,\lambda) = ||\Phi|| - \langle\lambda' H,\Phi\rangle + a'\lambda.$$

Hence,

$$\inf_{\Phi\in L_2} \mathcal{L}(\Phi,\lambda) = \begin{cases} a'\lambda & \text{if } ||\lambda' H|| \le 1 \\ -\infty & \text{otherwise.} \end{cases} \tag{7.19}$$

In fact, if $||\lambda' H|| \le 1$, then

$$||\Phi|| - \langle\lambda' H,\Phi\rangle \ge ||\Phi|| - ||\lambda' H||\,||\Phi|| = ||\Phi||\,(1 - ||\lambda' H||) \ge 0.$$

Trivially, $||0|| - \langle\lambda' H, 0\rangle = 0$.

Conversely, if $||\lambda' H|| > 1$, then, taking $\Phi = \alpha\lambda' H$, we have

$$||\Phi|| - \langle\lambda' H,\Phi\rangle = |\alpha|\,||\lambda' H|| - \alpha||\lambda' H||^2 = \alpha||\lambda' H||\,(1 - ||\lambda' H||) \to -\infty$$

as $\alpha \to \infty$. So, (7.16) follows from (7.19) and (5.7).

Let us now prove Assertions 2 and 3 of Lemma 7.2. The objective function of the dual problem is continuos and the constraint defines a compact set. Therefore the solution of (7.16) λ^0 exists.

Applying the Lagrange theorem [75], [51] (see also Chapter 2) to (7.16), we get that there exist $\alpha_0 \ge 0$ and $\alpha \ge 0$, not both zero, such that the Lagrange function

$$\tilde{\mathcal{L}}(\lambda) = \alpha_0 a'\lambda + \alpha\,(\lambda' W\lambda - 1) \tag{7.20}$$

attains its minimum at λ^0 (obviously, the power $\frac{1}{2}$ in the constraint can be omitted) and the complementary slackness condition is valid:

$$\alpha\,(\lambda' W \lambda - 1) = 0. \tag{7.21}$$

Moreover, if the constraint qualification hypothesis holds:

$$\text{there exists } \lambda \text{ such that } \lambda' W \lambda < 1, \tag{7.22}$$

then $\alpha_0 = 1$.

Obviously, (7.22) holds, e.g., with $\lambda = 0$; therefore $\alpha_0 = 1$. If $\alpha = 0$, then we have from (7.20) that

$$a'\lambda^0 \leq a'\lambda, \qquad \forall \lambda \in \mathbf{R}^n.$$

Clearly, the latter inequality holds only in the trivial case for $a = 0$. So, in (7.20) $\alpha_0 = 1$, $\alpha > 0$. Differentiating (7.20) with respect to λ, we get

$$\lambda^0 = \frac{1}{2\alpha} W^{-1} a. \tag{7.23}$$

Substituting (7.23) into (7.21) ($\alpha > 0$), we obtain (7.17). Formula (7.18) follows from (7.17) and (7.5). Lemma 7.2 is proved.

To obtain an upper estimate for Δ in (7.11), let us construct an estimate from below for I_0. By virtue of (7.14),

$$I_0 = I^0 \geq a'\lambda, \quad \text{for any } \lambda \text{ such that } \max_{[0,T]} |H'(t)\lambda| \leq 1. \tag{7.24}$$

So, to obtain a "good" estimate for I^0, we should find an "appropriate" feasible element of the dual problem (7.13). The best element for this purpose it the solution of the dual problem (7.13). But the dual problem is as difficult to solve as the primal one. The problem (7.9) aproximates (7.1), and (7.16) approximates (7.13). Therefore let us use the known solution of the dual problem (7.17) to find a "good" approximation for the unknown solution of (7.13). We shall search the approximating element along the direction determined by λ^0:

$$\lambda = \tilde{\nu}\lambda^0 = \nu W^{-1} a, \qquad \tilde{\nu}, \nu \in \mathbf{R}^1. \tag{7.25}$$

Then, taking into account (7.3), we see that for all ν such that

$$\max_{[0,T]} |\nu a' W^{-1} H(t)| \leq 1$$

the vectors $\lambda = \nu W^{-1} a$ and $-\lambda = -\nu W^{-1} a$ are feasible and

$$I_0 = I^0 \geq |a'\lambda| = |\nu|\, a' W^{-1} a = |\nu| \int_0^T \varphi^2(t)\, dt.$$

Hence, for $\nu \neq 0$,

$$\Delta \leq \frac{\int_0^T |\varphi(t)|\,dt}{|\nu| \int_0^T \varphi^2(t)\,dt}. \tag{7.26}$$

It follows from (7.3), (7.24)–(7.26) that the most successful value of ν is determined by the equality:

$$\max_{t\in[0,T]} |\nu a' W^{-1} H(t)| = 1 \quad \text{or} \quad \max_{[0,T]} |\nu\,\varphi(t)| = 1,$$

whence

$$\nu = \pm \left\{ \max_{[0,T]} |\varphi(t)| \right\}^{-1}. \tag{7.27}$$

Thus the estimate $\Delta \leq \Delta^0$ follows from (7.26) and (7.27).

Moreover, since φ is not an optimal guaranteed estimator,

$$\Delta = \frac{I(\varphi)}{I(\Phi^0)} \geq 1.$$

□

7.1. *Examples*

Let us consider simple examples.

EXAMPLE 7.1. Suppose that in the problem (7.1) $n = 1$ and $a = 1$, i.e., the problem has the form

$$\inf_{\Phi} \int_0^T |\Phi(t)|\,dt, \qquad \int_0^T H(t)\Phi(t)\,dt = 1, \tag{7.28}$$

where $H(t)$ is a given scalar continuous function.

Obviously, the solution of (7.28) is determined by the formula

$$\Phi^0(t) = H^{-1}(t^0)\,\delta(t - t^0), \qquad t^0 = \operatorname*{argmax}_{[0,T]} |H(t)|,$$

and the LSM-estimator has the form

$$\varphi(t) = H(t) \left(\int_0^T H^2(t)\,dt \right)^{-1}.$$

Consequently, in the scalar case,

$$\Delta = \Delta^0 = \frac{\displaystyle\int_0^T |H(t)|\, dt \cdot \max_{[0,T]} |H(t)|}{\displaystyle\int_0^T H^2(t)\, dt}. \tag{7.29}$$

Generally, Δ^0 can be arbitrarily large. In fact, consider a function

$$H^0(t) = \begin{cases} 1, & t \in [0, \epsilon^2] \\ \epsilon, & t \in (\epsilon^2, T]. \end{cases} \tag{7.30}$$

It is easy to compute that

$$\Delta^0 = \frac{\epsilon^2 + \epsilon(T - \epsilon^2)}{\epsilon^2 + \epsilon^2(T - \epsilon^2)} \sim \epsilon^{-1} \to \infty \quad \text{as} \quad \epsilon \to 0.$$

Though $H^0(t)$ defined by (7.30) is not continuous, the function $H^0(t)$ can be approximated by an appropriate continuous function with the same result. Moreover, it will be shown in Chapter 3 that Theorem 7.1 is valid for piecewise continuous functions as well.

The function (7.30) is quite exotic. Below we consider more regular functions.

EXAMPLE 7.2. Let us present the values of Δ^0 for certain elementary functions ($C = \text{const}$).

(a) $H(t) = C \cos t$, $T = 2\pi N$, N is an interger, $\qquad \Delta^0 = \dfrac{4}{\pi}$;

(b) $H(t) = C t^\alpha$, $\qquad \Delta^0 = \dfrac{2\alpha + 1}{\alpha + 1} \le 2$;

(c) $H(t) = C e^{\alpha t}$, $\qquad \Delta^0 = \begin{cases} \dfrac{2\left(e^{2\alpha T} - e^{\alpha T}\right)}{e^{2\alpha T} - 1} < 2 & \text{if } \alpha > 0 \\ \dfrac{2\left(1 - e^{\alpha T}\right)}{1 - e^{2\alpha T}} < 2 & \text{if } \alpha < 0. \end{cases}$

EXAMPLE 7.3. Let us consider the problem for the estimation of the Fourier coefficients. Suppose the measurements have the form:

$$z(t) = q_1 \cos t + q_2 \sin t + \ldots + q_{2p-1} \cos pt + q_{2p} \sin pt, \qquad t \in [0, T],$$

where $T = 2\pi N$ and p, N are integers. In our notation, $n = 2p$,

$$H(t) = \left(H_1(t), \ldots, H_{2p}(t)\right)' = \left(\cos t, \sin t, \ldots, \cos pt, \sin pt\right)',$$

$$a = e_k = (\underbrace{0, \ldots, 0, 1}_{k}, 0, \ldots, 0)'.$$

The LSM-estimator for $l_* = q_k$ is determined by (7.3):

$$\varphi_k(t) = e_k' W^{-1} H(t) = (\pi N)^{-1} H_k(t) = \begin{cases} (\pi N)^{-1} \cos \frac{(k+1)t}{2} & \text{if } k \text{ is odd} \\ (\pi N)^{-1} \sin \frac{kt}{2} & \text{if } k \text{ is even,} \end{cases}$$

$k = 1, \ldots, 2p$.

Then it follows from Theorem 7.1 that for $k = 1, \ldots, 2p$

$$\Delta_k \leq \Delta_k^0 = \frac{\displaystyle\int_0^{2\pi N} |H_k(t)|\, dt \cdot \max_{[0,T]} |H_k(t)|}{\displaystyle\int_0^{2\pi N} H_k^2(t)\, dt} = \frac{\displaystyle\int_0^{2\pi} |H_k(t)|\, dt}{\pi N \displaystyle\int_0^{2\pi} H_k^2(t)\, dt} = \frac{4}{\pi} \approx 1.28.$$

Note that the last estimate is valid for arbitrary p and N.

Thus the loss of optimality is at most 30 percent. So, our estimate of the nonoptimality level can be highly efficient.

EXAMPLE 7.4. Let us show that in some cases our estimate can be sharp for arbitrary n. Consider the problem with

$$H(t) = (\cos t, \ldots, \cos nt)', \quad T = 2\pi N, \qquad a = e_n = (0, \ldots, 0, 1)'. \quad (7.31)$$

We have shown in Section 5 that (7.1), (7.31) is equivalent to the following Chebyshev approximation problem

$$\min_{\{y_j\}_1^{n-1}} \max_{[0,2\pi N]} \Big|\cos nt + \sum_{j=1}^{n-1} y_j \cos jt\Big| = \left(\int_0^{2\pi N} |\Phi^0(t)|\, dt\right)^{-1} = I_0^{-1}. \quad (7.32)$$

It is proved in [15] (p. 145) that the optimal solution in (7.32) is attained at $y_j = 0$, $j = 1, \ldots, n-1$. So, $I_0 = 1$.

Obviously, $\varphi(t) = (\pi N)^{-1} \cos nt$ and $\Delta = \int_0^{2\pi N} |\varphi(t)|\, dt\, I_0^{-1} = 4/\pi$. Similar to Example 7.3, it is easy to compute that $\Delta^0 = 4/\pi$. Thus in our example $\Delta = \Delta^0$.

7.2. *Lidov's Estimate*

The approach proposed above can be used in many problems. Let us show that a very useful estimate of the nonoptimality level of an approximate impulse estimator [131] can be easily derived with the help of our approach.

Consider the problem of moments (3.15), (3.16). To solve this problem, we replace it by an appropriate discrete problem (1.10), (1.11) (with $\sigma_i = 1$, $i = 1, \ldots, N$). It is shown in Section 3 that the latter problem reduces to a linear programming problem (cf. (3.1), (3.2)), which can be resolved by the simplex method [46]. Suppose that, having solved the discrete problem or having stopped at a current iteration of the simplex method, we got an approximate impulse unbiased estimator of the form

$$\Phi(t) = \sum_{s=1}^{n} \Phi_{(s)}\delta(t - t_s), \qquad \sum_{s=1}^{n} H(t_s)\Phi_{(s)} = a,$$

$$\Phi_{(s)} \neq 0, \qquad \det\mathcal{H} \neq 0, \quad \mathcal{H} = (H(t_1), \dots, H(t_n)).$$

If $\Phi(t)$ would be optimal, then, in accordance with Theorem 4.1, Lemma 5.1, and the proof of Lemma 5.1, there exists a solution λ for the dual problem such that

$$\mathcal{H}'\lambda = e, \qquad e = \left(\operatorname{sign}\Phi_{(1)}, \dots, \operatorname{sign}\Phi_{(n)}\right)', \tag{7.33}$$

$$a'\lambda = \int_0^T |\Phi(t)|\, dt = \sum_{s=1}^{n} |\Phi_{(s)}|.$$

Since $\Phi(t)$ is not optimal, the vector λ formally determined from (7.33), generally, does not satisfy the constraints $|H'(t)\lambda| \leq 1$, $t \in [0, T]$, but the vector

$$\lambda_\alpha = \frac{\lambda}{1 + \alpha}, \qquad \alpha = \max_{t \in [0,T]} |e'\mathcal{H}^{-1}H(t)| - 1 \tag{7.34}$$

does. Obviously, $\alpha \geq 0$, and if Φ is optimal, then $\alpha = 0$; therefore α characterizes a "defect" of Φ.

By the duality relation (cf. Lemma 5.1),

$$\int_0^T |\Phi(t)|\, dt \geq I_0 = I^0 \geq a'\lambda_\alpha = (1 + \alpha)^{-1}\lambda' a = (1 + \alpha)^{-1} e'\mathcal{H}^{-1} a$$

$$= (1 + \alpha)^{-1} e' \left(\Phi_{(1)}, \dots, \Phi_{(n)}\right)' = (1 + \alpha)^{-1} \int_0^T |\Phi(t)|\, dt.$$

Hence,

$$1 \leq \frac{\int_0^T |\Phi(t)|\, dt}{\int_0^T |\Phi^0(t)|\, dt} \leq (1 + \alpha),$$

where α can be easily calculated from (7.34). Thus an appropriate approximation can be chosen *without the exact solution* of the initial variational problem.

8. Guaranteed Estimation with Anomalous Measurement Errors

In applied estimation problems a number of measurements can contain anomalous errors (outliers). Various heuristic approaches exist to filter out anomalous errors. One approach is to select and exclude outliers by appropriate threshold criterion (see, e.g., [62]). Another approach is to use the least absolute value method, which is known to be a robust estimation method (see, e.g., [45]). Howerever, in these approaches the estimation problems are not set as optimal ones. These approaches are effective only for outliers that stand out sharply against standard measurements. Moreover, these methods do not provide bounds for parameter estimation errors.

A lot of papers are devoted to statistical estimation problems in the presence of outliers. In a pioneer work by P. Huber [71] a shift parameter estimation problem under independent measurement errors with ϵ-polluted density was considered. This trend was extensively developed by P. Huber [72] and Ya.Z. Tsypkin [206]. The problem statement in [18] admits uncertain correlation between measurement errors, but in the paper [18] only linear estimators are considered.

In this section a new deterministic formalization for the guaranteeing parameter estimation problem with outliers in measurement errors is proposed. So, we generalize the "box scheme" [130], [137] discussed in Section 1.

8.1. *Single Outlier*

Let us consider N measurements of linear combinations of unknown vector parameter $q \in \mathbf{R}^n$

$$z = H'q + \varrho, \tag{8.1}$$

where $z = (z_1, \ldots, z_N)' \in \mathbf{R}^N$ is the measurement vector, $H = (H_1, \ldots, H_N)$ is a given $n \times N$ matrix, $N \gg n$, $\varrho = (\varrho_1, \ldots, \varrho_N)' \in \mathbf{R}^N$ is a vector of measurement errors. It is required to estimate a scalar quantity $l_* = a'q$, where $a \in \mathbf{R}^n$ is a given vector.

Let all errors, except one anomalous, be bounded:

$$\varrho \in \mathcal{P}(k), \qquad \mathcal{P}(k) = \left\{ \varrho \in \mathbf{R}^N \,\middle|\, |\varrho_i| \le \sigma_i, \quad i = 1, \ldots, N, \quad i \ne k \right\}, \tag{8.2}$$

where σ_i are given positive values, k is the index number of a measurement with an outlier. Assume that the outlier's index number and its range are unknown. Then the model for measurement errors with a single unbounded outlier takes the form

$$\varrho \in \mathcal{P}, \qquad \mathcal{P} = \bigcup_{k=1}^{N} \mathcal{P}(k). \tag{8.3}$$

Fig. 3. Set $\mathcal{P}$ for $N = 3$

Note that the set $\mathcal{P}$ is not convex and bounded in contrast to classical guaranteed estimation problem statements, where $|\varrho_i| \le \sigma_i$, $\forall i$ (see, [130], [138], [137]). This set is a sum of mutually orthogonal beams in $\mathbf{R}^N$ (see Figure 3). Without loss of generality it is assumed that $\sigma_i = 1$, $i = 1, \ldots, N$; otherwise one should make the substitution: $H_i' = \sigma_i^{-1} H_i$, $z_i' = \sigma_i^{-1} z_i$, $i = 1, \ldots, N$.

Let us consider all possible estimators $\hat{l} = s(z)$ for $l = l(q)$. Let us also introduce the set of parameters $q \in \mathbf{R}^n$ compatible with the given z:

$$Q(z) = \left\{ q \in \mathbf{R}^n \,\middle|\, H'q = z - \varrho, \quad \varrho \in \mathcal{P} \right\}.$$

Note that $Q(z)$ can be disconnected.

The guaranteed *a posteriori* estimation problem [123] is to find an estimator $s^0(z)$ such that for all other estimators $s(z)$

$$\max_{q \in Q(z)} |l(q) - s^0(z)| \le \max_{q \in Q(z)} |l(q) - s(z)|.$$

Note that in this section we admit nonlinear estimators $s(z)$ as well.

Maximal and minimal values of l are determined by the extremal problems

$$l_{\max} = \max_{q \in Q(z)} a'q, \qquad l_{\min} = \min_{q \in Q(z)} a'q. \tag{8.4}$$

Obviously, the optimal guaranteed *a posteriori* estimator has the form

$$s^0(z) = 1/2\,(l_{\max} + l_{\min}) \tag{8.5}$$

and the optimal guaranteed *a posteriori* estimation error is determined by the formula

$$\delta^0(z) = \max_{q \in Q(z)} |l(q) - s^0(z)| = 1/2\,(l_{\max} - l_{\min}). \tag{8.6}$$

A numerical realization of (8.4)–(8.6) is the main problem of the section. It should be emphasized that no statistical assumptions are used in the problem statement.

8.2. *Optimal Estimation Algorithm*

With an outlier in the measurement with the index number k, let us introduce the set of $q \in \mathbf{R}^n$ that are compatible with the vector z:

$$Q_k(z) = \left\{ q \in \mathbf{R}^n \,\middle|\, |z_i - H_i' q| \le 1, \quad i = 1, \dots, N, \quad i \neq k \right\}.$$

Then

$$Q(z) = \bigcup_{k=1}^{N} Q_k(z), \qquad l_{\max} = \max_{k=1,\dots,N} I_k(z), \quad l_{\min} = \min_{k=1,\dots,N} J_k(z), \quad (8.7)$$

$$I_k(z) = \max_{q \in Q_k(z)} a'q, \qquad J_k(z) = \min_{q \in Q_k(z)} a'q. \quad (8.8)$$

The dual problems [46] (see also Chapter 2) to the problems (8.8) have the form

$$I_k(z) = \min_{X_k} f(x), \qquad f(x) = \sum_{i=1}^{N} \left(|x_i| + z_i x_i \right), \quad (8.9)$$

$$J_k(z) = -\min_{X_k} g(x), \qquad g(x) = \sum_{i=1}^{N} \left(|x_i| - z_i x_i \right), \quad (8.10)$$

$$x = (x_1, \dots, x_N)', \qquad X_k = \left\{ x \in \mathbf{R}^N \,\middle|\, \sum_{i=1}^{N} H_i x_i = a, \; x_k = 0 \right\}.$$

It is assumed that $X_k \neq \emptyset$, $\quad k = 1, \dots, N$. Note that this condition is valid provided

$$\operatorname{rank}\{H_1, \dots, H_{k-1}, H_{k+1}, \dots, H_N\} = n, \qquad H_0 = H_{N+1} \triangleq 0,$$

$k = 1, \dots, N$.

Similar to Section 3, it can easily be checked that the problems (8.9), (8.10) are equivalent to the linear programming problems with x_i replaced by $u_i - v_i$ and $|x_i|$ replaced by $u_i + v_i$, $\quad u_i, v_i \ge 0$, $\quad i = 1, \dots, N$. Therefore they can be effectively solved, e.g., by the simplex method [46] (or by its modifications). The values of $I_k(z)$, $J_k(z)$ either are finite or $I_k(z) = -J_k(z) = -\infty$ (provided $Q_k(z) = \emptyset$) [46]. For at least one index k (that is equal to the true outlier number) $Q_k(z)$ is nonempty and the values $I_k(z)$ and $J_k(z)$ are finite.

Consequently $l_{\min}$ and $l_{\max}$ are also finite. In what follows the problems of (8.9), (8.10) type are called L-problems.

Thus the quantities $s^0(z)$ and $\delta^0(z)$ can be calculated by solving $2N$ L-problems. Clearly, under $N \gg n$ (just these problems are typical for applications) this direct algorithm is computationally tedious and not suitable for utilization. A more efficient algorithm is constructed below.

THEOREM 8.1. *In order to calculate $s^0(z)$ and $\delta^0(z)$ it is sufficient to solve at most $(2n+4)$ L-problems.*

PROOF. Consider the following L-problem (z is omitted):

$$I = \min_X f(x), \qquad X = \Big\{x \in \mathbf{R}^N \;\Big|\; \sum_{i=1}^{N} H_i x_i = a\Big\}. \tag{8.11}$$

Two cases are possible when solving (8.11).

Case A. The value of I is finite.

Then there exists a solution $x^0 \in X$ with at most n nonzero components $x^0_{i_s}$, $s = 1, \dots, n_{\mathrm{A}}$, $n_{\mathrm{A}} \le n$ [46]. The set of indices $\{i_1, \dots, i_{n_{\mathrm{A}}}\}$ is denoted by K_n.

Case B. The value of I is infinite.

Then there exists a sequence $x^{(m)} \subset X$, $m = 1, 2, \dots$, with at most $n+1$ nonzero components $x^{(m)}_{j_t}$, $t = 1, \dots, n_{\mathrm{B}}$, $n_{\mathrm{B}} \le n+1$ such that $f(x^{(m)}) \to -\infty$, $m \to \infty$ [46]. The set of indices $\{j_1, \dots, j_{n_{\mathrm{B}}}\}$ is denoted by K_{n+1}.

The realization of case B means that the system of inequalities $|z_k - H_k' q| \le 1$, $k \in K_{n+1}$, is inconsistent; consequently the index of the measurement with outlier belongs to K_{n+1}.

The sets K_n and K_{n+1} can be computed by the simplex method [46]. First let us study case A. With $k \notin K_n$ we have $I_k = I$. In fact, since $X_k \subset X$, we get $I_k \ge I$ for $k = 1, \dots, N$; at the same time, the optimal solution x^0 for problem (8.11) is an admissible one for problem (8.9) with $k \notin K_n$ and $I_k \le I$. Thus,

$$l_{\max} = \max\left(\max_{k \in K_n} I_k,\ \max_{k \notin K_n} I_k\right) = \max\left(\max_{k \in K_n} I_k,\ I\right) = \max_{k \in K_n} I_k \tag{8.12}$$

and in order to compute $l_{\max}$ it is sufficient to solve at most $n+1$ problems (one problem to determine K_n and n problems to find the maximum in (8.12)). The last equality in (8.12) provides the optimal estimation algorithm.

Let us now study case B. Similar to case A, it is not difficult to show that $I_k = -\infty$, $k \notin K_{n+1}$. Therefore the total number of L-problems does not exceed $n+2$. Since the outlier is located in K_{n+1}, $I_k > -\infty$ for at least one k.

The same arguments are valid for finding $l_{\min}$. Thus the total number of L-problems does not exceed $2n+4$. □

REMARK 8.1. It was noted above that in case B the index number of the outlier is located in K_{n+1}. In general case, the exact index of the measurement with outlier cannot be determined. For example, let us consider a simple problem with $n = 1$, $N = 3$:

$$z_1 = q + \varrho_1, \quad z_2 = q + \varrho_2, \quad z_3 = q + \varrho_3, \qquad \varrho_1 = -1,\ \varrho_2 = +1,\ \rho_3 = 3.$$

Obviously, it is impossible to determine from the data what measurement, z_1 or z_3, is anomalous.

8.3. *A priori Estimation Problem*

The *a priori* guaranteed parameter estimation problem [123] is to find an estimator $s_a(z)$ such that for any other estimator $s(z)$

$$\sup_{z\in\mathcal{Z}}\sup_{q\in Q(z)} |l(q) - s_a(z)| \le \sup_{z\in\mathcal{Z}}\sup_{q\in Q(z)} |l(q) - s(z)|,$$

$$\mathcal{Z} = \left\{ z \in \mathbf{R}^N \mid z = H'q + \varrho, \quad q \in \mathbf{R}^n, \varrho \in \mathcal{P} \right\}.$$

Obviously, it follows from the definitions that the optimal *a posteriori* estimator $s^0(z)$ is an optimal *a priori* estimator $s_a(z)$.

The guaranteed *a priori* estimation error is determined by the formula

$$\delta^a = \sup_{z\in\mathcal{Z}} \delta^0(z).$$

REMARK 8.2. The definition of the optimal *a priori* guaranteed estimator $s_a(z)$ can be also given in another form (cf. Section 1):

$$\sup_{q\in\mathbf{R}^n,\ \varrho\in\mathcal{P}} |l(q) - s_a(z)| \le \sup_{q\in\mathbf{R}^n,\ \varrho\in\mathcal{P}} |l(q) - s(z)|.$$

It can be easily proved that these definitions are equivalent and

$$\delta^a = \sup_{q\in\mathbf{R}^n,\ \varrho\in\mathcal{P}} |l(q) - s_a(z)|.$$

Let us introduce the function $f^0(x) = \sum_{i=1}^N |x_i|$ and the set

$$X_{\alpha,\beta} = \left\{ x \in \mathbf{R}^N \;\middle|\; \sum_{i=1}^N H_i x_i = a, \quad x_\alpha = x_\beta = 0 \right\}.$$

Assume in addition that $X_{\alpha,\beta} \ne \emptyset$, $\alpha, \beta = 1, \dots, N$.

THEOREM 8.2. *The a priori estimation error is finite and is restricted by the inequalities*

$$\delta_1 \le \delta^a \le \delta_2, \qquad \delta_1 = \max_{k=1,\dots,N} \min_{X_k} f^0(x), \qquad \delta_2 = \max_{\alpha,\,\beta=1,\dots,N} \min_{X_{\alpha,\beta}} f^0(x).$$

PROOF. By virtue of (8.4), (8.6) – (8.10),

$$2\delta^0(z) = \max_k I_k - \min_k J_k = \max_k \min_{X_k} f(x) - \min_k \left(- \min_{X_k} g(x) \right) \tag{8.13}$$

$$= \max_k \min_{X_k} f(x) + \max_k \min_{X_k} g(x).$$

Hence,

$$\delta^0(z) \ge 1/2 \max_k \left(\min_{X_k} f(x) + \min_{X_k} g(x) \right). \tag{8.14}$$

Since $z = 0$ is an admissible value of measurement vector (e.g., at $q = 0$, $\varrho = 0$), the lower estimate follows from (8.14):

$$\delta^a = \sup_z \delta^0(z) \ge \delta^0(0) \ge \max_k \min_{X_k} f^0(x) = \delta_1.$$

Let the first maximum in the last part of (8.13) be attained at $k = \alpha$, and the second maximum be attained at $k = \beta$. Then

$$\delta^0(z) = 1/2 \left(\min_{X_\alpha} f(x) + \min_{X_\beta} g(x) \right).$$

Since $X_{\alpha,\beta} \subset X_i$, $i = \alpha, \beta$,

$$\min_{X_{\alpha,\beta}} f(x) \ge \min_{X_\alpha} f(x), \qquad \min_{X_{\alpha,\beta}} g(x) \ge \min_{X_\beta} g(x).$$

Consequently,

$$\delta^0(z) \le 1/2 \left(\min_{X_{\alpha,\beta}} f(x) + \min_{X_{\alpha,\beta}} g(x) \right) \le 1/2 \min_{X_{\alpha,\beta}} \left(f(x) + g(x) \right) \tag{8.15}$$

$$\le \max_{\alpha,\beta} \min_{X_{\alpha,\beta}} f^0(x) = \delta_2.$$

The problem of the computation of δ_1 is a special case of the computation of $l_{\max}$ at $z = 0$ with single outlier and it can be solved by the method described above. We shall see below that the problem of the computation of δ_2 is a special case of the calculation of $l_{\max}$ at $z = 0$ with two outliers and it can be solved similarly. □

8.4. *Arbitrary Number of Outliers*

Let p outliers be admitted. The model of measurement errors has the form (cf. (8.2), (8.3))

$$\varrho \in \mathcal{P}^p, \qquad \mathcal{P}^p = \bigcup_{\omega\in\Omega} \mathcal{P}^p(\omega),$$

$$\mathcal{P}^p(\omega) = \left\{\varrho \in \mathbf{R}^N \mid |\varrho_i| \le 1, \quad i = 1,\dots,N, \quad i \notin \omega\right\},$$

where Ω is the set of all subsets $\omega = \{i_1,\dots,i_p\}$ $(i_1 < \dots < i_p)$ with p integers from the system of indices $\{1,\dots,N\}$.

The optimal guaranteed estimation problem reduces to the finding of $l_{\max}$ and $l_{\min}$ by the formulas (similar to (8.8)–(8.10)):

$$l_{\max} = \max_{\omega\in\Omega} I_\omega(z), \qquad I_\omega(z) = \min_{X_\omega} f(x), \tag{8.16}$$

$$l_{\min} = \min_{\omega\in\Omega} J_\omega(z), \qquad J_\omega(z) = -\min_{X_\omega} g(x),$$

$$X_\omega = \left\{x \in \mathbf{R}^N \;\middle|\; \sum_{i=1}^N h_i x_i = a, \quad x_j = 0, \quad j \in \omega\right\}. \tag{8.17}$$

Then the optimal estimator $s^0(z)$ and the estimation error $\delta^0(z)$ are determined by (8.5), (8.6).

In order to calculate $l_{\max}$ and $l_{\min}$ we have to solve $2\binom{N}{p}$ L-problems. Their number can be considerably reduced similar to Section 8.3. In what follows, it is assumed that

$$X_\omega \ne \emptyset, \quad \forall\, \omega \in \Omega \;\text{ and }\; X_{\omega^1 \bigcup \omega^2} \ne \emptyset, \quad \forall\, \omega^1, \omega^2 \in \Omega. \tag{8.18}$$

THEOREM 8.3. 1^0. *In order to determine the estimator $s^0(z)$ and the estimation error $\delta^0(z)$ under p outliers it is sufficient to solve at most*

$$d_{n,p} = 2n^{-1}[(n+1)^{p+1} - 1]$$

L-problems;
2^0. *The a priori estimation accuracy is finite and is determined by the inequalities:*

$$\delta_{1p} \le \sup_z \delta^0(z) \le \delta_{2p},$$

$$\delta_{1p} = \max_{\omega\in\Omega} \min_{X_\omega} f^0(x), \qquad \delta_{2p} = \max_{\omega^1,\omega^2\in\Omega} \min_{X_{\omega^1 \bigcup \omega^2}} f^0(x).$$

To compute δ_{1p} *and* δ_{2p} *it is sufficient to solve at most* $\frac{1}{2}d_{n-1,p}$ *and* $\frac{1}{2}d_{n-1,2p}$ *L-problems, respectively* ($d_{0,p} \triangleq 2(p+1)$).

PROOF. Consider the auxiliary problem (8.11). Let us first assume that (8.11) belongs to type A. Introduce the partitioning

$$\Omega = \Omega_1 \bigcup \Omega_2, \qquad \Omega_1 \bigcap \Omega_2 = \emptyset, \tag{8.19}$$

$$\Omega_1 = \{\omega \in \Omega \mid \omega \cap K_n = \emptyset\}, \qquad \Omega_2 = \{\omega \in \Omega \mid \omega \cap K_n \neq \emptyset\}.$$

Similar to the case of single outlier, it is easy to prove that

$$I_\omega = I, \qquad \omega \in \Omega_1 \tag{8.20}$$

and therefore (cf. (8.12))

$$l_{\max} = \max_{\omega \in \Omega_2} I_\omega. \tag{8.21}$$

Note that the excluded set Ω_1 contains "the bulk" of $\omega \in \Omega$ (for $n, p \ll N$). The set Ω_2 can be represented as a union of subsets ω that contain the first element from K_n, the second element from K_n, etc. Let us denote by $\Omega(1,s)$, $s \in K_n$, the set of all subsets $\omega(1)$ consisting of $p-1$ (distinct) elements that belong to the system $\{1, \dots, N\} \setminus s$. Then it follows from (8.16), (8.17), (8.21) that

$$l_{\max} = \max_{s \in K_n} l_{\max}^{(s)},$$

$$l_{\max}^{(s)} = \max_{\omega(1) \in \Omega(1,s)} I_{\omega(1)}^{(1,s)}, \qquad I_{\omega(1)}^{(1,s)} = \min_{X_{\omega(1) \cup s}} f(x). \tag{8.22}$$

So, the problem for the calculation of $l_{\max}$ reduces to the solving of (at most) n problems (8.22). By virtue of (8.16), (8.17), $l_{\max}^{(s)}$ is calculated as the exact upper bound for the parameter l in a problem with $p-1$ outliers in the measurements $\{z_i,\ i = 1, \dots, N,\ i \neq s\}$.

Thus the problem of finding $l_{\max}$ reduces to n problems of finding $l_{\max}$ with one potential outlier less than before. By executing this procedure $p-1$ times, we get a run of problems with less and less *a priori* number of outliers. Obviously, the number of L-problems in this branching process does not exceed

$$\sum_{i=0}^{p} n^i = (n^{p+1} - 1)/(n-1) = 1/2\, d_{n-1,p}.$$

In case B all arguments remain valid with K_n replaced by K_{n+1} and $d_{n-1,p}$ replaced by $d_{n,p}$. The same number of L-problems is needed for finding $l_{\min}$.

TABLE 1. Number of L-problems

n	p=1	p=2	p=3	p=4	p=5
1	6 (4)	14 (5)	30 (6)	62 (7)	126 (9)
2	8 (5)	26 (6)	80 (8)	242 (9)	728 (11)
3	10 (6)	42 (8)	170 (10)	682 (12)	2730 (14)
4	12 (7)	62 (9)	312 (11)	1562 (14)	7812 (16)

The inequalities for the *a priori* estimation error can be obtained just like that of Theorem 8.2. Since $f^0(x) > -\infty$, $x \in \mathbf{R}^N$, all L-problems for the finding of δ_{1p}, δ_{2p} have type A and their number does not exceed $1/2\, d_{n-1,p}$ and $1/2\, d_{n-1,2p}$, respectively.

The optimal estimation algorithm is generated by branching iterations of (8.22) type. □

The values for $d_{n,p}$ are presented in Table 1. The values of N, starting from which $2\binom{N}{p} > d_{n,p}$, are indicated in parentheses.

It follows from Table 1 that the upper bound for the number of L-problems grows rapidly with n and p, but remains considerably less than $2\binom{N}{p}$ under direct exhaustive search in (8.16) (with sufficiently large N). For example, $2\binom{50}{5} = 4237520$.

Let us construct a (nonoptimal) estimator that has a bounded *a priori* estimation error, but requires considerably fewer L-problems than those in Table 1 for its realization. In addition, it is further assumed that

$$X_c \neq \emptyset, \qquad C \text{ is any set of } 2p(n+1) \text{ indices from } \{1,\ldots,N\}.$$

THEOREM 8.4. *In order to construct an estimator with a bounded a priori estimation error under p outliers it is sufficient to solve* $2(p+1)$ *L-problems.*

PROOF. The main idea of the procedure, which generates this estimator, is that, instead of detailed sorting of the elements from K_n or K_{n+1} (as it was done for the optimal estimator), all of the set K_n (K_{n+1}) are erased at each procedure's step.

Let us construct an upper bound for $l_{\max}$. First consider that the L-problem (8.11) has type A.

LEMMA 8.1. *The following inequality holds:*

$$l_{\max} \leq l^{[1]}_{\max}, \qquad l^{[1]}_{\max} = \max_{\nu(1)\in\Omega(1)} I^{[1]}_{\nu(1)},$$

where $\Omega(1)$ *is the set of all subsets* $\nu(1)$ *from the system of numbers*

$$\{1,\ldots,N\} \setminus K_n$$

consisting of $p-1$ *elements, and*

$$I^{[1]}_{\nu(1)} = \min_{X_{\nu(1)\bigcup K_n}} f(x).$$

Proof of Lemma 8.1. Introduce the partitioning (8.19) and recall that (8.20) holds. Since $X_{\omega\bigcup K_n} \subset X_\omega \subset X$,

$$I_{\omega\bigcup K_n} \geq I_\omega \geq I. \tag{8.23}$$

It follows from (8.20) and (8.23) that

$$l_{\max} = \max\left(\max_{\omega\in\Omega_1} I_\omega,\ \max_{\omega\in\Omega_2} I_\omega\right) = \max_{\omega\in\Omega_2} I_\omega \leq \max_{\omega\in\Omega_2} I_{\omega\bigcup K_n}. \tag{8.24}$$

For every ω there exists ω' such that $\omega'\bigcup K_n \supset \omega\bigcup K_n$ and $\omega'\bigcap K_n$ contains exactly one element. Consequently $X_{\omega'\bigcup K_n} \subset X_{\omega\bigcup K_n}$, $I_{\omega'\bigcup K_n} \geq I_{\omega\bigcup K_n}$ and the maximum in (8.24) is attained at ω that has exactly one common element with K_n. Thus the right-hand expression in (8.24) is equal to $l^{[1]}_{\max}$. Lemma 8.1 is proved.

According to (8.16), (8.17), $l^{[1]}_{\max}$ is calculated as the exact upper bound for admissible values of l in a problem with $p-1$ outliers in the measurements $\{z_i,\ i=1,\ldots,N,\ i\notin K_n\}$.

By executing this procedure $p-1$ times more, we obtain the sequence of inequalities:

$$l_{\max} \leq l^{[1]}_{\max} \leq \cdots \leq l^{[p]}_{\max},$$

where $l^{[p]}_{\max}$ is a maximal value for l in a problem without outliers in the measurements $\{z_1,\ldots,z_N\}$ with corresponding exceptions.

Note that if the starting L-problem has type A, the next L-problems are all type A. It results from the fact that the corresponding sets become more and more restricted.

Let us now consider case B. The proof is the same as in case A with minor modifications. In particular: K_n should be replaced by K_{n+1} (including Lemma 8.1, which remains valid); in the proof of Lemma 8.1 I should be changed by $-\infty$.

Note that the $(p+1)$-th L-problem will be type A. Indeed, if any intermediate L-problem has type A, all the next problems will also be the same type. If all intermediate L-problems are type B, at each procedure's step at least one outlier will be located in the corresponding K_{n+1}, and up to the moment of solving the last $(p+1)$-th L-problem the outliers in the measurements will be absent.

To construct $l_{\min}^{[p]}$ (the lower bound for $l_{\min}$), we have to solve $(p+1)$ L-problems as well.

Let us construct the estimator

$$s^*(z) = 1/2\,(l_{\max}^{[p]} + l_{\min}^{[p]}).$$

As far as $l_{\min}^{[p]} \leq l_{\min} \leq l_{\max} \leq l_{\max}^{[p]}$, its estimation error

$$\delta(z) = \max_{q \in Q(z)} |l(q) - s^*(z)| \leq \delta^*(z) = 1/2\,(l_{\max}^{[p]} - l_{\min}^{[p]}).$$

The proof of the boundness of the *a priori* estimation error of $s^*(z)$ is the same as that for $s^0(z)$. Similar to (8.15),

$$\delta(z) \leq \max_C \min_{X_c} f^0(x), \quad \forall\, z.$$ □

8.5. *Examples*

Consider a number of examples.

EXAMPLE 8.1. Consider N measurements of a shift parameter q with a single outlier:

$$z_i = q + \varrho_i, \qquad i = 1, \ldots, N, \qquad N \geq 3. \tag{8.25}$$

With the preceding notation $n = 1,\ p = 1,\ H_i = \sigma_i = 1,\ i = 1, \ldots, N,\ a = 1$.

In accordance with (8.9), (8.10),

$$I_k = \min \sum_{i=1,\ldots,N,\, i\neq k} (|x_i| + z_i x_i)\,, \qquad \sum_{i=1,\ldots,N,\, i\neq k} x_i = 1, \tag{8.26}$$

$$J_k = -\min \sum_{i=1,\ldots,N,\, i\neq k} (|x_i| - z_i x_i)\,, \qquad \sum_{i=1,\ldots,N,\, i\neq k} x_i = 1. \tag{8.27}$$

Let us place the measurements in ascending order (with a corresponding renumbering)

$$z_1 \leq \ldots \leq z_N. \tag{8.28}$$

It is easy to show that the solutions of (8.26), (8.27) have the form

$$I_1 = \begin{cases} z_2 + 1 & \text{if the system } |z_i - q| \leq 1 \ (i=2,\ldots,N) \text{ is consistent} \\ -\infty & \text{otherwise;} \end{cases}$$

for $k = 2, \ldots, N$,

$$I_k = \begin{cases} z_1 + 1 & \text{if the system } |z_i - q| \leq 1 \ (i=1,\ldots,N,\ i\neq k) \text{ is consistent} \\ -\infty & \text{otherwise;} \end{cases}$$

for $s = 1, \ldots, N-1$,

$$J_s = \begin{cases} -z_N + 1 & \text{if the system } |z_i - q| \leq 1 \ (i=1,\ldots,N,\ i \neq s) \text{ is consistent} \\ -\infty & \text{otherwise;} \end{cases}$$

$$J_N = \begin{cases} -z_{N-1} + 1 & \text{if the system } |z_i - q| \leq 1 \ (i=1,\ldots,N-1) \text{ is consistent} \\ -\infty & \text{otherwise;} \end{cases}$$

Then, by virtue of (8.7),

$$l_{\max} = \begin{cases} z_2 + 1 & \text{if the system } |z_i - q| \leq 1 \ (i=2,\ldots,N) \text{ is consistent} \\ z_1 + 1 & \text{otherwise,} \end{cases}$$

$$l_{\min} = \begin{cases} z_{N-1} - 1 & \text{if the system } |z_j - q| \leq 1 \ (j=1,\ldots,N-1) \text{ is consistent} \\ z_N - 1 & \text{otherwise.} \end{cases}$$

The values of $s^0(z)$, $\delta^0(z)$ are determined by (8.5) and (8.6). We see that the optimal algorithm either cuts off the large outlier or cuts off the extreme measurements "just in case". Obviously, $\delta_1 = \delta_2 = 1$, and the *a priori* estimation error is equal to unity, i.e., it coincides with the *a priori* estimation error without outliers.

EXAMPLE 8.2. Let us consider the previous estimation problem with p outliers ($2p < N$ cf.(8.18)). In addition, it is assumed that the system

$$|z_i - q| \leq 1, \quad i = 1, \ldots, N \tag{8.29}$$

is consistent. It follows immediately from (8.16), (8.17) that in this case $l_{\max} = z_{p+1} + 1$, $l_{\min} = z_{N-p} - 1$, i.e., the optimal algorithm cuts off p left and p right measurements.

Note that, in spite of the consistency of (8.29), small outliers can be present. Their discount can worsen an estimate and can lead to extremely optimistic value of estimation error. For example, let the true shift parameter value $q = 0$, $p = 1$, and $z_i = \varrho_i = 0$ $(i = 1, \ldots, N-1)$, $z_N = \varrho_N = 2$ (anomalous error). Then $q = 1$ is the only common point of (8.29) and the ordinary guaranteed algorithm, which ignores outlier, yields the estimate $\hat{q} = 1$. Moreover, the ordinary algorithm states that this estimate has zero error. The guaranteed algorithm, which admits the presence of an outlier, gives another, delicate answer: $\hat{q} = s^0 = 0$ and $\delta^0 = 1$. So, small outliers are processed with care.

EXAMPLE 8.3. Consider the estimation problem for the initial angular velocity of the small oscillations of a pendulum. The motion of a pendulum is described by the equation

$$\ddot{x}(t) + x(t) = 0, \qquad t \in [0, T], \tag{8.30}$$

TABLE 2. One and two outliers; $p = 1$, $p = 2$

$l_{\min} = l_{\min}^{[p]}$	$l_{\max} = l_{\max}^{[p]}$	$s^0 = s^*$	$s^0 - l = s^* - l$	$\delta^0 = \delta^*$
13.48	13.61	13.54	-0.02	0.07

where $x(t)$ is the angular position of a pendulum, $\dot{x}(t)$ is its angular velocity. Assume the discrete measurements of the pendulum position are available:

$$z(t_i) = x(t_i) + \varrho(t_i), \quad t_i = \frac{T}{N-1}(i-1), \quad i = 1, \ldots, N, \tag{8.31}$$

where $\varrho(t_i)$ is a measurement noise. It is required to evaluate the initial angular velocity $\dot{x}(0)$. Thus (8.30), (8.31) reduce to (8.1) with

$$z = (z(t_1), \ldots, z(t_N))', \qquad H = \begin{pmatrix} \cos t_1 & \ldots & \cos t_N \\ \sin t_1 & \ldots & \sin t_N \end{pmatrix},$$

$$q = (x(0), \dot{x}(0))', \qquad \varrho = (\varrho(t_1), \ldots, \varrho(t_N))', \qquad a = (0, 1)'.$$

Further we put $T = 4\pi$, $N = 100$.

In our example, the measurement noise is formed by a "regular" component $\varrho_R(t_i)$, $|\varrho_R(t_i)| \leq 1$ with rare anomalous additions. Values of $\varrho_R(t_i)$ are simulated as a superposition of a deterministic function and pseudorandom numbers uniformly distributed in $[-1, +1]$ ($\max_i |\varrho_R(t_i)| = 0.9774$). The realization of $\varrho_R(t_i)$ is presented in Figure 4. The true value of q was set by $q^0 = (5.37,\ 13.56)'$.

Three cases are considered:

(1) the noise contains a single outlier in the measurement with index 47: $p = 1$, $\varrho(t_{47}) = \varrho_R(t_{47}) + 500$;

(2) the noise contains two outliers in the measurements with indices 34 and 71: $p = 2$, $\varrho(t_{34}) = \varrho_R(t_{34}) + 297$, $\varrho(t_{71}) = \varrho_R(t_{71}) + 123$;

(3) the noise contains six outliers in the measurements with indices 11, 23, 27, 34, 63, and 81: $p = 6$, and the corresponding additional summands are 10, 11, 10, 99, 101, and 10, respectively.

In cases (1) and (2), the optimal algorithm s^0 and the suboptimal algorithm s^* gave the same bounds, estimates, and estimation errors. The results of simulations are presented in Table 2.

In case (1), $\delta_1 = 1.000$, $\delta_2 = 1.001$, and therefore the *a priori* estimation error satisfies the inequalities: $1.000 \leq \delta^a \leq 1.001$.

In case (3), the suboptimal algorithm s^* only was utilized. The results are presented in Table 3.

It should be noted that even for six outliers the bounds, the estimate, and the estimation error of the suboptimal algorithm s^* are not far from those in cases (1) and (2).

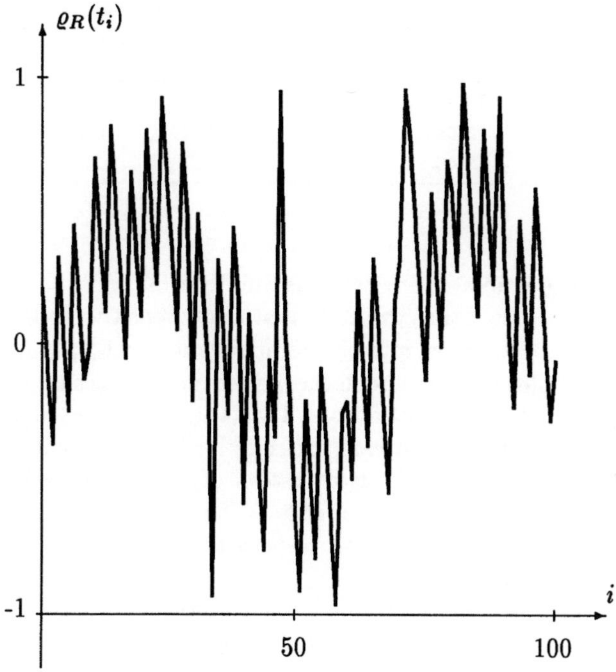

Fig. 4. "Regular" component of measurement noise

TABLE 3. Six outliers; $p = 6$

$l_{\min}^{[6]}$	$l_{\max}^{[6]}$	s^*	$s^* - l$	δ^*
13.27	13.65	13.46	-0.10	0.19

Now let us compare the proposed optimal guaranteeing method (OGM) and the suboptimal guaranteeing method (SGM) with classical estimation methods: the least squares method (LSM) and the least absolute value method (LAVM). We have

$$\hat{l}_{LSM} = a'\left(\arg\min_q \sum_1^N (z_i - h_i'q)^2\right),$$

$$\hat{l}_{LAVM} = a'\left(\arg\min_q \sum_1^N |z_i - h_i'q|\right).$$

TABLE 4. Comparison of estimation methods

	$s^0 = \hat{l}_{OGM}$	$s^* = \hat{l}_{SGM}$	$\hat{l}_{LSM}$	$\hat{l}_{LAVM}$	$\hat{l}_{LSM} - l$	$\hat{l}_{LAVM} - l$
Case 1	13.54	13.54	23.64	13.62	10.08	0.06
Case 2	13.54	13.54	20.76	13.62	7.20	0.06
Case 3	–	13.46	16.93	13.67	3.37	0.11

The corresponding estimates are presented in Table 4.

It follows from Table 4 that the LSM is strongly sensitive to large outliers: the estimates do not even get into the admissible interval $[l_{\min}, l_{\max}]$.

In contrast to LSM, LAVM cuts off the outliers and yields a good estimation performance (nevertheless, a little worse than the guaranteed algorithms).

The LAVM has standard software packages for its implementation and, at first sight, it can be successfully used instead of the guaranteed algorithms. But there exists an important peculiarity. The LAVM can fail when measurement errors are unbalanced with respect to zero. In fact, consider the following simple example.

Let the measurements be determined by (8.25) with the true shift parameter value $q^0 = 0$. It can easily be checked that in this case (with the renumbering of the measurements in the ascending order (8.28))

$$\hat{l}_{LAVM} = \begin{cases} z_{\frac{N+1}{2}} = \varrho_{\frac{N+1}{2}}, & N \text{ is odd} \\ \frac{1}{2}\left(z_{\frac{N}{2}} + z_{\frac{N}{2}+1}\right) = \frac{1}{2}\left(\varrho_{\frac{N}{2}} + \rho_{\frac{N}{2}+1}\right), & N \text{ is even.} \end{cases} \tag{8.32}$$

Assume that r components from $\{\varrho\}_1^N$ are equal to -1, p components are anomalous, and $N - p - r$ components are equal to $+1$. Then it follows from (8.32) that with $N > 2(p + r)$ we get $\hat{l}_{LAVM} = 1$. The optimal guaranteed method (with $r > p$) gives $\hat{l}_{OGM} = 0$ with $\delta^0(z) = 0$, i.e., the true value of q with zero guaranteed error.

Note that the classical methods do not provide the bounds for estimation errors (under our set-membership assumptions).

The obtained results show that the proposed guaranteed algorithms filter out large outliers, and they process measurements with small outliers carefully. Moreover, they provide correct bounds for estimation error.

Thus we proposed a new formalization for the guaranteed parameter estimation problem with anomalous measurements. Constructive estimation algorithms are obtained. They allow us to realize an optimal measurement processing with a calculation of the guaranteed estimation error. Numerical testing verifies the efficiency of the suggested algorithms. In particular, in some cases their accuracy exceeds the accurracy of the least-squares method and the least absolute value method.

9. Comments to Chapter 1

9.1. *Section 1*

Guaranteed estimation problems were studied by P. Huber [71], N.N. Krasovskii [113], [117], M.L. Lidov [130], F.C. Schweppe [189], H.S. Witsenhausen [215]. It is worth noting that the guaranteed approach was applied to the problem on *accumulation of perturbations* by B.V. Bulgakov in 1939 [42] (the "worst perturbation" was found from the set of uncertain bounded functions acting on a mechanical system). General concepts of observation and control under uncertainty basing on ideas of convex and functional analysis were developed by N.N. Krasovskii [117] and A.B. Kurzhanskii [123]. The problems of the ellipsoidal estimation were studied in detail by F.C. Schweppe [190], F.L. Chernousko [43], A.B. Kurzhanskii and I. Valyi [125]. The application of guaranteed approach to orbit determination is presented in [20] by P.E. Elyasberg, B.Ts. Bakhshiyan, and R.R. Nazirov. Spectral guaranteed estimation problems are considered in [120]. Statistical approach to robust estimaiton was discussed in [72], [206]. An elementary introduction to guaranteed estimation was presented by P.E. Elyasberg [54].

Various questions of guaranteed estimation are considered in [28], [11], [30], [184], [209], [210], [208], [174], [175], [211], [26], [172], [78], [181]. Let us cite the overviews [169], [138], [137]; they contain an extensive bibliography. The list of papers devoted to guaranteed estimation is continued in comments to other sections but this list is incomplete, certainly.

M.L. Lidov posed the "box scheme" problem and reduced it to the linear programming problem [130]. Thus he obtained an efficient numerical algorithm to solve guaranteed estimation problems. The "box scheme" problem ("worst correlation" problem) was used repeatedly for the analysis of applied problems (see, e.g., [55], [142], [22], [166], [38]).

The case of polyhedral constraints on measurement noise is considered in [141] (see also [18], [14]).

In this book we follow the trend initiated by M.L. Lidov [130], [138], [137].

9.2. *Section 2*

General guaranteed estimation problem for continuous case was studied by N.N. Krasovskii [113]. Continuous "worst correlation" problem was considered in [28].

The problem of moments is investigated, e.g., in [173], [118], [183]. A very clear and elementary introduction to this topic is given in [117]. Some applications of functional analysis to automatic control are presented in [24], [25].

9.3. *Section 3*

A tremendous amount of literature is devoted to linear programming. We cite only [86], [46], [60], [13]. Here we tried to give a brief and self-contained account of some basic results of linear programming that are necessary for our purposes.

9.4. *Section 4*

Necessary and sufficient conditions for guaranteed parameter estimation problem were obtained with the help of the infinite-dimensional Lagrange theorem [75], [51], [5]. This result is a special case of [133], [156].

9.5. *Section 5*

Theory of the Chebychev approximation is discussed in a lot of books (see. e.g., [129], [117], [49], [183]). An extensive survey on the approximation theory was written by V.M. Tikhomirov [204]. In [151], [166], [38] the Chebyshev approximation was applied to inertial navigation.

The dual problem and the Chebychev polynomial have clear geometric interpretation. A useful geometric interpretation for the optimal solution in terms of initial linear programming problem was given by M.L. Lidov and P.E. Elyasberg [54].

9.6. *Section 6*

In Section 6 we follow the paper by M.L. Lidov [132]. The discrete version of the problem in question was considered in [188] by J. Sacks and D. Ylvisaker. We pay a special attention for the derivation of the necessary condition of optimality (Lemma 6.1). In [132] this derivation is rather schematic. In [134] the same problem with small white noise component was treated: $r(t) = \epsilon\,\bar{r}(t)$, $\epsilon \ll 1$. An asymptotic in ϵ solution of the problem was obtained.

A very elegant approach for solving the main problem of Section 6 was developed by V.N. Soloviov [194]. He found that the corresponding dual problem is *an unconstrained n-dimensional smooth* problem that can be solved by well-known numerical methods. Using the solution of the dual problem, the optimal estimator was constructed with the help of iterations (see, also, [21], [193], [195], [196], [198], [199] and an excellent survey [200] devoted to new minimax theorems and minimax estimation problems). By now, this approach was developed only for static problems, i.e., for problems with constant state vector. The results by V.N. Soloviov can form the subject of a separate book. Since these results lie somewhat aside from the main line of our study, we do not discuss them here. In Chapter 3 of our book, we apply our method (see Section 7) to dynamic systems as well.

9.7. *Section 7*

Formula (7.12) was first obtained in [157]. The nonoptimality level of a given algorithm in multicriteria designing problems was studied in [29]. It was interesting to find out that the duality theory was also applied to vibroisolation problems [61].

The estimate of the proximity of a current iteration of the simplex method to the optimal solution of the linear programming program was first obtained by M.L. Lidov in [131], where a more traditional simplex method technique was used for proof.

9.8. *Section 8*

The results of this section completes the fundamental material of Section 1. These results are contained in the paper [167].

10. Exercises to Chapter 1

1. Assume that along with the measurements (1.1) an *a priori* information about unknown vector q is also available:
$$q \in Q, \qquad Q = \left\{ q \in \mathbf{R}^n \;\middle|\; |H'_{-s} q| \le \sigma_{-s}, \quad s = 1, \ldots, n \right\},$$
where $H_{-s} \in \mathbf{R}^n$ and $\sigma_{-s} \ge 0$, $s = 1, \ldots, n$, are given vectors and numbers, respectively.
Prove that, instead of (1.10), (1.11), the corresponding variational problem takes the form
$$\min_{\Phi_{-s},\,\Phi_i} \left(\sum_{s=1}^{n} \sigma_{-s} |\Phi_{-s}| + \sum_{i=1}^{N} \sigma_i |\Phi_i| \right), \qquad \sum_{s=1}^{n} H_{-s}\Phi_{-s} + \sum_{i=1}^{N} H_i \Phi_i = a.$$
What variational problem corresponds to the case of continuous measurements?
2. Consider the measurements
$$z = H'q + B\varrho,$$
where $B = (U, V)$ is an $N \times p$ matrix $(p \ge N)$, U is a nonsingular $N \times N$ matrix, and V is an $N \times (p - N)$ matrix. Assume that
$$\varrho \in \mathcal{G}' = \left\{ \varrho \in \mathbf{R}^p \;\middle|\; -\beta \le C\varrho \le \beta \right\},$$
where C is an $s \times p$ matrix $(s \ge p)$, $\beta \in \mathbf{R}^s$ is a vector with nonnegative elements. It is required to estimate $l_* = a'q$ with the help of linear estimators (cf. Section 1).
Construct a variational problem for finding an optimal estimator similar to (1.23) and (1.24).
3. Construct an example of linear programming problem with a nonunique solution.

4. Prove the assertion: if the optimal value of linear programming problem is equal to $-\infty$, then there exists a sequence of vectors $x^{(k)}$ with at most $n+1$ nonzero entries such that
$$\sum_{i=1}^{K} c_i x_i^{(k)} \to -\infty, \qquad \sum_{i=1}^{K} A_i x_i^{(k)} = a.$$
5. Construct an example of continuous linear programming problem (3.15), (3.16) with piecewise continuous $H(t)$ such that the assertion of Theorem 3.4 is not valid.
6. Consider the motion of a material point from Example 4.2. It is required to estimate the location of the point at $t = T$, i.e., to estimate the second component of the state vector. Construct the primal problem, the dual problem, and the generalized polynomial. Find the solutions of the primal and dual problems. Find the Chebyshev generalized polynomial.
7. Using the pattern (5.6), (5.7), construct the dual problem for the canonical linear programming problem (3.1), (3.2).
8. Construst the dual problem for the primal problem of the form
$$\inf_{\Phi \in \mathcal{F}} \left(\int_0^T |\Phi(t)|\, dt, \right)^2, \qquad \int_0^T H(t)\Phi(t)\, dt = a.$$
9. Prove that for any $\varphi \in C[0,T]$ the nonlinear functional $\Delta^0(\varphi) \geq 1$, where $\Delta^0(\varphi)$ is defined by (7.12).
10. Compute the value of Δ^0 for the problem (7.1) with $n = 2$, $H(t) = (1,t)'$, $a = (1,0)'$ or $a = (0,1)'$.
11. Compute the value of Δ^0 in Example 7.3 for the cases: (a) T is arbitrary; (b) $n = 2$, a is arbitrary.
12. Prove the assertion of Remark 8.2.
13. Analyze the combinations of the placement of z_k and $Q_k(z)$ for the case of $N = 3$.

Chapter 2

Guaranteed Estimation in Dynamic Systems

This chapter is devoted to dynamic systems. After a brief summary of the elements of convex analysis and theory of optimization, the optimal guaranteed estimation problem for dynamic systems that are described by linear ordinary differential equations is posed. The necessary and sufficient conditions for optimality of an estimator are obtained. Various examples are studied in detail. An important problem of the optimality of linear estimators is considered. Finally, the guaranteed estimation problem with phase constraints is discussed.

1. Lagrange Principle and Duality

This section summarizes the essentials of the theory of convex analysis and optimization theory. We ommit those points that we do not directly use for the analysis of estimation problems.

1.1. *Convex Analysis*

Let us discuss some elementary facts from convex analysis which we shall use further. Throughout this section X denotes a real normed vector space[1] and X^* denotes the dual space, i.e., the space of continuous linear functionals on X.

The set

$$[x_1, x_2] = \{x \in X \mid x = \alpha x_1 + (1 - \alpha)x_2, \quad 0 \le \alpha \le 1\}$$

is said to be the *interval* joining the points x_1 and x_2 in X. A subset A of the space X is said to be *convex* if it contains the interval joining any two of its points. The empty set is assumed to be convex by definition. A convex

[1] The general theory is developed for a separable locally convex topological vector space. In our book, all actual spaces are normed ones.

set A is said to be *balanced* if, for any $x \in A$, $\alpha x \in A$ for all $\alpha \in \mathbf{R}^1$ such that $|\alpha| \leq 1$.

A continuous linear functional x^* is said to separate the sets A and B if, for all $x \in A$, $y \in B$,

$$\langle x^*, x \rangle \leq \langle x^*, y \rangle,$$

where $\langle x^*, x \rangle$ denotes the value of the linear functional x^* on the element x.

THEOREM 1.1 [75], [109]. *Let A and B be convex sets of the space X and* $\text{int}A \neq \emptyset$. *Then A and B can be separated by a nonzero continuous linear functional x^* if and only if* $(\text{int}A) \cap B = \emptyset$.

The mapings into the extended straight line $\overline{\mathbf{R}} \equiv [-\infty, +\infty]$ are called *functions*. The following two sets can be associated with every function f given on X:

$$\text{dom } f = \left\{ x \in X \,\middle|\, f(x) < \infty \right\},$$

$$\text{epi } f = \left\{ (\alpha, x) \in \mathbf{R}^1 \times X \,\middle|\, \alpha \geq f(x) \right\}.$$

The first of these sets is called the *effective domain* of the function f, and the second one is called the *epigraph* of this function. A function is called *proper* function if $\text{dom } f \neq \emptyset$ and $f(x) > -\infty$ for all x.

A function $f : X \to [-\infty, \infty]$ is said to be *convex* if the set epi f is convex in the space $\mathbf{R}^1 \times X$. It is easy to prove that f is convex if and only if

$$f(\alpha x + (1-\alpha)y) \leq \alpha f(x) + (1-\alpha) f(y) \qquad \text{for all } \alpha \in [0,1], \tag{1.1}$$

whenever the right-hand side is defined.

A function $f : X \to \mathbf{R}^1$ is said to be *strictly convex* if it is convex and the strict equality holds in (1.1), $\forall\, x, y$, $x \neq y$, and $\forall\, \alpha \in (0,1)$.

Let f be a proper convex function on X. A functional $x^* \in X^*$ is said to be a *subgradient* of the function f at a point x if

$$f(z) - f(x) \geq \langle x^*, z - x \rangle \qquad \text{for all } z \in X.$$

The set of all subgradients of the function f at the point x is called the *subdifferential* of the function f at the point x, and it is denoted by $\partial f(x)$,

$$\partial f(x) = \left\{ x^* \in X^* \,\middle|\, f(z) - f(x) \geq \langle x^*, z - x \rangle,\ \forall\, z \in X \right\}.$$

If a convex function f has the Gâteaux derivative $F(x)$ at a point x, then $\partial f(x) = \{F(x)\}$.

By L_2 denote the Banach space of square integrable functions $\Phi\colon [0,T] \to \mathbf{R}^1$. Consider the functional $f\colon L_2 \to \mathbf{R}^1$,

$$f(\Phi) = \int_0^T \sigma(t)|\Phi(t)|\,dt, \qquad \sigma(t) > 0, \quad t \in [0,T], \tag{1.2}$$

$\sigma(t)$ is a piecewise continuous function.

LEMMA 1.1. *The subdifferential of f defined by* (1.2) *has the form*

$$\partial f(\Phi) = \left\{\varphi \in L_2 \,\middle|\, \varphi(t) = \sigma(t)\,\mathrm{sign}^*\Phi(t)\right\}, \tag{1.3}$$

where

$$\mathrm{sign}^*\Phi(t) = \begin{cases} +1, & \text{for } t \in [0,T]\colon \Phi(t) > 0 \\ -1, & \text{for } t \in [0,T]\colon \Phi(t) < 0 \\ e(t) \in L_2,\ |e(t)| \le 1, & \text{for } t \in [0,T]\colon \Phi(t) = 0. \end{cases}$$

PROOF. First recall that if $X = L_2$, then the dual space X^* may be identified with L_2 and

$$\langle x^*, x\rangle = \int_0^T x^*(t)x(t)\,dt \quad \text{for some} \quad x^*(\cdot) \in L_2.$$

Recall also the definition:

$$\operatorname*{ess\,sup}_{t\in[0,T]} \alpha(t) \triangleq \inf\left\{\sup_{t\in[0,T]} \tilde{\alpha}(t) \,\middle|\, \alpha(t) = \tilde{\alpha}(t) \text{ almost everywhere}\right\}.$$

Let us prove the following two assertions:
(1) If $\Phi = 0$, then

$$\partial f(0) = \left\{\varphi \in L_2 \,\middle|\, \operatorname*{ess\,sup}_{t\in[0,T]} \left|\frac{\varphi(t)}{\sigma(t)}\right| \le 1\right\}; \tag{1.4}$$

(2) If $\Phi \ne 0$, then

$$\partial f(0) = \left\{\varphi \in L_2 \,\middle|\, \operatorname*{ess\,sup}_{t\in[0,T]} \left|\frac{\varphi(t)}{\sigma(t)}\right| = 1, \right. \tag{1.5}$$

$$\left.\int_0^T \sigma(t)|\Phi(t)|\,dt = \int_0^T \varphi(t)\Phi(t)\,dt\right\}.$$

(1) Consider $\Phi = 0$. If $\operatorname{ess\,sup}_{t\in[0,T]} \left|\frac{\varphi(t)}{\sigma(t)}\right| \le 1$, then

$$\int_0^T \sigma(t)|\Phi(t)|\,dt \ge \int_0^T \frac{\varphi(t)}{\sigma(t)}\sigma(t)\Phi(t)\,dt, \qquad \forall\, \Phi \in L_2,$$

and $\varphi \in \partial f(0)$ by the definition of subgradient.

Conversely, if $\varphi \in \partial f(0)$, then, by the definition of subgradient,

$$\int_0^T \sigma(t)|\Phi(t)|\,dt \geq \int_0^T \varphi(t)\Phi(t)\,dt, \qquad \forall\,\Phi \in L_2. \tag{1.6}$$

Then $\operatorname{ess\,sup}_{t\in[0,T]}\left|\frac{\varphi(t)}{\sigma(t)}\right| \leq 1$. In fact, assume that on a set E of nonzero measure $|\varphi(t)| > \sigma(t)$. Let $\chi_E(t)$ be the indicator function of E, i.e.,

$$\chi_E(t) = \begin{cases} 1, & t \in E \\ 0, & t \notin E. \end{cases}$$

Taking $\Phi(t) = \chi_E(t)\operatorname{sign}\varphi(t)$, we have from (1.6)

$$\int_E \sigma(t)\,dt \geq \int_E \sigma(t)|\operatorname{sign}\varphi(t)|\,dt \geq \int_E |\varphi(t)|\,dt$$

$$= \int_E \left|\frac{\varphi(t)}{\sigma(t)}\right| \sigma(t)\,dt > \int_E \sigma(t)\,dt.$$

The last inequality is inconsistent. So, our assumption is false and (1.4) is proved.

(2) Consider $\Phi \neq 0$. Suppose

$$\operatorname*{ess\,sup}_{t\in[0,T]}\left|\frac{\varphi(t)}{\sigma(t)}\right| = 1 \quad \text{and} \quad \int_0^T \sigma(t)|\Phi(t)|\,dt = \int_0^T \varphi(t)\Phi(t)\,dt.$$

Then, for any $z \in L_2$,

$$\int_0^T \varphi(t)z(t)\,dt = \int_0^T \frac{\varphi(t)}{\sigma(t)}\sigma(t)z(t)\,dt \leq \int_0^T \left|\frac{\varphi(t)}{\sigma(t)}\right| \sigma(t)|z(t)|\,dt$$

$$\leq \int_0^T \sigma(t)|z(t)|\,dt.$$

Therefore, for any $z \in L_2$,

$$\int_0^T \sigma(t)|z(t)|\,dt - \int_0^T \sigma(t)|\Phi(t)|\,dt \geq \int_0^T \varphi(t)z(t)\,dt - \int_0^T \varphi(t)\Phi(t)\,dt$$

$$= \int_0^T \varphi(t)\left(z(t) - \Phi(t)\right)\,dt$$

and $\varphi \in \partial f(\Phi)$.

Conversely, if $\varphi \in \partial f(\Phi)$, then

$$-\int_0^T \sigma(t)|\Phi(t)|\,dt = \int_0^T \sigma(t)0\,dt - \int_0^T \sigma(t)|\Phi(t)|\,dt$$

$$\geq \int_0^T \varphi(t)\,(0-\Phi(t))\;dt = -\int_0^T \varphi(t)\Phi(t)\,dt.$$

Moreover,

$$\int_0^T \sigma(t)|\Phi(t)|\,dt = 2\int_0^T \sigma(t)|\Phi(t)|\,dt - \int_0^T \sigma(t)|\Phi(t)|\,dt$$

$$\geq \int_0^T \varphi(t)\,(2\Phi(t)-\Phi(t))\;dt = \int_0^T \varphi(t)\Phi(t)\,dt.$$

Hence,

$$\int_0^T \sigma(t)|\Phi(t)|\,dt = \int_0^T \varphi(t)\Phi(t)\,dt. \tag{1.7}$$

By the definition of subgradient, for any $z \in L_2$ and $M > 0$,

$$\int_0^T \sigma(t)|Mz(t)+\Phi(t)|\,dt - \int_0^T \sigma(t)|\Phi(t)|\,dt \geq \int_0^T \varphi(t)Mz(t)\,dt,$$

i.e.,

$$\int_0^T \sigma(t)\left|z(t)+\frac{\Phi(t)}{M}\right|dt - \int_0^T \sigma(t)\frac{|\Phi(t)|}{M}\,dt \geq \int_0^T \varphi(t)z(t)\,dt.$$

Hence, with $M \to \infty$,

$$\int_0^T \sigma(t)|z(t)|\,dt \geq \int_0^T \varphi(t)z(t)\,dt, \qquad \forall\, z \in L_2. \tag{1.8}$$

It follows from the last inequality that

$$\operatorname*{ess\,sup}_{t\in[0,T]}\left|\frac{\varphi(t)}{\sigma(t)}\right| \leq 1. \tag{1.9}$$

In fact, let us assume that on a set E of nonzero measure $|\varphi(t)| > |\sigma(t)|$. Taking $z(t) = \chi_E(t)\text{sign}\,\varphi(t)$, we have from (1.8)

$$\int_E \sigma(t)\,dt \geq \int_E \sigma(t)|\text{sign}\,\varphi(t)|\,dt \geq \int_E |\varphi(t)|\,dt = \int_E \left|\frac{\varphi(t)}{\sigma(t)}\right|\sigma(t)\,dt$$

$$> \int_E \sigma(t)\,dt.$$

This inequality is inconsistent. Thus (1.9) holds.

Let us now prove that (1.9) is a strict equality. In fact, assume that $\operatorname{ess\,sup}_{t\in[0,T]} \left|\frac{\varphi(t)}{\sigma(t)}\right| < 1$. Then, by virtue of (1.7),

$$\int_0^T \sigma(t)|\Phi(t)|\,dt = \int_0^T \frac{\varphi(t)}{\sigma(t)}\sigma(t)\Phi(t)\,dt \le \int_0^T \left|\frac{\varphi(t)}{\sigma(t)}\right|\sigma(t)|\Phi(t)|\,dt$$

$$= \int_P \left|\frac{\varphi(t)}{\sigma(t)}\right|\sigma(t)|\Phi(t)|\,dt < \int_P \sigma(t)|\Phi(t)|\,dt,$$

where

$$P = \left\{ t \in [0,T] \,\Big|\, |\Phi(t)| > 0 \right\}$$

(since $\Phi \neq 0$, the set P has a nonzero measure). This contradiction concludes the proof of (1.5).

Let us obtain (1.3) from (1.4) and (1.5). Note that φ from (1.4) can be represented in the form

$$\varphi(t) = \sigma(t)e(t), \qquad e \in L_2, \quad \operatorname*{ess\,sup}_{t\in[0,T]} |e(t)| \le 1. \tag{1.10}$$

It follows from (1.5) that

$$\int_0^T (\sigma(t)|\Phi(t)| - \varphi(t)\Phi(t)\,dt) = 0 \quad \text{and} \quad \sigma(t) \ge \varphi(t) \quad \text{a.e. on } [0,T].$$

Hence,

$$\sigma(t)|\Phi(t)| = \varphi(t)\Phi(t) \qquad \text{a.e. on } [0,T]$$

and

$$\varphi(t) = \frac{\sigma(t)|\Phi(t)|}{\Phi(t)} \quad \text{for almost all } t \text{ such that } \Phi(t) \neq 0. \tag{1.11}$$

Since the functions that are equal almost everywhere on $[0,T]$ are equivalent in L_2, we see that (1.3) follows from (1.4), (1.5), (1.10), and (1.11). □

THEOREM 1.2 (Moreau-Rockafellar) [75], [51]. *Let* f_1 and f_2 *be proper convex functions on* X. *Then*

$$\partial(f_1 + f_2)(x) \supset \partial f_1(x) + \partial f_2(x).$$

If one of these functions is continuous at a point belonging to the effective domain of the other, then

$$\partial(f_1 + f_2)(x) = \partial f_1(x) + \partial f_2(x)$$

for all x.

We remind the reader that $\partial f_1(x)$ and $\partial f_2(x)$ are sets in X^*, and the expression $\partial f_1(x) + \partial f_2(x)$ denotes the algebraic sum of the sets. This result can be considered as a generalization of the theorem from classical analysis about the derivative of the sum of two differentiable functions.

THEOREM 1.3 (Dubovitskii-Milyutin) [203], [44]. *Let $F: X \to \mathbf{R}^1$ be a convex function, let $g: \mathbf{R}^1 \to \mathbf{R}^1$ be a convex monotonically increasing function, and $f(x) = g(F(x))$. If F is continuous at a point x and g is continuous at the point $F(x)$, then*

$$\partial f(x) = \left\{ x^* \in X^* \,\middle|\, x^* = \alpha y^*,\ \alpha \in \partial g(F(x)),\ y^* \in \partial F(x) \right\},$$

i.e.,

$$\partial (g \circ F)(x) = \partial g(F(x)) \circ \partial F(x).$$

1.2. *Convex Problems. The Kuhn-Tucker Theorem*

Let us consider the problem of the unconditional minimum of a convex function f:

$$f(x) \to \inf.$$

THEOREM 1.4 [75], [51]. *For a function f to attain a minimum at a point $\bar{x}$, it is necessary and sufficient that the following relation holds:*

$$0 \in \partial f(\bar{x}).$$

The last relation is an analog of the Fermat theorem for smooth unconstrainted problems.

Let X be a linear space, let $f_0, \ldots, f_s$ be finite functions on X, and let A be a subset of the space X. We consider the problem

$$f_0(x) \to \inf; \tag{1.12}$$

$$f_1(x) \le 0, \ldots, f_s(x) \le 0, \tag{1.13}$$

$$x \in A. \tag{1.14}$$

The relations (1.13) are called *inequality constraints*. If the functions $f_0, \ldots, f_s$ and the set A are convex, then the problem (1.12)–(1.14) is called a convex programming problem.

The function

$$\mathcal{L}(x, \lambda_0, \ldots, \lambda_s) = \sum_{i=0}^{s} \lambda_i f_i(x), \qquad \lambda_i \in \mathbf{R}^1, \quad i = 0, \ldots, s,$$

is called the *Lagrange function*; (in English literature this function is usually called the "Lagrangian").

THEOREM 1.5 (Kuhn-Tucker) [75]. *Let the functions $f_0, \ldots, f_s$ and the set A be convex. Assume that a vector $\bar{x}$ satisfies the constraints* (1.13) *and* (1.14). *If $\bar{x}$ is a solution of the problem* (1.12) – (1.14), *then there exist Lagrange multipliers $\lambda_0 \geq 0, \ldots, \lambda_s \geq 0$, not all zero, such that*

$$\mathcal{L}(\bar{x}, \lambda_0, \ldots, \lambda_s) = \min_{x \in A} \mathcal{L}(x, \lambda_0, \ldots, \lambda_s) \tag{1.15}$$

and

$$\lambda_i f_i(\bar{x}) = 0 \quad \text{for} \quad i = 1, \ldots, s. \tag{1.16}$$

If, moreover, there exists a vector $x \in A$ such that $f_i(x) < 0$ for all $i = 1, \ldots, s$ (the Slater condition), then $\lambda_0 = 1$. In this case, the relations (1.15) *and* (1.16) *are sufficient for the point $\bar{x}$ satisfying the conditions* (1.13) *and* (1.14) *to be a solution of the problem* (1.12) – (1.14).

The relation (1.15) is called the *Kuhn-Tucker condition*, and the equations (1.16) are called the *conditions of complementary slackness.* The Kuhn-Tucker condition shows that, for convex programming problems, the *Lagrange principle* holds: *a solution of a problem with constraints, with an appropriate choice of Lagrange multipliers, yields a minimum for the problem $\mathcal{L} \to \inf$ with no constraints, or subject to the constraints not included in the Lagrange function.*

The Lagrange principle is valid for more general constraints. Consider the following problem:

$$f_0(x) \to \inf; \tag{1.17}$$

$$F(x) = 0, \tag{1.18}$$

$$f_1(x) \leq 0, \ldots, f_s(x) \leq 0, \tag{1.19}$$

$$x \in A, \tag{1.20}$$

where $F: X \to Y$, $f_0, \ldots, f_s$ are function on X, $A \subset X$, and Y is a Banach space. Let us construct the Lagrange function

$$\mathcal{L}(x, \lambda_0, \ldots, \lambda_s, y^*) = \sum_{i=0}^{s} \lambda_i f_i(x) + \langle y^*, F(x) \rangle, \qquad y^* \in Y^*.$$

THEOREM 1.6 [75]. *In the problem* (1.17) – (1.20), *let X be a linear space, let $f_0, \dots, f_s$ be convex functions on X, let F be affine mapping (i.e., $F(x) = y_0 + \Lambda x$, where $y_0 \in Y$ and $\Lambda: X \to Y$ is a linear operator), and let A be a convex set.*

If $\bar{x}$ is a solution of the problem (1.17) – (1.20), *then there exist Lagrange multipliers $\lambda_0 \geq 0, \dots, \lambda_s \geq 0$ and $y^* \in Y$, not all zero, such that*

$$\mathcal{L}(\bar{x}, \lambda_0, \dots, \lambda_s, y^*) = \sum_{i=0}^{s} \lambda_i f_i(\bar{x}) + \langle y^*, F(\bar{x}) \rangle = \min_{x \in A} \mathcal{L}(x, \lambda_0, \dots, \lambda_s, y^*)$$

and

$$\lambda_i f_i(\bar{x}) = 0 \quad \text{for} \quad i = 1, \dots, s.$$

If, moreover, the set A under the mapping $X \to Y$ contains a neighborhood of the origin of Y, and if there exists a vector $x \in A$ such that

$$F(x) = 0, \qquad f_i(x) < 0, \quad i = 1, \dots, s,$$

then $\lambda_0 \neq 0$ and one can set $\lambda_0 = 1$. In the last case, the relations written above are sufficient for the point $\bar{x}$ satisfying the conditions (1.18) – (1.20) *to be a solution of the problem* (1.17) – (1.20).

REMARK 1.1. If the problem (1.17) – (1.20) has no inequality conditions (1.19), then Theorem 1.6 remains valid with $\lambda_1 = \dots = \lambda_s = 0$. In fact, in this case we formally put $f_1(x) = \dots = f_s(x) \equiv -1$.

1.3. *Duality Theory*

First we formulate a result related to the existence of the minimum of a convex function.

Let $\mathcal{A}$ be a non-empty *closed convex* subset of X. We take a function f_0 of $\mathcal{A}$ into $\mathbf{R}^1$ and we assume that

$$f_0 \text{ is convex and continuous.} \tag{1.21}$$

We also concerned with the minization problem

$$\inf_{x \in \mathcal{A}} f_0(x). \tag{1.22}$$

Any element $x \in \mathcal{A}$ such that:

$$f_0(x) = \inf_{v \in \mathcal{A}} f_0(v)$$

is termed a *solution* of the problem.

LEMMA 1.2 [51]. *Let X be a reflexive Banach space. Let us assume, in addition to* (1.21), *that the set $\mathcal{A}$ is bounded, or that the function f_0 is coercive over $\mathcal{A}$, i.e. that:*

$$\lim f_0(x) = +\infty, \qquad \text{for } x \in \mathcal{A}, \qquad \|x\| \to \infty.$$

Then the problem (1.22) *has at least one solution. It has a unique solution if the function f_0 is strictly convex over $\mathcal{A}$.*

Let us associate to a minimization problem $(\mathcal{P})$ a maximization problem $(\mathcal{P}^*)$ termed the dual problem of $(\mathcal{P})$ and we shall examine the relationship between these two problems. Further we also suppose that (1.21) holds. The variational problems, which we consider in our book, possess this condition.

We discuss two variational problems.

Variational problem $\mathcal{P}_1$:

$$\inf_{x \in X} f_0(x) \quad \text{under constraint} \quad \Lambda x = 0, \tag{1.23}$$

where $\Lambda: X \to Y$ is an affine continuous operator, Y is a Banach space.

We also introduce the following *perturbed problem $\mathcal{P}_{1y}$*

$$\inf_{x \in X} f_0(x) \quad \text{under constraint} \quad \Lambda x = y, \quad y \in Y. \tag{1.24}$$

For $y \in Y$, let $h_1(y)$ be the optimal value of the perturbed problem (1.24). Clearly, $h_1(0)$ is the optimal value for the problem (1.23).

The Lagrange function is defined by the formula

$$\mathcal{L}_1(x, y^*) = f_0(x) + \langle y^*, \Lambda x \rangle, \qquad y^* \in Y^*. \tag{1.25}$$

Obviously, the primal problem $\mathcal{P}_1$ can be represented in the form

$$\inf_{x \in X} \left(\sup_{y^* \in Y^*} \mathcal{L}_1(x, y^*) \right).$$

The *dual problem $\mathcal{P}_1^*$* for the primal problem $\mathcal{P}_1$ is defined by the formula

$$\sup_{y^* \in Y^*} \left(\inf_{x \in X} \mathcal{L}_1(x, y^*) \right). \tag{1.26}$$

Variational problem $\mathcal{P}_2$:

$$\inf_{x \in X} f_0(x) \quad \text{under constraints} \quad f_1(x) \leq 0, \ldots, f_s(x) \leq 0, \tag{1.27}$$

where $f_i: X \to \mathbf{R}^1$, $i = 1, \ldots, s$, are convex continuous functions on X.

The perturbed problem $\mathcal{P}_{2y}$ is defined as follows:

$$\inf_{x \in X} f_0(x) \quad \text{under constraints} \quad f_i(x) \le y_i, \quad y_i \in \mathbf{R}^1, \quad i = 1, \dots, s. \tag{1.28}$$

For $y = (y_1, \dots, y_s)' \in \mathbf{R}^s$, let $h_2(y)$ be the optimal value of the perturbed problem (1.28). Clearly, $h_2(0)$ is the optimal value for the problem (1.27).

The Lagrange function is defined by the formula

$$\mathcal{L}_2(x, y^*) = f_0(x) + \sum_{i=1}^{s} \lambda_i f_i(x), \quad \lambda = (\lambda_1, \dots, \lambda_s)' \in \mathbf{R}^s, \quad \lambda \ge 0, \tag{1.29}$$

where $\lambda \ge 0$ means that $\lambda_i \ge 0$, $i = 1, \dots, s$.

The primal problem $\mathcal{P}_2$ can be represented in the form

$$\inf_{x \in X} \left(\sup_{\lambda \ge 0} \mathcal{L}_2(x, \lambda) \right).$$

The *dual problem* $\mathcal{P}_2^*$ for the primal problem $\mathcal{P}_2$ is defined by the formula

$$\sup_{\lambda \ge 0} \left(\inf_{x \in X} \mathcal{L}_2(x, \lambda) \right). \tag{1.30}$$

The following general inequalities hold.

LEMMA 1.3 [51], [5].

$$\sup_{y^* \in Y^*} \left(\inf_{x \in X} \mathcal{L}_1(x, y^*) \right) \le \inf_{x \in X} \left(\sup_{y^* \in Y^*} \mathcal{L}_1(x, y^*) \right),$$

$$\sup_{\lambda \ge 0} \left(\inf_{x \in X} \mathcal{L}_2(x, \lambda) \right) \le \inf_{x \in X} \left(\sup_{\lambda \ge 0} \mathcal{L}_2(x, \lambda) \right).$$

LEMMA 1.4 [51], [5]. *The functions $h_1(y) \colon Y^* \to \overline{\mathbf{R}}$ and $h_2(y) \colon \mathbf{R}^s \to \overline{\mathbf{R}}$ are convex.*

Convex problems possess an important feature that is used constantly in our book.

THEOREM 1.7 (Duality theorem) [75], [51], [5]. *If $h_1(y)$ is continuous at* 0, *then*

$$h_1(0) = \sup_{y^* \in Y^*} \left(\inf_{x \in X} \mathcal{L}_1(x, y^*) \right) = \inf_{x \in X} \left(\sup_{y^* \in Y^*} \mathcal{L}_1(x, y^*) \right). \tag{1.31}$$

If $h_2(y)$ is continuous at 0, *then*

$$h_2(0) = \sup_{\lambda \geq 0} \left(\inf_{x \in X} \mathcal{L}_2(x, \lambda) \right) = \inf_{x \in X} \left(\sup_{\lambda \geq 0} \mathcal{L}_2(x, \lambda) \right). \tag{1.32}$$

For convex problems, the Lagrange principle (cf. Theorems 1.5 and 1.6) can be strengthened.

A pair $(\bar{x}, \bar{y}^*)$ is called a *saddle point* of $\mathcal{L}_1$ if

$$\mathcal{L}_1(\bar{x}, y^*) \leq \mathcal{L}_1(\bar{x}, \bar{y}^*) \leq \mathcal{L}_1(x, \bar{y}^*), \qquad \forall\, x \in X, \quad \forall\, y^* \in Y^*.$$

Similarly, a pair $(\bar{x}, \bar{\lambda})$ is called a *saddle point* of $\mathcal{L}_2$ if

$$\mathcal{L}_2(\bar{x}, \lambda) \leq \mathcal{L}_2(\bar{x}, \bar{\lambda}) \leq \mathcal{L}_2(x, \bar{\lambda}), \qquad \forall\, x \in X, \quad \forall\, \lambda \geq 0. \tag{1.33}$$

Note that (1.33) implies the condition of complementary slackness:

$$\bar{\lambda}_i f_i(\bar{x}) = 0, \qquad i = 1, \ldots, s.$$

THEOREM 1.8 [51], [5]. *The following two conditions are equivalent to each other:*

(i) $(\bar{x}, \bar{y}^*)$ $((\bar{x}, \bar{\lambda}))$ *is a saddle point of* $\mathcal{L}_1$ $(\mathcal{L}_2)$,

(ii) $\bar{x}$ *is a solution of* $\mathcal{P}_1$ $(\mathcal{P}_2)$, $\bar{y}^*$ $(\bar{\lambda})$ *is a solution of* $\mathcal{P}_1^*$ $(\mathcal{P}_2^*)$, *and* (1.31) ((1.32)) *holds.*

THEOREM 1.9 [51], [5]. *Let us assume that $h_1(0)$ $(h_2(0))$ is finite and $h_1(y)$ $(h_2(y))$ is subdifferentiable at* 0, *then $\bar{x}$ is a solution of $\mathcal{P}_1$ $(\mathcal{P}_2)$ if and only if there exists $\bar{y}^*$ $(\bar{\lambda})$ such that $(\bar{x}, \bar{y}^*)$ $((\bar{x}, \bar{\lambda}))$ is a saddle point of $\mathcal{L}_1$ $(\mathcal{L}_2)$.*

To verify the conditions on $h_1(y)$ $(h_2(y))$ that are required in Theorems 1.7 and 1.9 the following lemmas can be useful.

LEMMA 1.5 [51], [5]. *If in the neighbourhood of a point $x \in X$, a convex function f is bounded above by a finite constant, then f is continuous at x.*

LEMMA 1.6 [51]. *Let f be a convex function of X into $\overline{\mathbf{R}}$, finite and continuous at the point $x \in X$. Then $\partial f(v) \neq \emptyset$ for all $v \in \operatorname{int} \operatorname{dom} f$, and in particular $\partial f(x) \neq \emptyset$.*

2. Uncertain Deterministic Disturbances

In this section we consider a generalization of continuous estimation problem discussed in Chapter 1 to the case of dynamic systems.

Consider a linear dynamic system

$$\dot{x}(t) = A(t)x(t) + B(t)u(t), \qquad t \in [0,T], \quad x(0) = x_0, \tag{2.1}$$

where $x(t) = (x_1(t), \dots, x_n(t))' \in \mathbf{R}^n$ is the state vector of the system, $A(t) \in \mathbf{R}^{n\times n}$, $B(t) \in \mathbf{R}^{n\times r}$ are given matrices, and $u(t) = (u_1(t), \dots, u_r(t))' \in \mathbf{R}^r$ is a perturbation vector. A prime denotes the transposition sign.

We assume that

$$|u_j(t)| \le \gamma_j(t), \qquad t \in [0,T], \quad j = 1, \dots, r, \tag{2.2}$$

where $\gamma_j(t)$ are given nonnegative functions.

REMARK 2.1. If equation (2.1) describes a mechanical system, then, usually, the disturbances $u(t) \in \mathbf{R}^r$ are unknown accelerations (forces acting on the system). The inequalities (2.2) give quite poor description of these accelerations. Therefore the system (2.1) sometimes is called the system with *unmodeled accelerations.*

Let measurements of the state vector be made on the segment $[0,T]$:

$$z(t) = H'(t)x(t) + \varrho(t), \qquad z(t), \varrho(t) \in \mathbf{R}^m, \tag{2.3}$$

where $z(t)$ are the measurements, $H(t)$ is a known $n \times m$ matrix, $\varrho(t)$ is a measurement noise. We assume that $\varrho(t) = (\varrho_1(t), \dots, \varrho_m(t))'$ is an unknown vector function with bounded components

$$|\varrho_l(t)| \le \sigma_l(t), \qquad l = 1, \dots, m, \tag{2.4}$$

where $\sigma_l(t)$, $l = 1, \dots, m$ are known nonnegative functions.

Moreover, the initial state of the system is assumed to belong to the parallelepiped

$$|x_d(0)| \le \sigma_{-d}, \qquad d = 1, \dots, n, \tag{2.5}$$

where σ_{-d}, $d = 1, \dots, n$, are known nonnegative numbers.

The entries of the matrices $A(t)$, $B(t)$, $H(t)$, and the functions $u_j(t)$, $\gamma_j(t)$, $\varrho_l(t)$, $\sigma_l(t)$ are assumed to be piecewise continuous. Furthermore, we use the notation $x(0) \in \Pi$, $u \in U$, and $\varrho \in \sigma$ for the sets (2.5), (2.2), and (2.4), respectively.

Our problem is to estimate a scalar quantity $l_* = a'x(T)$ (where $a \in \mathbf{R}^n$ is a given vector) by linear functionals

$$\hat{l} = \int_0^T \Phi'(t)z(t)\,dt, \qquad \Phi(t) = (\Phi_1(t), \dots, \Phi_m(t))',$$

$$\Phi(t) = \Phi_{pc}(t) + \sum_{s=1}^{M} \Phi_{(s)}\delta(t - t_s). \tag{2.6}$$

Here $\Phi_{pc}(t) \in \mathbf{R}^m$ is a piecewise continuous vector function, $\Phi_{(s)} \in \mathbf{R}^m$, $t_s \in [0,T]$, $s = 1,\ldots,M$, M is an arbitrary interger, and $\delta(t)$ is the Dirac delta-function. We denote the set (2.6) by $\mathcal{F}^m$ and refer to Φ also as an estimator. The estimate $\hat{l}$ includes both averaged and individual values of $z(t)$.

Optimal *a priori* guaranteed estimation problem is to find an estimator Φ^0 that minimizes the guaranteed estimation error

$$d(\Phi^0) = \inf_{\Phi\in\mathcal{F}^m} d(\Phi), \qquad d(\Phi) = \sup_{x(0)\in\Pi,\, u\in U,\, \varrho\in\sigma} |\hat{l} - l_*|.$$

To reduce the formulas let us introduce the notation

$$y(t) = X^{-1}(t)x(t) \in \mathbf{R}^n, \qquad h(t) = X'(t)H(t) \in \mathbf{R}^{n\times m},$$

$$b(t) = X^{-1}(t)B(t) \in \mathbf{R}^{n\times r}, \qquad \tilde{a} = X'(T)a,$$

$$h(t) = (h_1(t),\ldots,h_m(t)), \qquad h_l(t) \in \mathbf{R}^n, \quad l = 1,\ldots,m.$$

$$b(t) = (b_1(t),\ldots,b_r(t)), \qquad b_j(t) \in \mathbf{R}^n, \quad j = 1,\ldots,r,$$

where $X(t) \in \mathbf{R}^{n\times n}$ is a fundamental matrix of the homogeneous equation

$$\dot{x}(t) = A(t)x(t), \qquad X(0) = E_n, \quad E_n \text{ is the } n\times n \text{ identity matrix.}$$

Then,

$$\dot{y}(t) = b(t)u(t), \quad y(t) \in \mathbf{R}^n, \quad y(0) = x(0), \qquad t \in [0,T],$$

and

$$z(t) = h'(t)y(t) + \varrho(t).$$

With this notation, $l_* = \tilde{a}'y(T)$ and the guaranteed estimation error takes the form

$$d(\Phi) = \sup_{y(0)\in\Pi,\, u\in U,\, \varrho\in\sigma} |\hat{l} - l_*|$$

$$= \sup_{y(0)\in\Pi,\, u\in U,\, \varrho\in\sigma} \left| \left(\int_0^T h(t)\Phi(t) - \tilde{a} \right)' y(T) \right.$$

$$\left. + \int_0^T \Phi'(t)\Big(- h'(t) \int_t^T b(\tau)u(\tau)\, d\tau + \varrho(t)\Big)\, dt \right|,$$

$$y(T) = y(0) + \int_0^T b(\tau)u(\tau)\, d\tau.$$

Continuing the computation, we get

$$d(\Phi) = \sup_{y(0)\in\Pi,\, u\in U,\, \varrho\in\sigma} \left| \left(\int_0^T h(t)\Phi(t)\, dt - \tilde{a} \right)' y(0) + \int_0^T \Phi'(t)\varrho(t)\, dt \right.$$

$$+ \left(\int_0^T h(t)\Phi(t)\, dt - \tilde{a} \right)' \int_0^T b(\tau)u(\tau)\, d\tau$$

$$\left. - \int_0^T \Phi'(t)h'(t) \int_t^T b(\tau)u(\tau)\, d\tau\, dt \right|$$

$$= \sup_{y(0)\in\Pi,\, u\in U,\, \varrho\in\sigma} \left| - \Phi'_- y(0) + \int_0^T \Phi'(t)\varrho(t)\, dt - \Phi'_- \int_0^T b(\tau)u(\tau)\, d\tau \right.$$

$$\left. - \int_0^T \Phi'(t)h'(t) \int_t^T b(\tau)u(\tau)\, d\tau\, dt \right|,$$

where

$$\Phi_- + \int_0^T h(t)\Phi(t)\, dt = \tilde{a}. \tag{2.7}$$

The last equation is the analog of the unbiasness condition (cf. Section 2 of Chapter 1). We denote the set of (Φ_-, Φ) satisfying (2.7) by $\mathcal{N}$. Obviously, $\mathcal{N} \neq \emptyset$.

Then, taking into account that (cf. Section 2 of Chapter 1)

$$\sup_{y(0)\in\Pi} -\Phi'_-\, y(0) = \sum_{d=1}^{n} \sigma_{-d}|\Phi_{-d}|, \qquad \Phi_- = (\Phi_{-1}, \ldots, \Phi_{-n})',$$

$$\sup_{\varrho\in\sigma} \int_0^T \Phi'(t)\varrho(t)\, dt = \sum_{l=1}^{m} \int_0^T \sigma_l(t)|\Phi_l(t)|\, dt,$$

and that for any admissible $(y(0), u, \varrho)$ the triple $(-y(0), -u, -\varrho)$ is also admissible, we have

$$d(\Phi) = \sup_{u\in U} \mathcal{I}(\Phi_-, \Phi, u),$$

$$\mathcal{I}(\Phi_-, \Phi, u) = \sum_{d=1}^{n} \sigma_{-d}|\Phi_{-d}| + \sum_{l=1}^{m} \int_0^T \sigma_l(t)|\Phi_l(t)|\, dt \tag{2.8}$$

$$-\int_0^T \Phi'(t)h'(t)\int_t^T b(\tau)u(\tau)\,d\tau dt - \Phi'_-\int_0^T b(\tau)u(\tau)\,d\tau,$$

where $\Phi_- = \Phi_-(\Phi)$ is defined by (2.7).

Thus the solution of the optimal guaranteed estimation problem reduces to the solution of the following game problem

$$\inf_{(\Phi_-,\Phi)\in\mathcal{N}}\ \sup_{u\in U}\ \mathcal{I}(\Phi_-,\Phi,u). \tag{2.9}$$

To obtain the explicit formula for the supremum in (2.9) let us use the following

LEMMA 2.1. *Suppose $\Phi\in\mathcal{F}^m$; then the Dirichlet formula holds:*

$$\int_0^T\int_t^T \Phi'(t)h'(t)\,b(\tau)u(\tau)\,d\tau\,dt = \int_0^T\int_0^\tau \Phi'(t)h'(t)\,b(\tau)u(\tau)\,dt\,d\tau. \tag{2.10}$$

PROOF. The Dirichlet formula for the piecewise continuous component of Φ is well known from analysis. Therefore it is sufficient to prove (2.10) for impulse component only. By our definition, for an impulse function $\Phi(t) = \sum_{s=1}^M \Phi_{(s)}\delta(t-t_s)$ and a piecewise continuous function $f(t)$

$$\int_\alpha^\beta \Phi'(t)f(t)\,dt \triangleq \sum_{t_s\in[\alpha,\beta]} \Phi'_{(s)}f(t_s).$$

Hence,

$$\int_0^T\int_t^T \Phi'(t)h'(t)b(\tau)u(\tau)\,d\tau\,dt \triangleq \sum_{s=1}^M \Phi'_{(s)}h'(t_s)\int_{t_s}^T b(\tau)u(\tau)\,d\tau$$

$$= \sum_{s=1}^M \Phi'_{(s)}h'(t_s)\sum_{j=s}^M \int_{t_j}^{t_{j+1}} b(\tau)u(\tau)\,d\tau$$

$$= \sum_{s=1}^M\sum_{j=1}^M \chi_{sj}\Phi'_{(s)}h'(t_s)\int_{t_j}^{t_{j+1}} b(\tau)u(\tau)\,d\tau,$$

where

$$\chi_{sj} = \begin{cases} 1, & j\ge s \\ 0, & j<s, \end{cases} \qquad \text{and} \quad t_{M+1}=T.$$

Therefore,

$$\int_0^T\int_t^T \Phi'(t)h'(t)b(\tau)u(\tau)\,d\tau\,dt = \sum_{j=1}^M\sum_{s=1}^M \chi_{sj}\Phi'_{(s)}h'(t_s)\int_{t_j}^{t_{j+1}} b(\tau)u(\tau)\,d\tau$$

$$= \sum_{j=1}^{M} \left(\sum_{s=1}^{j} \Phi'_{(s)} h'(t_s) \right) \int_{t_j}^{t_{j+1}} b(\tau) u(\tau)\, d\tau$$

$$\triangleq \sum_{j=1}^{M} \left(\int_{0}^{t_j} \Phi'(t) h'(t)\, dt \right) \int_{t_j}^{t_{j+1}} b(\tau) u(\tau)\, d\tau$$

$$= \sum_{j=1}^{M} \int_{t_j}^{t_{j+1}} \left(\int_{0}^{\tau} \Phi'(t) h'(t)\, dt \right) b(\tau) u(\tau)\, d\tau$$

$$= \int_{0}^{T} \int_{0}^{\tau} \Phi'(t) h'(t)\, dt\, b(\tau) u(\tau)\, d\tau.$$

The next to the last equality holds due to the formula

$$\int_{0}^{\tau} \Phi'(t) h'(t)\, dt \triangleq \begin{cases} \int_0^{t_j} \Phi'(t) h'(t)\, dt, & \tau \in [t_j, t_{j+1}) \\ \int_0^{t_{j+1}} \Phi'(t) h'(t)\, dt, & \tau = t_{j+1}. \end{cases}$$

□

Using Lemma 2.1, the last integral term in the expression for $\mathcal{I}(\Phi_-, \Phi, u)$ can be represented in the form

$$\int_0^T \Phi'(t) h'(t) \int_t^T b(\tau) u(\tau)\, d\tau dt = \int_0^T \left(\int_0^{\tau} h(t) \Phi(t)\, dt \right)' b(\tau) u(\tau)\, d\tau.$$

If the supremum in (2.9) is taken over all measurable functions $u(t)$ restricted by (2.2), then this supremum is attained at $\bar{u}(t) = (\bar{u}_1(t), \ldots, \bar{u}_r(t))'$, where

$$\bar{u}_j(t) = -\gamma_j(t) \operatorname{sign} \left[b'_j(t) \left(\Phi_- + \int_0^t h(\tau) \Phi(\tau)\, d\tau \right) \right], \tag{2.11}$$

$j = 1, \ldots, r$, and, in this case,

$$d(\Phi) = I(\Phi_-, \Phi) \Big|_{(\Phi_-, \Phi) \in \mathcal{N}},$$

$$I(\Phi_-, \Phi) = \sum_{d=1}^{n} \sigma_{-d} |\Phi_{-d}| + \sum_{l=1}^{m} \int_0^T \sigma_l(t) |\Phi_l(t)|\, dt \tag{2.12}$$

$$+ \sum_{j=1}^{r} \int_0^T \gamma_j(t) \left| b'_j(t) \left(\Phi_- + \int_0^t h(\tau) \Phi(\tau)\, d\tau \right) \right| dt.$$

Since $\bar{u}(t)$ can be approximated by piecewise continuous functions restricted by inequalities (2.2), the supremum in (2.9) may be taken over piecewise continuous functions as well.

Therefore the optimal *a priori* guaranteed estimation problem reduces to the following variational problem

$$I_0 = \inf_{(\Phi_-,\Phi)\in\mathcal{N}} I(\Phi_-,\Phi). \tag{2.13}$$

The solution of (2.13) (Φ_-^0, Φ^0) determines the optimal estimator Φ^0.

3. Conditions for Optimality of Estimator

Necessary and sufficient conditions for the optimality of an estimator determine the structure of the solution. In some problems, they yield the solution in explicit form. In complex applied problems, the necessary and sufficient conditions allow a researcher to feel about the optimal estimator. Using these conditions, the researcher can construct an estimator that is close to the optimal one and then to modify it slightly in order to satisfy all conditions.

THEOREM 3.1. 1^0. *Let a pair* (Φ_-^0, Φ^0),

$$\Phi^0(t) = \Phi_{pc}^0(t) + \sum_{s=1}^{M} \Phi_{(s)}^0 \delta(t - t_s),$$

be a solution of the problem (2.13); *then there exist a vector* $\beta \in \mathbf{R}^n$ *and a measurable function* $u^0(t) = (u_1(t), \ldots, u_r(t))' \in \mathbf{R}^r$ *such that*

(a)

$$|u_j^0(t)| \le \gamma_j(t), \qquad t \in [0,T], \quad j = 1, \ldots, r; \tag{3.1}$$

(b)

$$u_j^0(t) = -\gamma_j(t) \operatorname{sign}\left(b_j'(t) p^0(t)\right) \;\; \textit{on the set} \;\; b_j'(t)p^0(t) \ne 0, \tag{3.2}$$

$$t \in [0,T], \quad j = 1, \ldots, r,$$

where

$$p^0(t) = \Phi_-^0 + \int_0^t h(s)\Phi^0(s)\, ds, \qquad \Phi_-^0 + \int_0^T h(s)\Phi^0(s)\, ds = \tilde{a};$$

(c)

$$|f_l^0(t)| \le \sigma_l(t), \qquad t \in [0,T], \quad l = 1, \ldots, m, \tag{3.3}$$

$$|f_{-d}^0| \le \sigma_{-d}, \qquad d = 1, \ldots, n,$$

where

$$f_l^0(t) = h_l'(t)\left(\beta + \int_t^T b(\tau)u^0(\tau)\,d\tau\right),$$

$$f_{-d}^0 = e_d'\left(\beta + \int_0^T b(\tau)u^0(\tau)\,d\tau\right), \qquad e_d = (\underbrace{0, \ldots, 0, 1}_{d}, \ldots, 0)';$$

(d)

$$\textit{On the set } \mathcal{T}_l^1 = \left\{t \;\middle|\; |f_l^0(t)| < \sigma_l(t)\right\}, \qquad \left(\Phi_{pc}^0\right)_l(t) = 0; \tag{3.4}$$

$$\textit{On the set } \mathcal{T}_l^2 = \left\{t \;\middle|\; |f_l^0(t)| = \sigma_l(t), \quad \sigma_l(t) > 0\right\},$$

$$\textit{either } \left(\Phi_{pc}^0\right)_l(t) = 0 \quad \textit{or} \quad \operatorname{sign}\left(\left(\Phi_{pc}^0\right)_l(t)\right) = \operatorname{sign} f_l^0(t), \quad l = 1, \ldots, m,$$

$$\Phi_{pc}^0(t) = \left(\left(\Phi_{pc}^0\right)_1(t), \ldots, \left(\Phi_{pc}^0\right)_m(t)\right)',$$

(possibly except for a finite number of discontinuity points of $\left(\Phi_{pc}^0\right)_l(t)$, $h_l(t)$, *and* $\sigma_l(t)$*).*

$$\textit{If } t_s \in \mathcal{T}_l^1, \textit{ then } \left(\Phi_{(s)}^0\right)_l = 0; \tag{3.5}$$

$$\textit{If } t_s \in \mathcal{T}_l^2, \textit{ then } \quad \textit{either } \left(\Phi_{(s)}^0\right)_l = 0 \quad \textit{or} \quad \operatorname{sign}\left(\Phi_{(s)}^0\right)_l = \operatorname{sign} f_l^0(t_s),$$

$s = 1, \ldots, M$, $l = 1, \ldots, m$,

$$\Phi_{(s)}^0 = \left(\left(\Phi_{(s)}^0\right)_1, \ldots, \left(\Phi_{(s)}^0\right)_m\right)';$$

$$\textit{On the set } \mathcal{T}_-^1 = \left\{d \;\middle|\; |f_{-d}^0| < \sigma_{-d}\right\}, \qquad \Phi_{-d}^0 = 0; \tag{3.6}$$

$$\textit{On the set } \mathcal{T}_-^2 = \left\{d \;\middle|\; |f_{-d}^0)| = \sigma_{-d}, \quad \sigma_{-d} > 0\right\},$$

$$\textit{either } \Phi_{-d}^0 = 0 \quad \textit{or} \quad \operatorname{sign}\left(\Phi_{-d}^0\right) = \operatorname{sign} f_{-d}^0, \quad d = 1, \ldots, n.$$

2^0. *Suppose a pair* (Φ_-^0, Φ^0), $\Phi^0 \in \mathcal{F}^m$, *a vector* $\beta \in \mathbf{R}^n$, *and a measurable function* $u^0(t) \in \mathbf{R}^r$ *satisfy conditions* (3.1) – (3.6); *then* (Φ_-^0, Φ^0) *is the solution of the problem* (2.13).

PROOF. Let L_1^r be the Banach space of absolutely integrable vector functions w with the norm:

$$\|w\|_{L_1^r} = \int_0^T \|w(t)\|_{\mathbf{R}^r}\, dt, \qquad \|w(t)\|_{\mathbf{R}^r} = \left(\sum_{i=1}^r w_i^2(t)\right)^{\frac{1}{2}}.$$

Consider the following extremal problem:

$$I_0 = \inf_{\Phi_- \in \mathbf{R}^n,\, \Phi \in \mathcal{F}^m,\, w \in L_1^r} I(\Phi_-, \Phi, w), \tag{3.7}$$

$$I(\Phi_-, \Phi, w) = \sum_{d=1}^n \sigma_{-d}|\Phi_{-d}| + \sum_{l=1}^m \int_0^T \sigma_l(t)|\Phi_l(t)|\, dt + \sum_{j=1}^r \int_0^T \gamma_j(t)|w_j(t)|\, dt,$$

$$b'(t)\left(\Phi_- + \int_0^t h(s)\Phi(s)\, ds\right) - w(t) = 0, \qquad t \in [0, T], \tag{3.8}$$

$$\Phi_- + \int_0^T h(t)\Phi(t)\, dt = \tilde{a}.$$

Clearly, if (Φ_-^0, Φ^0) is a solution of (2.13), then (Φ_-^0, Φ^0, w^0), where

$$w^0(t) = b'(t)\left(\Phi_-^0 + \int_0^t h(s)\Phi^0(s)\, ds\right), \qquad t \in [0, T],$$

is a solution of (3.7), (3.8). Conversely, if (Φ_-^0, Φ^0, w^0) is a solution of (3.7), (3.8), then (Φ_-^0, Φ^0) is a solution of (2.13).

LEMMA 3.1. 1^0. *Let* (Φ_-^0, Φ^0, w^0) *be a solution of* (3.7), (3.8). *Then there exist a vector* $\beta \in \mathbf{R}^n$ *and a bounded measurable function* $\eta(t) = (\eta_1(t), \ldots, \eta_r(t))' \in \mathbf{R}^r$ *such that*

$$\mathcal{L}(\Phi_-^0, \Phi^0, w^0; \beta, \eta) = \min_{\Phi_- \in \mathbf{R}^n,\, \Phi \in \mathcal{F}^m,\, w \in L_1^r} \mathcal{L}(\Phi_-, \Phi, w; \beta, \eta), \tag{3.9}$$

where

$$\mathcal{L}(\Phi_-, \Phi, w; \beta, \eta) = I(\Phi_-, \Phi, w)$$

$$+ \int_0^T \eta'(t)\left\{b'(t)\left(\Phi_- + \int_0^t h(s)\Phi(s)\, ds\right) - w(t)\right\} dt$$

$$+\beta'\left\{\tilde{a} - \Phi_- - \int_0^T h(t)\Phi(t)\,dt\right\}.$$

2^0. *Suppose a triple* (Φ_-^0, Φ^0, w^0) *satisfy* (3.8) *and there exist a vector* $\beta \in \mathbf{R}^n$, *and a bounded measurable function* $\eta(t) \in \mathbf{R}^r$ *such that* (3.9) *holds. Then* (Φ_-^0, Φ^0, w^0) *is the solution of the problem* (3.7), (3.8).

PROOF OF LEMMA 3.1. Similar to Section 2 of Chapter 1, let us equip the linear space $\mathcal{F}^m$ with the norm: for $\Phi(t) = \Phi_{pc}(t) + \sum_{s=1}^{M} \Phi_{(s)}\delta(t - t_s)$, we define

$$\|\Phi\|_{\mathcal{F}^m} = \int_0^T \|\Phi_{pc}(t)\|_{\mathbf{R}^m}\,dt + \sum_{s=1}^{M} \|\Phi_{(s)}\|_{\mathbf{R}^m}.$$

We introduce the spaces $\mathcal{X} = \mathbf{R}^n \times \mathcal{F}^m \times L_1^r$ and $\mathcal{Y} = L_1^r \times \mathbf{R}^n$ with the norm

$$\|x\|_{\mathcal{X}} = \max\left(\|\Phi_-\|_{\mathbf{R}^n}, \|\Phi\|_{\mathcal{F}^m}, \|w\|_{L_1^r}\right) \quad \text{for} \quad x = (\Phi_-, \Phi, w)$$

and the norm

$$\|y\|_{\mathcal{Y}} = \max\left(\|w\|_{L_1^r}, \|\Phi_-\|_{\mathbf{R}^n}\right) \quad \text{for} \quad y = (w, \Phi_-),$$

respectively.

Define an affine operator Λ: $\mathcal{X} \to \mathcal{Y}$ by the relations:

$$\Lambda(\Phi_-, \Phi, w) = (v, d), \qquad \Phi_-, d \in \mathbf{R}^n, \quad \Phi \in \mathcal{F}^m, \quad w, v \in L_1^r,$$

$$v(t) = b'(t)\left(\Phi_- + \int_0^t h(s)\Phi(s)\,ds\right) - w(t), \qquad t \in [0, T],$$

$$d = -\Phi_- - \int_0^T h(t)\Phi(t)\,dt + \tilde{a}.$$

Thus the initial extremal problem can be rewritten in the form

$$\inf_{(\Phi_-, \Phi, w) \in \mathcal{X}} I(\Phi_-, \Phi, w) \quad \text{under constraint} \quad \Lambda(\Phi_-, \Phi, w) = 0, \tag{3.10}$$

where the continuous functional I: $\mathcal{X} \to \mathbf{R}^1$ is convex and the continuous operator Λ: $\mathcal{X} \to \mathcal{Y}$ is affine.

Let us apply Theorem 1.6 to the problem (3.10) (see Remark 1.1). By virtue of Theorem 1.6, if (Φ_-^0, Φ^0, w^0) is the solution of (3.10), then there exist a number $\beta_0 \geq 0$ and a continuous linear functional y^* on $\mathcal{Y}$, not both zero, such that

$$\beta_0 I(\Phi_-^0, \Phi^0, w^0) + \langle y^*, \Lambda(\Phi_-^0, \Phi^0, w^0)\rangle \tag{3.11}$$

$$\leq \beta_0 I(\Phi_-, \Phi, w) + \langle y^*, \Lambda(\Phi_-, \Phi, w)\rangle, \qquad \forall\, (\Phi_-, \Phi, w) \in \mathcal{X}.$$

Since $\mathcal{Y}$ is a direct product of Banach spaces, the value of the continuous linear functional y^* on the element $y = (w, d) \in \mathcal{Y}$ is determined by the expression [5]

$$\langle y^*, y\rangle = \langle w^*, w\rangle + \langle d^*, d\rangle, \tag{3.12}$$

where w^* is a continuous linear functional on L_1^r, and d^* is a continuous linear functional on $\mathbf{R}^n$. The general form of these continuous linear functionals are determined by the formulas [75]

$$\langle w^*, w\rangle = \int_0^T \eta'(t)w(t)\,dt, \qquad \langle d^*, d\rangle = \beta' d, \tag{3.13}$$

where $\eta \in L_\infty^r$, $\beta \in \mathbf{R}^n$; L_∞^r is the Banach space of bounded measurable functions with the norm

$$||\eta||_{L_\infty^r} = \operatorname*{ess\,sup}_{t\in[0,T]} ||\eta(t)||_{\mathbf{R}^r}.$$

Recall that

$$\operatorname*{ess\,sup}_{t\in[0,T]} \alpha(t) = \inf\left\{ \sup_{t\in[0,T]} \tilde{\alpha}(t) \,\middle|\, \alpha(t) = \tilde{\alpha}(t) \text{ almost everywhere} \right\}.$$

Therefore the inequality (3.11) can be represented in the form

$$\beta_0 I(\Phi_-^0, \Phi^0, w^0) + \int_0^T \eta'(t)\left\{b'(t)\left(\Phi_-^0 + \int_0^t h(s)\Phi^0(s)\,ds\right) - w^0(t)\right\} dt$$

$$+ \beta'\left\{\tilde{a} - \Phi_-^0 - \int_0^T h(t)\Phi^0(t)\,dt\right\} \tag{3.14}$$

$$\leq \beta_0 I(\Phi_-, \Phi, w) + \int_0^T \eta'(t)\left\{b'(t)\left(\Phi_- + \int_0^t h(s)\Phi(s)\,ds\right) - w(t)\right\} dt$$

$$+ \beta'\left\{\tilde{a} - \Phi_- - \int_0^T h(t)\Phi(t)\,dt\right\}, \qquad \forall\, (\Phi_-, \Phi, w) \in \mathcal{X}.$$

Since the values of $\eta(t)$ on a set of zero measure has no affect on the integrals, we may consider $\eta(t)$ bounded everywhere on $[0, T]$.

Let us show that $\beta_0 > 0$. On the contrary, assume $\beta_0 = 0$. Putting $\Phi_- = 0$ and $\Phi(t) = 0$ in (3.14), we have

$$0 \leq -\int_0^T \eta'(t)w(t)\,dt + \beta'\tilde{a}, \qquad \forall\, w \in L_1^r.$$

If $\eta \neq 0$, then, taking $w(t) = c\,\eta(t)$, we get

$$0 \leq -c \int_0^T \|\eta(t)\|^2_{\mathbf{R}^n}\, dt + \beta' \tilde{a}, \qquad \forall\, c.$$

So, for sufficiently large c the latter inequality is invalid.

Suppose $\eta = 0$; then it follows from (3.14) that

$$\beta \neq 0, \qquad 0 \leq \beta' \left\{ \tilde{a} - \Phi_- - \int_0^T h(s)\Phi(s)\, ds \right\}, \quad \forall\, \Phi_- \text{ and } \Phi.$$

For the same reason, this inequality is also invalid.

Thus $\beta_0 > 0$ and we may put $\beta_0 = 1$. So, we proved the first assertion of the lemma.

Suppose the conditions of the second assertion hold. Then $\Lambda(\Phi^0_-, \Phi^0, w^0) = 0$. Moreover, in accordance with (3.12) and (3.13), the vector β and the function $\eta(t)$ define a continuous linear functional on $\mathcal{Y}$ and the inequality (3.11) is valid with $\beta_0 = 1$. Then it follows from the second assertion of Theorem 1.6 that (Φ^0_-, Φ^0, w^0) is the solution of (3.10).
Lemma 3.1 is proved.

Now let us prove the first assertion of Theorem 3.1. Let (Φ^0_-, Φ^0, w^0) be the solution of (3.7), (3.8). Then, by virtue of Lemma 3.1, relation (3.9) holds.

(1) Let us show that

$$|\eta_j(t)| \leq \gamma_j(t), \quad t \in [0, T], \quad \text{and} \tag{3.15}$$

$$\eta_j(t) = \gamma_j(t) \operatorname{sign} w^0_j(t) \quad \text{for almost every } t \text{ such that } w^0_j(t) \neq 0.$$

Setting $\Phi_l(t) = \Phi^0_l(t)$, $l = 1, \ldots, m$, and $\Phi_{-d} = \Phi^0_{-d}$, $d = 1, \ldots, n$, we have from (3.9) that

$$\omega^0 \triangleq \int_0^T \left(\sum_{j=1}^r \gamma_j(t)|w^0_j(t)| - \eta'(t) w^0(t) \right) dt \tag{3.16}$$

$$\leq \int_0^T \left(\sum_{j=1}^r \gamma_j(t)|w_j(t)| - \eta'(t) w(t) \right) dt, \qquad \forall\, w \in L^r_1.$$

Let us assume that, for a j^0, $|\eta_{j^0}(t)| > \gamma_{j^0}(t)$ on a set $\mathcal{M}$ of nonzero measure. Then it follows from (3.16) that for $w(t) = (0, \ldots, 0, w_{j^0}(t), 0, \ldots, 0)'$, where

$$\begin{cases} w_{j^0}(t) = c \operatorname{sign} \eta_{j^0}(t), & t \in \mathcal{M} \\ w_{j^0}(t) = 0, & t \notin \mathcal{M}, \end{cases}$$

the following inequality holds:

$$\omega^0 \leq c \int_{\mathcal{M}} \left(\gamma_{j^0}(t) - |\eta_{j^0}(t)| \right) dt, \qquad \forall\, c > 0. \tag{3.17}$$

Clearly, (3.17) is invalid for sufficiently large c. Therefore our assumption is false and

$$|\eta_j(t)| \leq \gamma_j(t) \quad \text{for almost all} \quad t \in [0,T], \quad j = 1, \ldots, r. \tag{3.18}$$

So, it follows from (3.16), (3.18) that $\omega^0 \geq 0$. Taking $w(t) = 0$, we get $\omega^0 \leq 0$. Consequently $\omega^0 = 0$ and

$$\gamma_j(t)|w_j^0(t)| - \eta_j(t)w_j^0(t) = 0 \quad \text{for almost all} \quad t \in [0,T], \quad j = 1, \ldots, r,$$

and

$$\eta_j(t) = \gamma_j(t) \operatorname{sign} w_j^0(t) \tag{3.19}$$

for almost every t such that $w_j^0(t) \neq 0$, $j = 1, \ldots, r$.

Since the values of $\eta(t)$ on sets of zero measure are not essential, we can consider that (3.18) and (3.19) hold everywhere on $[0,T]$.

(2) Let us prove that

$$|f_l(t)| \leq \sigma_l(t), \qquad t \in [0,T], \quad l = 1, \ldots, m, \tag{3.20}$$

$$|f_{-d}| \leq \sigma_{-d}, \qquad d = 1, \ldots, n,$$

where

$$f_l(t) = h_l'(t)\left(\beta - \int_t^T b(\tau)\eta(\tau)\,d\tau\right), \quad f_{-d} = e_d'\left(\beta - \int_0^T b(\tau)\eta(\tau)\,d\tau\right).$$

Moreover,

$$\int_0^T \left(\sigma_l(t)|\Phi_l^0(t)| - f_l(t)\Phi_l^0(t)\right)\,dt = 0, \qquad l = 1, \ldots, m, \tag{3.21}$$

$$\sigma_{-d}|\Phi_{-d}^0| - f_{-d}\Phi_{-d}^0 = 0, \qquad d = 1, \ldots, n.$$

Putting $w(t) = w^0(t)$ and $\Phi_- = \Phi_-^0$ in (3.9) and taking into account the Dirichlet formula (2.10)

$$\int_0^T \eta'(t)b'(t)\int_0^t h(s)\Phi(s)\,ds\,dt = \int_0^T \Phi'(s)h'(s)\int_s^T b(t)\eta(t)\,dt\,ds,$$

we get from (3.9) that

$$\omega_0 = \sum_{l=1}^m \int_0^T \left(\sigma_l(t)|\Phi_l^0(t)| - f_l(t)\Phi_l^0(t)\right)\,dt \tag{3.22}$$

$$\leq \sum_{l=1}^m \int_0^T \left(\sigma_l(t)|\Phi_l(t)| - f_l(t)\Phi_l(t)\right)\,dt, \qquad \forall\,\Phi \in \mathcal{F}^m.$$

Assume that there exist an l^0 and a $t^0 \in [0,T]$ such that $|f_{l^0}(t^0)| > \sigma_{l^0}(t^0)$. Setting in (3.22)

$$\Phi_{l^0}(t) = c \operatorname{sign} f_{l^0}(t^0)\, \delta(t-t^0), \quad c>0, \qquad \Phi_l(t)=0, \quad l \neq l^0,$$

we get that, for any $c > 0$,

$$\omega_0 \le c\left(\sigma_{l^0}(t^0) - |f_{l^0}(t^0)|\right).$$

For suficiently large c the latter inequality is invalid and our assumption is false. So, the first inequality from (3.20) holds and it follows from (3.22) that $\omega_0 \ge 0$. Putting $\Phi_l(t) = 0$, $l = 1,\dots,m$, in (3.22), we have $\omega_0 \le 0$ and then $\omega_0 = 0$. Consequently the first equality from (3.21) also holds. The rest relations from (3.20) and (3.21) can be proved similarly.

(3) The first formula of (3.21) is equivalent to the relation

$$\int_0^T \left(\sigma_l(t) \left| \left(\Phi^0_{pc}\right)_l(t)\right| - f_l(t)\left(\Phi^0_{pc}\right)_l(t)\right) dt \tag{3.23}$$

$$+\sum_{s=1}^{M}\left(\sigma_l(t_s)\left|\left(\Phi^0_{(s)}\right)_l\right| - f_l(t_s)\left(\Phi^0_{(s)}\right)_l\right) = 0, \qquad l=1,\dots,m.$$

We define

$$u^0(t) = -\eta(t), \qquad t\in[0,T]; \tag{3.24}$$

it follows from (2.11), (3.19), and (3.24) that $u^0(t)$ is the worst perturbation under the estimator (Φ^0_-, Φ^0). Then (3.15), (3.24), and (3.20) imply (3.1) – (3.3). The relation (3.23) imply (3.4) – (3.6) (see Section 4 of Chapter 1 for details).

The first assertion of Theorem 3.1 is proved.

Let the conditions of the second assertion of Theorem 3.1 hold. We define a function η by (3.24). Then (3.15) and (3.20) hold. Since

$$\mathcal{L}(\Phi_-, \Phi, w; \beta, \eta) = \sum_{d=1}^{n}\left(\sigma_{-d}|\Phi_{-d}| - f_{-d}\Phi_{-d}\right)$$

$$+\sum_{l=1}^{m}\int_0^T \left(\sigma_l(t)|\Phi_l(t)| - f_l(t)\Phi_l(t)\right) dt$$

$$+\sum_{j=1}^{r}\int_0^T \left(\gamma_j(t)|w_j(t)| - \eta_j(t)w_j(t)\right) dt + \beta'\tilde{a},$$

we get that

$$\min_{\Phi_-\in\mathbf{R}^n, \Phi\in\mathcal{F}^m, w\in L_1^r} \mathcal{L}(\Phi_-, \Phi, w; \beta, \eta) = \beta'\tilde{a}.$$

If (Φ_-^0, Φ^0, w^0) satisfies conditions (3.1) – (3.6), then $\mathcal{L}(\Phi_-^0, \Phi^0, w^0; \beta, \eta) = \beta'\tilde{a}$. Consequently, in accordance with the second assertion of Lemma 3.1, (Φ_-^0, Φ^0, w^0) is the solution for the problem (3.7), (3.8) and (Φ_-^0, Φ^0) is the solution of (2.13). □

REMARK 3.1. The set $\mathcal{T}_l^2$, $l = 1, \ldots, m$, consists of the maximum points of $f_l^0(t)$. In generic case, $\mathcal{T}_l^2$ contains only finite number of points. Therefore in "almost all" cases the piecewise component $\Phi_{pc}^0(t)$ equals to zero and the optimal estimator is described by impulse functions. This is the reason why we introduce the delta-functions into consideration.

Consider the game problem that is analogous to the problem (2.9)

$$\inf_{(\Phi_-,\Phi)\in\mathcal{N}} \sup_{u\in U^e} \mathcal{I}(\Phi_-, \Phi, u). \tag{3.25}$$

The distinction is that U^e is the set of all measurable (not only piecewise continuous) functions $u(t)$ such that

$$|u_j(t)| \leq \gamma_j(t), \qquad t \in [0, T], \quad j = 1, \ldots, r.$$

COROLLARY 3.1. *Let the pair (Φ_-^0, Φ^0) be a solution of the problem* (2.7), (2.13). *Then there exists a perturbation $u^0(t)$ satisfying* (3.1) – (3.6) *such that $\{(\Phi_-^0, \Phi^0); u^0\}$ is a saddle point for the game problem* (3.25):

$$\mathcal{I}(\Phi_-^0, \Phi^0, u) \leq \mathcal{I}(\Phi_-^0, \Phi^0, u^0) \leq \mathcal{I}(\Phi_-, \Phi, u^0), \qquad (\Phi_-, \Phi) \in \mathcal{N}, \; u \in U^e.$$

PROOF. Define $u^0(t)$ by (3.24). Then it follows from (2.11), (3.19), and (3.24) that $u^0(t)$ is the worst perturbation under the estimator Φ^0. Hence the inequality

$$\mathcal{I}(\Phi_-^0, \Phi^0, u) \leq \mathcal{I}(\Phi_-^0, \Phi^0, u^0), \qquad u \in U^e,$$

holds.

Let us prove the inequality

$$\mathcal{I}(\Phi_-^0, \Phi^0, u^0) \leq \mathcal{I}(\Phi_-, \Phi, u^0), \qquad (\Phi_-, \Phi) \in \mathcal{N}.$$

Using Lemma 3.1 with $w(t) = w^0(t)$ and $(\Phi_-, \Phi) \in \mathcal{N}$, we have

$$\sum_{d=1}^{n} \sigma_{-d}|\Phi_{-d}^0| + \sum_{l=1}^{m} \int_0^T \sigma_l(t)|\Phi_l^0(t)|\, dt \tag{3.26}$$

$$+ \int_0^T \eta'(t)b'(t)\left(\Phi_-^0 + \int_0^t h(s)\Phi^0(s)\, ds\right) dt$$

$$\leq \sum_{d=1}^{n} \sigma_{-d}|\Phi_{-d}| + \sum_{l=1}^{m} \int_0^T \sigma_l(t)|\Phi_l(t)|\, dt$$

$$+ \int_0^T \eta'(t)b'(t) \left(\Phi_- + \int_0^t h(s)\Phi(s)\, ds \right) dt,$$

where, by virtue of Lemma 2.1,

$$\int_0^T \eta'(t)b'(t) \left(\Phi_- + \int_0^t h(s)\Phi(s)\, ds \right) dt \tag{3.27}$$

$$= \Phi'_- \int_0^T b(t)\eta(t)\, dt + \int_0^T \Phi'(s)h'(s) \int_s^T b(t)\eta(t)\, dt\, ds.$$

Moreover, in accordance with (3.24),

$$\eta(t) = -u^0(t). \tag{3.28}$$

Then the desired inequality follows from (2.8), (3.26) – (3.28). □

Assume that no *a priori* information about the initial state vector is available: $x(0) \in \mathbf{R}^n$, i.e., $\sigma_{-d} = \infty$, $d = 1, \ldots, n$. It is not hard to prove that the corresponding guaranteed estimation error is finite if the *unbiasness condition* holds (cf. Section 2 of Chapter 1):

$$\int_0^T h(t)\Phi(t)\, dt = \tilde{a}. \tag{3.29}$$

In the latter case, the guaranteed estimation error is determined by the formula

$$d(\Phi) = \sup_{u \in U} \mathcal{I}(\Phi, u),$$

$$\mathcal{I}(\Phi, u) = \sum_{l=1}^{m} \int_0^T \sigma_l(t)|\Phi_l(t)|\, dt - \int_0^T \Phi'(t)h'(t) \int_t^T b(\tau)u(\tau)\, d\tau dt.$$

Besides, the solution of the corresponding optimal guaranteed estimation problem reduces to the solution of the following game problem

$$\inf_{\Phi \in \mathcal{N}_0} \sup_{u \in U} \mathcal{I}(\Phi, u), \tag{3.30}$$

where the set $\mathcal{N}_0$ is defined by the unbiasness condition (3.29).

Therefore, similar to (2.12) and (2.13), the optimal *a priori* guaranteed estimation problem reduces to the following variational problem

$$I_0 = \inf_{\Phi \in \mathcal{N}_0} I(\Phi), \tag{3.31}$$

where

$$I(\Phi) = \sum_{l=1}^{m} \int_0^T \sigma_l(t)|\Phi_l(t)|\,dt + \sum_{j=1}^{r} \int_0^T \gamma_j(t)\left| b_j'(t) \int_0^t h(\tau)\Phi(\tau)\,d\tau \right| dt.$$

The solution of (3.31) determines the optimal estimator Φ^0. In this case, Theorem 3.1 takes the form

COROLLARY 3.2. 1^0. *Let an estimator*

$$\Phi^0(t) = \Phi^0_{pc}(t) + \sum_{s=1}^{M} \Phi^0_{(s)}\delta(t - t_s)$$

be a solution of the problem (3.31); *then there exist a vector* $\beta \in \mathbf{R}^n$ *and a measurable function* $u^0(t) = (u_1(t), \ldots, u_r(t))' \in \mathbf{R}^r$ *such that*
(a1)

$$|u_j^0(t)| \le \gamma_j(t), \qquad t \in [0,T], \quad j = 1, \ldots, r; \tag{3.32}$$

(b1)

$$u_j^0(t) = -\gamma_j(t)\,\mathrm{sign}\,\big(b_j'(t)p^0(t)\big) \quad \textit{on the set} \quad b_j'(t)p^0(t) \neq 0, \tag{3.33}$$

$$t \in [0,T], \quad j = 1, \ldots, r,$$

where

$$p^0(t) = \int_0^t h(s)\Phi^0(s)\,ds, \qquad \int_0^T h(s)\Phi^0(s)\,ds = \tilde{a};$$

(c1)

$$|f_l^0(t)| \le \sigma_l(t), \qquad t \in [0,T], \quad l = 1, \ldots, m, \tag{3.34}$$

where

$$f_l^0(t) = h_l'(t)\left(\beta + \int_t^T b(\tau)u^0(\tau)\,d\tau\right);$$

(d1)

$$\textit{On the set } \mathcal{T}_l^1 = \left\{t \;\middle|\; |f_l^0(t)| < \sigma_l(t)\right\}, \qquad \left(\Phi^0_{pc}\right)_l(t) = 0; \tag{3.35}$$

$$\textit{On the set } \mathcal{T}_l^2 = \left\{t \;\middle|\; |f_l^0(t)| = \sigma_l(t), \quad \sigma_l(t) > 0\right\},$$

$$\textit{either } \left(\Phi^0_{pc}\right)_l(t) = 0 \quad \textit{or} \quad \mathrm{sign}\left(\left(\Phi^0_{pc}\right)_l(t)\right) = \mathrm{sign}\, f_l^0(t), \quad l = 1, \ldots, m,$$

(possibly except for a finite number of discontinuity points of $\left(\Phi^0_{pc}\right)_l(t)$, $h_l(t)$, *and* $\sigma_l(t)$*).*

$$\text{If } t_s \in \mathcal{T}_l^1, \text{ then } \left(\Phi^0_{(s)}\right)_l = 0; \tag{3.36}$$

$$\text{If } t_s \in \mathcal{T}_l^2, \text{ then either } \left(\Phi^0_{(s)}\right)_l = 0 \text{ or } \operatorname{sign}\left(\Phi^0_{(s)}\right)_l = \operatorname{sign} f_l^0(t_s),$$

$s = 1, \ldots, M, \ l = 1, \ldots, m.$

2^0. *Suppose an estimator* $\Phi^0 \in \mathcal{F}^m$, *a vector* $\beta \in \mathbf{R}^n$, *and a measurable function* $u^0(t) \in \mathbf{R}^r$ *satisfy conditions* (3.32) – (3.36); *then* Φ^0 *is the solution of the problem* (3.31).

The proof is similar to that of Theorem 3.1 (cf. Section 2 of Chapter 1).

3.1. *Discrete measurements*

Since we admit piecewise continuous matrices $H(t)$, the case of discrete measurements

$$z(t_i) = H'(t_i)x(t_i) + \varrho(t_i), \qquad |\varrho_l(t_i)| \le \sigma_l(t_i), \tag{3.37}$$

$$i = 0, \ldots, N, \qquad 0 \le t_0 < \ldots < t_N \le T, \qquad l = 1, \ldots, m,$$

is a special case of the problem under consideration. In fact, let us put

$$H(t) = \begin{cases} H(t_i) & \text{if } t = t_i \\ 0 & \text{otherwise} \end{cases} \quad \text{and} \quad \sigma_l(t) = \begin{cases} \sigma_l(t_i) & \text{if } t = t_i \\ 1 & \text{otherwise.} \end{cases}$$

Then the values of $\Phi(t)$ at $t \ne t_i$ do not influence on the third summand in the functional (2.12) and on the integral in the unbiasness condition (2.7) but on the second summand in (2.13). Therefore the optimal solution must possess the property $\Phi(t) = 0$, $t \ne t_i$, $i = 0, \ldots, N$. This means that the optimal estimator has the impulse component only:

$$\hat{l} = \int_0^T \Phi^{0\prime}(t)z(t)\,dt = \sum_{i=0}^N \Phi^{0\prime}(t_i)z(t_i), \qquad \Phi^0(t_i) = \left(\Phi^0_1(t_i), \ldots, \Phi^0_m(t_i)\right)'.$$

Moreover, the functional (2.12), the set $\mathcal{N}$ defined by (2.7), and the optimal problem (2.13) take the form

$$I_D(\Phi_-, \Phi) = \sum_{d=1}^n \sigma_{-d}|\Phi_{-d}| + \sum_{l=1}^m \sum_{i=0}^N \sigma_l(t_i)|\Phi_l(t_i)| \tag{3.38}$$

$$+\sum_{j=1}^{r}\int_0^T \gamma_j(t)\left|b_j'(t)\left(\Phi_- + \sum_{\{i\,|\,t_i\le t\}} h(t_i)\Phi(t_i)\right)\right|dt,$$

$$\mathcal{N}_D = \left\{(\Phi_-,\Phi)\;\middle|\; \Phi_- + \sum_{i=0}^{N} h(t_i)\Phi(t_i) = \tilde{a}\right\}, \tag{3.39}$$

$$I_{D0} = \inf_{(\Phi_-,\Phi)\in\mathcal{N}_D} I_D(\Phi_-,\Phi). \tag{3.40}$$

COROLLARY 3.3. 1^0. *Let a pair* (Φ_-^0, Φ^0), $\Phi^0 = \{\Phi^0(t_i)\}_0^N$, *be a solution of the problem* (3.40), (3.38), (3.39); *then there exist a vector* $\beta \in \mathbf{R}^n$ *and a measurable function* $u^0(t) = (u_1(t),\ldots,u_r(t))' \in \mathbf{R}^r$ *such that*

(a2)

$$|u_j^0(t)| \le \gamma_j(t), \qquad t\in[0,T], \quad j=1,\ldots,r; \tag{3.41}$$

(b2)

$$u_j^0(t) = -\gamma_j(t)\,\text{sign}\,\big(b_j'(t)p^0(t)\big) \quad \textit{on the set} \;\; b_j'(t)p^0(t)\ne 0, \tag{3.42}$$

$$t\in[0,T], \quad j=1,\ldots,r,$$

where

$$p^0(t) = \Phi_-^0 + \sum_{\{i\,|\,t_i\le t\}} h(t_i)\Phi^0(t_i), \qquad \Phi_-^0 + \sum_{s=0}^{N} h(t_s)\Phi^0(t_s) = \tilde{a};$$

(c2)

$$|f_l^0(t_i)| \le \sigma_l(t_i), \qquad i=0,\ldots,N, \quad l=1,\ldots,m, \tag{3.43}$$

$$|f_{-d}^0| \le \sigma_{-d}, \qquad d=1,\ldots,n,$$

where

$$f_l^0(t_i) = h_l'(t_i)\left(\beta + \int_{t_i}^T b(\tau)u^0(\tau)\,d\tau\right),$$

$$f_{-d}^0 = e_d'\left(\beta + \int_0^T b(\tau)u^0(\tau)\,d\tau\right), \qquad e_d = (\underbrace{0,\ldots,0,1}_{d},\ldots,0)';$$

(d2)

On the set $\mathcal{T}_l^{(1)} = \left\{t_i \;\middle|\; |f_l^0(t_i)| < \sigma_l(t_i)\right\}$, $\qquad \Phi_l^0(t_i) = 0;$ (3.44)

On the set $\mathcal{T}_l^{(2)} = \left\{ t \;\middle|\; |f_l^0(t_i)| = \sigma_l(t_i), \quad \sigma_l(t_i) > 0 \right\}$,

either $\Phi_l^0(t_i) = 0$ *or* $\operatorname{sign}\Phi_l^0(t_i) = \operatorname{sign} f_l^0(t_i), \quad l = 1, \ldots, m;$

$$\text{On the set } \mathcal{T}_-^{(1)} = \left\{ d \;\middle|\; |f_{-d}^0| < \sigma_{-d} \right\}, \qquad \Phi_{-d}^0 = 0; \tag{3.45}$$

On the set $\mathcal{T}_-^{(2)} = \left\{ d \;\middle|\; |f_{-d}^0| = \sigma_{-d}, \quad \sigma_{-d} > 0 \right\}$,

either $\Phi_{-d}^0 = 0$ *or* $\operatorname{sign}\Phi_{-d}^0 = \operatorname{sign} f_{-d}^0, \quad d = 1, \ldots, n.$

2^0. *Suppose a pair* (Φ_-^0, Φ^0), *a vector* $\beta \in \mathbf{R}^n$, *and a measurable function* $u^0(t) \in \mathbf{R}^r$ *satisfy conditions* (3.41) – (3.45); *then* (Φ_-^0, Φ^0) *is the solution of the problem* (3.40), (3.38), (3.39).

PROOF. The proof is similar to that of Theorem 3.1.

Let us now suppose that no *a priori* information about the initial state vector is available: $x(0) \in \mathbf{R}^n$, i.e., $\sigma_{-d} = \infty$, $d = 1, \ldots, n$. It is not hard to verify that the functional (3.38), the set $\mathcal{N}_\mathcal{D}$ defined by (3.39), and the optimal problem (3.40) take the form

$$I_D^0(\Phi) = \sum_{l=1}^{m} \sum_{i=0}^{N} \sigma_l(t_i)|\Phi_l(t_i)| + \sum_{j=1}^{r} \int_0^T \gamma_j(t) \left| b_j'(t) \sum_{\{i \,|\, t_i \leq t\}} h(t_i)\Phi(t_i) \right| dt, \tag{3.46}$$

$$\mathcal{N}_D^0 = \left\{ \Phi \;\middle|\; \sum_{i=0}^{N} h(t_i)\Phi(t_i) = \tilde{a} \right\}, \tag{3.47}$$

$$I_{D0}^0 = \inf_{\Phi \in \mathcal{N}_D^0} I_D^0(\Phi). \tag{3.48}$$

Note that if we formally put $\Phi_- = 0$ in (3.38) – (3.40), then we obtain (3.46) – (3.48).

COROLLARY 3.4. 1^0. *Let an estimator* $\Phi^0 = \left\{ \Phi^0(t_i) \right\}_0^N$ *be a solution of the problem* (3.48), (3.46), (3.47); *then there exist a vector* $\beta \in \mathbf{R}^n$ *and a measurable function* $u^0(t) = (u_1(t), \ldots, u_r(t))' \in \mathbf{R}^r$ *such that*
(a3)

$$|u_j^0(t)| \leq \gamma_j(t), \qquad t \in [0, T], \quad j = 1, \ldots, r; \tag{3.49}$$

(b3)

$$u_j^0(t) = -\gamma_j(t)\,\mathrm{sign}\,\big(b_j'(t)p^0(t)\big) \quad \textit{on the set} \;\; b_j'(t)p^0(t) \neq 0, \tag{3.50}$$

$$t \in [0,T], \quad j = 1,\dots,r,$$

where

$$p^0(t) = \sum_{\{i\,|\,t_i \le t\}} h(t_i)\Phi^0(t_i), \qquad \sum_{s=0}^{N} h(t_s)\Phi^0(t_s) = \tilde{a};$$

(c3)

$$|f_l^0(t_i)| \le \sigma_l(t_i), \qquad i = 0,\dots,N, \quad l = 1,\dots,m, \tag{3.51}$$

where

$$f_l^0(t_i) = h_l'(t_i)\left(\beta + \int_{t_i}^{T} b(\tau)u^0(\tau)\,d\tau\right);$$

(d3)

$$\textit{On the set } \mathcal{T}_l^{(1)} = \left\{t_i \;\middle|\; |f_l^0(t_i)| < \sigma_l(t_i)\right\}, \qquad \Phi_l^0(t_i) = 0; \tag{3.52}$$

$$\textit{On the set } \mathcal{T}_l^{(2)} = \left\{t_i \;\middle|\; |f_l^0(t_i)| = \sigma_l(t_i), \quad \sigma_l(t_i) > 0\right\},$$

$$\textit{either } \Phi_l^0(t_i) = 0 \quad \textit{or} \quad \mathrm{sign}\,\Phi_l^0(t_i) = \mathrm{sign}\,f_l^0(t_i), \quad l = 1,\dots,m.$$

2^0. *Suppose an estimator* Φ^0, *a vector* $\beta \in \mathbf{R}^n$, *and a measurable function* $u^0(t) \in \mathbf{R}^r$ *satisfy conditions* (3.49) – (3.52); *then* Φ^0 *is the solution of the problem* (3.48), (3.46), (3.47).

PROOF. The proof is similar to that of Theorem 3.1.

REMARK 3.2. Suppose that *a priori* information about the initial state vector is partial:

$$x_{d_1}(0) \in \mathbf{R}^1, \quad d_1 \in D^1, \qquad |x_{d_2}(0)| \le \sigma_{-d_2}, \quad d_2 \in D^2,$$

$$D^1 \cap D^2 = \emptyset, \qquad D^1 \cup D^2 = \{1,\dots,n\}.$$

By analogy with Corollaries 3.2 and 3.4 it is easy to proved that the necessary and sufficient conditions have the form of Corollary 3.3 where we formally set $\sigma_{-d_1} = \infty$, $d_1 \in D^1$.

By now there exists a number of numerical methods for solving the optimal guaranteed estimation problem (2.13), (2.7) [31], [133], [139], [19], [135], [214], [140], [136]. To implement these algorithms we have to solve linear programming problems with the number of equality constraints of order N, where N is the number of measurements. This can result in certain numerical difficulties. We shall proceed as follows. We shall use the classical Kalman filter for parameter estimation, which is easy to implement. Further we shall construct an estimate for the nonoptimality level of the Kalman filter in the optimal guaranteed estimation problem. This estimate can be efficiently computed as well. This approach is presented in Chapters 3 and 4 (see also Section 7 of Chapter 1 for simplest case).

4. Computation of Estimators

In this section, basing on general theorems of Section 3, we compute optimal estimators for certain dynamic systems.

EXAMPLE 4.1. Consider the motion of a material point of unit mass. The equation of the motion has the form

$$\frac{d^2S(\tau)}{d\tau^2} = W(\tau), \qquad \tau \in [0,T], \qquad |W(\tau)| \leq W, \tag{4.1}$$

where τ is the actual time, $S(\tau)$ is the position of the point, $W(\tau)$ is the acceleration of the point. We suppose that the acceleration is an unknown but bounded function; W is a given constant.

Let the position measurements are made on the segment $[0,T]$

$$Z(\tau) = S(\tau) + R(\tau), \qquad |R(\tau)| \leq \sigma, \tag{4.2}$$

where $R(\tau)$ is an unknown measurement error bounded with a given constant σ. Our problem is to evaluate the terminal velocity of the point, i.e., to evaluate the quantity

$$l_0 = \left(\frac{dS(\tau)}{d\tau}\right)\bigg|_{\tau=T}. \tag{4.3}$$

In order to simplify the computation let us introduce the new variables

$$\tau = Tt, \qquad s(t) = S(Tt), \quad \bar{u}(t) = W(Tt),$$

$$z(t) = Z(Tt), \quad \bar{\varrho}(t) = R(Tt).$$

Then the relations (4.1) – (4.3) take the form

$$\ddot{s}(t) = \bar{u}(t), \qquad |\bar{u}(t)| \leq \bar{\gamma}, \quad \bar{\gamma} = WT^2, \qquad (\dot{\ }) \triangleq \frac{d}{dt}, \quad t \in [0,1],$$

$$z(t) = s(t) + \bar{\varrho}(t), \qquad |\bar{\varrho}(t)| \leq \sigma,$$

and

$$l_0 = T^{-1}\dot{s}(1).$$

Introducing the vector $x(t) = (x_1(t), x_2(t))'$, where $x_1(t) = s(t)$, $x_2(t) = \dot{s}(t)$, we have

$$\dot{x}(t) = Ax(t) + B\bar{u}(t), \qquad t \in [0,1], \qquad |\bar{u}(t)| \leq \bar{\gamma} = WT^2, \tag{4.4}$$

$$A = \begin{pmatrix} 0 & 1 \\ 0 & 0 \end{pmatrix}, \qquad B = \begin{pmatrix} 0 \\ 1 \end{pmatrix}, \qquad H = \begin{pmatrix} 1 \\ 0 \end{pmatrix}, \tag{4.5}$$

$$z(t) = H'x(t) + \bar{\varrho}(t), \qquad |\bar{\varrho}(t)| \leq \sigma, \tag{4.6}$$

and

$$l_0 = a_T' x(1), \qquad a_T = T^{-1}(0,1)'. \tag{4.7}$$

PROPOSITION 4.1. *Let $\Phi^0(t)$ be the optimal estimator in the problem (2.1) – (2.6) for estimating the quantity $l_* = a'x(T)$. Then $\alpha\Phi^0(t)$ is the optimal estimator in the problem with $l_* = \alpha a'x(T)$, $\alpha \in \mathbf{R}^1$, $\alpha \neq 0$.*

PROOF. It was shown in Section 2 that the problem (2.1) – (2.6) with $l_* = a'x(t)$ reduces to the extremal problem (2.13), (2.7) with $\tilde{a} = X'(T)a$. The problem (2.1) – (2.6) with $l_* = \alpha a'x(T)$ reduces to (2.13), (2.7) where $\tilde{a}$ is replaced by $\alpha\tilde{a}$. Assume that $\alpha\Phi^0$ is not optimal, i.e., for a Φ^* such that

$$\int_0^T h(t)\Phi^*(t)\,dt = \alpha\tilde{a},$$

the following inequality holds:

$$I(\Phi^*) < I(\alpha\Phi^0) = |\alpha| I(\Phi^0) = |\alpha| I_0.$$

Then the estimator $\alpha^{-1}\Phi^*$ satisfies (2.7) and $I(\alpha^{-1}\Phi^*) < I(\Phi^0)$. So, Φ^0 is not optimal. This contradiction completes the proof. □

PROPOSITION 4.2. *Let $\Phi^0(t)$ be the optimal estimator in the problem (2.13), (2.7) with $\sigma(t) = \sigma =$ const and $\gamma(t) = \gamma =$ const. Then $\Phi^0(t)$ is the optimal estimator in the auxiliary problem with $\sigma(t) = 1$ and $\gamma(t) = \sigma^{-1}\gamma$. To compute the optimal guaranteed estimation error in the problem (2.13), (2.7), the optimal guaranteed estimation error in the auxiliary problem should be scaled by σ.*

PROOF. The proof is trivial.

In accordance with Propositions 4.1 and 4.2, our problem reduces to the problem

$$\frac{dx(t)}{dt} = Ax(t) + Bu(t), \qquad t \in [0,1], \quad |u(t)| \le \gamma = \frac{WT^2}{\sigma}, \tag{4.8}$$

$$z(t) = H'x(t) + \varrho(t), \qquad |\varrho(t)| \le 1, \tag{4.9}$$

and

$$l_* = a'x(1), \qquad a = (0,1)'. \tag{4.10}$$

Using the notation of Section 2, we get for system (4.8) – (4.10)

$$n = 2, \qquad r = 1, \qquad m = 1,$$

$$X(t) = \begin{pmatrix} 1 & t \\ 0 & 1 \end{pmatrix}, \qquad X^{-1}(t) = \begin{pmatrix} 1 & -t \\ 0 & 1 \end{pmatrix}, \qquad \tilde{a} = X'(1)a = (0,1)',$$

$$h(t) = X'(t)H = (1,t)', \qquad b(t) = X^{-1}(t)B = (-t,1)',$$

$$I(\Phi) = \int_0^1 |\Phi(t)|\, dt + \gamma \int_0^1 \left| b'(t) \int_0^t h(\tau)\Phi(\tau)\, d\tau \right| dt. \tag{4.11}$$

Let us apply Corollary 3.2. We shall search the optimal estimator in the form

$$\Phi^0(t) = \Phi^0_{(1)}\delta(t - t^*) + \Phi^0_{(2)}\delta(t-1), \qquad 0 \le t^* < 1, \tag{4.12}$$

where $\Phi_{(1)}$, $\Phi_{(2)}$, and t^* are to be determined.
In accordance with condition (b1) of Corollary 3.2,

$$p^0(t) = \int_0^t h(s)\Phi^0(s)\, ds = \begin{cases} 0, & 0 \le t < t^* \\ \Phi_{(1)} \begin{pmatrix} 1 \\ t^* \end{pmatrix}, & t^* \le t < 1 \\ \begin{pmatrix} 0 \\ 1 \end{pmatrix} = \tilde{a}, & t = 1. \end{cases}$$

The unbiasness condition $p^0(1) = \tilde{a}$ turns into the equations

$$\begin{cases} \Phi^0_{(1)} + \Phi^0_{(2)} = 0 \\ t^*\Phi^0_{(1)} + \Phi^0_{(2)} = 1, \end{cases}$$

whence,

$$\Phi^0_{(2)} = -\Phi^0_{(1)} = (1 - t^*)^{-1} > 0.$$

Furthermore,

$$b'(t)p^0(t) = \begin{cases} 0, & 0 \le t \le t^* \\ \Phi_{(1)}(t^* - t), & t^* < t < 1 \\ 1, & t = 1. \end{cases}$$

Then

$$u^0(t) = -\gamma, \qquad t^* < t \le 1,$$

and, taking into account condition (a1),

$$|u^0(t)| \le \gamma, \qquad 0 \le t \le t^*.$$

In accordance with condition (c1) of Corollary 3.2,

$$f^0(t) = \beta_1 + \beta_2 t$$

$$+\gamma \begin{cases} \frac{1}{2}\left(1 - t^{*2}\right) - t\left(1 - t^*\right) + \int_t^{t^*} (t - \tau) u^0(\tau)\, d\tau, & 0 \le t \le t^* \\ \frac{1}{2}\left(1 - t\right)^2, & t^* < t \le 1. \end{cases}$$

In accordance with condition (d1) of Corollary 3.2,

$$f^0(1) = \operatorname{sign} \Phi^0_{(2)} = 1, \qquad f^0(t^*) = \operatorname{sign} \Phi^0_{(1)} = -1,$$

i.e.,

$$\beta_1 + \beta_2 = 1, \qquad \beta_1 + \beta_2 t^* + \frac{1}{2}\gamma(1 - t^*)^2 = -1$$

whence

$$\beta_1 = -\frac{(1 + t^*) + \frac{1}{2}\gamma(1 - t^*)^2}{1 - t^*}, \qquad \beta_2 = \frac{2 + \frac{1}{2}\gamma(1 - t^*)^2}{1 - t^*}.$$

Then

$$g^0(t) \triangleq f^0(1 - t) = 1 - \frac{\left(2 + \frac{1}{2}\gamma K^2\right) t}{K}$$

$$+ \begin{cases} \frac{1}{2}\gamma K\,(2t - K) + \int_K^t (\tau - t) u^0(1 - \tau)\, d\tau, & K \le t \le 1 \\ \frac{1}{2}\gamma t^2, & 0 \le t < K, \end{cases}$$

where $K = 1 - t^*$, $0 < K \le 1$.

Note that $g^0(t)$ is continuously differentiable and

$$g^0(0) = 1, \qquad g^0(K) = -1. \tag{4.13}$$

Thus we should find K and $u^0(1-\tau)$ such that

$$|g^0(t)| \leq 1, \quad t \in [0,1], \qquad \text{and} \qquad |u^0(1-\tau)| \leq \gamma, \quad \tau \in [K,1]. \tag{4.14}$$

If (4.14) holds, then, due to Corollary 3.2, the estimator Φ^0 from (4.12) is optimal.

Assume that $0 < K < 1$. Then, taking into account (4.13), we have from (4.14) that

$$\left.\frac{dg^0(t)}{dt}\right|_{t=K} \equiv -\frac{2+\frac{1}{2}K^2}{K} + \gamma K = 0$$

whence $K = \frac{2}{\sqrt{\gamma}}$. Since $K < 1$ by assumption, we have $\gamma > 4$.

So, if $\gamma > 4$, then, putting $K = \frac{2}{\sqrt{\gamma}}$, we get

$$g^0(t) = \begin{cases} -1 + \int_K^t (\tau - t) u^0(1-\tau)\, d\tau, & K \leq t \leq 1 \\ 2\left(\dfrac{t}{K} - 1\right)^2 - 1, & 0 \leq t < K. \end{cases}$$

Obviously, with $u^0(1-\tau) = 0, \ \ K \leq \tau \leq 1$, we get that $|g^0(t)| \leq 1$; therefore if $\gamma > 4$ and $t^* = 1 - \frac{2}{\sqrt{\gamma}}$, then the estimator (4.12) is optimal.

Let us now consider the case $\gamma \leq 4$. Putting $K = 1$, we have

$$g^0(t) = 1 - \left(2 + \frac{\gamma}{2}\right) t + \frac{\gamma}{2} t^2.$$

Since

$$g(0) = 1, \quad g(1) = -1, \quad \text{and} \quad \frac{dg(t)}{dt} = -\left(2 + \frac{\gamma}{2}\right) + \gamma t \leq \frac{\gamma}{2} - 2 \leq 0,$$

we get that

$$|g(t)| \leq 1, \qquad t \in [0,1].$$

Thus, by virtue of Corollary 3.2, the sufficient conditions for optimality hold and the optimal estimator for the problem (4.8) – (4.10) is determined by the formula

$$\Phi^0(t) = \begin{cases} \delta(t-1) - \delta(t), & \gamma \leq 4 \\ \frac{\sqrt{\gamma}}{2}\left[\delta(t-1) - \delta(t - 1 + \frac{2}{\sqrt{\gamma}})\right], & \gamma > 4. \end{cases} \tag{4.15}$$

Moreover, the optimal value of the functional (4.11) takes the form

$$I(\Phi^0) = \begin{cases} 2 + \frac{1}{2}\gamma, & \gamma \leq 4 \\ 2\sqrt{\gamma}, & \gamma > 4. \end{cases}$$

By virtue of Propositions 4.1 and 4.2, the optimal estimator $\bar{\Phi}^0$ and the optimal guaranteed estimation error $\bar{I}(\bar{\Phi}^0)$ for the intermediate problem (4.4) – (4.7) take the form

$$\bar{\Phi}^0(t) = T^{-1}\Phi^0(t)$$

$$= \begin{cases} T^{-1}\left[\delta(t-1) - \delta(t)\right], & \sigma^{-1}\bar{\gamma} \le 4 \\ (2T)^{-1}\sqrt{\sigma^{-1}\bar{\gamma}}\left[\delta(t-1) - \delta(t-1+\frac{2}{\sqrt{\sigma^{-1}\bar{\gamma}}})\right], & \sigma^{-1}\bar{\gamma} > 4 \end{cases}$$

and

$$\bar{I}(\bar{\Phi}^0) = \sigma I(T^{-1}\Phi^0) = \sigma T^{-1} I(\Phi^0) = \begin{cases} 2T^{-1}\left(\sigma + \frac{1}{4}\bar{\gamma}\right), & \sigma^{-1}\bar{\gamma} \le 4 \\ 2T^{-1}\sqrt{\sigma\bar{\gamma}}, & \sigma^{-1}\bar{\gamma} > 4. \end{cases}$$

Thus the optimal estimate for the velocity (4.3) is determined by the relation

$$\hat{l}_0 = \begin{cases} T^{-1}\left[Z(T) - Z(0)\right], & \sigma^{-1}WT^2 \le 4 \\ \left(2\sqrt{W^{-1}\sigma}\right)^{-1}\left[Z(T) - Z(T - 2\sqrt{W^{-1}\sigma})\right], & \sigma^{-1}WT^2 > 4, \end{cases} \tag{4.16}$$

and the optimal guaranteed estimation error has the form

$$\bar{I}(\bar{\Phi}^0) = \begin{cases} 2\sigma T^{-1} + \frac{1}{2}WT, & \sigma^{-1}WT^2 \le 4 \\ 2\sqrt{\sigma W}, & \sigma^{-1}WT^2 > 4. \end{cases} \tag{4.17}$$

The relation (4.16) means that the optimal estimate is a divided difference of first order; in particular, the estimator is concentrated at two points. Moreover, for sufficiently large T such that

$$T > T^0, \quad \text{where} \quad T^0 = 2\sqrt{W^{-1}\sigma},$$

the measurements on the segment $[0, T^0)$ must be ignored. It follows from (4.17) that, for fixed W and σ, the optimal guaranteed estimation error is monotonely decreasing in T ($T \le T^0$). Nevertheless, if $T > T^0$, then the guaranteed estimation error is constant and can not be reduced. So, for arbitrary T, $2\sqrt{\sigma W}$ is the limiting value for estimation accuracy.

4.1. *Majorant Problem*

In order to clarify how we could feel about the type of the optimal estimator, let us consider the *majorant problem.*

Using the Dirichlet formula (cf. Lemma 2.1), it can easily be checked that

$$\int_0^1 \left| b'(t) \int_0^t h(\tau)\Phi(\tau)\, d\tau \right| dt \le \int_0^1 \int_0^t |b'(t)h(\tau)|\, |\Phi(\tau)|\, d\tau\, dt$$

$$= \int_0^1 \int_\tau^1 |b'(t)h(\tau)|\,|\Phi(\tau)|\,dt\,d\tau = \int_0^1 c_{\max}(\tau)|\Phi(\tau)|\,d\tau,$$

$$c_{\max}(\tau) = \int_\tau^1 |b'(t)h(\tau)|\,dt,$$

and

$$I(\Phi) \le I_{\max}(\Phi) = \int_0^1 (1 + \gamma c_{\max}(\tau))\,|\Phi(\tau)|\,d\tau.$$

Thus the majorant problem has the form

$$\inf_{\Phi\in\mathcal{F}} I_{\max}(\Phi) \quad \text{under constraint} \quad \int_0^1 h(\tau)\Phi(\tau)\,d\tau = \tilde{a}. \tag{4.18}$$

Since $(1 + \gamma c_{\max}(\tau))^{-1}\, h(\tau)$ is continuous on $[0,1]$, it follows from the results of Section 3 of Chapter 1 that the solution of (4.18) exists and, for Example 4.1, can be found in the form

$$\Phi(t) = \Phi_{(1)}\delta(t - t_1) + \Phi_{(2)}\delta(t - t_2).$$

Using the explicit expressions for $c_{\max}(\tau)$ and $h(\tau)$, the majorant problem may be rewritten in the form

$$\inf_{\Phi\in\mathcal{F}} \int_0^1 \left(1 + \gamma\frac{(1-\tau)^2}{2}\right)|\Phi(\tau)|\,d\tau$$

under constraints

$$\Phi_{(1)} + \Phi_{(2)} = 0, \qquad t_1\Phi_{(1)} + t_2\Phi_{(2)} = 1.$$

Obviously, $\Phi_{(2)} = -\Phi_{(1)} = (t_2 - t_1)^{-1}$ and the majorant problem reduces to the problem

$$\inf_{0\le t_1<t_2\le 1} \frac{4 + \gamma(1-t_1)^2 + \gamma(1-t_2)^2}{2(t_2 - t_1)}.$$

Clearly, $t_2 = 1$ and

$$\begin{cases} \text{if } \gamma \le 4, \text{ then} \quad t_1 = 0, \quad t_2 = 1, & \Phi^0_{(2)} = -\Phi^0_{(1)} = 1 \\ \text{if } \gamma > 4, \text{ then} \quad t_1 = 1 - \frac{2}{\sqrt{\gamma}}, \quad t_2 = 1, & \Phi^0_{(2)} = -\Phi^0_{(1)} = \frac{\sqrt{\gamma}}{2}. \end{cases} \tag{4.19}$$

So, we have obtained that the optimal estimator (4.19) for the majorant problem (4.18) has the type of (4.12) and coincides with the optimal estimator (4.15).

EXAMPLE 4.2. In the previous example the number of impulses in the optimal estimator is equal to the dimension of the state space. Let us show that, generally, this is not the case.

We consider the the problem

$$\frac{dx(t)}{dt} = Ax(t) + Bu(t), \qquad |u(t)| \le \gamma(t), \quad t \in [0,3], \tag{4.20}$$

where

$$\gamma(t) = \begin{cases} \gamma_1, & 0 \le t < 1 \\ \gamma_2, & 1 \le t \le 2 \\ 0, & 2 < t \le 3, \end{cases}$$

The measurements are made at three instants:

$$z(i) = H'x(i) + \varrho(i), \qquad |\varrho(i)| \le 1, \qquad i = 0,1,2, \tag{4.21}$$

where the matrices A, B, and H are defined by (4.5), and

$$l_* = a'x(3), \qquad a = (1, -\alpha)', \quad 0 \le \alpha < 1, \tag{4.22}$$

where α is a fixed parameter.

Using the notation of Sections 2 and 3, we get for system (4.20) – (4.22)

$$n = 2, \qquad r = 1, \qquad m = 1, \qquad N = 2, \qquad t_i = i, \quad i = 0,1,2.$$

$$X(t) = \begin{pmatrix} 1 & t \\ 0 & 1 \end{pmatrix}, \quad X^{-1}(t) = \begin{pmatrix} 1 & -t \\ 0 & 1 \end{pmatrix}, \quad \tilde{a} = X'(3)a = (1, 3-\alpha)',$$

$$h(t) = X'(t)H = (1,t)', \qquad b(t) = X^{-1}(t)B = (-t, 1)',$$

$$I_D^0(\Phi) = \sum_{i=0}^{2} |\Phi(i)| + \int_0^2 \gamma(t) \left| b'(t) \sum_{\{i \mid i \le t\}} h(i)\Phi(i) \right| dt,$$

$$\mathcal{N}_D^0 = \left\{ \Phi \,\middle|\, \sum_{i=0}^{2} h(i)\Phi(i) = \tilde{a} \right\},$$

$$I_{D0}^0 = \inf_{\Phi \in \mathcal{N}_D^0} I_D^0(\Phi).$$

Let us apply Corollary 3.4.

In accordance with condition (b3) of Corollary 3.4,

$$p^0(t) = \sum_{\{i \mid i \le t\}} h(i)\Phi(i), \qquad p^0(3) = \sum_{i=0}^{2} h(i)\Phi(i) = \tilde{a},$$

$$b'(t)p^0(t) = \begin{cases} -t\Phi(0), & 0 \le t < 1 \\ -t\Phi(0) + (1-t)\Phi(1), & 1 \le t < 2 \\ -t\Phi(0) + (1-t)\Phi(1) + (2-t)\Phi(2), & 2 \le t \le 3. \end{cases}$$

Then, resolving the unbiasness condition $p^0(3) = \tilde{a}$, we have

$$\Phi(1) = -\xi - 2\Phi(0), \ \ \Phi(2) = 1 + \xi + \Phi(0), \ \ \xi = 1 - \alpha, \ 0 < \xi \le 1, \tag{4.23}$$

$$b'(t)p^0(t) = \begin{cases} -\Phi(0)t, & 0 \le t < 1 \\ (\Phi(0) + \xi)t - (\xi + 2\Phi(0)), & 1 \le t < 2 \\ 2 - t + \xi, & 2 \le t \le 3. \end{cases} \tag{4.24}$$

Let us consider three types of estimators with the impulses $\Phi(0)$, $\Phi(1)$, and $\Phi(2)$.

Case 1. Suppose $\Phi(0) = 0$; then it follows from (4.23) and (4.24) that

$$\Phi(1) = -\xi, \quad \Phi(2) = 1 + \xi, \quad \text{i.e.,} \quad \operatorname{sign}\Phi(1) = -1, \ \operatorname{sign}\Phi(2) = 1, \tag{4.25}$$

and

$$b'(t)p^0(t) = \begin{cases} 0, & 0 \le t < 1 \\ \xi(t-1), & 1 \le t < 2 \\ 2 - t + \xi, & 2 \le t \le 3. \end{cases} \tag{4.26}$$

By virtue of conditions (a3) and (b3) of Corollary 3.4 and (4.26), we have

$$|u^0(t)| \le \gamma_1, \quad 0 \le t < 1, \qquad |u^0(1)| \le \gamma_2, \tag{4.27}$$

$$u^0(t) = \begin{cases} -\gamma_2, & 1 < t \le 2 \\ 0, & 2 < t \le 3. \end{cases}$$

Let us use conditions (c3) and (d3) of Corollary 3.4 taking into account (4.25). Then, for

$$f^0(t) = h'(t)\left(\beta + \int_t^3 b(\tau)u^0(\tau)\, d\tau\right),$$

we have

$$|f^0(0)| \le 1, \qquad f^0(1) = -1, \qquad f^0(2) = 1. \tag{4.28}$$

It follows from (4.27) and (4.28) that

$$\left|\beta_1 + \frac{3}{2}\gamma_2 - \int_0^1 \tau u^0(\tau)\, d\tau\right| \le 1, \qquad |u^0(\tau)| \le \gamma_1,$$

$$\beta_1 + \beta_2 + \frac{1}{2}\gamma_2 = -1, \qquad \beta_1 + 2\beta_2 = 1,$$

whence,

$$\left|3 - \frac{1}{2}\gamma_2 + \int_0^1 \tau u^0(\tau)\, d\tau\right| \le 1, \quad \beta_1 = -3 - \gamma_2, \quad \beta_2 = 2 + \frac{1}{2}\gamma_2. \tag{4.29}$$

Since $|u^0(\tau)| \le \gamma_1$, $0 \le \tau < 1$, the set of values of $\int_0^1 \tau u^0(\tau)\, d\tau$ coincides with the segment $[-\frac{1}{2}\gamma_1, \frac{1}{2}\gamma_1]$ and the first inequality from (4.29) is equivalent to the relation

$$\left[-\frac{1}{2}\gamma_1, \frac{1}{2}\gamma_1\right] \cap \left[-4 + \frac{1}{2}\gamma_2, -2 + \frac{1}{2}\gamma_2\right] \ne \emptyset,$$

i.e.,

$$\begin{cases} \gamma_1 \ge 4 - \gamma_2 \\ \gamma_1 \ge \gamma_2 - 8. \end{cases} \tag{4.30}$$

So, if (4.30) holds, then, by virtue of the sufficient conditions of Corollary 3.4, the estimator determined by (4.25) is optimal.

Case 2. Suppose $\Phi(1) = 0$; then it follows from (4.23) and (4.24) that

$$\Phi(0) = -\frac{1}{2}\xi, \quad \Phi(2) = 1 + \frac{1}{2}\xi, \text{ i.e., } \operatorname{sign}\Phi(0) = -1, \ \operatorname{sign}\Phi(2) = 1, \tag{4.31}$$

and

$$b'(t)p^0(t) = \begin{cases} \frac{1}{2}\xi t, & 0 \le t < 1 \\ \frac{1}{2}\xi t, & 1 \le t < 2 \\ 2 - t + \xi, & 2 \le t \le 3. \end{cases} \tag{4.32}$$

By virtue of conditions (a3) and (b3) of Corollary 3.4 and (4.32),

$$u^0(t) = \begin{cases} -\gamma_1, & 0 < t < 1 \\ -\gamma_2, & 1 \le t \le 2 \\ 0, & 2 < t \le 3. \end{cases} \tag{4.33}$$

Let us use conditions (c3) and (d3) of Corollary 3.4 taking into account (4.31). Then

$$f^0(0) = -1, \qquad |f^0(1)| \le 1, \qquad f^0(2) = 1. \tag{4.34}$$

It follows from (4.33) and (4.34) that

$$\beta_1 + \frac{1}{2}\gamma_1 + \frac{3}{2}\gamma_2 = -1, \qquad \left|\beta_1 + \beta_2 + \frac{1}{2}\gamma_2\right| \le 1, \qquad \beta_1 + 2\beta_2 = 1,$$

whence

$$\beta_1 = -1 - \frac{1}{2}(\gamma_1 + 3\gamma_2), \quad \frac{1}{4}|\gamma_1 + \gamma_2| \le 1, \quad \beta_2 = 1 + \frac{1}{4}(\gamma_1 + 3\gamma_2),$$

and, consequently,

$$\gamma_1 \leq 4 - \gamma_2. \tag{4.35}$$

So, if (4.35) holds, then, by virtue of the sufficient conditions of Corollary 3.4, the estimator determined by (4.31) is optimal.

Case 3. Suppose $\Phi(2) = 0$; then it follows from (4.23) and (4.24) that

$$\Phi(0) = -1 - \xi, \;\; \Phi(1) = 2 + \xi, \;\; \text{i.e., } \operatorname{sign}\Phi(0) = -1, \;\; \operatorname{sign}\Phi(1) = 1, \tag{4.36}$$

and

$$b'(t)p^0(t) = \begin{cases} (1+\xi)t, & 0 \leq t < 1 \\ 2 - t + \xi, & 1 \leq t \leq 3. \end{cases} \tag{4.37}$$

By virtue of conditions (a3) and (b3) of Corollary 3.4 and (4.37),

$$u^0(t) = \begin{cases} -\gamma_1, & 0 < t < 1 \\ -\gamma_2, & 1 \leq t \leq 2 \\ 0, & 2 < t \leq 3. \end{cases} \tag{4.38}$$

Let us use conditions (c3) and (d3) of Corollary 3.4 taking into account (4.36). Then

$$f^0(0) = -1, \qquad f^0(1) = 1, \qquad |f^0(2)| \leq 1. \tag{4.39}$$

It follows from (4.38) and (4.39) that

$$\beta_1 = -1 - \frac{1}{2}(\gamma_1 + 3\gamma_2), \quad \beta_2 = 2 + \frac{1}{2}(\gamma_1 + 2\gamma_2), \quad |\beta_1 + 2\beta_2| \leq 1,$$

and, consequently,

$$\left|3 + \frac{1}{2}(\gamma_1 + \gamma_2)\right| \leq 1.$$

Obviously, the last inequality is invalid and the estimator determined by (4.36) is not optimal.

In cases 1 – 3, the value of the first impulse is nonpositive: $\Phi(0) \leq 0$. Let us consider the contrary case.

Case 4. Suppose $\Phi(0) > 0$; then it follows from (4.23) that

$$\operatorname{sign}\Phi(0) = 1, \qquad \operatorname{sign}\Phi(1) = -1, \qquad \operatorname{sign}\Phi(2) = 1. \tag{4.40}$$

By virtue of conditions (a3) and (b3) of Corollary 3.4 and (4.24),

$$u^0(t) = \begin{cases} \gamma_1, & 0 < t < 1 \\ \gamma_2, & 1 \leq t < \tilde{z} \\ -\gamma_2, & \tilde{z} < t \leq 2 \\ 0, & 2 < t \leq 3, \end{cases} \qquad \tilde{z} = \frac{\xi + 2\Phi(0)}{\xi + \Phi(0)}, \quad 1 < \tilde{z} < 2. \tag{4.41}$$

Let us use conditions (c3) and (d3) of Corollary 3.4 taking into account (4.40). Then

$$f^0(0) = 1, \qquad f^0(1) = -1, \qquad f^0(2) = 1. \tag{4.42}$$

It follows from (4.41) and (4.42) that

$$\beta_1 - \frac{1}{2}\gamma_1 - \gamma_2\left(z^2 + 2z - \frac{3}{2}\right) = 1, \qquad \beta_1 + \beta_2 - \frac{1}{2}\gamma_2\left(2z^2 - 1\right) = -1,$$

$$\text{and} \quad \beta_1 + 2\beta_2 = 1, \quad \text{where} \quad z = \tilde{z} - 1 = \frac{\Phi(0)}{\xi + \Phi(0)}, \quad 0 < z < 1.$$

Solving this system, we get

$$\beta_1 = 2\gamma_2\left(z^2 - \frac{1}{2}\right) - 3, \qquad \beta_2 = 2 - \gamma_2\left(z^2 - \frac{1}{2}\right), \quad \text{and}$$

$$2z^2 - 4z + (1 - \kappa) = 0, \quad 0 < z < 1, \quad \text{where} \quad \kappa = \frac{\gamma_1 + 8}{\gamma_2} > 0. \tag{4.43}$$

It can easily be checked that if $\kappa < 1$, i.e.,

$$\gamma_1 < \gamma_2 - 8, \tag{4.44}$$

then the quadratic equation from (4.43) has one root on $(0,1)$: $z^0 = 1 - \sqrt{\frac{1}{2}(1+\kappa)}$ and, consequently,

$$\Phi(0) = \frac{z^0\xi}{1 - z^0}, \quad \Phi(1) = -\xi\left(\frac{1 + z^0}{1 - z^0}\right), \quad \Phi(2) = 1 + \left(\frac{\xi}{1 - z^0}\right). \tag{4.45}$$

So, if (4.44) holds, then, by virtue of the sufficient conditions of Corollary 3.4, the estimator determined by (4.45) is optimal. If (4.44) does not hold, then the quadratic equation has no root on $(0,1)$.

Thus we have constructed the optimal estimators in the domains (4.30), (4.35), (4.44). Clearly, these domains cover the quadrant $\{\gamma_1 \geq 0,\ \gamma_2 \geq 0\}$ (see Figure 1).

Therefore we solved our problem completely. Note that if $\gamma_1 < \gamma_2 - 8$, then the optimal estimator must have three nonzero impulses (that is greater than the state space dimension).

EXAMPLE 4.3. Let us consider a problem of the computation of a quadrature. Suppose we have the values of an absolutely continuous function $\mu(t)$ at certain instants:

$$z(t_i) = \mu(t_i) + \varrho(t_i), \qquad t_i = i\Delta t, \quad i = 0, \ldots, N, \qquad t_0 = 0, \quad t_N = T,$$

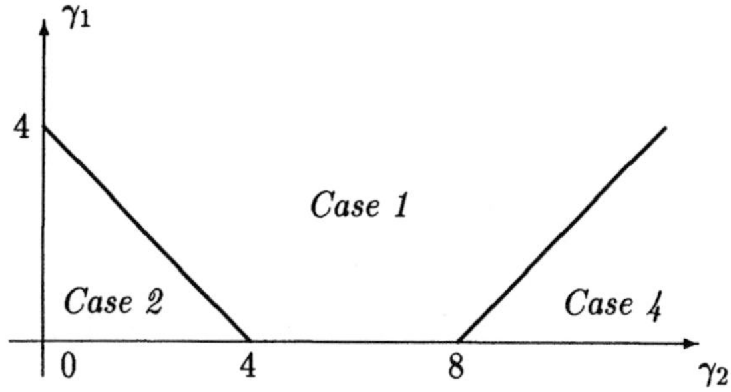

Fig. 1. Domains of (γ_1, γ_2) for Example 4.2

where the set $\{z(t_i)\}_0^N$ describes the available information about $\mu(t)$ and the set $\{\varrho(t_i)\}_0^N$ describes the errors of the information.

We assume that the function $\mu(t)$ has a bounded derivative and the errors are also bounded:

$$\left|\frac{d\mu(t)}{dt}\right| \leq \gamma, \quad t \in [0, T], \qquad |\varrho(t_i)| \leq \sigma, \quad i = 0, \ldots, N,$$

where γ and σ are given constants.

Our problem is to construct an estimate $\Gamma(T; z(t_0), \ldots, z(t_N))$ for the integral $\int_0^T \mu(t)\, dt$ that is optimal on the class of the functions with bounded derivative. In other words, we have to solve the following minimax problem:

$$\inf_{\Gamma} \sup_{|\dot{\mu}(t)| \leq \gamma, |\varrho(t_i)| \leq \sigma} \left|\Gamma(T; z(t_0, \ldots, z(t_N)) - \int_0^T \mu(t)\, dt\right|. \tag{4.46}$$

Obviously, this problem reduces to the general guaranteed estimation problem for the system

$$\frac{dx(t)}{dt} = Ax(t) + Bu(t), \qquad |u(t)| \leq \gamma, \quad t \in [0, T],$$

where

$$x(t) = \begin{pmatrix} x_1(t) \\ x_2(t) \end{pmatrix}, \qquad A = \begin{pmatrix} 0 & 1 \\ 0 & 0 \end{pmatrix}, \qquad B = \begin{pmatrix} 0 \\ 1 \end{pmatrix},$$

$$|x_1(0)| \leq \sigma_{-1}, \qquad \sigma_{-1} = 0, \qquad x_2(0) \in \mathbf{R}^1, \tag{4.47}$$

$$l_* = a'x(T), \qquad a = (1, 0)',$$

$$z(t_i) = H'x(t_i) + \varrho(t_i), \qquad H = (0, 1)', \qquad |\varrho(t_i)| \leq \sigma, \qquad i = 0, \ldots, N.$$

We shall consider a slightly modified constraints on $x(0)$:

$$|x_1(0)| \le \sigma_{-1}, \quad |x_2(0)| \le \sigma_{-2}, \qquad \sigma_{-1} \ge 0, \quad \sigma_{-2} \ge \sigma. \tag{4.48}$$

Using the notation of Sections 2 and 3, we get for our system (with (4.48)) (cf. Examples 4.1 and 4.2)

$$n = 2, \qquad r = 1, \qquad m = 1,$$

$$X(t) = \begin{pmatrix} 1 & t \\ 0 & 1 \end{pmatrix}, \quad X^{-1}(t) = \begin{pmatrix} 1 & -t \\ 0 & 1 \end{pmatrix}, \quad \tilde{a} = X'(T)a = (1, T)',$$

$$h(t) = X'(t)H = (0, 1)', \qquad b(t) = X^{-1}(t)B = (-t, 1)',$$

$$I_D(\Phi_-, \Phi) = \sigma_{-1}|\Phi_{-1}| + \sigma_{-2}|\Phi_{-2}| + \sigma \sum_{i=0}^{N} |\Phi(t_i)|$$

$$+ \gamma \int_0^T \left| -t\Phi_{-1} + \Phi_{-2} + \sum_{\{i \mid t_i \le t\}} \Phi(t_i) \right| dt,$$

$$\mathcal{N}_D = \left\{ (\Phi_-, \Phi) \,\middle|\, \Phi_{-1} = 1, \quad \Phi_{-2} + \sum_{i=0}^{N} \Phi(t_i) = T \right\},$$

$$I_{D0} = \inf_{(\Phi_-, \Phi) \in \mathcal{N}_D} I_D(\Phi_-, \Phi). \tag{4.49}$$

Proposition 4.3. *The optimal estimator*

$$\Gamma(T; z(t_1), \ldots, z(_N)) = \sum_{i=0}^{N} \Phi^0(t_i) z(t_i)$$

is determined by the trapezoid formula

$$\Phi^0(t_0) = \Phi^0(t_N) = \frac{1}{2}\Delta t, \qquad \Phi^0(t_i) = \Delta t, \quad i = 1, \ldots, N-1, \tag{4.50}$$

$$\Phi^0_{-1} = 1, \qquad \Phi^0_{-2} = 0.$$

Proof. Let us apply Corollary 3.3.
In accordance with condition (b2) of Corollary 3.3,

$$p^0(t) = (1, \sum_{\{i \mid t_i \le t\}} \Phi^0(t_i))', \qquad p^0(T) = (1, \sum_{i=0}^{N} \Phi^0(t_i))' = (1, T)'.$$

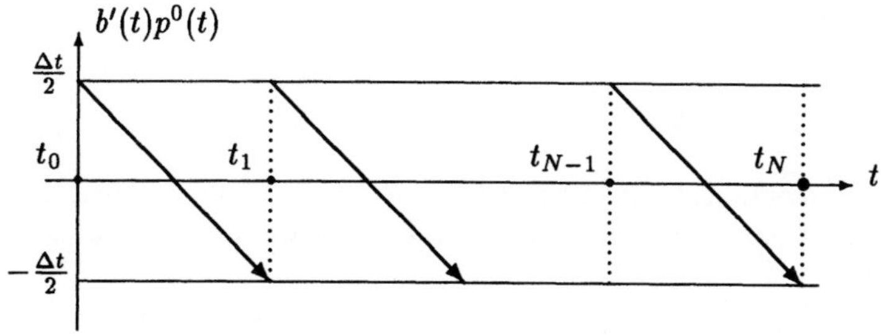

Fig. 2. The function for switching (Example 4.3)

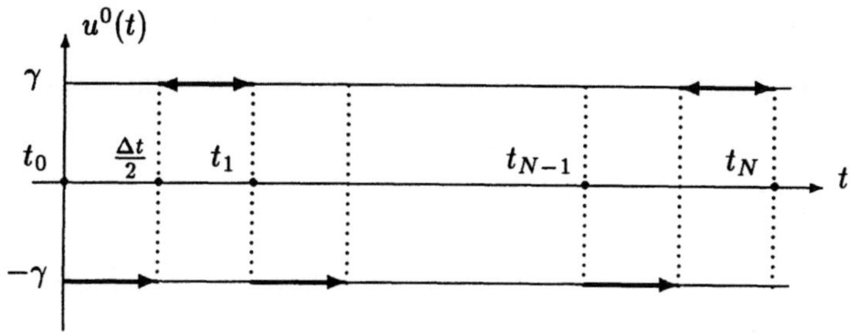

Fig. 3. Worst perturbation (Example 4.3)

Then

$$b'(t)p^0(t) = \sum_{\{i \,|\, t_i \le t\}} \Phi^0(t_i) - t$$

and

$$u^0(t) = \begin{cases} \gamma, & \frac{1}{2}\Delta t + t_k < t < t_{k+1} \\ -\gamma, & t_k \le t < t_k + \frac{1}{2}\Delta t, \qquad k = 0, \dots, N-1. \end{cases}$$

Clearly, $u^0(t)$ satisfies condition (a2) of Corollary 3.2. The functions $b'(t)p^0(t)$ and $u^0(t)$ are shown in Figures 2 and 3.

Let us check the conditions (c2) and (d2) of Corollary 3.2. Since all impulses from (4.50) are positive, these conditions take the form

$$f^0(t_i) = \beta_2 + \int_{t_i}^{T} u^0(\tau)\, d\tau = \sigma, \qquad i = 0, \dots, N, \tag{4.51}$$

$$f^0_{-1} = \beta_1 - \frac{1}{4}\gamma T\Delta t = \sigma_{-1}, \qquad |f^0_{-2}| = \left|\beta_2 + \int_0^T u^0(\tau)\, d\tau\right| \le \sigma_{-2},$$

whence we have

$$\beta_1 = \sigma_{-1} + \frac{1}{4}\gamma T\Delta t, \qquad \beta_2 = \sigma, \qquad \sigma \le \sigma_{-2}.$$

Since restrictions (4.48) hold, relations (4.51) also hold. Then, by virtue of the sufficient conditions of Corollary 3.2, the estimator determined by (4.50) is optimal. □

Substituting the estimator (4.50) into the functional (4.49), we compute that the optimal guaranteed estimation error is equal to

$$I_{D0} = I_D(\Phi^0_-, \Phi^0) = \sigma_{-1} + \sigma T\Delta t + \frac{1}{4}\gamma T\Delta t.$$

Note that if we use the simplest estimator

$$\Phi^R(t_i) = \Delta t, \quad i = 0, \dots, N, \qquad \Phi^R_{-1} = 1, \quad \Phi^R_{-2} = 0,$$

which corresponds to the rectangular formula, then the guaranteed estimation error is equal to

$$I_D(\Phi^R_-, \Phi^R) = \sigma_1 + \sigma T\Delta t + \frac{1}{2}\gamma T\Delta t.$$

Finally, consider the case (4.47). In accordance with Remark 3.2, we may formally put

$$\sigma_{-1} = 0, \qquad \sigma_{-2} = \infty;$$

so, we obtain that the trapezoidal formula is optimal for the problem (4.46) and

$$I_{D0} = \sigma T + \frac{1}{4}\gamma T\Delta t.$$

EXAMPLE 4.4. Let us study a guaranteed estimation problem that has its origin in inertial navigation. Suppose an object with the on-board inertial navigation system moves along Equator. The equations for the errors of the inertial navigation system in fore-and-aft direction have the form [177]

$$\dot{x}_1 = x_2, \quad \dot{x}_2 = -x_1 - x_3 + u(t), \quad \dot{x}_3 = x_4, \quad \dot{x}_4 = 0, \quad |u(t)| \le \gamma, \quad (4.52)$$

where $x_1(t)$ is the positioning error, $x_2(t)$ is the velocity error, $x_3(t)$ is the kinematic error, $x_4(t)$ is the gyroscopic drift, $u(t)$ is the accelerometer error that is assumed to be unknown but bounded. The information about the initial state vector have the form

$$|x_d(0)| \le \sigma_{-d}, \qquad d = 1, \dots, 4. \qquad (4.53)$$

In order to compensate the navigation errors we have to evaluate the terminal positioning error $x_1(T)$. For this purpose, we use the measurements of the actual velocity and transform them into the measurements of the velocity error:

$$z(t) = x_2(t) + \varrho(t), \qquad |\varrho(t)| \le \sigma, \qquad t \in [0,T], \tag{4.54}$$

where $z(t)$ are the measurements, $\varrho(t)$ is the measurement noise that is also assumed to be unknown but bounded.

In general notation of Section 2,

$$n = 4, \qquad m = 1, \qquad r = 1, \qquad x(t) = (x_1(t), \dots, x_4(t))',$$

$$A = \begin{pmatrix} 0 & 1 & 0 & 0 \\ -1 & 0 & -1 & 0 \\ 0 & 0 & 0 & 1 \\ 0 & 0 & 0 & 0 \end{pmatrix}, \quad X(t) = \begin{pmatrix} \cos t & \sin t & \cos t - 1 & \sin t - t \\ -\sin t & \cos t & -\sin t & \cos t - 1 \\ 0 & 0 & 1 & t \\ 0 & 0 & 0 & 1 \end{pmatrix},$$

$$B = (0,1,0,0)', \quad H = (0,1,0,0)', \quad a = (1,0,0,0)',$$

$$b(t) = (-\sin t,\ \cos t,\ 0,\ 0)', \quad h(t) = (-\sin t,\ \cos t, -\sin t, \cos t - 1)',$$

$$\tilde{a} = (\ \cos T,\ \sin T, \cos T - 1, \sin T - T)'.$$

By analogy with Example 4.3 we shall prove that the estimator

$$\Phi^0_{-1} = 1, \qquad \Phi^0_{-d} = 0, \quad d = 2,3,4, \qquad \Phi^0(t) = 1, \quad t \in [0,T], \tag{4.55}$$

is optimal.

Proposition 4.4. *The estimator* (4.55) *is optimal if and only if there exist constants* C_1 and C_2 *such that the following inequalities hold*

$$\sigma \le \sigma_{-2}, \qquad |C_1| \le \gamma, \qquad |C_1 + \sigma_{-1}| \le \sigma_{-3}, \tag{4.56}$$

$$|C_2| \le \sigma_{-4}, \qquad |(\sigma + C_2)T - C_1| \le \gamma.$$

Proof. Let us apply Theorem 3.1 to the problem (4.52) – (4.54). According to conditions (b) of Theorem 3.1,

$$p^0(t) = (\cos t, \sin t, \cos t - 1, \sin t - t,)',$$

and the condition $p^0(T) = \tilde{a}$ holds. Furthermore,

$$b'(t)p^0(t) \equiv 0, \qquad t \in [0,T],$$

i.e., the problem is degenerate.

In accordance with conditions (a), (c), and (d),

$$f^0(t) \equiv h'(t)\beta + \int_t^T \cos(t-\tau)u^0(\tau)\,d\tau = \sigma, \qquad t \in [0,T], \tag{4.57}$$

$$f^0_{-1} = \beta_1 - \int_0^T \sin\tau\, u^0(\tau)\,d\tau = \sigma_{-1}.$$

$$|f^0_{-2}| = \left|\beta_2 + \int_0^T \cos\tau\, u^0(\tau)\,d\tau\right| \le \sigma_{-2},$$

$$|\beta_3| \le \sigma_{-3}, \qquad |\beta_4| \le \sigma_{-4}, \qquad |u^0(t)| \le \gamma, \quad t \in [0,T].$$

Differentiating the first relation from (4.57) with respect to t, we get

$$\frac{d\,f^0(t)}{dt} \equiv -(\beta_1+\beta_3)\cos t - (\beta_2+\beta_4)\sin t - u^0(t) \tag{4.58}$$

$$-\int_t^T \sin(t-\tau)u^0(\tau)\,d\tau = 0,$$

$$\frac{d^2 f^0(t)}{dt^2} \equiv (\beta_1+\beta_3)\sin t - (\beta_2+\beta_4)\cos t$$

$$-\int_t^T \cos(t-\tau)u^0(\tau)\,d\tau - \frac{du^0(t)}{dt} \equiv -f^0(t) - \frac{du^0(t)}{dt} - \beta_4 = 0.$$

Summing the first relation from (4.57) with the second relation from (4.58), we have

$$\frac{du^0(t)}{dt} = -(\sigma+\beta_4), \qquad t \in [0,T].$$

We may assume from the first equality of (4.58) that $u^0(t)$ is absolutely continuous. Hence,

$$u^0(t) = -(\sigma+\beta_4)\,t + c, \qquad c = \text{const}. \tag{4.59}$$

Substituting (4.59) into (4.57), we can compute that the first two relations from (4.57) are equivalent to the following equalities

$$\beta_1 = c\,(1-\cos T) + \sigma_{-1} + (\beta_4+\sigma)\,(T\cos T - \sin T),$$

$$\beta_2 = -\beta_4 - c\sin T + (\beta_4+\sigma)\,(T\sin T + \cos T),$$

$$\beta_3 = -c - \sigma_{-1}.$$

The latter four inequalities turn into the following inequalities

$$\sigma \le \sigma_{-2}, \qquad |c+\sigma_{-1}| \le \sigma_{-3}, \qquad |\beta_4| \le \sigma_{-4}, \tag{4.60}$$

$$|c| \le \gamma, \qquad |(\sigma+\beta_4)T - c| \le \gamma.$$

Consequently if (4.60) hold, then, by virtue of the sufficient conditions of Theorem 3.1, the estimator (4.55) is optimal. Note that (4.60) is equivalent to (4.56) with $C_1 = c$, $C_2 = \beta_4$. This concludes the proof. □

It can easily be checked that the optimal value of the guaranteed estimation error is equal to

$$I_0 = I(\Phi^0_-, \Phi^0) = \sigma_{-1} + \sigma T. \tag{4.61}$$

PROPOSITION 4.5. *If* $\sigma_{-1} \le \sigma_{-3}$, $\sigma \le \sigma_{-2}$, *and* $\sigma \le \sigma_{-4}$, *then the optimal estimator* (4.55) *is unique.*

Note that for actual navigation systems the condition of Proposition 4.5 is valid.

PROOF. If the condition of Proposition 4.5 holds, then (4.56) is valid with $C_1 = 0$ and $C_2 = -\sigma$. Consequently the estimator (4.55) is optimal and the value (4.61) is the minimal guaranteed estimation error.

Assume that there exists another optimal estimator (Φ_-, Φ), $\Phi_- \in \mathbf{R}^4$, $\Phi \in \mathcal{F}^1$, such that

$$\sigma_{-1} + \sigma T = \sum_{d=1}^{4} \sigma_{-d}|\Phi_{-d}| + \sigma \int_0^T |\Phi(t)|\, dt + \gamma S, \tag{4.62}$$

where

$$S = \int_0^T \left| b'(t) \left(\int_0^t h(\tau)\Phi(\tau)\, d\tau + \Phi_- \right) \right| dt.$$

The analog of the unbiasness condition (2.7) takes the form

$$\Phi_{-1} - \int_0^T \sin t\, \Phi(t)\, dt = \cos T, \tag{4.63}$$

$$\Phi_{-2} + \int_0^T \cos t\, \Phi(t)\, dt = \sin T,$$

$$\Phi_{-3} - \int_0^T \sin t\, \Phi(t)\, dt = \cos T - 1,$$

$$\Phi_{-4} + \int_0^T (\cos t - 1)\Phi(t)\,dt = \sin T - T.$$

Consequently,

$$\Phi_{-1} - \Phi_{-3} = 1, \qquad \Phi_{-2} - \Phi_{-4} + \int_0^T \Phi(t)\,dt = T, \tag{4.64}$$

whence,

$$|\Phi_{-1}| + |\Phi_{-3}| \geq 1, \qquad |\Phi_{-2}| + |\Phi_{-4}| + \int_0^T |\Phi(t)|\,dt \geq T. \tag{4.65}$$

It follows from (4.62), (4.64), and (4.65) that

$$\sigma_{-1} + \sigma T \geq \sigma_{-1}\left(|\Phi_{-1}| + |\Phi_{-3}|\right) + \sigma\left(|\Phi_{-2}| + |\Phi_{-4}|\right)$$

$$+ \sigma \int_0^T |\Phi(t)|\,dt + \gamma S \geq \sigma_{-1} + \sigma T + \gamma S.$$

Hence,

$$S = 0, \qquad \Phi \in \mathcal{F}^1. \tag{4.66}$$

The function $b'(t)\left(\int_0^t h(\tau)\Phi(\tau)\,d\tau + \Phi_-\right)$ is continuous from the right. Therefore (4.66) is equivalent to the equality

$$\int_0^t \cos(t-\tau)\Psi(\tau)\,d\tau = 0, \qquad t \in [0,T), \tag{4.67}$$

$$\Psi(\tau) = \Phi(\tau) + \Phi_{-2}\delta(\tau) - \Phi_{-1}, \qquad \Psi \in \mathcal{F}^1.$$

Let us show that (4.67) implies the identity

$$\Psi(t) \equiv 0, \qquad t \in [0,T). \tag{4.68}$$

In fact, since $\Psi \in \mathcal{F}^1$, we have $\Psi(t) = \Psi_{pc}(t) + \sum_{s=1}^{M} \Psi_{(s)}\delta(t - t_s)$, where $\Psi_{pc}(t)$ is a piecewise continuous component of Ψ.

Without loss of generality we suppose that at discontinuity points the values of $\Psi_{pc}(t)$ coincide with the corresponding limits from the left or from the right.

Suppose $t_1 = 0$; then (4.67) implies the equality

$$\cos t\,\Psi_{(1)} + \int_0^t \cos(t-\tau)\Psi_{pc}(\tau)\,d\tau = 0, \qquad t \in [0, t_2),$$

whence $\Psi_{(1)} = 0$ and

$$\int_0^t \cos(t-\tau)\Psi_{pc}(\tau)\,d\tau = 0, \qquad t \in [0, t_2). \tag{4.69}$$

Differentiating identity (4.69) at any continuity point of $\Psi(t)$ on the interval $(0, t_2)$, we obtain the relation

$$\Psi_{pc}(t) - \int_0^t \sin(t-\tau)\Psi_{pc}(\tau)\,d\tau = 0. \tag{4.70}$$

It follows from (4.70) that $\Psi_{pc}(t)$ is absolutely continuous and differentiable on $[0, t_2)$. Moreover,

$$\frac{d\Psi_{pc}(t)}{dt} - \int_0^t \cos(t-\tau)\Psi_{pc}(\tau)\,d\tau = 0, \qquad t \in [0, t_2). \tag{4.71}$$

Summing (4.69) and (4.71), we get

$$\frac{d\Psi_{pc}(t)}{dt} = 0, \qquad t \in [0, t_2). \tag{4.72}$$

Proceeding to the limit in (4.70) at the point $t = 0$, we get $\Psi_{pc}(0) = \lim_{s\to 0+} \Psi(s) = 0$ and, by virtue of (4.72),

$$\Psi_{pc}(t) = 0, \qquad t \in [0, t_2).$$

Arguing as above, we can extend the last equality to the whole of $[0, T)$ and obtain the identity (4.68). In the case $t_1 > 0$, the result is the same.

Since $\Psi \in \mathcal{F}^1$, it follows from (4.68) that

$$\Phi(t) = \Phi_{-1} - \Phi_{-2}\,\delta(t) + \Phi_T\delta(T-t), \quad \Phi_T \in \mathbf{R}^1, \qquad t \in [0, T]. \tag{4.73}$$

Then relations (4.63) turn into the following equalities:

$$(\Phi_{-1} - 1)\cos T - \Phi_T \sin T = 0,$$

$$(\Phi_{-1} - 1)\sin T + \Phi_T \cos T = 0,$$

$$\Phi_{-3} + (\cos T - 1)\Phi_{-1} - \Phi_T \sin T = \cos T - 1,$$

$$\Phi_{-4} + (\sin T - T)\Phi_{-1} + (\cos T - 1)\Phi_T = \sin T - T,$$

whence $\Phi_{-1} = 1$, $\Phi_T = 0$, $\Phi_{-3} = 0$, and $\Phi_{-4} = 0$. Substituting the estimator (4.73) into (4.62), it can easily be checked that $\Phi_{-2} = 0$ and $\Phi(t) \equiv 1$. □

Let us consider a limiting case that the inequalities (4.56) do not hold:

$$\sigma > 0, \qquad \gamma = 0, \qquad \sigma_{-4} = 0, \tag{4.74}$$

i.e., the case that the sensors are precise ($\gamma = 0,\ \sigma_{-4} = 0$) and the errors are caused only by the inaccuracy of initial data ($x_1(0) \neq 0,\ x_2(0) \neq 0$) and by the levelling error $(x_1(0)+x_3(0)) \neq 0$. By virtue of (4.74), the equations take the form

$$\dot{x}_1 = x_2, \qquad \dot{x}_2 = -x_1 - x_3, \qquad \dot{x}_3 = 0. \tag{4.75}$$

The necessary conditions for optimality yield the estimator

$$\Phi_{-1} = 1, \quad \Phi_{-2} = 0, \quad \Phi_{-3} = 0, \qquad \Phi(t) = 2\sin\left(\frac{T}{2}\right)\delta\left(t - \frac{T}{2}\right), \tag{4.76}$$

and the parameters

$$\beta_1 = \sigma_{-1}, \qquad \beta_2 = -\sigma\cos\frac{T}{2}, \qquad \beta_3 = \sigma\sin\frac{T}{2} - \sigma_{-1}.$$

By virtue of the sufficient conditions for optimality it can easily be checked that if $\sigma \leq \sigma_{-2}$ and $\sigma + \sigma_{-1} \leq \sigma_{-3}$, then the estimator (4.76) is optimal.

Note that such a simple form of estimator (4.76) has a simple explanation. Indeed, resolving system (4.75), we get the identity

$$x_1(t) \equiv 2\sin\left(\frac{t}{2}\right)x_2\left(\frac{t}{2}\right) + x_1(0).$$

Thus if the inequalities (4.56) are not valid, then the optimal measurement strategy may be quite different from (4.55): we should make a unique measurement at the midpoint of the time segment.

5. Optimality of Linear Estimators

In this section we show that the optimal estimator for the *a priori* guaranteed estimation problem can be searched on the class of linear estimators.

In the previous sections we considered linear estimators of the form

$$\hat{l} = \int_0^T \Phi'(t)z(t)\,dt = \int_0^T \Phi'_{pc}(t)z(t)\,dt + \sum_{s=1}^{M} \Phi'_{(s)}z(t_s). \tag{5.1}$$

Let us introduce nonlinear estimators

$$\hat{l} = \mathcal{S}(z(\cdot)), \qquad \mathcal{S}\colon \mathcal{Z} \to \mathbf{R}^1, \tag{5.2}$$

where $\mathcal{Z}$ is the set of all samples $z(\cdot)$ from (2.3).

The optimal *a priori* guaranteed estimation problem on the class of nonlinear estimators (5.2) is to find an estimator $\mathcal{S}^0$ such that

$$\inf_{\mathcal{S}} \sup_{x(0)\in\Pi,\, u\in U,\, \varrho\in\sigma} |\mathcal{S}(z(\cdot)) - l_*| = \sup_{x(0)\in\Pi,\, u\in U,\, \varrho\in\sigma} |\mathcal{S}^0(z(\cdot)) - l_*|. \tag{5.3}$$

Obviously, the class of nonlinear estimators contains the linear estimators from (5.1). Therefore,

$$\inf_{\mathcal{S}} \sup_{x(0)\in\Pi,\, u\in U,\, \varrho\in\sigma} |\, \mathcal{S}(z(\cdot)) - l_* \,| \le I_0,$$

where I_0 is the best accuracy achievable with linear estimators (5.1); I_0 is determined by (2.13).

5.1. *General Estimation Problem*

Let $\mathcal{X}$ and $\mathcal{Y}$ be normed vector spaces with elements $x \in \mathcal{X}$ and $y \in \mathcal{Y}$ and the norms $\|\cdot\|_{\mathcal{X}}$ and $\|\cdot\|_{\mathcal{Y}}$, respectively. Suppose $\mathcal{V} \subset \mathcal{X}$ and $\mathcal{W} \subset \mathcal{Y}$ are convex balanced sets.

The problem is to estimate the value $l(x)$, $x \in \mathcal{V}$, of a continuous linear functional $l\colon \mathcal{X} \to \mathbf{R}^1$ by the measurements

$$\tilde{L}(x,\xi) = L(x) + \xi,$$

where $L\colon \mathcal{X} \to \mathcal{Y}$ is a continuous linear operator, $\xi \in \mathcal{W} \subset \mathcal{Y}$ is a measurement error. Let $\mathcal{S}\colon \mathcal{Y} \to \mathbf{R}^1$ be an estimator, i.e., $\mathcal{S}(\tilde{L}(x,\xi))$ is an estimate for $l(x)$.

The optimal estimation problem is to construct an estimator that minimizes the guaranteed estimation error:

$$\kappa(\mathcal{S}) = \sup_{x\in\mathcal{V},\, \xi\in\mathcal{W}} |\, \mathcal{S}(\tilde{L}(x,\xi)) - l(x) \,| \to \inf_{\mathcal{S}}.$$

LEMMA 5.1. *For any estimator* $\mathcal{S}$, $\kappa(\mathcal{S}) \ge \mathcal{D}^0$, *where*

$$\mathcal{D}^0 \triangleq \sup_{x\in\mathcal{V},\, L(x)\in\mathcal{W}} |l(x)|. \tag{5.4}$$

PROOF. Let λ be a realization of the measurement and let $\mathcal{V}(\lambda)$ is a set of x that are compatible with this realization:

$$\mathcal{V}(\lambda) = \Big\{ x \in \mathcal{V} \,\Big|\, \lambda - L(x) \in \mathcal{W} \Big\}.$$

The set $\mathcal{V}(\lambda)$ is convex. Really, if

$$\lambda - L(x_i) = \xi_i \in \mathcal{W}, \qquad x_i \in \mathcal{V}, \qquad i = 1, 2,$$

then

$$\alpha x_1 + (1-\alpha)x_2 \in \mathcal{V}, \qquad \alpha \in [0,1] \qquad (\mathcal{V} \text{ is convex})$$

and, since L is linear and $\mathcal{W}$ is convex,

$$\lambda - L(\alpha x_1 + (1-\alpha)x_2) = \alpha(\lambda - L(x_1)) + (1-\alpha)(\lambda - L(x_2))$$

$$= \alpha\xi_1 + (1-\alpha)\xi_2 \in \mathcal{W}.$$

Obviously,

$$\kappa(\mathcal{S}) = \sup_{x\in\mathcal{V},\,\xi\in\mathcal{W}} |\,\mathcal{S}(\tilde{L}(x,\xi)) - l(x)\,| \tag{5.5}$$

$$\geq \sup_{x\in\mathcal{V}(\lambda),\,\xi=\lambda-L(x)} |\,\mathcal{S}(\tilde{L}(x,\xi)) - l(x)\,| = \sup_{x\in\mathcal{V}(\lambda)} |\mathcal{S}(\lambda) - l(x)|.$$

The image of $\mathcal{V}(\lambda)$ under the linear map $l(x)$ is a convex set of real axis. This image determines the values of $l(x)$ that are compatible with the measurement λ. For fixed λ, the value $\mathcal{S}(\lambda)$ is a number. Hence,

$$\sup_{x\in\mathcal{V}(\lambda)} |\mathcal{S}(\lambda) - l(x)| \geq \frac{1}{2}\left(\sup_{x\in\mathcal{V}(\lambda)} l(x) - \inf_{x\in\mathcal{V}(\lambda)} l(x)\right). \tag{5.6}$$

Relations (5.5) and (5.6) are valid for any admissible λ, in particular, for $\lambda = 0$. Moreover, since $\mathcal{V}$ and $\mathcal{W}$ are balanced, $\mathcal{V}(0)$ is also balanced and

$$\inf_{x\in\mathcal{V}(0)} l(x) = -\sup_{x\in\mathcal{V}(0)} l(x).$$

Thus it follows from (5.5) and (5.6) that

$$\kappa(\mathcal{S}) \geq \sup_{x\in\mathcal{V}(0)} l(x) = \mathcal{D}^0.$$

□

THEOREM 5.1. *Suppose $\mathcal{W}$ contains a neighborhood of zero, i.e., $\mathcal{W}$ is a convex body in $\mathcal{Y}$. Then there exists a continuous linear functional $\mathcal{S}^0\colon \mathcal{Y} \to \mathbf{R}^1$ such that*

$$\kappa(\mathcal{S}^0) = \inf_{\mathcal{S}} \kappa(\mathcal{S}) = \mathcal{D}^0. \tag{5.7}$$

PROOF. Lemma 5.1 provides a lower bound for estimation accuracy. Let us prove that for a certain continuous linear functional $\mathcal{S}^0$

$$\kappa(\mathcal{S}^0) \leq \mathcal{D}^0. \tag{5.8}$$

If $\mathcal{D}^0 = \infty$, then (5.8) holds. So, suppose that $\mathcal{D}^0 < \infty$. Following [149], let us introduce the augmented vector space $\tilde{\mathcal{Y}} = \mathbf{R}^1 \times \mathcal{Y}$ with elements $\tilde{y} = (y^0, y)$, where $y^0 \in \mathbf{R}^1$, $y \in \mathcal{Y}$, and the norm $\|\tilde{y}\|_{\tilde{\mathcal{Y}}} = \max\left\{|y^0|, \|y\|_{\mathcal{Y}}\right\}$.

Consider the set

$$\mathcal{E} = \left\{e \in \tilde{\mathcal{Y}} \,\middle|\, e = (y^0, y),\quad y^0 = l(x),\ y = L(x) + \xi,\quad x \in \mathcal{V},\ \xi \in \mathcal{W}\right\}.$$

Note that the section of $\mathcal{E}$ with fixed y determines the set of the values of $l(x)$ compatible with y.

Let us prove the three auxiliary assertions:

(a) $\mathcal{E}$ is convex and balanced;

(b) $0 \in \operatorname{int} \mathcal{E}$;

(c) the point $v = (\mathcal{D}^0, 0) \notin \operatorname{int} \mathcal{E}$.

Proof of Assertion (a). Suppose $e_i = (y_i^0, y_i) \in \mathcal{E}$, $i = 1, 2$, are defined by the equalities:

$$y_i^0 = l(x_i), \quad y_i = L(x_i) + \xi_i, \qquad x_i \in \mathcal{V}, \quad \xi_i \in \mathcal{W}.$$

Then, since l, L are linear and $\mathcal{V}$, $\mathcal{W}$ are convex, we have that, for $\alpha \in [0, 1]$,

$$\begin{aligned}&\alpha e_1 + (1-\alpha)e_2 \\ &= \Big(l(\alpha x_1 + (1-\alpha)x_2),\ L(\alpha x_1 + (1-\alpha)x_2) + \alpha\xi_1 + (1-\alpha)\xi_2\Big) \\ &= \big(l(x),\ L(x) + \xi\big) \quad \text{for some} \quad x \in \mathcal{V} \quad \text{and} \quad \xi \in \mathcal{W}.\end{aligned}$$

If $e \in \tilde{\mathcal{Y}}$ for $x \in \mathcal{V}$, $\xi \in \mathcal{W}$, then $\alpha e = (l(\alpha x), L(\alpha x) + \alpha\xi)$. Since $\mathcal{V}$ and $\mathcal{W}$ are balanced, $\mathcal{E}$ is balanced.

Proof of Assertion (b). Let $\tilde{y} = (y^0, y) \in \tilde{\mathcal{Y}}$ be any point of the neighborhood $\|\tilde{y}\|_{\tilde{\mathcal{Y}}} < \epsilon$, i.e., $|y^0| < \epsilon$ and $\|y\|_{\mathcal{Y}} < \epsilon$. Let us show that if ϵ is sufficiently small, then for any $|y^0| < \epsilon$, $\|y\|_{\mathcal{Y}} < \epsilon$ there exist $x \in \mathcal{V}$ and $\xi \in \mathcal{W}$ such that $y^0 = l(x)$ and $y = L(x) + \xi$, i.e., $0 \in \operatorname{int} \mathcal{E}$.

If $l(x) = 0\ \forall x \in \mathcal{V}$, then we should take $\mathcal{S}^0 \equiv 0$. In this case, $\kappa(\mathcal{S}^0) = 0$ and $\mathcal{D}^0 = 0$. So, Theorem 5.1 is proved.

If $l(x) \not\equiv 0$ on $\mathcal{V}$, then $l(\bar{x}) = \nu \neq 0$ for an $\bar{x} \in \mathcal{V}$. Since $\mathcal{V}$ is balanced, we may suppose $\nu > 0$.

Denote by r the radius of a spherical neighbourhood of zero that is contained in $\mathcal{W}$ by the condition of Theorem 5.1. Let $\|L\|$ be the norm of the operator L. Choose

$$\epsilon < \min\left\{\nu,\ r\left(1 + \frac{\|L\|\,\|\bar{x}\|_{\mathcal{X}}}{\nu}\right)^{-1}\right\} \tag{5.9}$$

and construct $x = (\nu^{-1}y^0)\bar{x}$ and $\xi = y - L(x)$. Since $\mathcal{V}$ is balanced, $|y^0| < \epsilon$, and (5.9) holds, we see that $x \in \mathcal{V}$. Moreover, since $\|Y\|_{\mathcal{Y}} < \epsilon$ and (5.9) holds, we have

$$\begin{aligned}&\|\xi\|_{\mathcal{Y}} \le \|y\|_{\mathcal{Y}} + \|L\|\,\|x\|_{\mathcal{X}} = \|y\|_{\mathcal{Y}} + \nu^{-1}y^0\|L\|\,\|\bar{x}\|_{\mathcal{X}} \\ &\le \epsilon\left(1 + \nu^{-1}\|L\|\,\|\bar{x}\|_{\mathcal{X}}\right) < r \quad \text{and therefore} \quad \xi \in \mathcal{W}.\end{aligned}$$

Thus $\big(l(x),\ L(x) + \xi\big) = (y^0, y)$, i.e., $(y^0, y) \in \mathcal{E}$ and $0 \in \operatorname{int} \mathcal{E}$.

Proof of Assertion (c). If $v \in \operatorname{int}\mathcal{E}$, then $(\mathcal{D}^0, 0) + (\eta, 0) \in \mathcal{E}$ for an $\eta > 0$. Hence there exist $x_0 \in \mathcal{V}$ and $\xi_0 \in \mathcal{W}$ such that

$$l(x_0) = \mathcal{D}^0 + \eta, \qquad L(x_0) + \xi_0 = 0$$

and, consequently, $L(x_0) \in \mathcal{W}$. Therefore,

$$\sup_{x \in \mathcal{V},\, L(x) \in \mathcal{W}} |l(x)| \geq l(x_0) = \mathcal{D}^0 + \eta.$$

The last inequality contradicts the definition of $\mathcal{D}^0$ from (5.4). So, Assertions (a) – (c) are proved.

By Theorem 1.1 (separation theorem), there exist a nonzero continuous linear functional $\Omega(\tilde{y})$ over $\tilde{\mathcal{Y}}$ such that

$$\sup_{e \in \mathcal{E}} \Omega(e) \leq \Omega(v).$$

Since Ω is linear, we have

$$\Omega(\tilde{y}) = \Omega((y^0, y)) = \Omega((y^0, 0) + (0, y)) = \Omega((y^0, 0)) + \Omega((0, y))$$

$$= n_0 y^0 + n(y),$$

where $n_0 \in \mathbf{R}^1$ and $n(y)$ is a continuous linear functional over $\mathcal{Y}$.
Then

$$\sup_{(y^0, y) \in \mathcal{E}} (n_0 y^0 + n(y)) \leq n_0 \mathcal{D}^0. \tag{5.10}$$

Since $\mathcal{E}$ is balanced, it follows from (5.10) that

$$\sup_{(y^0, y) \in \mathcal{E}} |n_0 y^0 + n(y)| \leq n_0 \mathcal{D}^0. \tag{5.11}$$

Assume $n_0 \mathcal{D}^0 = 0$. Then from (5.11) it follows that $\Omega(e) = 0 \;\; \forall e \in \mathcal{E}$. By virtue of Assertion (b), $0 \in \operatorname{int}\mathcal{E}$ and $\Omega(\tilde{y}) = 0$ for all $\tilde{y} \in \tilde{\mathcal{Y}}$. The latter relation contradicts Theorem 1.1. Thus $n_0 \mathcal{D}^0 \neq 0$ and from (5.11) it follows that $n_0 > 0$ and

$$\sup_{(y^0, y) \in \mathcal{E}} |y^0 - \mathcal{S}^0(y)| \leq \mathcal{D}^0, \qquad \mathcal{S}^0(y) = -n_0^{-1} n(y).$$

Finally,

$$\sup_{x \in \mathcal{V},\, \xi \in \mathcal{W}} |\mathcal{S}^0(L(x) + \xi) - l(x)| = \sup_{(y^0, y) \in \mathcal{E}} |\mathcal{S}^0(y) - y^0| \leq \mathcal{D}^0.$$

Thus (5.7) follows from Lemma 5.1 and (5.8). □

5.2. *Optimality of Linear Estimators in Dynamic Systems*

Let us show that the guaranteed estimation problem for dynamic system that was studied in Section 2 is a special case of the general estimation problem considered above.

Recall that our dynamic system was represented in the form (see Section 2 for details)

$$\dot{y}(t) = b(t)u(t), \qquad y(0) = x(0), \qquad y(t) \in \mathbf{R}^n, \qquad t \in [0,T],$$

$$b(t) = \begin{pmatrix} \bar{b}_1'(t) \\ \vdots \\ \bar{b}_n'(t) \end{pmatrix}, \qquad \bar{b}_i(t) \in \mathbf{R}^r, \qquad i = 1,\dots,n,$$

$$z(t) = h'(t)y(t) + \varrho(t), \qquad z(t), \varrho(t) \in \mathbf{R}^m,$$

$$h'(t) = \begin{pmatrix} h_1'(t) \\ \vdots \\ h_m'(t) \end{pmatrix}, \qquad h_l(t) \in \mathbf{R}^n, \qquad l = 1,\dots,m,$$

$$l_* = \tilde{a}'y(T), \qquad y(T) = y(0) + \int_0^T b(\tau)u(\tau)\,d\tau,$$

where $b(t)$, $h(t)$, and $\tilde{a}$ are specified quantities and

$$|y_{-d}(0)| \le \sigma_{-d}, \quad d = 1,\dots,n, \qquad |u_j(t)| \le \gamma_j(t), \quad j = 1,\dots,r,$$

$$|\varrho_l(t)| \le \sigma_l(t), \quad l = 1,\dots,m.$$

Denote by $\mathcal{C}^n$ the linear space of piecewise continuous vector functions $v(t) = (v_1(t),\dots,v_n(t))' \in \mathbf{R}^n$ over the closed interval $[0,T]$ with the norm

$$\|v(\cdot)\|_{\mathcal{C}^n} \triangleq \sup_{t\in[0,T]} \|v(t)\|_{\mathbf{R}^n} = \sup_{t\in[0,T]} \sqrt{\sum_{i=1}^n v_i^2(t)}.$$

Thus, in the notation of the general estimation problem,

$$\mathcal{X} = \mathbf{R}^n \times \mathcal{C}^r, \qquad \mathcal{Y} = \mathcal{C}^m,$$

$$x = \big(y(0), u(\cdot)\big), \qquad y(0) \in \mathbf{R}^n, \quad u(\cdot) \in \mathcal{C}^r,$$

$$\|x\|_{\mathcal{X}} = \max\big\{\|y(0)\|_{\mathbf{R}^n}, \|u(\cdot)\|_{\mathcal{C}^r}\big\},$$

$$l(x) = \tilde{a}'\left(y(0) + \int_0^T b(\tau)u(\tau)\,d\tau\right), \tag{5.12}$$

$$L(x) = h'(t)\left(y(0) + \int_0^t b(\tau)u(\tau)\,d\tau\right), \qquad \xi = \varrho(\cdot), \tag{5.13}$$

$$\mathcal{V} = \left\{(y(0), u(\cdot)) \in \mathbf{R}^n \times \mathcal{C}^r \,\Big|\, y(0) \in \Pi,\ u \in U\right\}, \tag{5.14}$$

$$\mathcal{W} = \left\{\varrho(\cdot) \in \mathcal{C}^m \,\Big|\, \varrho \in \sigma\right\}. \tag{5.15}$$

Let us show that the fuctional $l(x)$ and the operator L defined by (5.12) and (5.13), respectively, are continuous. Really, it can easily be checked by standard computation that

$$|l(x)| \le \|\tilde{a}\|_{\mathbf{R}^n}\|y(0)\|_{\mathbf{R}^n} + \|\tilde{a}\|_{\mathbf{R}^n}\|v(T)\|_{\mathbf{R}^n},$$

$$\|L(x)\|_{\mathcal{Y}} \le \|h'(\cdot)y(0)\|_{\mathcal{Y}} + \|h'v(\cdot)\|_{\mathcal{Y}}, \qquad v(t) = \int_0^t b(\tau)u(\tau)\,d\tau,$$

where

$$\|v(T)\|_{\mathbf{R}^n} \le \sqrt{n}\,T \max_{1\le i\le n} \|\bar{b}_i(\cdot)\|_{\mathcal{C}^r}\, \|u(\cdot)\|_{\mathcal{C}^r},$$

$$\|h'(\cdot)y(0)\|_{\mathcal{Y}} \le \sqrt{m} \max_{1\le l\le m} \|h_l(\cdot)\|_{\mathcal{C}^n}\|y(0)\|_{\mathbf{R}^n},$$

$$\|h'v(\cdot)\|_{\mathcal{Y}} \le \sqrt{mn}\,T \max_{1\le l\le m} \|h_l(\cdot)\|_{\mathcal{C}^n} \max_{1\le i\le n} \|\bar{b}_i(\cdot)\|_{\mathcal{C}^r}\, \|u(\cdot)\|_{\mathcal{C}^r}.$$

Thus,

$$|l(x)| \le C_1\|x\|_{\mathcal{X}}, \qquad \|L(x)\|_{\mathcal{Y}} \le C_2\|x\|_{\mathcal{X}}, \tag{5.16}$$

where

$$C_1 = \|\tilde{a}\|_{\mathbf{R}^n}\left(1 + \sqrt{n}\,T \max_{1\le i\le n} \|\bar{b}_i(\cdot)\|_{\mathcal{C}^r}\right),$$

$$C_2 = \sqrt{m} \max_{1\le l\le m} \|h_l(\cdot)\|_{\mathcal{C}^n}\left(1 + \sqrt{n}\,T \max_{1\le i\le n} \|\bar{b}_i(\cdot)\|_{\mathcal{C}^r}\right).$$

COROLLARY 5.1. *Let*

$$\sigma_l(t) \ge \sigma^0 > 0, \qquad t \in [0,T], \qquad l = 1,\dots,m.$$

Then there exists a continuous linear functional $\mathcal{S}^0\colon \mathcal{C}^m \to \mathbf{R}^1$ *such that*

$$\sup_{x(0)\in\Pi,\, u\in U,\, \varrho\in\sigma} |\mathcal{S}^0(z(\cdot)) - l_*| \tag{5.17}$$

$$= \inf_{\mathcal{S}} \sup_{x(0)\in\Pi,\, u\in U,\, \varrho\in\sigma} |\, \mathcal{S}(z(\cdot)) - l_* \,| = \mathcal{D}^0.$$

Moreover,

$$\mathcal{D}^0 = \sup |\tilde{a}' y(T)| = \sup \tilde{a}' y(T) \tag{5.18}$$

under constraints

$$|h'(t)y(t)| \le \sigma_l(t), \qquad t \in [0,T], \quad l = 1,\ldots,m, \tag{5.19}$$

$$|y_{-d}(0)| \le \sigma_{-d}, \qquad d = 1,\ldots,n,$$

$$|u_j(t)| \le \gamma_j(t), \qquad t \in [0,T], \quad j = 1,\ldots,r,$$

where

$$\dot{y}(t) = b(t)u(t).$$

PROOF. Obviously, the sets (5.14) and (5.15) are convex and balanced. According to (5.16), $l(x)$ and $L(x)$ are continuous. It follows from the condition of Corollary 5.1 that the set (5.15) contains a neighbourhood of zero in $\mathcal{Y} = \mathcal{C}^m$. Recall also that $y(0) = x(0)$. Then, by virtue of Theorem 5.1, the relations (5.17) – (5.19) hold. The second equality in (5.18) is valid because the set (5.19) is balanced. □

A general form of a continuous linear functional on $\mathcal{C}^m$ is quite complex. For example, the functional

$$\mathcal{S}(z(\cdot)) = \int_0^T z(t)\, d\mu,$$

where μ is a Lebesgue-Stieltjes measure, is a linear continuous functional on $\mathcal{C}^m$. Nevertheless, if μ is a singular measure, then $\mathcal{S}$ can not be represented by the estimators of type (5.1). To emphasize the distinction between the general continuous linear estimators and the estimators of type (5.1) we call the estimators of type (5.1) $\mathcal{F}$-estimators.

In any case, Corollary 5.1 provides an important result: the minimal guaranteed estimation error is equal to $\mathcal{D}^0$.

5.3. *Optimality of $\mathcal{F}$-estimators*

Now we show that the solution of optimal guaranteed estimation problem for dynamic systems may be searched on the class of $\mathcal{F}$-estimators. Recall that if we use the $\mathcal{F}$-estimators, then the optimal guaranteed estimation problem reduce to to the following variational problem (see Section 2 for details)

$$I_0 = \inf_{(\Phi_-,\Phi)\in\mathcal{N}} I(\Phi_-,\Phi), \tag{5.20}$$

where

$$I(\Phi_-,\Phi) = \sum_{d=1}^{n} \sigma_{-d}|\Phi_{-d}| + \sum_{l=1}^{m} \int_0^T \sigma_l(t)|\Phi_l(t)|\,dt \tag{5.21}$$

$$+\sum_{j=1}^{r} \int_0^T \gamma_j(t)\left| b_j'(t)\left(\Phi_- + \int_0^t h(\tau)\Phi(\tau)\,d\tau\right)\right| dt$$

and the constraints $\mathcal{N}$ are defined by the equality

$$\Phi_- + \int_0^T h(t)\Phi(t)\,dt = \tilde{a}. \tag{5.22}$$

The solution of (5.20) (Φ_-^0, Φ^0) determines the optimal estimator Φ^0.

THEOREM 5.2. *Let* $\mathcal{D}^0$ *is defined by* (5.18), (5.19) *and* I_0 *is defined by* (5.20) - (5.22). *Then* $I_0 = \mathcal{D}^0$.

PROOF. First let us introduce auxiliary structures that are identical to that of the proof of Theorem 3.1. We shall repeat some fragments from the proof of Theorem 3.1 to make the proof of Theorem 5.2 self-contained.

Let L_1^r be the Banach space of absolutely integrable vector functions w with the norm:

$$\|w\|_{L_1^r} = \int_0^T \|w(t)\|_{\mathbf{R}^r}\,dt, \qquad \|w(t)\|_{\mathbf{R}^r} = \left(\sum_{i=1}^{r} w_i^2(t)\right)^{\frac{1}{2}}.$$

Recall that the norm in $\mathcal{F}^m$ is defined by

$$\|\Phi\|_{\mathcal{F}^m} = \int_0^T \|\Phi_{pc}(t)\|_{\mathbf{R}^m}\,dt + \sum_{s=1}^{M} \|\Phi_{(s)}\|_{\mathbf{R}^m}.$$

Consider the following extremal problem, which is equivalent to the problem (5.20) – (5.22):

$$I_0 = \inf_{\Phi_-\in\mathbf{R}^n,\,\Phi\in\mathcal{F}^m,\,w\in L_1^r} I(\Phi_-,\Phi,w),$$

$$I(\Phi_-, \Phi, w) = \sum_{d=1}^{n} \sigma_{-d}|\Phi_{-d}| + \sum_{l=1}^{m} \int_0^T \sigma_l(t)|\Phi_l(t)|\,dt + \sum_{j=1}^{r} \int_0^T \gamma_j(t)|w_j(t)|\,dt,$$

$$b'(t)\left(\Phi_- + \int_0^t h(s)\Phi(s)\,ds\right) - w(t) = 0, \qquad t \in [0,T],$$

$$\Phi_- + \int_0^T h(t)\Phi(t)\,dt = \tilde{a}.$$

We also introduce the spaces $\mathcal{X} = \mathbf{R}^n \times \mathcal{F}^m \times L_1^r$ and $\mathcal{Y} = L_1^r \times \mathbf{R}^n$ with the norm

$$\|x\|_{\mathcal{X}} = \max\left(\|\Phi_-\|_{\mathbf{R}^n}, \|\Phi\|_{\mathcal{F}^m}, \|w\|_{L_1^r}\right) \quad \text{for} \quad x = (\Phi_-, \Phi, w)$$

and the norm

$$\|y\|_{\mathcal{Y}} = \max\left(\|w\|_{L_1^r}, \|\Phi_-\|_{\mathbf{R}^n}\right) \quad \text{for} \quad y = (w, \Phi_-),$$

respectively.

We define the affine operator Λ: $\mathcal{X} \to \mathcal{Y}$ by the relations:

$$\Lambda(\Phi_-, \Phi, w) = (v, d), \qquad \Phi_-, d \in \mathbf{R}^n, \quad \Phi \in \mathcal{F}^m, \quad w, v \in L_1^r,$$

$$v(t) = b'(t)\left(\Phi_- + \int_0^t h(s)\Phi(s)\,ds\right) - w(t), \qquad t \in [0,T],$$

$$d = -\Phi_- - \int_0^T h(t)\Phi(t)\,dt + \tilde{a}.$$

Thus the extremal problem (5.20) – (5.22) can be rewritten in the form

$$\inf_{(\Phi_-, \Phi, w) \in \mathcal{X}} I(\Phi_-, \Phi, w) \quad \text{under constraint} \quad \Lambda(\Phi_-, \Phi, w) = 0, \tag{5.23}$$

where the functional I: $\mathcal{X} \to \mathbf{R}^1$ is continuous and convex and the operator Λ: $\mathcal{X} \to \mathcal{Y}$ is continuous and affine. The proof of continuity is similar to the proof of inequalities (5.16). The convexity of I is obvious.

So, the problem (5.23) is a special case of the variational problem $\mathcal{P}_1$ posed in Section 1.

Let us define the perturbation of the extremal problem (5.23) by the variation of the constraint:

$$I_0(\delta w, \delta\tilde{a}) = \inf_{\mathcal{X}} I(\Phi_-, \Phi, w)$$

under constraint

$$\Lambda(\Phi_-, \Phi, w) = \begin{pmatrix} \delta w \\ \delta\tilde{a} \end{pmatrix}, \qquad \delta w \in L_1^r, \quad \delta\tilde{a} \in \mathbf{R}^n. \tag{5.24}$$

If $\delta w = 0$ and $\delta\tilde{a} = 0$, then we get the initial problem (5.23).

PROPOSITION 5.1. *For any neighbourhood of zero $\mathcal{U} \subset \mathcal{Y}$, the function $I_0(\delta w, \delta\tilde{a})$ is bounded:*

$$0 \le I_0(\delta w, \delta\tilde{a}) \le C \quad \forall\ (\delta w, \delta\tilde{a}) \in \mathcal{U}, \qquad C = C(\mathcal{U}) = \text{const}.$$

PROOF. Obviously, $I_0(\delta w, \delta\tilde{a})$ is nonnegative. Let us construct an upper bound for $I_0(\delta w, \delta\tilde{a})$. The triple

$$(\bar{\Phi}_-, \bar{\Phi}, \bar{w}) = (\tilde{a} - \delta\tilde{a},\ 0,\ b'(t)(\tilde{a} - \delta\tilde{a}) - \delta w)$$

satisfies (5.24). Moreover,

$$I_0(\bar{\Phi}_-, \bar{\Phi}, \bar{w}) = \sum_{d=1}^{n} \sigma_{-d}|\tilde{a}_d - (\delta\tilde{a})_d|$$

$$+ \sum_{j=1}^{r} \int_0^T \gamma_j(t)\big|b_j'(t)(\tilde{a} - \delta\tilde{a}) - (\delta w)_j(t)\big|\, dt$$

$$\le \left(\sqrt{\sum_{d=1}^{n} \sigma_{-d}^2} + \sum_{j=1}^{r} \int_0^T \gamma_j(t)\|b_j(t)\|_{\mathbf{R}^n}\, dt\right)\left(\|\tilde{a}\|_{\mathbf{R}^n} + \|\delta\tilde{a}\|_{\mathbf{R}^n}\right)$$

$$+ \sup_{t\in[0,T]} \sqrt{\sum_{j=1}^{r} \gamma_j^2(t)}\ \|\delta w\|_{L_1^r} \le C_3 \max\left(\|\delta\tilde{a}\|_{\mathbf{R}^n},\ \|\delta w\|_{L_1^r}\right) + C_4,$$

where C_3, C_4 are constants. Since $0 \le I_0(\delta w, \delta\tilde{a}) \le I(\bar{\Phi}_-, \bar{\Phi}, \bar{w})$, we see that $I_0(\delta w, \delta\tilde{a})$ is also bounded. □

By virtue of Proposition 5.1, Lemma 1.4 and Lemma 1.5, $I_0(\delta w, \delta\tilde{a})$ is convex and continuous at the point $(0, 0)$. Then, in accordance with Theorem 1.7 (taking into account that $\mathcal{Y}^* = L_\infty^r \times \mathbf{R}^n$), the duality relation holds:

$$I_0 = I^0 \triangleq \sup_{p\in L_\infty^r,\ \lambda\in\mathbf{R}^n} \left\{ \inf_{\Phi_-\in\mathbf{R}^n, \Phi\in\mathcal{F}^m, w\in L_1^r} \mathcal{L}(\Phi_-, \Phi,\ w; p, \lambda) \right\}, \tag{5.25}$$

where L_∞^r is the space of measurable essentially bounded functions $p(t) \in \mathbf{R}^r$ with the norm $\|p(\cdot)\|_{L_\infty^r} = \text{ess sup}_{t\in[0,T]} \|p(t)\|_{\mathbf{R}^r}$ and

$$\mathcal{L}(\Phi_-, \Phi,\ w; p, \lambda) = I(\Phi_-, \Phi,\ w)$$

$$+\int_0^T p'(t)\left[b'(t)\left(\int_0^t h(s)\Phi(s)\,ds+\Phi_-\right)-w(t)\right]dt$$

$$+\lambda'\left(\tilde a-\int_0^T h(t)\Phi(t)\,dt-\Phi_-\right).$$

Explicit computation of the infimum in (5.25) results in the problem (5.18), (5.19). To perform the corresponding computations let us consider the following three auxiliary problems.

(a)

$$\mathcal{L}_{(1)}=\inf_{\Phi\in\mathcal{F}^m}\mathcal{L}(0,\Phi,0,p,\lambda)=\inf_{\Phi\in\mathcal{F}^m}\left\{\int_0^T\sum_{l=1}^m\sigma_l(t)|\Phi_l(t)|\,dt\right.$$

$$\left.+\int_0^T p'(t)b'(t)\int_0^t h(s)\Phi(s)\,ds\,dt-\int_0^T\lambda' h(t)\Phi(t)\,dt+\lambda'\tilde a\right\}$$

$$=\inf_{\Phi\in\mathcal{F}^m}\left\{\int_0^T\sum_{l=1}^m\sigma_l(t)|\Phi_l(t)|\,dt\right.$$

$$\left.-\int_0^T\left(\lambda'-\int_s^T p'(t)b'(t)\,dt\right)h(s)\Phi(s)\,ds\right\}+\lambda'\tilde a.$$

Clearly, $\mathcal{L}_{(1)}=\lambda'\tilde a$ if

$$\left|h_l'(t)\left(\lambda-\int_t^T b(s)p(s)\,ds\right)\right|\le\sigma_l(t),\quad t\in[0,T],\quad l=1,\dots,m; \qquad (5.26)$$

otherwise, $\mathcal{L}_{(1)}=-\infty$.

In fact, assume that (5.26) is not valid for $t=t_0$ and $l=l_0$. Putting

$$\Phi_{l_0}(t)=C\mathrm{sign}\left[h_{l_0}'(t_0)\left(\lambda-\int_{t_0}^T b(s)p(s)\,ds\right)\right]\delta(t-t_0),$$

$$\Phi_l(t)\equiv 0,\qquad l\neq l_0,$$

we get that $\mathcal{L}(0,\Phi,0,p,\lambda)\to-\infty$ as $C\to\infty$.

(b)

$$\mathcal{L}_{(2)}=\inf_{w\in L_1^r}\mathcal{L}(0,0,w,p,\lambda)$$

$$= \inf_{w \in L_1^r} \left\{ \int_0^T \sum_{j=1}^r \gamma_j(t)|w_j(t)| - \int_0^T p'(t)w(t)\,dt \right\} + \lambda'\tilde{a}.$$

Clearly, $\mathcal{L}_{(2)} = \lambda'\tilde{a}$ if

$$\operatorname*{ess\,sup}_{t\in[0,T]} \Big(|p_j(t)| - \gamma_j(t) \Big) \le 0, \qquad j = 1, \dots, r, \tag{5.27}$$

and $\mathcal{L}_{(2)} = -\infty$ if (5.27) is not valid. In fact, assume that (5.27) does not hold, i.e., there exist a set E of nonzero measure and a number j_0 such that

$$|p_{j_0}(t)| > \gamma_{j_0}(t), \qquad t \in E.$$

Setting

$$w_{j_0}(t) = \begin{cases} C \operatorname{sign} p_{j_0}(t), & t \in E \\ 0, & t \notin E, \end{cases} \qquad \text{and} \quad w_j(t) \equiv 0, \quad j \ne j_0,$$

we get that $\mathcal{L}(0, 0, w, p, \lambda) \to -\infty$ as $C \to \infty$.

(c)

$$\mathcal{L}_{(3)} = \inf_{\Phi_- \in \mathbf{R}^n} \mathcal{L}(\Phi_-, 0, 0, p, \lambda)$$

$$= \inf_{\Phi_- \in \mathbf{R}^n} \left\{ \sum_{d=1}^n \sigma_{-d}|\Phi_{-d}| - \left(\lambda' - \int_0^T p'(t)b'(t)\,dt \right) \Phi_- \right\} + \lambda'\tilde{a}.$$

Clearly, $\mathcal{L}_{(3)} = \lambda'\tilde{a}$ if

$$\left| e_d' \left(\lambda - \int_0^T b(t)p(t)\,dt \right) \right| \le \sigma_{-d}, \qquad d = 1, \dots, n, \tag{5.28}$$

$$e_d = (\underbrace{0, \dots, 1}_{d}, 0, \dots, 0)',$$

and $\mathcal{L}_{(3)} = -\infty$ if (5.28) is not valid. In fact, assume that (5.28) does not hold for $d = d_0$. Choosing

$$\Phi_{-d_0} = C \operatorname{sign} \left[e_{d_0}' \left(\lambda - \int_0^T b(t)p(t)\,dt \right) \right], \qquad \Phi_{-d} = 0, \qquad d \ne d_0,$$

we get that $\mathcal{L}(\Phi_-, 0, 0, p, \lambda) \to -\infty$ as $C \to \infty$.

It follows from (a) – (c) that

$$I^0 = \sup_{p \in L_1^r,\, \lambda \in \mathbf{R}^n} \lambda'\tilde{a} \qquad \text{under constraints} \quad (5.26) - (5.28).$$

Obviously, the quantity

$$\lambda - \int_t^T b(s)p(s)\,ds$$

can be treated as the solution of the equation

$$\dot{y}(t) = b(t)u(t), \qquad y(T) = \lambda, \quad u(t) = p(t).$$

Then (5.26) – (5.28) are equivalent to (5.19). Since (5.19) defines a balanced set, we have that

$$\sup_{(5.19)} |\tilde{a}'y(T)| = \sup_{(5.19)} \tilde{a}'y(T) = \sup_{(5.26)-(5.28)} \lambda'\tilde{a}.$$

□

Thus if a solution of extremal problem (5.20) – (5.22) exists, then it determines the optimal estimator over the class of nonlinear estimators. If the extremal problem (5.20) – (5.22) has no solution, then we can choose a minimizing sequence of linear estimators such that the estimation accuracy is arbitrarily close to $\mathcal{D}^0$.

5.4. *Optimality of Linear Estimators in Discrete Case*

Let the dynamic system be still continuous and let the measurements be discrete:

$$z(t_i) = H'(t_i)x(t_i) + \varrho(t_i), \qquad |\varrho_l(t_i)| \le \sigma_l(t_i), \qquad z(t_i), \varrho(t_i) \in \mathbf{R}^m,$$

$$i = 0, \ldots, N, \qquad l = 1, \ldots, m, \qquad 0 \le t_0 < \ldots < t_N \le T.$$

Similar to the case of continuous measurements, it can easily be checked that our problem is a special case of the general estimation problem; in particular, $\mathcal{S}\colon \mathbf{R}^{(N+1)m} \to \mathbf{R}^1$.

It is known (see, e.g., [109]) that the general form of the continuous linear functional on $\mathbf{R}^{(N+1)m}$ is determined by the formula

$$\mathcal{S}(z) = \mathcal{S}(z(t_0), \ldots, z(t_N)) = \sum_{i=0}^{N} \Phi'(t_i)z(t_i). \tag{5.29}$$

Therefore, in contrast to the case of continuous measurements, Theorem 5.1 provides the final result:

COROLLARY 5.2. *Let*

$$\sigma_l(t_i) > 0, \qquad i = 0, \ldots, N, \qquad l = 1, \ldots, m.$$

Then there exists a continuous linear functional $\mathcal{S}^0$ of the form (5.29) *such that*

$$\sup_{x(0)\in\Pi,\, u\in U,\, \varrho\in\sigma} |\mathcal{S}^0(z(t_0),\ldots,z(t_N)) - l_*|$$

$$= \inf_{\mathcal{S}} \sup_{x(0)\in\Pi,\, u\in U,\, \varrho\in\sigma} |\mathcal{S}(z(t_0),\ldots,z(t_N)) - l_*|.$$

Thus in the discrete case the finding of the optimal estimator (among linear and nonlinear estimators) is determined by the extremal problem (3.40), (3.38), and (3.39). Moreover, Corollary 5.2 guarantees that the solution for this problem exists.

5.5. *Optimality of Linear Estimators in the "Worst Correlation" Problem*

Let us consider the "worst correlation" problem described in details in Chapter 1. Recall that in this problem the mesurements have the form

$$z_i = H_i' q + \varrho_i, \qquad i = 1,\ldots,N,$$

where z_i are scalar measurements, $q \in \mathbf{R}^n$ is an unknown deterministic vector parameter, $H_i \in \mathbf{R}^n$ are given vectors, ϱ_i are random measurement errors. The measurement errors are supposed to have zero means, bounded variances, and unknown covariances:

$$\mathbf{E}\varrho_i = 0, \quad \mathbf{E}\varrho_i^2 \le \sigma_i^2, \quad |\mathbf{E}\varrho_i\varrho_j| \le (\mathbf{E}\varrho_i^2)^{\frac{1}{2}}(\mathbf{E}\varrho_j^2)^{\frac{1}{2}}, \quad i,j = 1,\ldots,N. \tag{5.30}$$

This set of $\varrho = (\varrho_1,\ldots,\varrho_N)'$ satisfying (5.30) was denoted by $\mathcal{P}_S$

The problem is to estimate a parameter, where $a \in \mathbf{R}^n$ is a given vector.

It was shown in Chapter 1 that if we use linear estimators

$$\hat{l} = \sum_{i=1}^{N} \Phi_i z_i, \tag{5.31}$$

then the estimation problem reduces to the following extremal problem

$$\sum_{i=1}^{N} \sigma_i |\Phi_i| \to \min_{\Phi} \tag{5.32}$$

subject to the unbiasness condition

$$\sum_{i=1}^{N} H_i \Phi_i = a. \tag{5.33}$$

Now, instead of linear estimates (5.31), let us consider all Borel functions $\hat{l} = \mathcal{S}(z_1, \dots, z_N)$. Our problem is to find an estimator $\mathcal{S}^0$ such that for any admissible $\mathcal{S}$

$$\sup_{q \in \mathbf{R}^n,\, \varrho \in \mathcal{P}_S} \mathbf{E}\big|\mathcal{S}^0(z_1, \dots, z_N) - l_*\big|^2 \le \sup_{q \in \mathbf{R}^n,\, \varrho \in \mathcal{P}_S} \mathbf{E}\big|\mathcal{S}(z_1, \dots, z_N) - l_*\big|^2.$$

By $s^0(z_1, \dots, z_n)$ denote the optimal linear estimator

$$s^0(z_1, \dots, z_n) = \sum_{i=1}^{N} \Phi_i^0 z_i, \tag{5.34}$$

where $\Phi^0 = (\Phi_1^0, \dots, \Phi_N^0)'$ is determined by the extremal problem (5.32), (5.33).

THEOREM 5.3. *The linear estimator s^0 determined by (5.34) is optimal on the class of all Borel functions (including nonlinear functions).*

PROOF. The guaranteed value for the second moment of estimation error for the estimator s^0 is determined by the formula

$$\sup_{q \in \mathbf{R}^n,\, \varrho \in \mathcal{P}_S} \mathbf{E}\big|s^0(z_1, \dots, z_N) - l_*\big|^2 = I_0^2,$$

where

$$I_0 = \sum_{i=1}^{N} \sigma_i |\Phi_i^0|.$$

By virtue of Corollary 3.4 there exists a vector β such that

$$\sum_{i=1}^{N} H_i'\beta\, \Phi_i^0 = \sum_{i=1}^{N} \sigma_i |\Phi_i^0| = I_0 \quad \text{and} \quad |H_i'\beta| \le \sigma_i, \quad i = 1, \dots, N. \tag{5.35}$$

Consider a stochastic process $\{\varrho_i^0\}_1^N$ that has two equiprobable samples (with probability 1/2)

$$\varrho_{1i}^0 = H_i'\beta \qquad \text{and} \qquad \varrho_{2i}^0 = -\, H_i'\beta, \qquad\qquad i = 1, \dots, N.$$

Obviously, $\varrho^0 = (\varrho_1^0, \dots, \varrho_N^0)' \in \mathcal{P}_S$.

Let us compute the value of the second moment of estimation error for any admissible estimator $\mathcal{S}$ under $\varrho = \varrho^0$. For brevity we shall write $\mathcal{S}([z_i])$ instead of $\mathcal{S}(z_1, \dots, z_N)$. So, we have

$$\mathcal{K}(q) \triangleq \mathbf{E}\big|\mathcal{S}([z_i]) - l_*\big|^2 \tag{5.36}$$

$$= \frac{1}{2}\big|\mathcal{S}([H_i'q + \varrho_{1i}^0]) - a'q\big|^2 + \frac{1}{2}\big|\mathcal{S}([H_i'q + \varrho_{2i}^0]) - a'q\big|^2.$$

Let us introduce the difference of the estimators $\chi = \mathcal{S} - s^0$. It follows from (5.36) that

$$\mathcal{K}(q) = \frac{1}{2}\left|\sum_{i=1}^{N} H_i'(q+\beta)\Phi_i^0 + \chi([H_i'(q+\beta)]) - a'q\right|^2$$

$$+ \frac{1}{2}\left|\sum_{i=1}^{N} H_i'(q-\beta)\Phi_i^0 + \chi([H_i'(q-\beta)]) - a'q\right|^2.$$

By virtue of (5.33) and (5.35), we get

$$\mathcal{K}(q) = \frac{1}{2}\left|I_0 + \chi([H_i'(q+\beta)])\right|^2 + \frac{1}{2}\left|I_0 - \chi([H_i'(q-\beta)])\right|^2. \tag{5.37}$$

Let us choose $q = c\beta$, $c \in \mathbf{R}^1$ and consider the quantity $\nu(c) = \chi([H_i' c\beta)])$. Then it follows from (5.37) that

$$\mathcal{K}(c\beta) = \frac{1}{2}\left|I_0 + \nu(c+1)\right|^2 + \frac{1}{2}\left|I_0 - \nu(c-1)\right|^2 \tag{5.38}$$

$$= I_0^2 + I_0\left(\nu(c+1) - \nu(c-1)\right) + \frac{1}{2}\left(\nu^2(c+1) + \nu^2(c-1)\right).$$

Assume that $\sup_{q\in\mathbf{R}^n} \mathcal{K}(q) < I_0^2$. Then

$$\sup_{c\in\mathbf{R}^1} \mathcal{K}(c\beta) < I_0^2. \tag{5.39}$$

LEMMA 5.2. *It (5.39) holds, then*

$$\sup_{c\in\mathbf{R}^1} |\nu(c)| \le \left(\sqrt{2} - 1\right) I_0, \quad \text{and} \quad \sup_{c\in\mathbf{R}^1} \left(\nu(c+1) - \nu(c-1)\right) < 0. \tag{5.40}$$

PROOF. It follows from (5.38) and (5.39) that

$$I_0\left(\nu(c+1) - \nu(c-1)\right) + \frac{1}{2}\left(\nu^2(c+1) + \nu^2(c-1)\right) < 0, \quad \forall c \in \mathbf{R}^1. \tag{5.41}$$

Since

$$-\frac{I_0^2}{2} \le \pm I_0\nu(c) + \frac{1}{2}\nu^2(c), \qquad \forall c \in \mathbf{R}^1,$$

we have that, for every c,

$$I_0\left(\nu(c+1) - \nu(c-1)\right) + \frac{1}{2}\left(\nu^2(c+1) + \nu^2(c-1)\right)$$

$$\geq -\frac{I_0^2}{2} - I_0\nu(c-1) + \frac{1}{2}\nu^2(c-1)$$

and

$$I_0\left(\nu(c+1) - \nu(c-1)\right) + \frac{1}{2}\left(\nu^2(c+1) + \nu^2(c-1)\right)$$

$$\geq -\frac{I_0^2}{2} + I_0\nu(c+1) + \frac{1}{2}\nu^2(c+1).$$

Consequently, inequality (5.41) implies the following inequalities:

$$-\frac{I_0^2}{2} \pm I_0\nu(c) + \frac{1}{2}\nu^2(c) < 0, \qquad \forall c \in \mathbf{R}^1.$$

Then $|\nu(c)| < \left(\sqrt{2}-1\right) I_0$ for all c. The second inequality of (5.40) directly follows from (5.41). Lemma 5.2 is proved.

Conditions (5.40) are inconsistent. In fact, the second inequality of (5.40) implies that

$$\nu(c+1) - \nu(c-1) < -c_0, \qquad c_0 = \text{const} > 0, \qquad \forall c \in \mathbf{R}^1,$$

whence $\inf_{c\in\mathbf{R}^1} \nu(c) = -\infty$. This relation contradicts the first inequality of (5.40). Thus $\sup_{q\in\mathbf{R}^n} \mathcal{K}(q) \geq I_0^2$ and

$$\sup_{q\in\mathbf{R}^n,\, \varrho\in\mathcal{P}_S} \mathbf{E}\left|\mathcal{S}(z_1,\dots,z_N) - l_*\right|^2 \geq \sup_{q\in\mathbf{R}^n} \mathcal{K}(q) \geq I_0^2, \qquad \forall \mathcal{S}.$$

□

6. Phase Constraints in Guaranteed Estimation Problem

Certain estimation problems contain linear phase constraints. In this Section we show that they are equivalent to additional measurements.

Consider the dynamic system (2.1), (2.2) and the measurements (2.3), (2.4). The problem is to evaluate the quantity $l_* = a'x(T)$. Suppose the phase coordinates are subjected to the additional inequalities

$$|G_s'(t)x(t)| \leq M_s(t), \qquad t \in [0,T], \qquad s = 1,\dots,k, \tag{6.1}$$

where $G_s(t) \in \mathbf{R}^n$ and $M_s(t) \in \mathbf{R}^1$ are given piecewise continuous functions. For the sake of simplicity, assume that there is no information about initial state:

$$-\infty < x_d(0) < \infty, \qquad d = 1,\dots,n.$$

Note that estimation problems with additional constraints on the derivatives of perturbations reduce to the problem (2.1), (2.3), (6.1). In fact, suppose that a certain component of the system perturbations, say $u_1(t)$, is an absolutely continuous function with a piecewise continuous derivative and the following additional restriction holds:

$$\left|\frac{du_1(t)}{dt}\right| \le L(t),$$

where $L(t)$ is a given scalar function. Introducing the auxiliary state vector $\bar{x}(t) = (u_1(t), x'(t))' \in \mathbf{R}^{n+1}$, we obtain the augmented system

$$\frac{d\bar{x}(t)}{dt} = \bar{A}(t)\bar{x}(t) + \bar{B}(t)\bar{u}(t),$$

where

$$\bar{A}(t) = \begin{pmatrix} 0 & 0_n' \\ B_1(t) & A(t) \end{pmatrix}, \qquad \bar{B}(t) = \begin{pmatrix} 1 & 0 & \dots & 0 \\ 0_n & B_2(t) & \dots & B_r(t) \end{pmatrix},$$

$B_j(t)$ is the j-th column of $B(t)$,

$$\bar{u}(t) = (v(t), u_2(t), \dots, u_r(t))' \in \mathbf{R}^r,$$

$$|v(t)| \le L(t), \qquad |u_j(t)| \le \gamma_j(t), \quad j = 2, \dots, r,$$

$$|G'(t)\bar{x}(t)| \le \gamma_1(t), \qquad G(t) = (1, 0_n')', \qquad 0_n \in \mathbf{R}^n \text{ is zero vector.}$$

The main result is as follows. Consider the estimation problem for the system (2.1), (2.2), without phase constraints (6.1) but with extended measurement vector:

$$z(t) = H'(t)x(t) + \varrho(t), \qquad |\varrho_l(t)| \le \sigma_l(t), \qquad l = 1, \dots, m, \tag{6.2}$$

$$\kappa(t) = G'(t)x(t) + \zeta(t), \qquad |\zeta_s(t)| \le M_s(t), \qquad s = 1, \dots, k,$$

$$G(t) = (G_1(t), \dots, G_k(t))', \qquad \zeta(t) = (\zeta_1(t), \dots, \zeta_k(t))'.$$

The estimator for this auxiliary estimation problem has the form

$$\hat{l}_e = \int_0^T \Phi'(t)z(t)\,dt + \int_0^T \varphi'(t)\kappa(t)\,dt, \qquad \Phi \in \mathcal{F}^m, \quad \varphi \in \mathcal{F}^k. \tag{6.3}$$

THEOREM 6.1. *Let* (Φ^0, φ^0) *be the solution of the problem* (2.1), (2.2), (6.2), (6.3). *Then the estimator* Φ^0 *is the solution of the guaranteed estimation problem with phase constraints* (2.1) – (2.4), (6.1).

PROOF. Let us construct the variational problem for the estimation problem with phase constraints. For brevity, we assume $m = 1$, $r = 1$, and $k = 1$. The general case can be studied similarly.

As before, introduce the vectors

$$y(t) = X^{-1}(t)x(t), \qquad b(t) = X^{-1}(t)B(t),$$

$$h(t) = X'(t)H(t), \qquad g(t) = X'(t)G(t), \qquad \tilde{a} = X'(T)a.$$

Then the equations for the system, measurements, the estimated parameter, and constraints can be rewritten in the form

$$\dot{y}(t) = b(t)u(t), \qquad |u(t)| \le \gamma(t),$$

$$z(t) = h'(t)y(t) + \varrho(t), \qquad |\varrho(t)| \le \sigma(t),$$

$$\left|g'(t)\Big(y(T) - \int_t^T b(s)u(s)\,ds\Big)\right| \le M(t), \qquad l_* = \tilde{a}'y(T),$$

$$y(t), b(t), h(t), g(t) \in \mathbf{R}^n, \qquad u(t), \gamma(t), z(t), \varrho(t), \sigma(t), M(t) \in \mathbf{R}^1.$$

We denote the phase constraint by the inclusion $(y(T), u) \in M$.

The estimation error is determined by the expression

$$\hat{l} - l_* = \Big(\int_0^T h(t)\Phi(t)\,dt - \tilde{a}\Big)' y(T) + \int_0^T \Phi(t)\varrho(t)\,dt \tag{6.4}$$

$$- \int_0^T \Phi(t)h'(t)\int_t^T b(s)u(s)\,ds\,dt.$$

Hence the guaranteed estimation error under estimator Φ is defined by the formula

$$I(\Phi) = \sup_{u\in U,\,\varrho\in\sigma,\,(y(T),u)\in M} |\hat{l} - l_*| = \sup_{u\in U,\,\varrho\in\sigma,\,(y(T),u)\in M} (\hat{l} - l_*). \tag{6.5}$$

The last equality holds because the set $\{u \in U,\ \varrho \in \sigma,\ (y(T), u) \in M\}$ is balanced.

In contrast to the case without phase consraint, we fail to obtain the explicit expression for $I(\Phi)$. But using the duality theory, we can reduce the computation of $I(\Phi)$ to a minimization problem.

LEMMA 6.1. *The following equality holds:*

$$I(\Phi) = \inf_{\varphi\in\mathcal{F}^1} \mathcal{O}(\Phi, \varphi),$$

where

$$\mathcal{O}(\Phi,\varphi) = \int_0^T M(t)|\varphi(t)|\,dt + \int_0^T \sigma(t)|\Phi(t)|\,dt \tag{6.6}$$

$$+ \int_0^T \gamma(t)\left|b'(t)\int_0^t \big(h(s)\Phi(s) + g(s)\varphi(s)\big)\,ds\right|dt,$$

under constraint

$$\int_0^T \big(h(t)\Phi(t) + g(t)\varphi(t)\big)\,dt = \tilde{a}. \tag{6.7}$$

PROOF OF LEMMA 6.1. Let us represent the extremal problem

$$\mathcal{O}(\Phi) = \inf_{\varphi\in\mathcal{F}^1} \mathcal{O}(\Phi,\varphi) \quad \text{under constraint (6.7)}$$

in the form

$$\mathcal{O}(\Phi) = \inf_{\varphi\in\mathcal{F}^1,\, w\in L_1} \mathcal{O}(\Phi,\varphi,w), \tag{6.8}$$

$$\mathcal{O}(\Phi,\varphi,w) = \int_0^T M(t)|\varphi(t)|\,dt + \int_0^T \sigma(t)|\Phi(t)|\,dt + \int_0^T \gamma(t)|w(t)|\,dt,$$

under the conditions

$$w(t) - b'(t)\int_0^t g(s)\varphi(s)\,ds = \Gamma(t), \qquad t \in [0,T], \tag{6.9}$$

$$\int_0^T g(t)\varphi(t)\,dt = c, \tag{6.10}$$

where

$$\Gamma(t) = b'(t)\int_0^t h(s)\Phi(s)\,ds, \qquad c = \tilde{a} - \int_0^T h(s)\Phi(s)\,ds.$$

Note that the problem (6.8) – (6.10) has the type of variational problem $\mathcal{P}_1$ of Section 1.

Assume the dimension of the subspace spanned on $\{g(t)\}$, $t \in [0,T]$, is equal to $d < n$. Then there exists a nonsingular $n \times n$ marix W such that

$$Wg(t) = \begin{pmatrix} \tilde{g}(t) \\ 0_{n-d} \end{pmatrix}, \qquad \tilde{g}(t) \in \mathbf{R}^d, \qquad 0_{n-d} \in \mathbf{R}^{n-d}. \tag{6.11}$$

Multiplying the equality (6.10) by W, we obtain the equivalent system of equalities

$$\begin{aligned} \int_0^T \tilde{g}(t)\varphi(t)\,dt &= \tilde{c}_1, \qquad \tilde{c}_1 \in \mathbf{R}^d, \\ 0_{n-d} &= \tilde{c}_2, \qquad \tilde{c}_2 \in \mathbf{R}^{n-d}, \qquad Wc = (\tilde{c}_1', \tilde{c}_2')'. \end{aligned} \tag{6.12}$$

Two cases are possible: (a) $\tilde{c}_2 \neq 0$ and (b) $\tilde{c}_2 = 0$.

In case (a), conditions (6.9), (6.10) are inconsistent and $\mathcal{O}(\Phi) = \infty$. Let us show that $I(\Phi) = \infty$. Really, let us put

$$u(s) = 0, \quad \varrho(t) = 0, \quad y(T) = -\alpha W' \begin{pmatrix} 0_d \\ \tilde{c}_2 \end{pmatrix}, \quad \alpha > 0.$$

Then, by virtue of (6.11), the phase constraints are satisfied: $|0| \leq M(t)$ and

$$\hat{l} - l_* = \alpha ||\tilde{c}_2||^2_{\mathbf{R}^{n-d}}.$$

So, we have $\hat{l} - l_* \to \infty$ as $\alpha \to \infty$ and $I(\Phi) = \infty$.

In case (b), conditions (6.9), (6.10) are resolvable with respect to (φ, w) for arbitrary $\Gamma(t)$ and $\tilde{c}_1$. Then, using Theorem 1.7[2], we obtain from the duality theory that

$$\mathcal{O}(\Phi) = \sup_{\tilde{\lambda}_1 \in \mathbf{R}^d,\, p \in L_\infty} \ \inf_{\varphi \in \mathcal{F}^1,\, w \in L_1} \mathcal{L}(\varphi, w, p, \tilde{\lambda}_1), \tag{6.13}$$

where L_∞ is the space of essentially bounded functions on $[0, T]$, L_1 is the space of absolutely integrable functions, and the Lagrange function has the form

$$\mathcal{L}(\varphi, w, p, \tilde{\lambda}_1) = \mathcal{O}(\Phi, \varphi, w) + \int_0^T p(t)\left[\Gamma(t) - w(t) + b'(t)\int_0^t g(s)\varphi(s)\,ds\right]dt$$

$$+ \tilde{\lambda}_1'\left[-\tilde{c}_1 + \int_0^T \tilde{g}(t)\varphi(t)\,dt\right].$$

Taking into account Lemma 2.1 and (6.11), this expression can be transformed to the formula

$$\mathcal{L}(\varphi, w, p, \tilde{\lambda}_1) = \int_0^T \sigma(t)|\Phi(t)|\,dt \tag{6.14}$$

$$+ \int_0^T \left\{ M(t)|\varphi(t)| + g'(t)\left[\int_t^T b(s)p(s)\,ds + W'\begin{pmatrix} \tilde{\lambda}_1 \\ 0_{n-d} \end{pmatrix}\right]\varphi(t) \right\} dt$$

[2] the details are similar to these of the proof of Theorem 5.2

$$+\int_0^T h'(t)\Phi(t)\left(\int_t^T b(s)p(s)\,ds\right)dt+\int_0^T\Big(\gamma(t)|w(t)|-p(t)w(t)\Big)\,dt-\tilde{\lambda}_1'\tilde{c}_1.$$

Then the explicit computation of the infimum in (6.13) (or in (6.14)) results in the relation

$$\inf_{\varphi,w}\mathcal{L}=\int_0^T\sigma(t)|\Phi(t)|\,dt+\int_0^T h'(t)\Phi(t)\left(\int_t^T b(s)p(s)\,ds\right)dt \tag{6.15}$$

$$-c'W'\begin{pmatrix}\tilde{\lambda}_1\\ 0_{n-d}\end{pmatrix}$$

under constraints

$$|p(t)|\le\gamma(t),\qquad t\in[0,T], \tag{6.16}$$

$$\left|g'(t)\left[W'\begin{pmatrix}\tilde{\lambda}_1\\ 0_{n-d}\end{pmatrix}+\int_t^T b(s)p(s)\,ds\right]\right|\le M(t),\quad t\in[0,T]$$

and

$$\inf_{\varphi,w}\mathcal{L}=-\infty, \tag{6.17}$$

if constraints (6.16) are violated.

According to (6.11), (6.12), for any $\tilde{\lambda}_2\in\mathbf{R}^{n-d}$

$$g'(t)W'\begin{pmatrix}\tilde{\lambda}_1\\ 0_{n-d}\end{pmatrix}=g'(t)W'\begin{pmatrix}\tilde{\lambda}_1\\ \tilde{\lambda}_2\end{pmatrix},\quad c'W'\begin{pmatrix}\tilde{\lambda}_1\\ 0_{n-d}\end{pmatrix}=c'W'\begin{pmatrix}\tilde{\lambda}_1\\ \tilde{\lambda}_2\end{pmatrix}.$$

Then we get from (6.13), (6.15) – (6.17) that

$$\mathcal{O}(\Phi)=\sup_{p\in L_\infty,\,\mu\in\mathbf{R}^n}\left\{\int_0^T\sigma(t)|\Phi(t)|\,dt\right. \tag{6.18}$$

$$\left.+\int_0^T h'(t)\Phi(t)\left(\int_t^T b(s)p(s)\,ds\right)dt-c'\mu\right\},\qquad \mu=W'\begin{pmatrix}\tilde{\lambda}_1\\ \tilde{\lambda}_2\end{pmatrix}$$

under constraints

$$|p(t)|\le\gamma(t),\qquad t\in[0,T], \tag{6.19}$$

$$\left|g'(t)\left(\mu+\int_t^T b(s)p(s)\,ds\right)\right|\le M(t),\qquad t\in[0,T].$$

Since

$$\int_0^T h(t)\Phi(t)\,dt-\tilde{a}=-c\quad\text{and}\quad\sup_{\varrho\in\sigma}\int_0^T\varrho(t)\Phi(t)\,dt=\int_0^T\sigma(t)|\Phi(t)|\,dt,$$

the maximization problem (6.4), (6.5) coincides with the maximization problem (6.18), (6.19) if we put

$$y(T) = \mu \quad \text{and} \quad u(s) = -p(s).$$

So, for $d < n$ Lemma 6.1 is proved.

If $d = n$, then in order to prove the lemma we should repeat the previous arguments without decomposition (6.11).

Lemma 6.1 is proved completely.

In accordance with Lemma 6.1 the variational prolem that corresponds to (2.1) – (2.4), and (6.1) takes the form

$$\inf_{\Phi} I(\Phi) = \inf_{\Phi} \left\{ \inf_{\varphi \in (6.7)} \mathcal{O}(\Phi, \varphi) \right\} = \inf_{\Phi, \varphi \in (6.7)} \mathcal{O}(\Phi, \varphi), \quad \Phi, \varphi \in \mathcal{F}^1. \tag{6.20}$$

It is easy to show by direct computation (just as in Section 2) that $\mathcal{O}(\Phi, \varphi)$ under (6.7) coincides with the guaranteed value of the estimation error for the auxiliary problem (2.1), (2.3), (6.2), (6.3). Thus the solving of the problem (6.20), (6.6), and (6.7) yields the optimal estimator (Φ^0, φ^0) for the auxiliary problem as well as the optimal estimator Φ^0 for the original problem. □

It follows from (6.20) that if $(\Phi_\varepsilon, \varphi_\varepsilon)$ is an ε-optimal estimator for the auxiliary problem, then Φ_ε is an ε-optimal estimator for the problem with phase constraints.

REMARK 6.1. It can easily be checked (similar to Section 5) that the use of nonlinear estimators in the guaranteed estimation problem with phase constraints will not improve the estimation accuracy.

EXAMPLE 6.1. Consider the measurements of the form

$$z(t) = H'(t)q + \varrho(t), \qquad t \in [0, T], \tag{6.21}$$

where $z(t), \varrho(t) \in \mathbf{R}^1$, $H(t), q \in \mathbf{R}^n$, $H(t)$ is a continuously differentiable vector function.

Suppose the measurement noise is a bounded absolutely continuous function with a bounded piecewise continuous derivative:

$$|\varrho(t)| \le 1, \qquad |\dot{\varrho}(t)| \le \nu, \qquad t \in [0, T], \tag{6.22}$$

where ν is a given constant. The problem is to evaluate the quantity $l_* = a'q$, $a \in \mathbf{R}^n$.

In accordance with Theorem 6.1, this problem is equivalent to the following estimation problem without phase constraints

$$\begin{cases} \dot{\varrho}(t) = \nu(t), & |\nu(t)| \le \nu, \\ \dot{q}(t) = 0, & \end{cases}$$

$$\begin{cases} z(t) = \varrho(t) + H'(t)q(t) + \tilde{\varrho}(t), & |\tilde{\varrho}(t)| \le 0, \\ \kappa(t) = \varrho(t) + \zeta(t), & |\zeta(t)| \le 1, \end{cases}$$

where

$$l_* = (0, a')(\varrho(T), q(T))' = a'q.$$

The corresponding variational problem has the form

$$I_0 = \inf_{\Phi,\varphi\in\mathcal{F}^1} \left\{ \int_0^T |\varphi(t)|\, dt + \nu \int_0^T \left| \int_0^t (\Phi(s) + \varphi(s))\, ds \right| dt \right\} \tag{6.23}$$

under constrtaints

$$\int_0^T (\Phi(t) + \varphi(t))\, dt = 0, \qquad \int_0^T H(t)\Phi(t)\, dt = a. \tag{6.24}$$

Let us reduce the problem (6.23), (6.24) to a simpler form. First note that, without loss of generality, the infimum in (6.23) can be taken over piecewise continuous functions from $\mathcal{C}^1$. Introduce the absolutely continuous function

$$\Psi(t) = -\int_0^t (\Phi(s) + \varphi(s))\, ds. \tag{6.25}$$

It follows from (6.24), (6.25) that (for piecewise continuous functions)

$$\Psi(0) = \Psi(T) = 0, \qquad \Phi(t) = -\varphi(t) - \dot{\Psi}(t). \tag{6.26}$$

Using the substitution (6.25) and taking into account (6.26), we get that the problem (6.23), (6.24) is equivalent to the problem

$$I_0 = \inf_{\varphi,\Psi} \left\{ \int_0^T |\varphi(t)|\, dt + \nu \int_0^T |\Psi(t)|\, dt \right\} \tag{6.27}$$

under constraint

$$\int_0^T \dot{H}(t)\Psi(t)\, dt - \int_0^T H(t)\varphi(t)\, dt = a, \tag{6.28}$$

$$\varphi \in \mathcal{C}^1, \quad \Psi(T) = \Psi(0) = 0, \quad \Psi \text{ is absolutely continuous and has a piecewise continuous derivative.} \tag{6.29}$$

Moreover, the estimator takes the form

$$\hat{l} = \int_0^T \Psi(t)\dot{z}(t)\, dt - \int_0^T \varphi(t)z(t)\, dt. \tag{6.30}$$

Since $\Psi(t)$ in (6.27) and (6.28) is contained in integrals only and $\dot{H}(t)$ is continuous, the condition (6.29) may be relaxed:

$$\varphi,\ \Psi \in \mathcal{C}^1. \tag{6.31}$$

Thus, having an ε-optimal solution of the moment problem (6.27), (6.28), (6.31), we can construct a 2ε-optimal solution of (6.27) – (6.29)[3]. So, an ε-optimal solution of (6.27), (6.28), (6.31) determines a 2ε-optimal estimator for the initial problem (6.23), (6.24).

Note that the solution of the moment problem (6.27), (6.28) is not generally attained on the set (6.31); it is attained on impulse functions from $\mathcal{F}^1$. Therefore the solution of the initial problem (6.23), (6.24) may not exist (ε-optimal solution exists). If we admit the exact differentiation, then it follows from (6.30) that the optimal estimate exists and is determined by (6.27), (6.28).

The problem (6.27), (6.28) formally corresponds to the estimation problem for the measurments

$$\begin{cases} z(t) = H'(t)q + \alpha(t), & |\alpha(t)| \le 1, \\ \dot{z}(t) = \dot{H}'(t)q + \beta(t), & |\beta(t)| \le \nu. \end{cases}$$

Nevertheless, the noises $\alpha(t) = \varrho(t)$ and $\beta(t) = \dot{\varrho}(t)$ are related. Therefore the equivalence of the initial problem to the moment problem (6.27), (6.28) is not so trivial (the similar problem with discrete measurements is discussed in Chapter 1).

7. Comments to Chapter 2

7.1. *Section 1*

This auxiliary part of our book is devoted to some basic results of functional analysis, convex analysis, and optimization theory. We directly cited the corresponding sections and theorems from the classical monographs [75], [51], [44], and from a brilliant overview [203]. For in-depth studying these fields we could also recommend the well-known books [185], [5], [182], [183]. An interesting geometrical illustration of the duality is given in [128].

7.2. *Section 2*

Guaranteed estimation problem is posed for dynamic systems. The estimation problem reduces to a nonsmooth variational problem. We follow the paper

[3]To guarantee this feature, it is sufficient to suppose that the usual condition $\int_0^T H(t)H'(t)\,dt > 0$ holds.

[133]. A new detail is the presence of the additional information about initial state.

7.3. *Section 3*

Necessary and sufficient conditions for guaranteed estimation problem were obtained by M.L. Lidov [133] (see also [123] for more general case). M.L. Lidov used the theory of differential games [77]. We follow [156], where the general Lagrange principle [75], [51], [5] was applied to a nonsmooth variational problem with extended constraints and, thus, the necessary conditions for optimality were obtained more rigorously than in [133].

7.4. *Section 4*

In this section certain examples are studied in detail. They show that the necessary and sufficient conditions for optimality can be quite constructive. Examples 4.1 and 4.2 were investigated by M.L. Lidov in [133].

7.5. *Section 5*

The problem for the optimality of linear estimators has a long history. Our investigation is based on a result by S.A. Smolyak [192]. This result was further developed in [23], [149], [168]. We generalize the theorem by A.G. Marchuk and K.Yu. Osipenko [149] to the case of continuous measurements [153], [154]. Some results on the optimality of linear estimators are contained in the classical monograph [123]. We also discuss the optimality of linear estimators in the stochastic "worst correlation problem" [155]. An interesting minimax feature of nonlinear stochastic estimation problems is presented in a dramatic note by V.D. Vandelinde [207]. The optimality of linear estimators in general stochastic problems were investigated recently by V.N. Soloviov in [197], [199], [200]; these works are essentially based on the fine results of the conjugate duality theory.

7.6. *Section 6*

B.Ts. Bakhshiyan called the author's attention to the guaranteed estimation problems with restrictions to noise derivative. Here we presented a slightly modified version of [162].

8. Exercises to Chapter 2

1. Obtain the variational problem that corresponds to the guaranteed estimation problem with $l_* = a'x(t^0)$, $0 \le t^0 < T$.

2. Prove that if in the problem (4.8) – (4.11) $H = H(t) = (1, t^2 - t)'$, i.e., $h(t) = (1, t^2)'$, then the optimal estimator may contain three impulses.
3. Suppose $\{(\Phi_-^0, \Phi^0), u^0\}$ with $(\Phi_-^0, \Phi^0) \in \mathcal{N}$, $u^0 \in U^e$ is a saddle point for the game problem (3.25). Prove that (Φ_-^0, Φ^0) is the solution for (2.13) and u^0 satisfies the conditions (3.2) – (3.6) from Theorem 3.1.
4. Obtain the necessary and sufficient conditions for optimality in the guaranteed estimation problem with an interval $(t_\alpha, t_\beta) \subset [0, T]$ that is "forbidden" for measurements.
5. Formulate the necessary and sufficient conditions for optimality in the guaranteed estimation problem with $l_* = a'x(t^0)$, $0 \le t^0 < T$.
6. Suppose the observable system has the form
$$\dot{x}(t) = u(t), \qquad x(t), u(t) \in \mathbf{R}^1,$$
$$z(t) = x(t) + \varrho(t), \qquad z(t), \varrho(t) \in \mathbf{R}^1,$$
where $t \in [0, T]$, $|u(t)| \le \gamma$, $|\varrho(t)| \le \sigma$, and $l_* = x(T)$. Find the optimal guaranteed estimator.
7. Consider the problem of Example 4.1. Is there exist an optimal estimator with a piecewise continuous component only?
8. We say that the matrix $b(t) = X^{-1}(t)B(t)$ is analytical if there exists a domain $\mathcal{B}$ in the plane of complex variable w such that (a) $[0, T] \subset \mathcal{B}$; (b) the functions $b_{ij}(w)$, $i = 1, \ldots, n$, $j = 1, \ldots, r$, are analytical in $\mathcal{B}$. Prove the following lemma.
LEMMA. Let the matrix $b(t)$ be analytical and $\sigma_l(t_i) > 0$, $i = 0, \ldots, N$, $l = 1, \ldots, m$. Then the analog of Corollary 3.1 for the discrete case is valid with $u^0 \in U \subset U^e$, i.e., with piecewise continuous u^0.
9. Consider the guaranteed estimation problem with phase constraints on the class of nonlinear estimators. Prove the assertion of Remark 6.1, i.e., prove that the nonlinear estimators do not improve the estimation accuracy as compared to linear ones.
10. Consider the measurements of the form (6.21), (6.22) with $n = 2$, $H(t) = (t, 1)'$, $t \in [0, T]$. Solve the estimation problem with the bounded amplitude and the derivative of the measurement noise for $a = (1, 0)'$ and $a = (0, 1)'$.
11. Consider the guaranteed estimation problem that is described in Exercise 6 with the additional condition $|\dot{u}(t)| \le \nu$, $t \in [0, T]$. Find the optimal estimator.

Chapter 3

Kalman Filter in Guaranteed Estimation Problem

In applied problems there exist at least two reasons why we use nonoptimal estimators. The first reason is that the finding of optimal estimator can result in serious numerical problems. The second reason is that an algorithm that is optimal for one model of noises sometimes must operate under another assumptions about noises. Moreover, the notion of the optimal estimator for actual system is conventional because the mathematical models are not quite accurate. Therefore the design decision is often based on the orders of quantities that describe the system performance but not on the exact characteristics of inaccurate models.

Thus in many cases, instead of searching for an optimal estimator, it is more preferable to use the existing (simpler) estimation algorithms. In such an approach, however, we must evaluate *without finding the optimal algorithm* the increment in estimation error due to the rejection of optimal filter. In this chapter, we shall obtain such an estimate for the deterministic systems that are described by linear ordinary differential equations.

Since we prefer to make the presentation of different chapters more self-contained, we have to repeat some basic constructions.

1. Level of Nonoptimality for Kalman Filter

In this section we recall the statement of the optimal guaranteed estimation problem that was formulated in Section 2 of Chapter 2 in detail. Then we shall define the level of nonoptimality of simplified estimation algorithms.

Consider a linear dynamic system

$$\dot{x}(t) = A(t)x(t) + B(t)u(t), \qquad t \in [0,T], \quad x(0) = x_0, \tag{1.1}$$

where $x(t) = (x_1(t),\ldots,x_n(t))' \in \mathbf{R}^n$ is the state vector, $A(t) \in \mathbf{R}^{n\times n}$, $B(t) \in \mathbf{R}^{n\times r}$ are given matrices, and $u(t) = (u_1(t),\ldots,u_r(t))' \in \mathbf{R}^r$ is a perturbation vector.

We assume that

$$|u_j(t)| \leq \gamma_j(t), \qquad t \in [0,T], \quad j = 1, \dots, r, \tag{1.2}$$

where $\gamma_j(t)$ are given nonnegative functions.

Let measurements of the state vector be available on the segment $[0,T]$:

$$z(t) = H'(t)x(t) + \varrho(t), \qquad z(t), \varrho(t) \in \mathbf{R}^m, \tag{1.3}$$

where $z(t)$ are the measurements, $H(t)$ is a known $n \times m$ matrix, $\varrho(t)$ is a measurement noise. We assume that $\varrho(t) = (\varrho_1(t), \dots, \varrho_m(t))'$ is an unknown vector function with bounded components

$$|\varrho_l(t)| \leq \sigma_l(t), \qquad l = 1, \dots, m, \tag{1.4}$$

where $\sigma_l(t)$, $l = 1, \dots, m$ are known nonnegative functions.

Moreover, the initial state of the system is assumed to belong to the parallelepiped

$$|x_d(0)| \leq \sigma_{-d}, \qquad d = 1, \dots, n, \tag{1.5}$$

where σ_{-d}, $d = 1, \dots, n$, are known nonnegative numbers.

We suppose that the entries of the matrices $A(t)$, $B(t)$, $H(t)$, and the functions $u_j(t)$, $\gamma_j(t)$, $\varrho_l(t)$, $\sigma_l(t)$ are piecewise continuous. Moreover, in order to simplify the resulting formulas we also suppose that these functions at the discontinuity points are continuous from the right, possibly, except for the point T at which the functions are supposed to be continuous from the left. So, the functions have no isolated jumps. In addition, for technical reasons we assume that the uncertainty bounds are restricted from below:

$$\gamma_j(t) \geq \gamma^0 > 0, \quad j = 1, \dots, r; \qquad \sigma_l(t) \geq \sigma^0 > 0, \quad l = 1, \dots, m; \tag{1.6}$$

$$\sigma_{-d} > 0, \quad d = 1, \dots, n.$$

Note that the first restriction is not essential because instead of the column $B_j(t)u_j(t)$, $|u_j(t)| \leq \gamma_j(t)$, we can consider the column $\tilde{B}_j(t)\tilde{u}_j(t)$, $|\tilde{u}_j(t)| \leq 1$, where $\tilde{B}_j(t) = B_j(t)\gamma_j(t)$.

As in Chapter 2 we use the notation $x(0) \in \Pi$, $u \in U$, and $\varrho \in \sigma$ for the sets (1.5), (1.2), and (1.4), respectively.

The problem is to evaluate a scalar quantity $l_* = a'x(T)$ (where $a \in \mathbf{R}^n$ is a given vector) by linear functionals

$$\hat{l} = \int_0^T \Phi'(t)z(t)\,dt, \qquad \Phi \in \mathcal{F}^m, \tag{1.7}$$

i.e., $\Phi(t) = (\Phi_1(t), \dots, \Phi_m(t))'$ and $\Phi(t) = \Phi_{pc}(t) + \sum_{s=1}^M \Phi_{(s)}\delta(t - t_s)$, where $\Phi_{pc}(t) \in \mathbf{R}^m$ is a piecewise continuous vector function, $\Phi_{(s)} \in \mathbf{R}^m$, $t_s \in [0,T]$,

$s = 1, \ldots, M$, M is an arbitrary interger, and $\delta(t)$ is the Dirac delta-function. Obviously, we may consider $a \neq 0$; otherwise, the estimation problem is trivial.

The optimal *a priori* guaranteed estimation problem is to find an estimator Φ^0 that minimizes the guaranteed estimation error

$$d(\Phi^0) = I_0 \triangleq \inf_{\Phi \in \mathcal{F}^m} d(\Phi), \qquad d(\Phi) = \sup_{x(0)\in\Pi,\, u\in U,\, \varrho\in\sigma} |\hat{l} - l_*|.$$

To reduce the further computation we introduce the notation

$$y(t) = X^{-1}(t)x(t) \in \mathbf{R}^n, \qquad h(t) = X'(t)H(t) \in \mathbf{R}^{n\times m},$$

$$b(t) = X^{-1}(t)B(t) \in \mathbf{R}^{n\times r}, \qquad \tilde{a} = X'(T)a,$$

$$h(t) = (h_1(t), \ldots, h_m(t)), \qquad h_l(t) \in \mathbf{R}^n, \quad l = 1, \ldots, m,$$

$$b(t) = (b_1(t), \ldots, b_r(t)), \qquad b_j(t) \in \mathbf{R}^n, \quad j = 1, \ldots, r,$$

where $X(t) \in \mathbf{R}^{n\times n}$ is a fundamental matrix of the homogeneous equation

$$\dot{x}(t) = A(t)x(t), \qquad X(0) = E_n, \quad E_n \text{ is identity matrix.} \tag{1.8}$$

Then,

$$\dot{y}(t) = b(t)u(t), \quad y(t) \in \mathbf{R}^n, \quad y(0) = x(0), \qquad t \in [0, T], \tag{1.9}$$

and

$$z(t) = h'(t)y(t) + \varrho(t). \tag{1.10}$$

With this notation, $l_* = \tilde{a}'y(T)$ and it can be verified (see Section 2 of Chapter 2) that the guaranteed estimation error is determined by the formula

$$d(\Phi) = \sup_{y(0)\in\Pi,\, u\in U,\, \varrho\in\sigma} |\hat{l} - l_*| = \sup_{u\in U} \mathcal{I}(\Phi_-, \Phi, u), \tag{1.11}$$

where

$$\mathcal{I}(\Phi_-, \Phi, u) = \sum_{d=1}^{n} \sigma_{-d}|\Phi_{-d}| + \sum_{l=1}^{m} \int_0^T \sigma_l(t)|\Phi_l(t)|\, dt$$

$$- \int_0^T \Phi'(t)h'(t) \int_t^T b(\tau)u(\tau)\, d\tau dt - \Phi'_- \int_0^T b(\tau)u(\tau)\, d\tau,$$

and $\Phi_- = \Phi_-(\Phi)$ is defined by the analog of the unbiasness condition:

$$\Phi_- + \int_0^T h(t)\Phi(t)\, dt = \tilde{a}. \tag{1.12}$$

Computing the supremum in (1.11) with the help of the Dirichlet formula, we obtain that

$$d(\Phi) = I(\Phi_-, \Phi)\Big|_{(\Phi_-,\Phi)\in(1.12)}, \tag{1.13}$$

where

$$I(\Phi_-, \Phi) = \sum_{d=1}^{n} \sigma_{-d}|\Phi_{-d}| + \sum_{l=1}^{m} \int_0^T \sigma_l(t)|\Phi_l(t)|\,dt \tag{1.14}$$

$$+ \sum_{j=1}^{r} \int_0^T \gamma_j(t)\left| b_j'(t)\left(\Phi_- + \int_0^t h(\tau)\Phi(\tau)\,d\tau\right)\right| dt.$$

Thus the optimal *a priori* guaranteed estimation problem reduces to the solution of the following nonsmooth variational problem

$$I_0 = \inf_{\Phi_-\in\mathbf{R}^n,\, \Phi\in\mathcal{F}^m} I(\Phi_-, \Phi) \quad \text{under constraint (1.12).} \tag{1.15}$$

The solution of (1.15) (Φ_-^0, Φ^0) determines the optimal estimator Φ^0.

The solving of (1.15) is a difficult numerical problem [31], [133], [139], [19], [135], [214], [140], [136]. In order to resolve this problem approximately let us use the classical Kalman filter [79], [47].

Note that the term *Kalman filter* means a specific algorithm for measurement data processing regardless of the actual noise models.

The level of the nonoptimality of the Kalman filter in the guaranteed estimation prblem we define by the ratio

$$\Delta = \frac{\sup_{x(0)\in\Pi,\, u\in U,\, \varrho\in\sigma} |\hat{l}_\varphi - l_*|}{\sup_{x(0)\in\Pi,\, u\in U,\, \varrho\in\sigma} |\hat{l}_0 - l_*|} = \frac{\sup_{y(0)\in\Pi,\, u\in U,\, \varrho\in\sigma} |\hat{l}_\varphi - l_*|}{\sup_{y(0)\in\Pi,\, u\in U,\, \varrho\in\sigma} |\hat{l}_0 - l_*|}, \tag{1.16}$$

where

$$\hat{l}_0 = \int_0^T \Phi^{0\prime}(t)z(t)\,dt \quad \text{and} \quad \hat{l}_\varphi = \int_0^T \varphi'(t)z(t)\,dt$$

are the unknown optimal estimate and the Kalman filter estimate, respectively. Obviously, $\Delta \geq 1$.

Using the first equality from (1.11), we see that

$$1 \leq \Delta = \frac{d(\varphi)}{d(\Phi^0)} = \frac{d(\varphi)}{I_0}. \tag{1.17}$$

If the optimal estimator Φ^0 does not exist, then the second equality from (1.17) should be used for the definition of Δ.

The Kalman filter estimator φ can be easily computed and the numerator $d(\varphi)$ is a known quantity. Since the optimal estimator is unknown, the denominator $d(\Phi^0)$ is unknown and Δ is unknown.

Our idea is to obtain an upper bound Δ^0 for Δ that can be efficiently computed *without solving the original nonsmooth problem* (1.15). If the value of Δ^0 is not large, then the use of the Kalman filter will be justified. The results of Section 7 of Chapter 1 for simplest static systems allows us to hope that this approach will be successful for dynamic systems as well.

2. Bound for the Level of Nonoptimality

Let us present the main result of this chapter.

The Kalman filter for the system (1.9), (1.10) is described by the following relations [84], [41], [79], [47], [187]:

$$\hat{l}_\varphi = \tilde{a}'\hat{y}(T), \qquad \dot{\hat{y}}(t) = \tilde{K}(t)\big(z(t) - h'(t)\hat{y}(t)\big), \quad \hat{y}(0) = 0, \tag{2.1}$$

$$\tilde{K}(t) = \tilde{P}(t)h(t)R^{-1}(t),$$

$$\dot{\tilde{P}}(t) = -\tilde{P}(t)h(t)R^{-1}(t)h'(t)\tilde{P}(t) + b(t)Q(t)b'(t), \quad \tilde{P}(0) = P_0, \tag{2.2}$$

$$\hat{y}(t) \in \mathbf{R}^n, \quad \tilde{K}(t) \in \mathbf{R}^{n\times m}, \quad \tilde{P}(t) \in \mathbf{R}^{n\times n},$$

$$R(t) \in \mathbf{R}^{m\times m}, \quad Q(t) \in \mathbf{R}^{r\times r}, \qquad t \in [0, T].$$

The symmetric matrices $P_0 \geq 0$, $R(t) \geq R_0 > 0$, and $Q(t) \geq 0$ [1] are specified by a researcher; P_0 characterizes the uncertainty in the initial state, $R(t)$ and $Q(t)$ characterize the uncertainty in the measurements and in the system. Equation (2.2) is the *Riccati equation.*

Note again that for the guaranteed estimation problems discussed in this book the Kalman filter is only the algorithm to process the measurements.

The estimate (2.1) can be presented in the form (1.7):

$$\hat{l}_\varphi = \int_0^T \tilde{a}'\tilde{\theta}(T)\tilde{\theta}^{-1}(t)\tilde{K}(t)z(t)\,dt, \tag{2.3}$$

where $\tilde{\theta}(t) \in \mathbf{R}^{n\times n}$ satisfies the homogeneous equation

$$\dot{\tilde{\theta}}(t) = -\tilde{K}(t)h'(t)\tilde{\theta}(t), \qquad \tilde{\theta}(0) = E_n. \tag{2.4}$$

Thus, for the Kalman filter,

$$\Phi(t) = \varphi(t) = \tilde{K}'(t)\left(\tilde{\theta}^{-1}(t)\right)'\tilde{\theta}'(T)\tilde{a}. \tag{2.5}$$

[1] $C > 0$ ($C \geq 0$) means that $z'Cz > 0$ ($z'Cz \geq 0$), $\forall z$.

We call φ the *Kalman estimator*.

Let L_2^m be the Hilbert space of square integrable vector functions $\Phi(t) = (\Phi_1(t), \dots, \Phi_m(t))'$ with the inner product

$$\langle \Phi, \Psi \rangle_{L_2^m} \triangleq \int_0^T \Phi'(t)\Psi(t)\, dt, \qquad \Phi, \Psi \in L_2^m.$$

THEOREM 2.1. 1^0. *The Kalman estimator φ is determined by the solution (φ_-, φ) of the following variational problem:*

$$J_0 = \inf_{\Phi_- \in \mathbf{R}^n, \Phi \in L_2^m} J(\Phi_-, \Phi) \quad \textit{under constraint} \quad (1.12), \tag{2.6}$$

where

$$J(\Phi_-, \Phi) = \Bigg\{ \Phi_-' P_0 \Phi_- + \int_0^T \Phi'(t) R(t) \Phi(t)\, dt \tag{2.7}$$

$$+ \int_0^T \left(\Phi_- + \int_0^t h(\tau)\Phi(\tau)\, d\tau \right)' b(t)Q(t)b'(t) \left(\Phi_- + \int_0^t h(\tau)\Phi(\tau)\, d\tau \right) dt \Bigg\}^{\frac{1}{2}}.$$

2^0. *The solution of the problem* (2.6) (φ_-, φ) *is determined by the solution of the boundary value problem:*

$$\varphi_- = \tilde{\xi}(0), \qquad \varphi(t) = R^{-1}(t)h'(t)\tilde{\eta}(t) = \tilde{K}'(t)\tilde{\xi}(t), \tag{2.8}$$

where

$$\dot{\tilde{\xi}}(t) = h(t)R^{-1}(t)h'(t)\tilde{\eta}(t), \qquad \tilde{\xi}(T) = \tilde{a}, \tag{2.9}$$

$$\dot{\tilde{\eta}}(t) = b(t)Q(t)b'(t)\tilde{\xi}(t), \qquad \tilde{\eta}(0) = P_0\tilde{\xi}(0).$$

Moreover,

$$\tilde{\eta}(t) = \tilde{P}(t)\tilde{\xi}(t), \qquad t \in [0, T], \tag{2.10}$$

$$\dot{\tilde{\xi}}(t) = -\left(-\tilde{K}(t)h'(t)\right)'\tilde{\xi}(t), \qquad \tilde{\xi}(T) = \tilde{a}, \tag{2.11}$$

and

$$J_0 = \left(\tilde{a}'\tilde{P}(T)\tilde{a}\right)^{\frac{1}{2}}, \tag{2.12}$$

where $\tilde{K}(t)$ and $\tilde{P}(t)$ are defined by (2.1) *and* (2.2).

The proof of Theorem 2.1 is given in Section 3.

It follows from (2.8) and (2.9) that

$$\tilde{\xi}(t) = \int_0^t h(\tau)\varphi(\tau)\,d\tau + \varphi_- = \tilde{a} - \int_t^T h(\tau)\varphi(\tau)\,d\tau. \tag{2.13}$$

Let us put

$$P_0 = \mathrm{diag}(\sigma_{-1}^2, \ldots, \sigma_{-n}^2), \quad R(t) = \beta_1 \mathrm{diag}(\sigma_1^2(t), \ldots, \sigma_m^2(t)), \tag{2.14}$$

$$Q(t) = \beta_2 \mathrm{diag}(\gamma_1^2(t), \ldots, \gamma_r^2(t)),$$

where β_1, β_2 are specified positive constants: $\beta_1, \beta_2 \in \mathbf{R}^1$, $\beta_1, \beta_2 > 0$.

It follows from Theorem 2.1 that the use of the Kalman filter is equivalent to changing $I(\Phi_-, \Phi)$ to $J(\Phi_-, \Phi)$.

Note that the functionals $I(\Phi_-, \Phi)$ and $J(\Phi_-, \Phi)$ defined by (1.14) and (2.7) have a similar structure. In fact, consider the functional

$$F_\alpha(\Phi_-, \Phi) = \Bigg\{ \sum_{d=1}^n \left|\sigma_{-d}\Phi_{-d}\right|^\alpha + \sum_{l=1}^m \beta_1^{-1} \int_0^T \left|\beta_1 \sigma_l(t)\Phi_l(t)\right|^\alpha dt$$

$$+ \sum_{j=1}^r \beta_2^{-1} \int_0^T \left|\beta_2 \gamma_j(t) b_j'(t)\left(\Phi_- + \int_0^t h(\tau)\Phi(\tau)\,d\tau\right)\right|^\alpha dt \Bigg\}^{\frac{1}{\alpha}}, \qquad \alpha > 0.$$

We see that $F_1(\Phi_-, \Phi) = I(\Phi_-, \Phi)$ and $F_2(\Phi_-, \Phi) = J(\Phi_-, \Phi)$. This fact allows us to hope that the suboptimal quality of the Kalman filter will be satisfactory.

There is another point in favour of the link between $I(\Phi_-, \Phi)$ and $J(\Phi_-, \Phi)$. Using the Cauchy-Schwarz-Bunyakovskii inequality, it can easily be checked that

$$I(\Phi_-, \Phi) \le \sqrt{n + m\beta_1^{-1}T + r\beta_2^{-1}T}\; J(\Phi_-, \Phi).$$

Thus the minimization of $J(\Phi_-, \Phi)$ can lead to the minimization of $I(\Phi_-, \Phi)$.

The main result of this chapter is described by the following theorem.

THEOREM 2.2. *Let φ be the Kalman estimator and the pair (φ_-, φ) be the solution of the problem* (2.6). *Then*

$$1 \le \Delta \le \Delta^0, \qquad \Delta^0 = \frac{\mathcal{N}_1 \cdot \mathcal{N}_2}{\mathcal{N}_\infty^2}, \tag{2.15}$$

$$\mathcal{N}_1 = \sum_{d=1}^n |\Psi_{-d}| + \sum_{l=1}^m \beta_1^{-1} \int_0^T |\Psi_l(t)|\,dt + \sum_{j=1}^r \beta_2^{-1} \int_0^T |\chi_j(t)|\,dt, \tag{2.16}$$

$$\mathcal{N}_2 = \left(\sum_{d=1}^{n} \Psi_{-d}^2 + \sum_{l=1}^{m} \beta_1^{-1} \int_0^T \Psi_l^2(t)\, dt + \sum_{j=1}^{r} \beta_2^{-1} \int_0^T \chi_j^2(t)\, dt \right)^{\frac{1}{2}},$$

$$\mathcal{N}_\infty = \max\left\{ \max_{1 \le d \le n} |\Psi_{-d}|, \sup_{1 \le l \le m,\ t \in [0,T]} |\Psi_l(t)|, \sup_{1 \le j \le r,\ t \in [0,T]} |\chi_j(t)| \right\},$$

where

$$\Psi_{-d} = \sigma_{-d}\varphi_{-d}, \quad d = 1, \ldots, n, \qquad \Psi_l(t) = \beta_1 \sigma_l(t) \varphi_l(t), \quad l = 1, \ldots, m,$$

$$\chi_j(t) = \beta_2 \gamma_j(t) b_j'(t) \tilde{\xi}(t), \quad j = 1, \ldots, r,$$

$$\varphi_- = (\varphi_{-1}, \ldots, \varphi_{-n})', \qquad \varphi(t) = (\varphi_1(t), \ldots, \varphi_m(t))',$$

and $\tilde{\xi}(t)$ *is determined by* (2.11) *or by relations* (2.13).

The proof of Theorem 2.2 is given in Section 3.

It follows from Theorem 2.2 that the value of Δ^0 is determined by the characteristics of the Kalman filter only. These characteristics can be easily computed. Therefore the estimate of the nonoptimality level Δ^0 can be efficiently computed without solving the original nonsmooth variational problem (1.15).

If Δ^0 is not large, then we do not need to resolve the difficult nonsmooth problem (1.15) and the Kalman filter can be used in the optimal guaranteed estimation probem as well.

Let us return to the initial variables. Consider the system (1.1), (1.3) again. Introduce the new variables

$$K(t) = X(t)\tilde{K}(t), \qquad P(t) = X(t)\tilde{P}(t)X'(t),$$

$$\xi(t) = \left(X^{-1}(t)\right)' \tilde{\xi}(t), \qquad \eta(t) = X(t)\tilde{\eta}(t).$$

Then the Kalman filter for the system (1.1), (1.3) takes the standard form

$$\hat{l}_\varphi = a'\hat{x}(T),$$

$$\dot{\hat{x}}(t) = A(t)\hat{x}(t) + K(t)\big(z(t) - H'(t)\hat{x}(t)\big), \qquad \hat{x}(0) = 0,$$

$$K(t) = P(t)H(t)R^{-1}(t), \tag{2.17}$$

$$\dot{P}(t) = A(t)P(t) + P(t)A'(t) - P(t)H(t)R^{-1}(t)H'(t)P(t) \tag{2.18}$$

$$+ B(t)Q(t)B'(t), \qquad P(0) = P_0,$$

$$\hat{x}(t) \in \mathbf{R}^n, \quad K(t) \in \mathbf{R}^{n\times m}, \quad P(t) \in \mathbf{R}^{n\times n}, \qquad t \in [0,T].$$

In these variables, the Kalman estimator is determined by the formula

$$\varphi(t) = K'(t)\left(\theta^{-1}(t)\right)'\theta'(T)a, \qquad \varphi_- = \theta'(T)a,$$

where

$$\dot{\theta}(t) = \big(A(t) - K(t)H'(t)\big)\theta(t), \qquad \theta(0) = E_n.$$

It can easily be checked that

$$\varphi_- = \xi(0), \qquad \varphi(t) = R^{-1}(t)H'(t)\eta(t) = K'(t)\xi(t), \tag{2.19}$$

and the corresponding boundary value problem turns into the system

$$\dot{\xi}(t) = -A'(t)\xi(t) + H(t)R^{-1}(t)H'(t)\eta(t), \qquad \xi(T) = a, \tag{2.20}$$

$$\dot{\eta}(t) = B(t)Q(t)B'(t)\xi(t) + A(t)\eta(t), \qquad \eta(0) = P_0\xi(0).$$

Moreover,

$$\eta(t) = P(t)\xi(t), \qquad t \in [0,T], \tag{2.21}$$

and

$$\dot{\xi}(t) = -\left(A(t) - K(t)H'(t)\right)'\xi(t), \qquad \xi(T) = a, \tag{2.22}$$

where $K(t)$ and $P(t)$ are defined by (2.17) and (2.18).

It follows from (2.19) and (2.20) that

$$\dot{\xi}(t) = -A'(t)\xi(t) + H(t)\varphi(t) \tag{2.23}$$

with the initial value $\xi(0) = \varphi_-$ or with the terminal value $\xi(T) = a$.

The only modification in Theorem 2.2 is that the quantity $\chi(t)$ is computed now by the formula

$$\chi_j(t) = \beta_2\gamma_j(t)B_j'(t)\xi(t), \qquad B(t) = \left(B_1(t), \ldots, B_r(t)\right), \quad j = 1, \ldots, r,$$

and $\xi(t)$ is determined by (2.22) or by relation (2.23).

2.1. *Computational Schemes*

The elements $\{\varphi_-, \varphi(t), \xi(t)\}$, $t \in [0,T]$, can be computed by various ways. Let us discuss two computational schemes.

The first scheme is as follows. The Kalman gain $K(t)$ is computed by (2.17), (2.18) on $[0,T]$ from left to right. Then the matrix $A(t) - K(t)H'(t)$

is known and $\xi(t)$ is computed by (2.22) from right to left. Given $\xi(t)$, the pair $(\varphi_-, \varphi(t))$ is determined by (2.19).

The second scheme is iterative. This scheme is natural when Δ^0 is calculated for a set of consecutive values of T. Denote by $(\varphi_-^{(T_L)}, \varphi^{(T_L)}(t))$, $t \in [0, T_L]$, the pair that corresponds to the estimation of the quantity $l_*^{(T_L)} = a'x(T_L)$ with the help of the measurements on $[0, T_L]$, $0 \leq T_L \leq T$. Obviously,

$$\varphi_-^{(T_L)} = \left(a'\varpi_-^{(T_L)}\right)', \qquad \varpi_-^{(T_L)} = \theta(T_L),$$

$$\varphi^{(T_L)}(t) = \left(a'\varpi^{(T_L)}(t)\right)', \qquad \varpi^{(T_L)}(t) = \theta(T_L)\theta^{-1}(t)K(t), \quad t \in [0, T_L].$$

Then, for a T_M such that $T_L < T_M \leq T$, we have

$$\varpi_-^{(T_M)} = \left[\theta(T_M)\theta^{-1}(T_L)\right] \varpi_-^{(T_L)},$$

$$\varpi^{(T_M)}(t) = \left[\theta(T_M)\theta^{-1}(T_L)\right] \varpi^{(T_L)}(t), \qquad t \in [0, T_L],$$

$$\varpi^{(T_M)}(t) = \left[\theta(T_M)\theta^{-1}(t)\right] K(t), \qquad t \in (T_L, T_M].$$

So, having computed the pair $(\varphi_-^{(T_L)}, \varphi^{(T_L)}(t))$ on the segment $[0, T_L]$, we can convert it into the pair $(\varphi_-^{(T_M)}, \varphi^{(T_M)}(t))$ on the segment $t \in [0, T_M]$ with the help of the transition matrix $\theta(T_M)\theta^{-1}(t)$, $t \in [T_L, T_M]$. This transition matrix is uniquely determined by the matrix $A(t) - K(t)H'(t)$ on the interval $[T_L, T_M]$ only. The iteration process with increasing T_M is initiated by $\varpi_-^{(0)} = E_n$, $\varpi^{(0)}(0) = K(0)$ and is finished when $T_M = T$.

Using these iteration process, we obtain the pair $(\varphi_-, \varphi(t))$, $t \in [0, T]$. Then $\xi(t)$ is computed by (2.23).

2.2. *Prediction Problem*

Let us consider a prediction problem for the system (1.1) – (1.6). The problem is to evaluate the quantity [139]

$$l_* = a'x(T + \kappa), \qquad \kappa > 0, \qquad a \in \mathbf{R}^n.$$

Formally, this problem reduces to the filtering problem with

$$H(t) \equiv 0, \qquad t \in (T, T + \kappa].$$

It can easily be checked that the optimal guaranteed prediction problem takes the form (cf. Section 1)

$$I_0^\kappa = \inf I^\kappa(\Phi_-, \Phi)$$

under the constraint

$$\Phi_- + \int_0^T h(t)\Phi(t)\,dt = \tilde{a},$$

where

$$I^\kappa(\Phi_-, \Phi) = I(\Phi_-, \Phi) + F^\kappa, \qquad F^\kappa = \sum_{j=1}^{r} \int_T^{T+\kappa} \gamma_j(t)|b_j'(t)\tilde{a}|\,dt, \tag{2.24}$$

$I(\Phi_-, \Phi)$ is defined by (1.14),

$$h(t) = X'(t)H(t), \qquad b(t) = X^{-1}(t)B(t), \qquad \tilde{a} = X'(T+\kappa)a,$$

and $X(t)$ is defined by (1.8).

Therefore the optimal guaranteed estimator in the prediction problem coincides with the optimal guaranteed estimator in a filterting problem for the quantity

$$l_* = a_\kappa' x(T), \qquad a_\kappa = \left(X^{-1}(T)\right)' X'(T+\kappa)a. \tag{2.25}$$

It is not hard to prove that the Kalman estimator in the prediction problem coincides with the Kalman estimator in the filtering problem under condition (2.25). Denote by Δ_κ the level of nonoptimality of the Kalman filter in the prediction problem (that is analogous to the quantity Δ from (1.16)).

THEOREM 2.3. *The following inequalities hold:*

$$1 \le \Delta_\kappa \le \Delta_\kappa^0, \qquad \Delta_\kappa^0 = \frac{\mathcal{N}_1 + F^\kappa}{\mathcal{N}_2^2\mathcal{N}_\infty^{-1} + F^\kappa},$$

where $\mathcal{N}_1$, $\mathcal{N}_2$, $\mathcal{N}_\infty$, *and* F^κ *are defined by* (2.16) *and* (2.24), *respectively.*

Theorem 2.3 can be proved similarly to Theorem 2.2.

2.3. *Poor Initial Information*

In some problems the *a priori* information about the initial state $x(0)$ is poor:

$$\sigma_{-d} = \sigma_{-d}(\varepsilon) \equiv \varepsilon^{-1}\bar{\sigma}_{-d}, \quad \bar{\sigma}_{-d} = \text{const}, \quad \varepsilon \ll 1, \quad d = 1, \dots, n. \tag{2.26}$$

Denote by $(\varphi_-(\varepsilon), \varphi(t,\varepsilon))$ and by $\tilde{\xi}(t,\varepsilon)$, $\tilde{\eta}(t,\varepsilon)$ the solutions of the corresponding variational problem and the boundary value problem, respectively.

Let the system (1.9), (1.10) ((1.1), (1.3)) be completely observable on the segment $[0,T]$, i.e., the matrix

$$\mathcal{O} = \int_0^T h(t)R^{-1}(t)h'(t)\,dt \quad \text{is nonsingular.} \tag{2.27}$$

Lemma 2.1. *Suppose conditions* (2.26) and (2.27) *hold; then there exist the limits* $\varphi_\infty(t)$, $\tilde{\xi}_\infty(t)$, $\tilde{\eta}_\infty(t)$ *such that*

$$\lim_{\varepsilon\to 0+} \varphi_-(\varepsilon) = 0, \qquad \lim_{\varepsilon\to 0+} \sigma_{-d}(\varepsilon)\varphi_{-d}(\varepsilon) = 0, \qquad d = 1,\dots,n, \tag{2.28}$$

$$\lim_{\varepsilon\to 0+} \|\varphi(\cdot,\varepsilon) - \varphi_\infty(\cdot)\|_{L_2^m} = 0,$$

$$\lim_{\varepsilon\to 0+} \max_{t\in[0,T]} \|\tilde{\xi}(t,\varepsilon) - \tilde{\xi}_\infty(t)\|_{\mathbf{R}^n} = 0, \tag{2.29}$$

$$\lim_{\varepsilon\to 0+} \max_{t\in[0,T]} \|\tilde{\eta}(t,\varepsilon) - \tilde{\eta}_\infty(t)\|_{\mathbf{R}^n} = 0,$$

where $\varphi_\infty(t)$ *is a unique solution of the variational problem*

$$\min_{\Phi\in L_2^m} G_\infty(\Phi) \quad \textit{under the constraint} \quad \int_0^T h(s)\Phi(s)\,ds = \tilde{a}, \tag{2.30}$$

$$G_\infty(\Phi) = \int_0^T \Phi'(t)R(t)\Phi(t)\,dt$$

$$+ \int_0^T \left(\int_0^t h(s)\Phi(s)\,ds\right)' b(t)Q(t)b'(t)\left(\int_0^t h(s)\Phi(s)\,ds\right) dt.$$

The proof of Lemma 2.1 is given in Section 3.

In accordance with Lemma 2.1, if $A(t) = 0$ and $B(t) = 0$, then Theorem 2.2 turns into the main result of Section 7 of Chapter 1.

3. Derivation of Main Result

Let us decompose the proof of the main results into three parts.

3.1. *Proof of Theorem 2.1*

First let us prove Assertion 1 of Theorem 2.1. We have obtained above that if the initial state, the perturbations in the dynamic system, and the measurement noise are denoted by $y(0) = x(0)$, $u(t)$, and $\varrho(t)$, respectively, then the estimation error under the estimator $\Phi(t)$ takes the form

$$\hat{l} - l_* = \int_0^T \Phi'(t)z(t)\,dt - \tilde{a}'y(T) = \left(\int_0^T h(t)\Phi(t) - \tilde{a}\right)' y(T)$$

$$+\int_0^T \Phi'(t)\Big(-h'(t)\int_t^T b(\tau)u(\tau)\,d\tau + \varrho(t)\Big)\,dt$$

$$= \left(\int_0^T h(t)\Phi(t)\,dt - \tilde{a}\right)' y(0) + \int_0^T \Phi'(t)\varrho(t)\,dt$$

$$+ \left(\int_0^T h(t)\Phi(t)\,dt - \tilde{a}\right)' \int_0^T b(\tau)u(\tau)\,d\tau$$

$$- \int_0^T \Phi'(t)h'(t)\int_t^T b(\tau)u(\tau)\,d\tau\,dt$$

$$= -\Phi'_-\,y(0) + \int_0^T \Phi'(t)\varrho(t)\,dt - \Phi'_-\int_0^T b(\tau)u(\tau)\,d\tau$$

$$- \int_0^T \Phi'(t)h'(t)\int_t^T b(\tau)u(\tau)\,d\tau\,dt,$$

where

$$\Phi_- + \int_0^T h(t)\Phi(t)\,dt = \tilde{a}. \tag{3.1}$$

Using the Dirichlet formula (cf. Section 2 of Chapter 2), we have

$$\hat{l} - l_* = -\Phi'_-\,y(0) + \int_0^T \Phi'(t)\varrho(t)\,dt \tag{3.2}$$

$$- \int_0^T \left(\int_0^t h(\tau)\Phi(\tau)\,d\tau + \Phi_-\right)' b(t)u(t)\,dt.$$

Let $y(0) = x(0)$ be a random variable, let $u(t)$ and $\varrho(t)$ be Gaussian white noise processes:

$$\mathbf{E}y(0) = 0, \qquad \mathbf{E}y(0)y(0)' = P_0, \tag{3.3}$$

$$\mathbf{E}u(t) = 0, \qquad \mathbf{E}u(t)u'(s) = Q(t)\delta(t-s),$$

$$\mathbf{E}\varrho(t) = 0, \qquad \mathbf{E}\varrho(t)\varrho'(s) = R(t)\delta(t-s).$$

It follows directly from (3.1) – (3.3) that

$$\mathbf{E}(\hat{l} - l) = 0, \qquad \mathbf{E}(\hat{l} - l_*)^2 = J^2(\Phi_-, \Phi).$$

Thus the solution of (2.6) yields the linear estimator with minimal variance. Then it follows from the general filtering theory [79], [47], [187] that the Kalman estimator is determined by the solution of the problem (2.6). So, Assertion 1 of Theorem 2.1 is proved. Note that Assertion 1 of Theorem 2.1 also follows directly from (2.4), (2.5), (2.8), and (2.11). In fact, it can easily be checked that the solution of (2.11) is determined by the formula

$$\tilde{\xi}(t) = \left(\tilde{\theta}^{-1}(t)\right)' \tilde{\theta}'(T)\tilde{a},$$

where $\tilde{\theta}(t)$ is given by (2.4).

Let us now prove Assertion 2 of Theorem 2.1. Introduce two Hilbert spaces: the space $\mathcal{H} = \mathbf{R}^n \times L_2^m \times L_2^r$ with the elements $\Gamma = (\Phi_-, \Phi, w)$ and the inner product

$$\langle \Gamma_{(1)}, \Gamma_{(2)} \rangle_{\mathcal{H}} = \Phi'_{-(1)}\Phi_{-(2)} + \int_0^T \Phi'_{(1)}(t)\Phi_{(2)}(t)\,dt + \int_0^T w'_{(1)}(t)w_{(2)}(t)\,dt,$$

$$\Gamma_{(i)} \in \mathcal{H}, \qquad \Phi_{-(i)} \in \mathbf{R}^n, \qquad \Phi_{(i)} \in L_2^m, \qquad w_{(i)} \in L_2^r, \qquad i = 1, 2,$$

and the space $\tilde{\mathcal{H}} = L_2^r \times \mathbf{R}^n$ with the elements $\tilde{\Gamma} = (G, g)$ and the inner product

$$\langle \tilde{\Gamma}_{(1)}, \tilde{\Gamma}_{(2)} \rangle_{\tilde{\mathcal{H}}} = \int_0^T G'_{(1)}(t)G_{(2)}(t)\,dt + g'_{(1)}g_{(2)},$$

$$\tilde{\Gamma}_{(i)} \in \tilde{\mathcal{H}}, \qquad G_{(i)} \in L_2^r, \qquad g_{(i)} \in \mathbf{R}^n, \qquad i = 1, 2.$$

Let us represent the problem (2.6) in the form

$$J_0 = \inf_{\Phi_- \in \mathbf{R}^n, \Phi \in L_2^m, w \in L_2^r} J(\Phi_-, \Phi, w), \tag{3.4}$$

where

$$J(\Phi_-, \Phi, w) \tag{3.5}$$

$$= \left\{ \Phi'_- P_0 \Phi_- + \int_0^T \Phi'(t)R(t)\Phi(t)\,dt + \int_0^T w'(t)Q(t)w(t)\,dt \right\}^{\frac{1}{2}},$$

under the constraints

$$b'(t)\left(\int_0^t h(s)\Phi(s)\,ds + \Phi_-\right) - w(t) = 0, \tag{3.6}$$

$$\tilde{a} - \int_0^T h(s)\Phi(s)\,ds - \Phi_- = 0.$$

Thus the variational problem (3.4) – (3.6) takes the form

$$J_0 = \inf_{\mathcal{H}} J(\Phi_-, \Phi, w) \quad \text{under the constraint} \quad \Lambda(\Phi_-, \Phi, w) = 0. \tag{3.7}$$

The functional $J(\Phi_-, \Phi, w): \mathcal{H} \to \mathbf{R}^1$ defined by (3.5) is a convex continuous functional, the operator $\Lambda: \mathcal{H} \to \tilde{\mathcal{H}}$ defined by (3.6) is a continuous affine operator. So, the problem (3.7) is a variational problem of the type $\mathcal{P}_1$ from Section 1 of Chapter 2.

Let us prove the existence and uniqueness of the solution for the problem (3.7). The space $\mathcal{H}$ is a Hilbert space; consequently it is a reflexive Banach space. By virtue of restrictions (1.6) and (2.14), P_0, $R(t)$, and $Q(t)$ are nonsingular. Then it can easily be checked that the functional $J(\Phi_-, \Phi, w)$ is coercive and strictly convex. Since the operator Λ is continuous and affine, the set $\Lambda(\Phi_-, \Phi, w) = 0$ is an affine closed manifold. Then, by virtue of Lemma 1.2 of Chapter 2, the problem (3.7) has a unique solution $(\varphi_-, \varphi, w_\varphi)$.

Let us turn to the dual problem. Since the space $\tilde{\mathcal{H}}$ is Hilbert, the dual space can be identified with the original space: $\tilde{\mathcal{H}}^* = \tilde{\mathcal{H}}$. So, the Lagrange function has the form

$$\mathcal{L}_{\mathcal{H}}(\Phi_-, \Phi, w; p, \lambda) = J(\Phi_-, \Phi, w) \tag{3.8}$$

$$+ \int_0^T p'(t) \left[b'(t) \left(\int_0^t h(s)\Phi(s)\, ds + \Phi_- \right) - w(t) \right] dt$$

$$+ \lambda' \left(\tilde{a} - \int_0^T h(t)\Phi(t)\, dt - \Phi_- \right).$$

The perturbation of problem (3.7) has the form (cf. Section 1 of Chapter 2)

$$J_0(\delta w, \delta\tilde{a}) = \inf_{\mathcal{H}} J(\Phi_-, \Phi, w)$$

under the constraint

$$\Lambda(\Phi_-, \Phi, w) = \begin{pmatrix} \delta w \\ \delta\tilde{a} \end{pmatrix}, \qquad \delta w \in L_2^r, \quad \delta\tilde{a} \in \mathbf{R}^n.$$

With $\delta w = 0$ and $\delta\tilde{a} = 0$ we get the initial problem (3.4) – (3.6).

Similar to the proof of Theorem 5.2 of Chapter 2 we can verify that $J_0(\delta w, \delta\tilde{a})$ is bounded in a neighborhood of the point $(0,0)$ (cf. Proposition 5.1 of Chapter 2). Then, using Lemmas 1.4 – 1.6 of Chapter 2, we can establish that $J_0(\delta w, \delta\tilde{a})$ is convex, and at the point $(0,0)$ $J_0(\delta w, \delta\tilde{a})$ is continuous and has a non-empty subdifferential. Then, by virtue of Theorems 1.8 and 1.9 of Chapter 2, the dual problem

$$J^0 = \sup_{p \in L_2^r, \lambda \in \mathbf{R}^n} \left\{ \inf_{\Phi_- \in \mathbf{R}^n, \Phi \in L_2^m, w \in L_2^r} \mathcal{L}_{\mathcal{H}}(\Phi_-, \Phi, w; p, \lambda) \right\} \tag{3.9}$$

has a solution (p^0, λ^0), the Lagrange function has a saddle point:

$$\mathcal{L}_{\mathcal{H}}(\varphi_-, \varphi, w_\varphi; p, \lambda) \le \mathcal{L}_{\mathcal{H}}(\varphi_-, \varphi, w_\varphi; p^0, \lambda^0) \le \mathcal{L}_{\mathcal{H}}(\Phi_-, \Phi, w; p^0, \lambda^0),$$

$$\forall\ \Phi_-, \lambda \in \mathbf{R}^n, \quad \Phi \in L_2^m, \quad w, p \in L_2^r,$$

and the duality relation holds:

$$J_0 = J^0. \tag{3.10}$$

In order to compute the infimum in (3.9) and to obtain the explicit relation between the solutions of the primal and dual problems, let us simplify the expression for the Lagrange function. Using the structure of Hilbert space and applying the Dirichlet formula (Lemma 2.1 of Chapter 2), we represent the Lagrange function (3.8) in another form

$$\mathcal{L}_{\mathcal{H}}(\Phi_-, \Phi, w; p, \lambda) = \bar{\mathcal{L}}_{\mathcal{H}}(\Gamma; \Xi) \triangleq \|\Gamma\|_{\mathcal{H}} - \langle \Xi, \Gamma \rangle_{\mathcal{H}} + \lambda' \tilde{a}, \tag{3.11}$$

where[2]

$$\Gamma = \Gamma(\Phi_-, \Phi, w) = \left(P_0^{\frac{1}{2}} \Phi_-,\ R^{\frac{1}{2}}(t)\Phi(t),\ Q^{\frac{1}{2}}(t) w(t) \right),$$

$$\Xi = \Xi(p, \lambda) = \Bigg(P_0^{-\frac{1}{2}} \left(\lambda - \int_0^T b(t) p(t)\, dt \right),$$
$$R^{-\frac{1}{2}}(t) h'(t) \left(\lambda - \int_t^T b(s) p(s)\, ds \right),\ Q^{-\frac{1}{2}}(t) p(t) \Bigg).$$

With these new variables, the dual problem is represented in the form

$$J^0 = \sup_{\Xi(p,\lambda)} \left(\inf_{\Gamma \in \mathcal{H}} \bar{\mathcal{L}}_{\mathcal{H}}(\Gamma; \Xi) \right) \tag{3.12}$$

and has a solution $\Xi^0 = \Xi(p^0, \lambda^0)$. The saddle point inequalities take the form

$$\bar{\mathcal{L}}_{\mathcal{H}}(\Gamma^0; \Xi) \le \bar{\mathcal{L}}_{\mathcal{H}}(\Gamma^0; \Xi^0) \le \bar{\mathcal{L}}_{\mathcal{H}}(\Gamma; \Xi^0), \qquad \Gamma \in \mathcal{H}, \quad \Xi = \Xi(p, \lambda). \tag{3.13}$$

Computing the infimum in (3.12), we get

$$\inf_{\Gamma \in \mathcal{H}} \bar{\mathcal{L}}_{\mathcal{H}}(\Gamma, \Xi) = \begin{cases} \tilde{a}'\lambda & \text{if } \|\Xi\|_{\mathcal{H}} \le 1 \\ -\infty & \text{otherwise.} \end{cases}$$

[2] For a symmetric matrix $A \ge 0$, $A^{\frac{1}{2}}$ is a symmetric matrix such that $A^{\frac{1}{2}} A^{\frac{1}{2}} = A$; $A^{-\frac{1}{2}} \triangleq \left(A^{\frac{1}{2}} \right)^{-1}$.

In fact, if $||\Xi|| \leq 1$, then

$$||\Gamma||_{\mathcal{H}} - \langle \Xi, \Gamma \rangle_{\mathcal{H}} \geq ||\Gamma||_{\mathcal{H}} - ||\Xi||_{\mathcal{H}}\, ||\Gamma||_{\mathcal{H}} = ||\Gamma||_{\mathcal{H}}\, (1 - ||\Xi||_{\mathcal{H}}) \geq 0.$$

Obviously, $||0||_{\mathcal{H}} - \langle \Xi, 0 \rangle_{\mathcal{H}} = 0$.

Conversely, if $||\Xi||_{\mathcal{H}} > 1$, then, taking $\Gamma = \alpha\Xi$, $\alpha > 0$, we have

$$||\Gamma||_{\mathcal{H}} - \langle \Xi, \Gamma \rangle_{\mathcal{H}} = |\alpha|\, ||\Xi||_{\mathcal{H}} - \alpha ||\Xi||^2_{\mathcal{H}} = \alpha ||\Xi||_{\mathcal{H}}\, (1 - ||\Xi||_{\mathcal{H}}) \to -\infty$$

as $\alpha \to \infty$.

Hence the dual problem is determined by the relation:

$$\sup_{\lambda \in \mathbf{R}^n,\, p \in L_2^r} \tilde{a}'\lambda \quad \text{under the constraint} \quad ||\Xi(p, \lambda)||_{\mathcal{H}} \leq 1. \tag{3.14}$$

The second inequality of (3.13) can be written in the explicit form

$$||\Gamma^0||_{\mathcal{H}} - \langle \Xi^0, \Gamma^0 \rangle_{\mathcal{H}} \leq ||\Gamma||_{\mathcal{H}} - \langle \Xi^0, \Gamma \rangle_{\mathcal{H}}, \qquad \forall \Gamma \in \mathcal{H}. \tag{3.15}$$

Since $||\Xi^0||_{\mathcal{H}} \leq 1$, inequality (3.15) implies

$$||\Xi^0||_{\mathcal{H}} ||\Gamma^0||_{\mathcal{H}} \leq ||\Gamma^0||_{\mathcal{H}} \leq \langle \Xi^0, \Gamma^0 \rangle_{\mathcal{H}} \leq ||\Xi^0||_{\mathcal{H}} ||\Gamma^0||_{\mathcal{H}},$$

whence,

$$\langle \Xi^0, \Gamma^0 \rangle_{\mathcal{H}} = ||\Xi^0||_{\mathcal{H}} ||\Gamma^0||_{\mathcal{H}}.$$

Therefore,

$$\Gamma^0 = k\, \Xi^0 \qquad \text{for a certain} \quad k \geq 0. \tag{3.16}$$

Returning to the initial variables, we obtain from (3.14) and (3.16) that the dual problem has the form

$$\sup_{p \in L_2^r,\, \lambda \in \mathbf{R}^n} \tilde{a}'\lambda \quad \text{under the constraint} \quad S(p, \lambda) \leq 1, \tag{3.17}$$

where

$$S(p, \lambda) = \left(\lambda - \int_0^T b(s)p(s)\, ds\right)' P_0^{-1} \left(\lambda - \int_0^T b(s)p(s)\, ds\right) \tag{3.18}$$

$$+ \int_0^T \left[h'(t)\left(\lambda - \int_t^T b(s)p(s)\, ds\right)\right]' R^{-1}(t) \left[h'(t)\left(\lambda - \int_t^T b(s)p(s)\, ds\right)\right] dt$$

$$+ \int_0^T p'(t) Q^{-1}(t) p(t)\, dt.$$

Moreover, for a certain $k \geq 0$,

$$\varphi_- = kP_0^{-1}\left(\lambda^0 - \int_0^T b(t)p^0(t)\,dt\right), \tag{3.19}$$

$$\varphi(t) = kR^{-1}(t)h'(t)\left(\lambda^0 - \int_t^T b(s)p^0(s)\,ds\right),$$

$$w_\varphi(t) = k\,Q^{-1}(t)p^0(t),$$

where $(p^0(t), \lambda^0)$ is the solution of the dual problem (3.17).

Let us now resolve the dual problem (3.17). First note that since $\sup \tilde{a}'\lambda = -\inf(-\tilde{a}'\lambda)$, the problem (3.17) has the type $\mathcal{P}_2$ from Section 1 of Chapter 2 with the linear functional

$$f_0(p,\lambda) \equiv -\tilde{a}'\lambda, \qquad f_0 : \tilde{\mathcal{H}} \to \mathbf{R}^1,$$

and the convex scalar constraint

$$f_1(p,\lambda) \equiv S(p,\lambda) - 1 \leq 0.$$

The space $\tilde{\mathcal{H}}$ is Hilbert and the constraint defines a non-empty convex bounded closed subset. Then, in accordance with Lemma 1.2 of Chapter 2, a solution of (3.17) exists.

By virtue of the Kuhn-Tucker theorem (Theorem 1.5 of Chapter 2)[3], if (p^0, λ^0) is the solution of (3.17), then (p^0, λ^0) yields an unconstrained minimum to the Lagrange function

$$\mathcal{L}^0(p,\lambda;\kappa) = -\tilde{a}'\lambda + \kappa\,(S(p,\lambda) - 1) \qquad \text{for a certain} \quad \kappa \geq 0.$$

Moreover, the condition of complementary slackness holds:

$$\kappa\,(S(p,\lambda) - 1) = 0.$$

Without loss of generality we may put $\kappa > 0$. In fact, if $\kappa = 0$, then it follows from the Kuhn-Tucker theorem that

$$\tilde{a}'\lambda^0 \leq \tilde{a}'\lambda \qquad \text{for all} \quad \lambda \in \mathbf{R}^n.$$

The latter inequality is valid only if $\tilde{a} = a = 0$. In this case, all variational problems associated with the estimation problem can be resolved immediately.

[3] Arguing as above, we could also consider the corresponding perturbed problem and apply Lemmas 1.4 – 1.6 and Theorem 1.9 of Chapter 2. Nevertheless, in this specific case, the use of the simpler Kuhn-Tucker theorem will provide the desired result.

In particular, the optimal estimator $\varphi(t) \equiv 0$. So, omitting this trivial case,[4] we may put $\kappa > 0$; hence

$$S(p^0, \lambda^0) = 1. \tag{3.20}$$

Since $\mathcal{L}^0(p, \lambda)$ is a smooth convex function, the necessary and sufficient condition for its minimum are determined by the relations (see Theorem 1.4 of Chapter 2)

$$\frac{\partial \mathcal{L}^0}{\partial p} = 0, \qquad \frac{\partial \mathcal{L}^0}{\partial \lambda} = 0, \tag{3.21}$$

where $\frac{\partial \mathcal{L}^0}{\partial p}$ is a Frechét derivative. Differentiating $\mathcal{L}^0$ with resprect to p and λ and equating them to zero, we obtain from (3.21) that

$$p(t) = Q(t)b'(t)\left\{ P_0^{-1}\left(\lambda - \int_0^T b(s)p(s)\,ds\right)\right.$$

$$\left. + \int_0^t h(s)R^{-1}(s)h'(s)\left(\lambda - \int_s^T b(\tau)p(\tau)\,d\tau\right) ds\right\}$$

and

$$P_0^{-1}\left(\lambda - \int_0^T b(s)p(s)\,ds\right)$$

$$+ \int_0^T h(s)R^{-1}h'(s)\left(\lambda - \int_s^T b(\tau)p(\tau)\,d\tau\right) ds = \frac{\tilde{a}}{2\kappa}.$$

The last relations imply that

$$p(t) = Q(t)b'(t)\bar{\xi}(t), \tag{3.22}$$

where

$$\bar{\xi}(t) = P_0^{-1}\left(\lambda - \int_0^T b(s)Q(s)b'(s)\bar{\xi}(s)\,ds\right)$$

$$+ \int_0^t h(s)R^{-1}(s)h'(s)\left(\lambda - \int_s^T b(\tau)Q(\tau)b'(\tau)\bar{\xi}(\tau)\,d\tau\right) ds,$$

and

$$P_0^{-1}\left(\lambda - \int_0^T b(s)Q(s)b'(s)\bar{\xi}(s)\,ds\right) \tag{3.23}$$

[4] Obviously, Theorem 2.1 is valid for $\tilde{a} = a = 0$ as well.

$$+\int_0^T h(s)R^{-1}(s)h'(s)\left(\lambda-\int_s^T b(\tau)Q(\tau)b'(\tau)\bar{\xi}(\tau)\,d\tau\right)ds=\frac{\tilde{a}}{2\kappa}.$$

Let us introduce an auxiliary variable

$$\bar{\eta}(s)=\bar{\eta}(T)-\int_s^T b(\tau)Q(\tau)b'(\tau)\bar{\xi}(\tau)\,d\tau,\qquad \bar{\eta}(T)=\lambda. \tag{3.24}$$

It follows from (3.22) – (3.24) that

$$\dot{\bar{\xi}}(t)=h(t)R^{-1}(t)h'(t)\bar{\eta}(t),\qquad \bar{\xi}(T)=\frac{\tilde{a}}{2\kappa}, \tag{3.25}$$

$$\dot{\bar{\eta}}(t)=b(t)Q(t)b'(t)\bar{\xi}(t),\qquad \bar{\eta}(0)=P_0\bar{\xi}(0).$$

Since the problem (3.17) has a solution, the boundary value problem (3.25) must have a solution. Moreover, this solution is unique. In fact, if there are two solutions $(\bar{\xi}_{(1)}(t),\bar{\eta}_{(1)}(t))$ and $(\bar{\xi}_{(2)}(t),\bar{\eta}_{(2)}(t))$, then the non-zero functions

$$\delta\xi(t)=\bar{\xi}_{(2)}(t)-\bar{\xi}_{(1)}(t),\qquad \delta\eta(t)=\bar{\eta}_{(2)}(t)-\bar{\eta}_{(1)}(t)$$

satisfy the equations

$$\dot{\delta\xi}(t)=h(t)R^{-1}(t)h'(t)\delta\eta(t),\qquad \delta\xi(T)=0, \tag{3.26}$$

$$\dot{\delta\eta}(t)=b(t)Q(t)b'(t)\delta\xi(t),\qquad \delta\eta(0)=P_0\delta\xi(0).$$

It follows immediately from (3.26) that

$$\frac{d}{dt}\left(\delta\eta'(t)\delta\xi(t)\right)=\delta\eta'(t)h(t)R^{-1}(t)h'(t)\delta\eta(t)+\delta\xi'(t)b(t)Q(t)b'(t)\delta\xi(t),$$

whence,

$$\delta\xi'(0)P_0\delta\xi(0)+\int_0^T\Big(\delta\eta'(t)h(t)R^{-1}(t)h'(t)\delta\eta(t)$$

$$+\,\delta\xi'(t)b(t)Q(t)b'(t)\delta\xi(t)\Big)\,dt=0.$$

Consequently,

$$P_0^{\frac{1}{2}}\delta\xi(0)=0,\qquad R^{-\frac{1}{2}}(t)h'(t)\delta\eta(t)\equiv0,\qquad Q^{\frac{1}{2}}(t)b'(t)\delta\xi(t)\equiv0.$$

Then we get from the first equation of (3.26) that $\delta\xi(t)\equiv0$ and therefore $\delta\eta(t)\equiv0$. Clearly, the solutions of (2.9) and (3.25) differ only by the scale 2κ.

Thus, the boundary value problem (2.9) has the unique solution $(\tilde{\xi}(t), \tilde{\eta}(t))$ and the solution of (3.17) has the form

$$p^0(t) = (2\kappa)^{-1} Q(t)b'(t)\tilde{\xi}(t), \qquad \lambda^0 = (2\kappa)^{-1}\tilde{\eta}(T). \tag{3.27}$$

Moreover, using the boundary value problem, the solution (3.19) of (3.7) may be rewritten in the form

$$\varphi_- = \frac{k}{2\kappa}\tilde{\xi}(0), \qquad \varphi(t) = \frac{k}{2\kappa}R^{-1}(t)h'(t)\tilde{\eta}(t), \tag{3.28}$$

$$w_\varphi(t) = \frac{k}{2\kappa}b'(t)\tilde{\xi}(t).$$

Substituting (3.28) into (1.12), we have

$$\frac{k}{2\kappa}\left(\tilde{\xi}(0) + \int_0^T h(t)R^{-1}(t)h'(t)\tilde{\eta}(t)\,dt\right) = \frac{k}{2\kappa}\tilde{\xi}(T) = \frac{k}{2\kappa}\tilde{a} = \tilde{a},$$

whence $\tilde{a} \neq 0$,

$$k = 2\kappa$$

and

$$\varphi_- = \tilde{\xi}(0), \qquad \varphi(t) = R^{-1}(t)h'(t)\tilde{\eta}(t), \qquad w_\varphi(t) = b'(t)\tilde{\xi}(t). \tag{3.29}$$

So, the first two equalities of (2.8) hold.

Substituting (3.27) into constraint (3.20), we get

$$\tilde{\eta}'(0)P_0^{-1}\tilde{\eta}(0) + \int_0^T \tilde{\eta}'(t)h(t)R^{-1}(t)h'(t)\tilde{\eta}(t)\,dt$$

$$+ \int_0^T \tilde{\xi}'(t)b(t)Q(t)b'(t)\tilde{\xi}(t)\,dt = 4\kappa^2.$$

In accordance with (3.29), the second boundary condition from (2.9), and (3.4) – (3.6), the last equality takes the form $J_0^2 = 4\kappa^2$, whence,

$$2\kappa = J_0.$$

Finally, the solution of the dual problem (3.17) has the form

$$p^0(t) = J_0^{-1}Q(t)b'(t)\tilde{\xi}(t), \qquad \lambda^0 = J_0^{-1}\tilde{\eta}(T). \tag{3.30}$$

It can be easily checked by direct differentiation that the function $\tilde{P}(t)\xi(t)$ satisfies the second equation of (2.9). Then, by virtue of the uniquness of a solution, (2.10) holds. In particular,

$$\varphi(t) = R^{-1}(t)h'(t)\tilde{\eta}(t) = \tilde{K}'(t)\tilde{\xi}(t)$$

and the last equality of (2.8) is valid. Relation (2.11) follows from (2.9) and (2.10).

It follows directly from (2.9), (2.10), (3.10), and (3.30) that

$$J_0 = J^0 = \tilde{a}'\lambda^0 = J_0^{-1}\tilde{a}'\tilde{\eta}(T) = J_0^{-1}\tilde{a}'\tilde{P}(T)\tilde{a}. \tag{3.31}$$

So, relation (2.12) also holds. □

3.2. *Proof of Theorem 2.1 for Singular P_0 and $Q(t)$*

Strictly speaking, Theorem 2.1 was proved only for $P_0 > 0$ and $Q(t) > 0$ that is consistent with (1.6). Here we shall study the singular case and prove that Theorem 2.1 is still valid.

Let P_0 and $Q(t)$ be singular. Consider the perturbed nonsingular matrices

$$P_0(\varepsilon) = P_0 + \varepsilon E_n, \qquad Q(t,\varepsilon) = Q(t) + \varepsilon E_r \tag{3.32}$$

and the corresponding ε-perturbed variational problem (1.12), (2.6), (2.7).

Let us denote by $\varphi_-(\varepsilon)$, $\varphi(t,\varepsilon)$, $\tilde{\xi}(t,\varepsilon)$, $\tilde{\eta}(t,\varepsilon)$, $J(\Phi_-,\Phi;\varepsilon)$ the corresponding elements of the ε-perturbed problem. Obviously, Assertion 1^0 of Theorem 2.1 remains valid. So, we focus our attention on Assertion 2^0.

Using condition (1.12), let us represent the ε-perturbed problem (1.12), (2.6), (2.7), (3.32) as an unconstrained problem of the form

$$\inf_{\Phi \in L_2^m} F(\Phi,\varepsilon), \tag{3.33}$$

where

$$F(\Phi,\varepsilon) = \left(\tilde{a} - \int_0^T h(s)\Phi(s)\,ds\right)' P_0(\varepsilon)\left(\tilde{a} - \int_0^T h(s)\Phi(s)\,ds\right)$$

$$+ \int_0^T \Phi'(t)R(t)\Phi(t)\,dt$$

$$+ \int_0^T \left(\tilde{a} - \int_t^T h(s)\Phi(s)\,ds\right)' b(t)Q(t,\varepsilon)b'(t)\left(\tilde{a} - \int_t^T h(s)\Phi(s)\,ds\right) dt.$$

For brevity, we shall write $F(\Phi)$ instead of $F(\Phi,0)$, i.e., $F(\Phi) \triangleq F(\Phi,0)$.

PROPOSITION 3.1. *Suppose $R(t) \geq R_0 > 0$; then the following assertions hold.*

1^0. *The functional $F(\Phi)\colon L_2^m \to \mathbf{R}^1$ is coercive:*

$$F(\Phi) \to \infty \quad \text{as} \quad \|\Phi\|_{L_2^m} \to \infty.$$

2^0. *The functional* $F(\Phi)$ *is strongly convex:*

$$F(\alpha\Phi + (1-\alpha)\Psi) \leq \alpha F(\Phi) + (1-\alpha)F(\Psi) - \alpha(1-\alpha)\|\Phi - \Psi\|^2_{L_2^m} \quad (3.34)$$

for any $\alpha \in [0, 1]$.
3^0. *The problem* (3.33) *with* $\varepsilon = 0$ *has a unique solution* $\varphi(t)$[5].
4^0. *For any* $\Phi \in L_2^m$,

$$\|\Phi - \varphi\|^2_{L_2^m} \leq \frac{2}{c}(F(\Phi) - F(\varphi)), \qquad c > 0. \quad (3.35)$$

PROOF. Denoting by $c > 0$ the minimal eigenvalue of R_0, we have

$$F(\Phi) \geq \int_0^T \Phi'(t)R(t)\Phi(t)\,dt \geq \int_0^T \Phi'(t)R_0\Phi(t)\,dt \geq c\,\|\Phi\|^2_{L_2^m}. \quad (3.36)$$

Hence the functional is coercive.

Consider the functional $F_R(\Phi) = \int_0^T \Phi'(t)R(t)\Phi(t)\,dt$. It is easy to verify by direct computation that

$$F_R(\alpha\Phi + (1-\alpha)\Psi) = \alpha F_R(\Phi) + (1-\alpha)F_R(\Psi) - \alpha(1-\alpha)F_R(\Phi - \Psi)$$

$$\leq \alpha F_R(\Phi) + (1-\alpha)F_R(\Psi) - \alpha(1-\alpha)c\,\|\Phi - \Psi\|^2_{L_2^m}, \qquad \alpha \in [0, 1].$$

So, $F_R(\Phi)$ is strongly convex. Since the other two summands in $F(\Phi)$ are convex, the functional $F(\Phi)$ is strongly convex.

We have proved that the functional $F(\Phi)$ is coercive and strongly convex (consequently $F(\Phi)$ is strictly convex). The space L_2^m is a Hilbert one. Obviously, $F(\Phi)$ is continuous. Then, by virtue of Lemma 1.2 of Chapter 2, a solution of (3.33) for $\varepsilon = 0$ exists; moreover, it is unique.

With $\alpha = \frac{1}{2}$, it follows directly from (3.34) that

$$F(\tfrac{1}{2}\Phi + \tfrac{1}{2}\varphi) \leq \tfrac{1}{2}F(\Phi) + \tfrac{1}{2}F(\varphi) - \frac{c}{4}\|\Phi - \varphi\|^2_{L_2^m}. \quad (3.37)$$

Since φ yields a minimum to $F(\Phi)$, we have

$$F(\varphi) \leq F(\tfrac{1}{2}\Phi + \tfrac{1}{2}\varphi).$$

Hence (3.37) implies (3.35). Proposition 3.1 is proved.

PROPOSITION 3.2. *Suppose* $\varphi(t, \varepsilon)$ *and* $\varphi(t) = \varphi(t, 0)$ *are the solutions of the problem* (3.33) *for* $\varepsilon > 0$ *and for* $\varepsilon = 0$ *(or of* (2.6)*), respectively; then*

$$\lim_{\varepsilon \to 0+} \|\varphi(\cdot, \varepsilon) - \varphi(\cdot)\|_{L_2^m} = 0. \quad (3.38)$$

[5] It was already proved above that (3.33) has a unique solution for $\varepsilon > 0$.

Proof. For brevity, we denote $\varphi(\cdot,\varepsilon)$, $F(\cdot,\varepsilon)$, and $F(\cdot,0)$ by φ^ε, F^ε, and F, respectively.

As $F(\varphi)\le F(\varphi^\varepsilon)$ and $F^\varepsilon(\varphi^\varepsilon)\le F^\varepsilon(\varphi)$, we have

$$0\le F(\varphi^\varepsilon)-F(\varphi)=\left[F(\varphi^\varepsilon)-F^\varepsilon(\varphi^\varepsilon)\right]+\left[F^\varepsilon(\varphi^\varepsilon)-F(\varphi)\right] \tag{3.39}$$

$$\le\left[F(\varphi^\varepsilon)-F^\varepsilon(\varphi^\varepsilon)\right]+\left[F^\varepsilon(\varphi)-F(\varphi)\right].$$

Using the Cauchy-Swartz-Bunyakovskii inequality, it is not hard to prove that

$$|F(\Phi)-F^\varepsilon(\Phi)|\le\varepsilon(c_1+c_2\|\Phi\|_{L_2^m})^2 \quad \text{for some positive constants } c_1,\ c_2.$$

Similar to (3.36),

$$\|\varphi^\varepsilon\|^2_{L_2^m}\le c^{-1}F^\varepsilon(\varphi^\varepsilon)\le c^{-1}F^\varepsilon(0)\le\varepsilon\, c_3+c_4,\quad c_3,\ c_4=\text{const}>0. \tag{3.40}$$

It follows from (3.35), (3.39), and (3.40) that, for an $\varepsilon^0>0$ and a constant c_5,

$$\|\varphi^\varepsilon-\varphi\|^2_{L_2^m}\le\varepsilon\, c_5,\qquad 0\le\varepsilon\le\varepsilon^0,$$

and (3.38) is valid. Proposition 3.2 is proved.

In accordance with (2.13),

$$\tilde\xi(t,\varepsilon)=\tilde a-\int_t^T h(s)\varphi(t,\varepsilon)\,ds.$$

By virtue of Proposition 3.2, $\varphi(\cdot,\varepsilon)\to\varphi(\cdot)$; hence,

$$\tilde\xi(t,\varepsilon)\to\tilde\xi(t)\triangleq\tilde a-\int_t^T h(s)\varphi(s)\,ds.$$

In fact, using the Cauchy-Schwarz-Bunyakovskii inequality, it is not hard to prove that

$$\|\tilde\xi(t,\varepsilon)-\tilde\xi(t)\|_{\mathbf{R}^n}\le c_6\,\|\varphi(\cdot,\varepsilon)-\varphi(\cdot)\|_{L_2^m}\qquad \text{for all}\quad t\in[0,T]$$

and therefore,

$$\lim_{\varepsilon\to 0+}\max_{t\in[0,T]}\|\tilde\xi(t,\varepsilon)-\tilde\xi(t)\|_{\mathbf{R}^n}=0. \tag{3.41}$$

By virtue of (2.8) and (2.9),

$$\varphi_-(\varepsilon)=\tilde\xi(0,\varepsilon), \tag{3.42}$$

$$\tilde\eta(t,\varepsilon)=P_0(\varepsilon)\tilde\xi(0,\varepsilon)+\int_0^t b(s)Q(s,\varepsilon)b'(s)\tilde\xi(s,\varepsilon)\,ds.$$

Since (3.41) and (3.42) hold, it can easily be shown that $\varphi_-(\varepsilon)$ and $\tilde{\eta}(t,\varepsilon)$ also have the limits φ_- and $\tilde{\eta}(t)$ such that

$$\lim_{\varepsilon\to 0+}\max_{t\in[0,T]}\|\tilde{\eta}(t,\varepsilon)-\tilde{\eta}(t)\|_{\mathbf{R}^n}=0. \tag{3.43}$$

According to (2.8) and (2.9),

$$\varphi(t,\varepsilon)=R^{-1}(t)h(t)\tilde{\eta}(t,\varepsilon) \tag{3.44}$$

and

$$\tilde{\xi}(t,\varepsilon)=\tilde{a}-\int_t^T h(s)R^{-1}(s)h'(s)\tilde{\eta}(s,\varepsilon)\,ds. \tag{3.45}$$

Proceeding to limit in (3.42), (3.44), and (3.45), we obtain that

$$\varphi_-=\tilde{\xi}(0),\qquad \varphi(t)=R^{-1}(t)h(t)\tilde{\eta}(t),$$

$$\tilde{\xi}(t)=\tilde{a}-\int_t^T h(s)R^{-1}(s)h'(s)\tilde{\eta}(s)\,ds,$$

and

$$\tilde{\eta}(t)=P_0\tilde{\xi}(0)+\int_0^t b(s)Q(s)b'(s)\tilde{\xi}(s)\,ds.$$

Consequently the limiting functions $\tilde{\xi}(t)$ and $\tilde{\eta}(t)$ satisfy the boundary value problem (2.9). Relation (2.10) follows immediately from (2.9). Moreover, the pair (φ_-,φ) satisfies the condition (1.12), the pair (φ_-,φ) is a solution of the variational problem (2.6), and (2.8) is also valid.

It follows from the equality (3.31) for the ε-perturbed problem that

$$J(\varphi_-(\varepsilon),\varphi(\cdot,\varepsilon);\varepsilon)=J^{-1}(\varphi_-(\varepsilon),\varphi(\cdot,\varepsilon);\varepsilon)\,\tilde{a}'\tilde{\eta}(T,\varepsilon).$$

Proceeding to limit in this equality, we get

$$J_0^2=\tilde{a}'\tilde{\eta}(T).$$

By virtue of (2.10), the last assertion of Theorem 2.1 also holds. Thus Theorem 2.1 is valid for any positive semi-definite matrices P_0 and $Q(t)$. □

3.3. *Proof of Theorem 2.2*

In accordance with formulas (1.12) – (1.15) and (1.17),

$$\Delta=\frac{I(\varphi_-,\varphi)}{I_0}, \tag{3.46}$$

where (φ_-, φ) are subjected to (1.12) and φ is the Kalman estimator. The numerator of (3.46) can be easily computed. The denominator is unknown; so, Δ is unknown. In order to obtain the upper bound for Δ we have to evaluate I_0 from below.

Similar to (3.4) – (3.6), the problem (1.15) can be represented in the form (see the proof of Theorem 5.2 of Chapter 2)

$$I_0 = \inf_{\Phi_- \in \mathbf{R}^n,\, \Phi \in \mathcal{F}^m,\, w \in L_1^r} I(\Phi_-, \Phi, w), \tag{3.47}$$

$$I(\Phi_-, \Phi, w) = \sum_{d=1}^{n} \sigma_{-d} |\Phi_{-d}| + \sum_{l=1}^{m} \int_0^T \sigma_l(t) |\Phi_l(t)|\, dt + \sum_{j=1}^{r} \int_0^T \gamma_j(t) |w_j(t)|\, dt,$$

under constraints (3.6).

By virtue of Theorem 2.1, the pair (φ_-, φ) is the solution of the problem (2.6). This problem can be represented in the form (3.4) – (3.6).

Let us extract the dual problems for the variational problems (3.47), (3.6) and (3.4), (3.6) from the proofs developed above.

THEOREM 3.1. *The problem dual to* (3.47), (3.6) *has the form*

$$I^0 = \sup_{p \in L_\infty^r,\, \lambda \in \mathbf{R}^n} \lambda' \tilde{a} \tag{3.48}$$

under the constraints

$$\sigma_l^{-1}(t) \left| h_l'(t) \left(\lambda - \int_t^T b(s) p(s)\, ds \right) \right| \le 1, \quad t \in [0,T], \quad l = 1, \ldots, m, \tag{3.49}$$

$$\gamma_j^{-1}(t)\, |p_j(t)| \le 1, \qquad t \in [0,T], \qquad j = 1, \ldots, r,$$

$$\sigma_{-d}^{-1} \left| e_d' \left(\lambda - \int_0^T b(t) p(t)\, dt \right) \right| \le 1, \qquad d = 1, \ldots, n,$$

$$e_d = (\underbrace{0, \ldots, 1}_{d}, 0, \ldots, 0)'.$$

Moreover,

$$I_0 = I^0. \tag{3.50}$$

PROOF. The proof of Theorem 3.1 is a part of the proof of Theorem 5.2 of Chapter 2. Since the function $p(t)$ enters in the integrals, the second condition in (3.49) may not contain the sign of essential supremum (as in Chapter 2) and is defined on the whole of the interval $[0, T]$.

THEOREM 3.2. *The problem dual to* (3.4), (3.6) *has the form*

$$J^0 = \sup_{p \in L_2^r, \lambda \in \mathbf{R}^n} \tilde{a}'\lambda \tag{3.51}$$

under the constraint

$$S(p, \lambda) \leq 1, \tag{3.52}$$

where

$$S(p,\lambda) = \left(\lambda - \int_0^T b(s)p(s)\,ds\right)' P_0^{-1}\left(\lambda - \int_0^T b(s)p(s)\,ds\right)$$

$$+ \int_0^T \left[h'(t)\left(\lambda - \int_t^T b(s)p(s)\,ds\right)\right]' R^{-1}(t)\left[h'(t)\left(\lambda - \int_t^T b(s)p(s)\,ds\right)\right] dt$$

$$+ \int_0^T p'(t)Q^{-1}(t)p(t)\,dt.$$

The solution of the dual problem is determined by the formulas

$$p^0(t) = J_0^{-1}Q(t)b'(t)\tilde{\xi}(t), \qquad \lambda^0 = J_0^{-1}\tilde{\eta}(T). \tag{3.53}$$

Moreover,

$$J_0 = J^0.$$

PROOF. Theorem 3.2 follows directly from (3.10), (3.17), (3.18), and (3.30).

Dual problem (3.48), (3.49) is as difficult as the primal one (3.47), (3.6). Nevertheless, with the help of the dual problem, taking into account the equality (3.50), we can obtain the lower bound for I_0. To do this, we should select a "proper" feasible pair $(\hat{p}, \hat{\lambda})$ of the dual problem. Then

$$I_0 = I^0 \geq \tilde{a}'\hat{\lambda}.$$

The problem (3.51), (3.52), in some sence, approximates the problem (3.48), (3.49). So, the solution of (3.51), (3.52) prompts us the search direction. In accordance with our approach, we take as the "proper" pair the pair that is a solution of a restricted problem (3.48), (3.49) along the direction determined by the solution (p^0, λ^0) of the problem (3.51), (3.52). In other words, we shall find the solution of (3.48), (3.49) along the line $\nu(p^0, \lambda^0)$, $\nu \in \mathbf{R}^1$.

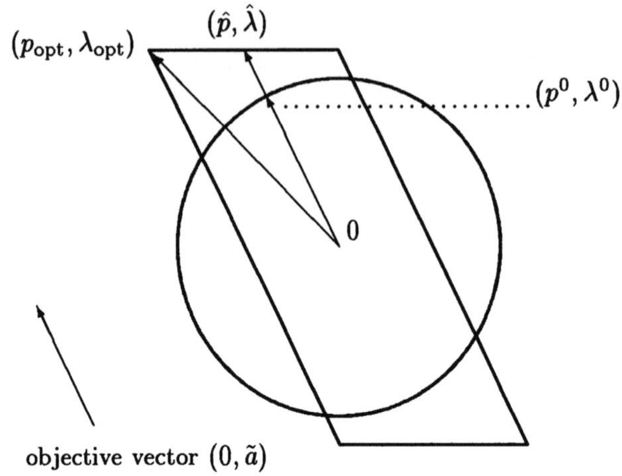

Fig. 1. **Geometric illustration**

Our approach has a simple geometric illustration (see Figure 1). A parallelogram in Figure 1 displays the domain (3.49) and a circle displays the domain (3.52). The dual problem (3.48), (3.49) (or (3.51), (3.52)) is to find a feasible pair (p, λ) that has a maximal projection on the common objective vector $(0, \tilde{a})$. In Figure 1 there are shown: the solution of (3.48), (3.49) $(p_{\rm opt}, \lambda_{\rm opt})$; the solution of (3.51), (3.52) (p^0, λ^0); the feasible pair $(\hat{p}, \hat{\lambda})$, which approximates the solution $(p_{\rm opt}, \lambda_{\rm opt})$.

Let us perform the corresponding computations. In accordance with (3.49), (3.53), the pair $\nu\,(p^0, \lambda^0)$, $\nu \in \mathbf{R}^1$, is feasible if and only if

$$\mu\sigma_l^{-1}(t)\left| h_l'(t)\left(\tilde{\eta}(T) - \int_t^T b(s)Q(s)b'(s)\tilde{\xi}(s)\,ds\right)\right| \le 1, \qquad t \in [0,T], \quad l = 1,\dots,m, \tag{3.54}$$

$$\mu\gamma_j^{-1}(t)\left|\left(Q(t)b'(t)\tilde{\xi}(t)\right)_j\right| \le 1, \qquad t \in [0,T], \quad j = 1,\dots,r,$$

$$\mu\sigma_{-d}^{-1}\left| e_d'\left(\tilde{\eta}(T) - \int_0^T b(s)Q(s)b'(s)\tilde{\xi}(s)\,ds\right)\right| \le 1, \qquad d = 1,\dots,n,$$

$$\mu = |\nu| J_0^{-1}.$$

In accordance with (2.8), (2.9), and (2.14), conditions (3.54) are equivalent to the system of inequalities

$$\mu \left| \beta_1 \sigma_l(t) \varphi_l(t) \right| \leq 1, \qquad t \in [0,T], \qquad l = 1, \ldots, m, \tag{3.55}$$

$$\mu \left| \beta_2 \gamma_j(t) b_j'(t) \tilde{\xi}(t) \right| \leq 1, \qquad t \in [0,T], \qquad j = 1, \ldots, r,$$

$$\mu \left| \sigma_{-d} \varphi_{-d} \right| \leq 1, \qquad d = 1, \ldots, n.$$

Consequently the pair $\nu(p^0, \lambda^0)$ is admissible if and only if

$$|\nu| \leq \nu^0 = J_0 \mathcal{N}_\infty^{-1},$$

where $\mathcal{N}_\infty$ is defined by (2.16). By virtue of (2.9), (2.10), (2.12), and (3.53), the value of the objective function $\tilde{a}'\lambda$ along the chosen direction is equal to

$$\nu \tilde{a}' \lambda^0 = \nu J_0^{-1} \tilde{a}' \tilde{\eta}(T) = \nu J_0^{-1} \tilde{a}' \tilde{P}(T) \tilde{a} = \nu J_0.$$

Therefore the pair $\nu(p^0, \lambda^0)$ provides the maximal value to the objective function at $\nu = \nu^0$ and we should take $(\hat{p}, \hat{\lambda}) = \nu^0(p^0, \lambda^0)$. By virtue of (2.13),

$$\int_0^t h(s) \varphi(s)\, ds + \varphi_- = \tilde{\xi}(t);$$

hence, taking into account, (2.14), (1.14), and (2.7), we see that

$$I(\varphi_-, \varphi) = \mathcal{N}_1 \quad \text{and} \quad J_0 = J(\varphi_-, \varphi) = \mathcal{N}_2,$$

where $\mathcal{N}_1$ and $\mathcal{N}_2$ are defined by (2.16). Thus the desired estimate from below for I_0 has the form

$$I_0 = I^0 \geq \nu^0 J_0 = J_0^2 \mathcal{N}_\infty^{-1} = \mathcal{N}_2^2 \mathcal{N}_\infty^{-1}. \tag{3.56}$$

Finally, the estimate (2.15) follows from (3.46) and (3.56). Theorem 2.2 is proved. □

3.4. *Proof of Theorem 2.2 for Non-Negative σ_{-d} and $\gamma_j(t)$*

Theorem 2.2 was proved under assumptions (1.6). In contrast to (1.6), let us consider the degenerate case. Suppose that some bounds for uncertainties in initial system state and in system dynamics may attain zero values:

$$\gamma_j(t) \geq \varepsilon \gamma^0, \qquad \gamma^0 > 0, \qquad j \in \mathcal{J}_1, \qquad \varepsilon \geq 0, \tag{3.57}$$

$$\gamma_j(t) \geq \gamma^0 > 0, \qquad j \in \mathcal{J}_2, \qquad \mathcal{J}_1 \cup \mathcal{J}_2 = \{1, \ldots, r\};$$

$$\sigma_l(t) \geq \sigma^0 > 0, \qquad l = 1, \ldots, m;$$

$$\sigma_{-d} \geq \varepsilon\sigma^0, \qquad \sigma^0 > 0, \qquad d \in \mathcal{D}_1,$$

$$\sigma_{-d} > 0, \qquad d \in \mathcal{D}_2, \qquad \mathcal{D}_1 \cup \mathcal{D}_2 = \{1, \ldots, n\}.$$

Let us consider the limiting case: $\varepsilon \to 0^+$. The auxiliary matrices for the linear-quadratic problem (2.6) are still defined by (2.14). Though the auxiliary dual problem (3.51), (3.52) has no explicit limiting form as $\varepsilon \to 0^+$, it follows from the proof of Theorem 2.1 for singular matrices P_0 and $Q(t)$ that the solution of the auxiliary dual problem (p^0, λ^0) does have a limit, which is determined by (3.53).

The idea of the proof is the same as above. Arguing as in the proof of Theorem 2.2 for the case of strictly positive σ_{-d} and $\gamma_j(t)$, we consider the dual problem for the original nonsmooth variational problem (3.47), (3.6). Note that this dual problem was obtained in Theorem 5.2 of Chapter 2 in the form

$$I^0 = \sup_{p \in L_1^r, \lambda \in \mathbf{R}^n} \lambda' \tilde{a}$$

under the constraints

$$\left| h_l'(t) \left(\lambda - \int_t^T b(s)p(s)\, ds \right) \right| \leq \sigma_l(t), \quad t \in [0, T], \quad l = 1, \ldots, m, \tag{3.58}$$

$$|p_j(t)| \leq \gamma_j(t), \qquad t \in [0, T], \qquad j = 1, \ldots, r,$$

$$\left| e_d' \left(\lambda - \int_0^T b(t)p(t)\, dt \right) \right| \leq \sigma_{-d}, \qquad d = 1, \ldots, n,$$

$$e_d = (\underbrace{0, \ldots, 1}_{d}, 0, \ldots, 0)'.$$

Substituting the pair $\nu(p^0, \lambda^0)$ into constraints (3.58), we get the analog of (3.55):

$$\mu \left| \beta_1 \sigma_l^2(t) \varphi_l(t) \right| \leq \sigma_l(t), \qquad t \in [0, T], \qquad l = 1, \ldots, m,$$

$$\mu \left| \beta_2 \gamma_j^2(t) b_j'(t) \tilde{\xi}(t) \right| \leq \gamma_j(t), \qquad t \in [0, T], \qquad j = 1, \ldots, r,$$

$$\mu \left| \sigma_{-d}^2 \varphi_{-d} \right| \leq \sigma_{-d}, \qquad d = 1, \ldots, n.$$

So, these constraints for zero σ_{-d}, $\gamma_j(t)$ hold trivially. The final part of the proof coincides with that of the nondegenerate case. Thus Theorem 2.2 is valid for degenerate case (3.57) as well. □

3.5. *Proof of Lemma 2.1*

Consider the problem (1.9), (1.10), (2.6), (2.14), (2.26), (2.27). Obviously, the pair

$$\bar{\Phi}_- = 0, \qquad \bar{\Phi}(t) = R^{-1}(t) h'(t) \mathcal{O}^{-1} \tilde{a}$$

satisfies (1.12). Hence,

$$J(\varphi_-(\varepsilon), \varphi(\varepsilon)) \leq J(\bar{\Phi}_-, \bar{\Phi}) = L = \text{const}, \quad \forall \varepsilon.$$

Therefore, for any $\varepsilon > 0$,

$$\varphi_-'(\varepsilon) P_0(\varepsilon) \varphi_-(\varepsilon) = \sum_{d=1}^{n} \sigma_{-d}^2(\varepsilon) \varphi_-^2(\varepsilon) \leq L$$

and

$$\int_0^T \varphi'(t, \varepsilon) R(t) \varphi(t, \varepsilon)\, dt \leq L.$$

Hence,

$$\varphi_-(\varepsilon) \to 0 \quad \text{as} \quad \varepsilon \to 0^+ \qquad \text{(the first equality of (2.28))}$$

and, since $R(t) \geq R_0 > 0$, we get

$$c \, \|\varphi(\cdot, \varepsilon)\|_{L_2^m}^2 \leq \int_0^T \varphi'(t, \varepsilon) R_0 \varphi(t, \varepsilon)\, dt \leq \int_0^T \varphi'(t, \varepsilon) R(t) \varphi(t, \varepsilon)\, dt \leq L,$$

where $c > 0$ is the minimal eigenvalue of R_0. Consequently the set $\varphi(\cdot, \varepsilon)$ is bounded:

$$\|\varphi(\cdot, \varepsilon)\|_{L_2^m}^2 \leq c^{-1} L \qquad \forall \varepsilon. \tag{3.59}$$

Using the unbiasness condition, we rewrite the problem (2.30) in the form

$$\min_{\Phi \in L_2^m} G(\Phi) \quad \text{under the constraint} \quad \int_0^T h(s) \Phi(s)\, ds = \tilde{a}, \tag{3.60}$$

where

$$G(\Phi) = \int_0^T \Phi'(t)R(t)\Phi(t)\,dt$$

$$+ \int_0^T \left(\tilde{a} - \int_t^T h(s)\Phi(s)\,ds\right)' b(t)Q(t)b'(t)\left(\tilde{a} - \int_t^T h(s)\Phi(s)\,ds\right) dt.$$

The pair $(\varphi_-(\varepsilon), \varphi(t,\varepsilon))$ is the solution of (1.12), (2.6), (2.7), (2.14), and (2.26). Therefore, for the fixed $\varphi_-(\varepsilon)$, the function $\varphi(t,\varepsilon)$ is the solution of the problem

$$\min_{\Phi \in L_2^m} G(\Phi) \quad \text{under the constraint} \quad \int_0^T h(s)\Phi(s)\,ds = \tilde{a} - \varphi_-(\varepsilon). \qquad (3.61)$$

PROPOSITION 3.3. *Suppose* $R(t) \geq R_0 > 0$. *Then the following assertions hold.*
1^0. *The functional* $G(\Phi)\colon L_2^m \to \mathbf{R}^1$ *is coercive:*

$$G(\Phi) \to \infty \quad \text{as} \quad \|\Phi\|_{L_2^m} \to \infty.$$

2^0. *The functional* $G(\Phi)$ *is strongly convex:*

$$G(\alpha\Phi + (1-\alpha)\Psi) \leq \alpha G(\Phi) + (1-\alpha)G(\Psi) - \alpha(1-\alpha)\|\Phi - \Psi\|^2_{L_2^m}$$

for any $\alpha \in [0,1]$.
3^0. *The functional* $G(\Phi)$ *is Lipschitzian on any bounded set* $\mathcal{B}$:

$$|G(\Phi) - G(\Psi)| \leq C(\mathcal{B})\|\Phi - \Psi\|_{L_2^m}, \qquad \forall\, \Phi, \Psi \in \mathcal{B}. \qquad (3.62)$$

4^0. *Suppose* Φ *satisfies the unbiasness condition from* (3.60); *then*

$$\|\Phi - \varphi_\infty\|^2_{L_2^m} \leq \frac{2}{c}(G(\Phi) - G(\varphi_\infty)). \qquad (3.63)$$

PROOF. Assertions 1^0, 2^0, and 4^0 can be verified similar to these of Proposition 3.1. Let us prove Assertion 3^0.

It can easily be checked that

$$G(\Phi) - G(\Psi) = \int_0^T (\Phi(t) - \Psi(t))' R(t)\Phi(t)\,dt$$

$$+ \int_0^T \Psi'(t)R(t)(\Phi(t) - \Psi(t))\,dt$$

$$+\int_0^T \left(\int_0^t h(s)\big(\Phi(s)-\Psi(s)\big)\,ds\right)' b(t)Q(t)b'(t)\left(\int_0^t h(s)\Phi(s)\,ds\right) dt$$

$$+\int_0^T \Psi'(t)b(t)Q(t)b'(t)\left(\int_0^t h(s)\big(\Phi(s)-\Psi(s)\big)\,ds\right) dt$$

$$-\,2\tilde{a}'\int_0^T b(t)Q(t)b'(t)\int_0^t h(s)\big(\Phi(s)-\Psi(s)\big)\,ds\,dt.$$

Thus, using the Cauchy-Schwarz-Bunyakovskii inequality, we get

$$|G(\Phi)-G(\Psi)| \le \bar{C}_1\|\Phi\|_{L_2^m}\|\Phi-\Psi\|_{L_2^m} + \bar{C}_2\|\Psi\|_{L_2^m}\|\Phi-\Psi\|_{L_2^m}$$

$$+\,\bar{C}_3\|\Phi-\Psi\|_{L_2^m}.$$

Hence (3.62) holds. Proposition 3.1 is proved.

PROPOSITION 3.4. *The solutions of the problems* (3.60) *and* (3.61) *exist and are unique. Moreover,*

$$\lim_{\varepsilon\to 0+}\|\varphi(\cdot,\varepsilon)-\varphi_\infty(\cdot)\|_{L_2^m}=0. \tag{3.64}$$

PROOF. In accordance with Proposition 3.3, the functional $G(\Phi)$ is continuous, coercive, and strongly convex (consequently $G(\Phi)$ is strictly convex). Obviously, the constraints in the problems (3.60) and (3.61) define convex and closed affine manifolds. The space L_2^m is Hilbert. Then, by virtue of Lemma 1.2 of Chapter 2, the solutions of (3.60) and (3.61) exist and are unique.

For brevity, we denote by φ^ε, φ_∞ the functions $\varphi(t,\varepsilon)$, $\varphi_\infty(t)$, respectively. Introduce the functions

$$\delta\varphi^\varepsilon=\delta\varphi(t,\varepsilon)=R^{-1}(t)h'(t)\mathcal{O}^{-1}\varphi_-(\varepsilon)\quad\text{and}\quad \bar{\varphi}^\varepsilon=\bar{\varphi}(t,\varepsilon)=\varphi^\varepsilon+\delta\varphi^\varepsilon.$$

Note that $\delta\varphi^\varepsilon\to 0$ as $\varepsilon\to 0^+$.

The element $\bar{\varphi}^\varepsilon=\varphi^\varepsilon+\delta\varphi^\varepsilon$ is admissible for (3.60); the functional $G(\Phi)$ is Lipschitzian. Hence, taking into account (3.59),

$$G(\varphi_\infty)\le G(\bar{\varphi}^\varepsilon)=G(\varphi^\varepsilon+\delta\varphi^\varepsilon)=G(\varphi^\varepsilon)+c_1(\varepsilon),\qquad c_1(\varepsilon)\to 0.$$

The element $\tilde{\varphi}^\varepsilon=\varphi_\infty-\delta\varphi^\varepsilon$ is admissible for (3.61). Hence,

$$G(\varphi^\varepsilon)\le G(\tilde{\varphi}^\varepsilon)=G(\varphi_\infty-\delta\varphi^\varepsilon)=G(\varphi_\infty)+c_2(\varepsilon),\qquad c_2(\varepsilon)\to 0.$$

Consequently,

$$G(\varphi_\infty)-c_1(\varepsilon)\le G(\varphi^\varepsilon)\le G(\varphi_\infty)+c_2(\varepsilon),$$

whence,

$$G(\varphi^{\varepsilon}) \to G(\varphi_{\infty}) \quad \text{as} \quad \varepsilon \to 0^{+}.$$

Since G is Lipschitzian, $\delta\varphi^{\varepsilon} \to 0$, and $\bar{\varphi}^{\varepsilon}$, φ^{ε} are bounded for all ε, we have

$$G(\bar{\varphi}^{\varepsilon}) = G(\varphi^{\varepsilon}) + (G(\bar{\varphi}^{\varepsilon}) - G(\varphi^{\varepsilon})) \to G(\varphi_{\infty}).$$

Then it follows from (3.63) that

$$\|\varphi^{\varepsilon} - \varphi_{\infty}\|_{L_2^m} = \|\bar{\varphi}^{\varepsilon} - \varphi_{\infty}\|_{L_2^m} + \|\varphi^{\varepsilon} - \bar{\varphi}^{\varepsilon}\|_{L_2^m} \to 0 \quad \text{as} \quad \varepsilon \to 0^{+}.$$

Thus (3.64) holds. Proposition 3.4 is proved. The third relation from (2.28) is equivalent to (3.64).

In accordance with (2.13) and (3.64),

$$\tilde{\xi}(t,\varepsilon) \to \tilde{\xi}_{\infty}(t) \triangleq \tilde{a} - \int_t^T h(s)\varphi_{\infty}(s)\,ds \qquad \text{uniformly in } t.$$

In fact,

$$\tilde{\xi}(t,\varepsilon) - \tilde{\xi}_{\infty}(t) = -\int_t^T h(s)\big(\varphi(s,\varepsilon) - \varphi_{\infty}(s)\big)\,ds.$$

Hence, using the Cauchy-Schwarz-Bunyakovskii inequality,

$$\|\tilde{\xi}(t,\varepsilon) - \tilde{\xi}_{\infty}(t)\|_{\mathbf{R}^n} \le c_3\,\|\varphi(\cdot,\varepsilon) - \varphi_{\infty}(\cdot)\|_{L_2^m} \qquad \text{for all} \quad t \in [0,T]$$

and the first equality from (2.29) is proved.

By virtue of (2.8) and (2.9),

$$\varphi_{-}(\varepsilon) = \tilde{\xi}(0,\varepsilon) = \tilde{a} - \int_0^T h(s)R^{-1}(s)h'(s)$$

$$\times\left[\tilde{\eta}(0,\varepsilon) + \int_0^s b(\tau)Q(\tau)b'(\tau)\tilde{\xi}(\tau,\varepsilon)\,d\tau\right] ds = -\mathcal{O}\tilde{\eta}(0,\varepsilon) + v(\varepsilon),$$

where

$$v(\varepsilon) = \tilde{a} - \int_0^T h(s)R^{-1}(s)h'(s)\int_0^s b(\tau)Q(\tau)b'(\tau)\tilde{\xi}(\tau,\varepsilon)\,d\tau\,ds.$$

Thus,

$$\tilde{\eta}(0,\varepsilon) = -\mathcal{O}^{-1}\varphi_{-}(\varepsilon) + \mathcal{O}^{-1}v(\varepsilon).$$

Since $\varphi_{-}(\varepsilon) \to 0$ and $\tilde{\xi}(t,\varepsilon) \to \tilde{\xi}_{\infty}(t)$, we see that $\tilde{\eta}(0,\varepsilon)$ has a limit:

$$\tilde{\eta}(0,\varepsilon) \to \tilde{\eta}_{\infty}(0). \tag{3.65}$$

According to (2.8), boundary conditions from (2.9), and (2.26),

$$\tilde{\eta}(0,\varepsilon) = P_0(\varepsilon)\varphi_-(\varepsilon) \tag{3.66}$$

$$= \varepsilon^{-1}\big(\sigma_{-1}(\varepsilon)\varphi_{-1}(\varepsilon)\bar{\sigma}_{-1},\ldots,\sigma_{-n}(\varepsilon)\varphi_{-n}(\varepsilon)\bar{\sigma}_{-n},\big)'.$$

Then from (3.65) and (3.66) we obtain the second relation from (2.28). Note also that if $\sigma_{-d}\varphi_{-d}(\varepsilon) \to 0\ \forall d$, then $\varphi'_-(\varepsilon)P_0(\varepsilon)\varphi_-(\varepsilon) \to 0$.

Finally, we get from (2.9)

$$\tilde{\eta}(t,\varepsilon) = \tilde{\eta}(0,\varepsilon) + \int_0^t b(s)Q(s)b'(s)\tilde{\xi}(s,\varepsilon)\,ds.$$

Since $\tilde{\eta}(0,\varepsilon) \to \tilde{\eta}_\infty(0)$ and $\tilde{\xi}(t,\varepsilon) \to \tilde{\xi}_\infty(t)$, we get the second relation of (2.29). □

4. Kalman Filter with Discrete Measurements

In almost all applications, the measurements are carried out at discrete instants. Therefore, in this section, we study the case of discrete measurements in detail.

Consider a continuous linear dynamic system

$$\dot{x}(t) = A(t)x(t) + B(t)u(t), \qquad t \in [0,T], \quad x(0) = x_0, \tag{4.1}$$

where $x(t) \in \mathbf{R}^n$ is the state vector, $u(t) \in \mathbf{R}^r$ is a perturbation vector.

We assume that

$$|u_j(t)| \le \gamma_j(t), \qquad t \in [0,T], \quad j = 1,\ldots,r. \tag{4.2}$$

Let the discrete measurements of the state vector are available on the segment $[0,T]$:

$$z(t_i) = H'(t_i)x(t_i) + \varrho(t_i), \qquad z(t_i), \varrho(t_i) \in \mathbf{R}^m, \tag{4.3}$$

$$|\varrho(t_i)| \le \sigma_l(t_i), \tag{4.4}$$

$$i = 0,\ldots,N, \qquad l = 1,\ldots,m, \qquad t_0 = 0, \qquad t_N = T,$$

where $z(t_i)$ are the measurements, $\varrho(t_i)$ is a measurement noise.

Moreover, the initial state of the system is assumed to belong to the parallelepiped

$$|x_d(0)| \le \sigma_{-d}, \qquad d = 1,\ldots,n. \tag{4.5}$$

We suppose again that the entries of the matrices $A(t)$, $B(t)$, and the functions $u_j(t)$, $\gamma_j(t)$ are piecewise continuous. We also suppose that these functions at the discontinuity points are continuous from the right, possibly, except for the point T at which the functions are supposed to be continuous from the left. In addition, we also assume that the uncertainty bounds are restricted from below:

$$\gamma_j(t) \geq \gamma^0 > 0, \quad j = 1, \dots, r, \tag{4.6}$$

$$\sigma_l(t_i) \geq \sigma^0 > 0, \quad l = 1, \dots, m, \quad i = 0, \dots, N,$$

$$\sigma_{-d} > 0, \quad d = 1, \dots, n.$$

Similar to the case of continuous measurements, we shall use the notation $x(0) \in \Pi$, $u \in U$, and $\varrho \in \sigma_D$ for the sets (4.5), (4.2), and (4.4), respectively.

The problem is to estimate a scalar quantity $l_* = a'x(T)$ with the help of linear functionals

$$\hat{l} = \sum_{i=0}^{N} \Phi'(t_i)z(t_i), \qquad \Phi(t_i) \in \mathbf{R}^m. \tag{4.7}$$

As above, we introduce the notation

$$y(t) = X^{-1}(t)x(t) \in \mathbf{R}^n, \qquad h(t_i) = X'(t_i)H(t_i) \in \mathbf{R}^{n\times m},$$

$$b(t) = X^{-1}(t)B(t) \in \mathbf{R}^{n\times r}, \qquad \tilde{a} = X'(T)a,$$

$$h(t_i) = (h_1(t_i), \dots, h_m(t_i)), \qquad h_l(t_i) \in \mathbf{R}^n, \quad l = 1, \dots, m.$$

$$b(t) = (b_1(t), \dots, b_r(t)), \qquad b_j(t) \in \mathbf{R}^n, \quad j = 1, \dots, r,$$

where the matrix $X(t) \in \mathbf{R}^{n\times n}$ satisfies the homogeneous equation

$$\dot{X}(t) = A(t)X(t), \qquad X(0) = E_n.$$

Then,

$$\dot{y}(t) = b(t)u(t), \quad y(t) \in \mathbf{R}^n, \quad y(0) = x(0), \qquad t \in [0, T], \tag{4.8}$$

$$z(t_i) = h'(t_i)y(t_i) + \varrho(t_i), \tag{4.9}$$

and

$$l_* = \tilde{a}'y(T).$$

It can easily be verified (see Chapter 2) that the guaranteed estimation error under the estimator $\Phi = \{\Phi(t_i)\}_0^N$ is determined by the expression

$$\sup_{x(0)\in\Pi,\, u\in U,\, \varrho\in\sigma_D} |\hat{l} - l_*| = I_D(\Phi_-, \Phi) \tag{4.10}$$

$$= \sum_{d=1}^{n} \sigma_{-d}|\Phi_{-d}| + \sum_{l=1}^{m}\sum_{i=0}^{N} \sigma_l(t_i)|\Phi_l(t_i)|$$

$$+ \sum_{j=1}^{r} \int_0^T \gamma_j(t) \left| b_j'(t) \left(\Phi_- + \sum_{\{i|t_i \le t\}} h(t_i)\Phi(t_i) \right) \right| dt,$$

where

$$\Phi_- + \sum_{i=0}^{N} h(t_i)\Phi(t_i) = \tilde{a}. \tag{4.11}$$

Note that the last term in the expression for the guaranteed estimation error can be represented in a more standard form

$$\sum_{j=1}^{r} \int_0^T \gamma_j(t) \left| b_j'(t) \left(\Phi_- + \sum_{\{i|t_i \le t\}} h(t_i)\Phi(t_i) \right) \right| dt$$

$$= \sum_{j=1}^{r} \sum_{i=0}^{N-1} \int_{t_i}^{t_{i+1}} \gamma_j(t) \left| b_j'(t) \left(\Phi_- + \sum_{s=0}^{i} h(t_s)\Phi(t_s) \right) \right| dt.$$

The optimal *a priori* guaranteed estimation problem is to find the estimator that yields the minimal value to the guaranteed estimation error. In other words, we have to solve the following variational problem:

$$I_{D0} = \inf_{\substack{\Phi_- \in \mathbf{R}^n,\ \Phi(t_i) \in \mathbf{R}^m, \\ i=0,\dots,N}} I_D(\Phi_-, \Phi) \quad \text{under constraint (4.11)}. \tag{4.12}$$

Note that relations (4.10), (4.11) can be obtained from (1.14), (1.12) if we formally set

$$\Phi(t) = \sum_{i=0}^{N} \Phi(t_i)\delta(t - t_i).$$

The solution of (4.12) (Φ_-^0, Φ^0) determines the optimal estimator Φ^0.

The solving of (4.12) is a difficult numerical problem [31], [133], [139], [19], [135], [214], [140], [136]. In order to solve this problem approximately let us use the discrete Kalman filter [84], [79].

It should be emphasized that, in this book, the "Kalman filter" means a numerical algorithm for data processing (regardless of the actual noise models).

The level of the nonoptimality of the Kalman filter in the guaranteed estimation problem we define by the ratio

$$\Delta_D = \frac{\sup_{x(0)\in\Pi,\, u\in U,\, \varrho\in\sigma_D} |\hat{l}_\varphi - l_*|}{\sup_{x(0)\in\Pi,\, u\in U,\, \varrho\in\sigma_D} |\hat{l}_0 - l_*|} = \frac{\sup_{y(0)\in\Pi,\, u\in U,\, \varrho\in\sigma_D} |\hat{l}_\varphi - l_*|}{\sup_{y(0)\in\Pi,\, u\in U,\, \varrho\in\sigma_D} |\hat{l}_0 - l_*|},$$

where

$$\hat{l}_0 = \sum_{i=0}^{N} \Phi^{0\prime}(t_i)z(t_i) \quad \text{and} \quad \hat{l}_\varphi = \sum_{i=0}^{N} \varphi'(t_i)z(t_i)$$

are the unknown optimal estimator and the discrete Kalman filter estimator, respectively. Obviously,

$$1 \le \Delta_D = \frac{I_D(\varphi_-, \varphi)}{I_{D0}},$$

where (φ_-, φ) satisfies (4.11).

Our aim is to evaluate Δ_D from above (cf. Section 2).

Let us cite the necessary facts from the Kalman filter theory for discrete measurements.

The Kalman filter for the system (4.8), (4.9) is described by the following relations [84], [79], [47], [187]:

$$\hat{l}_\varphi = \tilde{a}'\hat{y}(T|N) = \tilde{a}'\hat{y}(t_N|N),$$

$$\hat{y}(t_i|i) = \hat{y}(t_{i-1}|i-1) + \tilde{K}(t_i)\big(z(t_i) - h'(t_i)\hat{y}(t_{i-1}|i-1)\big),$$

$$\hat{y}(t_{-1}|-1) = 0,$$

$$\tilde{K}(t_i) = \tilde{P}(t_i|i-1)h(t_i)\big(h'(t_i)\tilde{P}(t_i|i-1)h(t_i) + R(t_i)\big)^{-1}, \tag{4.13}$$

$$\tilde{P}(t_i|i) = \tilde{P}(t_i|i-1) - \tilde{K}(t_i)h'(t_i)\tilde{P}(t_i|i-1), \tag{4.14}$$

$$\tilde{P}(t_0|-1) = P_0, \qquad i = 0, \dots, N,$$

$$\tilde{P}(t_i|i-1) = \tilde{P}(t_{i-1}|i-1) + \int_{t_{i-1}}^{t_i} b(\tau)Q(\tau)b'(\tau)\,d\tau, \qquad i = 1, \dots, N.$$

The symmetric matrices $P_0 \ge 0$, $R(t_i) > 0$, and $Q(t) \ge 0$ are specified by a researcher; P_0 characterizes the uncertainty in the initial state, $R(t_i)$ and $Q(t)$ characterize the uncertainty in the measurements and in the system.

The Kalman estimate $\hat{l}_\varphi$ can be presented in the form (4.7):

$$\hat{l}_\varphi = \sum_{i=0}^{N} \tilde{a}'\tilde{\Theta}(T, t_i)\tilde{K}(t_i)z(t_i), \tag{4.15}$$

where

$$\tilde{\Theta}(T, t_i) = \prod_{j=i+1}^{N} \tilde{\Theta}(t_j, t_{j-1}) \triangleq \tilde{\Theta}(t_N, t_{N-1}) \cdot \ldots \cdot \tilde{\Theta}(t_{i+1}, t_i),$$

$$\tilde{\Theta}(t_i, t_{i-1}) = E_n - \tilde{K}(t_i)h'(t_i), \qquad \tilde{\Theta}(T,T) = \tilde{\Theta}(t_N, t_N) = E_n.$$

Thus, for the Kalman filter,

$$\Phi(t_i) = \varphi(t_i) = \tilde{K}'(t_i)\tilde{\Theta}'(T, t_i)\tilde{a}, \qquad i = 0, \ldots, N.$$

We call $\varphi = \{\varphi(t_i)\}_0^N$ the *Kalman estimator.*

THEOREM 4.1. 1^0. *The Kalman estimator* φ *is determined by the solution* (φ_-, φ) *of the following variational problem:*

$$J_{D0} = \inf_{\substack{\Phi_- \in \mathbf{R}^n,\, \Phi(t_i) \in \mathbf{R}^m, \\ i=0,\ldots,N}} J_D(\Phi_-, \Phi) \quad \textit{under constraint} \quad (4.11), \tag{4.16}$$

where

$$J_D(\Phi_-, \Phi) = \left\{ \Phi'_- P_0 \Phi_- + \sum_{i=0}^{N} \Phi'(t_i) R(t_i) \Phi(t_i) \right. \tag{4.17}$$

$$\left. + \int_0^T \Big(\Phi_- + \sum_{\{i | t_i \le t\}} h(t_i)\Phi(t_i) \Big)' b(t)Q(t)b'(t) \Big(\Phi_- + \sum_{\{i | t_i \le t\}} h(t_i)\Phi(t_i) \Big)\, dt \right\}^{\frac{1}{2}}.$$

2^0. *The solution of the problem* (4.16) (φ_-, φ) *is determined by the following formulas:*

$$\varphi_- = \big(E_n - \tilde{K}(t_0)h'(t_0)\big)'\tilde{\xi}(t_0) \tag{4.18}$$

$$= \big(E_n + h(t_0)R^{-1}(t_0)h'(t_0)P_0\big)^{-1}\tilde{\xi}(t_0),$$

$$\varphi(t_i) = R^{-1}(t_i)h'(t_i)\tilde{\eta}(t_i) = \tilde{K}'(t_i)\tilde{\xi}(t_i),$$

where

$$\tilde{K}(t_i) = \tilde{P}(t_i | i-1)h(t_i)\big(h'(t_i)\tilde{P}(t_i | i-1)h(t_i) + R(t_i)\big)^{-1} \tag{4.19}$$

$$= \tilde{P}(t_i | i)h(t_i)R^{-1}(t_i),$$

$$i = 0, \ldots, N,$$

and $\tilde{\xi}(t_i)$, $\tilde{\eta}(t_i)$ *satisfy the boundary value problem*

$$\tilde{\xi}(t_{i+1}) = \tilde{\xi}(t_i) + h(t_{i+1})R^{-1}(t_{i+1})h'(t_{i+1})\,\tilde{\eta}(t_{i+1}), \tag{4.20}$$

$$\tilde{\eta}(t_{i+1}) = \tilde{\eta}(t_i) + \int_{t_i}^{t_{i+1}} b(t)Q(t)b'(t)\, dt\; \tilde{\xi}(t_i),$$

$$i = 0, \ldots, N-1,$$

with the boundary conditions

$$\tilde{\xi}(t_N) = \tilde{a}, \tag{4.21}$$

and

$$\tilde{\eta}(t_0) = P(t_0|0)\tilde{\xi}(t_0) = P_0\big(E_n - \tilde{K}(t_0)h'(t_0)\big)'\tilde{\xi}(t_0)$$

$$= P_0\big(E_n + h(t_0)R^{-1}(t_0)h'(t_0)P_0\big)^{-1}\tilde{\xi}(t_0)$$

$$= \big(E_n + P_0h(t_0)R^{-1}(t_0)h'(t_0)\big)^{-1}P_0\tilde{\xi}(t_0).$$

Moreover,

$$\tilde{\eta}(t_i) = \tilde{P}(t_i|i)\tilde{\xi}(t_i), \qquad i = 0, \ldots, N, \tag{4.22}$$

$$\tilde{\xi}(t_i) = \big(E_n - \tilde{K}(t_{i+1})h'(t_{i+1})\big)'\tilde{\xi}(t_{i+1}), \qquad i = 0, \ldots, N-1, \tag{4.23}$$

$$\tilde{\xi}(t_N) = \tilde{a},$$

and

$$J_{D0} = \Big(\tilde{a}'\tilde{P}(t_N|N)\tilde{a}\Big)^{\frac{1}{2}}, \tag{4.24}$$

where $\tilde{K}(t_i)$ *and* $\tilde{P}(t_i|i)$ *are defined by* (4.13) *and* (4.14).

The proof of Theorem 4.1 is given in Section 5.

It follows from (4.18), (4.20), and (4.21) that

$$\tilde{\xi}(t_i) = \sum_{j=0}^{i} h(t_j)\varphi(t_j) + \varphi_- = \tilde{a} - \sum_{j=i+1}^{N} h(t_j)\varphi(t_j). \tag{4.25}$$

Let us put

$$P_0 = \operatorname{diag}\big(\sigma_{-1}^2, \ldots, \sigma_{-n}^2\big), \quad R(t_i) = \beta_1 \operatorname{diag}\big(\sigma_1^2(t_i), \ldots, \sigma_m^2(t_i)\big), \tag{4.26}$$

$$Q(t) = \beta_2 \operatorname{diag}\big(\gamma_1^2(t), \ldots, \gamma_r^2(t)\big),$$

where β_1, β_2 are specified positive constants: $\beta_1, \beta_2 \in \mathbf{R}^1$, $\beta_1, \beta_2 > 0$.

Note that the functionals $I_D(\Phi_-, \Phi)$ and $J_D(\Phi_-, \Phi)$ defined by (4.10) and (4.17) have the similar structure.

The main result for the case of discrete measurements is the following theorem.

THEOREM 4.2. *Let φ be the Kalman estimator and the pair (φ_-, φ) be the solution of the problem* (4.16). *Then*

$$1 \leq \Delta_D \leq \Delta_D^0, \qquad \Delta_D^0 = \frac{\mathcal{N}_{D1} \cdot \mathcal{N}_{D2}}{\mathcal{N}_{D\infty}^2}, \tag{4.27}$$

$$\mathcal{N}_{D1} = \sum_{d=1}^{n} |\Psi_{-d}| + \sum_{l=1}^{m} \beta_1^{-1} \sum_{i=0}^{N} |\Psi_l(t_i)| + \sum_{j=1}^{r} \beta_2^{-1} \int_0^T |\chi_j(t)|\, dt, \tag{4.28}$$

$$\mathcal{N}_{D2} = \left(\sum_{d=1}^{n} \Psi_{-d}^2 + \sum_{l=1}^{m} \beta_1^{-1} \sum_{i=0}^{N} \Psi_l^2(t) + \sum_{j=1}^{r} \beta_2^{-1} \int_0^T \chi_j^2(t)\, dt \right)^{\frac{1}{2}},$$

$$\mathcal{N}_{D\infty} = \max\left\{ \max_{1 \leq d \leq n} |\Psi_{-d}|, \sup_{1 \leq l \leq m,\ 0 \leq i \leq N} |\Psi_l(t_i)|, \sup_{1 \leq j \leq r,\ t \in [0,T)} |\chi_j(t)| \right\},$$

where

$$\Psi_{-d} = \sigma_{-d}\varphi_{-d}, \qquad d = 1, \ldots, n,$$

$$\Psi_l(t_i) = \beta_1 \sigma_l(t_i) \varphi_l(t_i), \qquad l = 1, \ldots, m, \qquad i = 0, \ldots, N,$$

$$\chi_j(t) = \beta_2 \gamma_j(t) b_j'(t) \tilde{\xi}(t), \qquad j = 1, \ldots, r, \qquad t \in [0, T),$$

$$\tilde{\xi}(t) = \tilde{\xi}(t_i), \qquad t_i \leq t < t_{i+1}, \qquad i = 0, \ldots, N-1,$$

and $\tilde{\xi}(t_i)$ is determined by (4.23) *or* (4.25);

$$\varphi_- = (\varphi_{-1}, \ldots, \varphi_{-n})', \qquad \varphi(t_i) = (\varphi_1(t_i), \ldots, \varphi_m(t_i))'.$$

The proof of Theorem 4.2 is given in Section 5.

It follows from Theorem 4.2 that the value of Δ_D^0 is determined by the characteristics of the discrete Kalman filter only. These characteristics can be computed easily. Therefore the estimate of the nonoptimality level Δ^0 can be efficiently computed without solving the original nonsmooth variational problem (4.12).

If Δ_D^0 is not large, then we do not need to resolve the difficult nonsmooth problem (4.12) and the discrete Kalman filter can be used in the optimal guaranteed estimation probem as well.

Let us turn back to the original variables. Consider the system (4.1), (4.3) again. For convenience, we define the solution of the boundary value problem on the whole of $[0, T]$

$$\tilde{\xi}(t) = \tilde{\xi}(t_i), \qquad \tilde{\eta}(t) = \tilde{\eta}(t_i) + \int_{t_i}^{t} b(\tau)Q(\tau)b'(\tau)\tilde{\xi}(\tau)\,d\tau,$$

$$t_i \le t < t_{i+1}, \qquad i = 0, \ldots, N-1.$$

Introduce new variables

$$K(t_i) = X(t_i)\tilde{K}(t_i),$$

$$P(t_i|i) = X(t_i)\tilde{P}(t_i)X'(t_i), \qquad P(t_i|i-1) = X(t_i)\tilde{P}(t_i|i-1)X'(t_i),$$

$$\xi(t) = \left(X^{-1}(t)\right)'\tilde{\xi}(t), \qquad \eta(t) = X(t)\tilde{\eta}(t).$$

Obviously, $\tilde{\eta}(t)$, $\eta(t)$ are continuous and $\tilde{\xi}(t)$, $\xi(t)$ are piecewise continuous (continuous from the right).

Then the Kalman filter for the system (4.1), (4.3) takes the standard form

$$\hat{l}_\varphi = a'\hat{x}(T|N) = a'\hat{x}(t_N|N),$$

$$\hat{x}(t_i|i) = \hat{x}(t_i|i-1) + K(t_i)\left(z(t_i) - H'(t_i)\hat{x}(t_i|i-1)\right), \qquad i = 0, \ldots, N,$$

$$\hat{x}(t_i|i-1) = \Upsilon(t_i, t_{i-1})\hat{x}(t_{i-1}|i-1), \qquad i = 1, \ldots, N,$$

$$\hat{x}(t_0|-1) = 0,$$

where $\Upsilon(t_i|t_{i-1}) = X(t_i)X^{-1}(t_{i-1})$ is the transition matrix of the homogeneous system $\dot{x}(t) = A(t)x(t)$,

$$K(t_i) = P(t_i|i-1)H(t_i)\left(H'(t_i)P(t_i|i-1)H(t_i) + R(t_i)\right)^{-1} \tag{4.29}$$

$$= P(t_i|i)H(t_i)R^{-1}(t_i),$$

$$P(t_i|i) = P(t_i|i-1) - K(t_i)H'(t_i)P(t_i|i-1), \tag{4.30}$$

$$P(t_0|-1) = P_0, \qquad i = 0, \ldots, N,$$

$$P(t_i|i-1) = \Upsilon(t_i, t_{i-1})P(t_{i-1}|i-1)\Upsilon'(t_i, t_{i-1})$$

$$+ \int_{t_{i-1}}^{t_i} \Upsilon(t_i, \tau)B(\tau)Q(\tau)B'(\tau)\Upsilon'(t_i, \tau)\,d\tau, \qquad i = 1, \ldots, N.$$

In these variables, the discrete Kalman estimator is determined by the formula

$$\varphi(t_i) = K'(t_i)\Theta'(T, t_i)a, \qquad i = 0, \ldots, N, \tag{4.31}$$

$$\varphi_- = \big(E_n - K(t_0)H'(t_0)\big)'\Theta'(T, 0)a,$$

where

$$\Theta(T, t_i) = \prod_{j=i+1}^{N} \Theta(t_j, t_{j-1}) \triangleq \Theta(t_N, t_{N-1}) \cdot \ldots \cdot \Theta(t_{i+1}, t_i),$$

$$\Theta(t_i, t_{i-1}) = \big(E_n - K(t_i)H'(t_i)\big)\Upsilon(t_i, t_{i-1}), \qquad i = 1, \ldots, N,$$

$$\Theta(T, T) = \Theta(t_N, t_N) = E_n.$$

On the other hand, the Kalman estimator can be expressed by the solution of the boundary value problem:

$$\varphi(t_i) = R^{-1}(t_i)H'(t_i)\eta(t_i) = K'(t_i)\xi(t_i), \tag{4.32}$$

$$\varphi_- = \xi^-(t_0) \triangleq \big(E_n - K(t_0)H'(t_0)\big)'\xi(t_0),$$

where the piecewise continuous (continuous from the right) function $\xi(t)$ and the continuous function $\eta(t)$ satisfy the boundary value problem

$$\dot{\xi}(t) = -A'(t)\xi(t), \qquad t_i \leq t < t_{i+1}, \qquad i = 0, \ldots, N-1,$$

$$\dot{\eta}(t) = B(t)Q(t)B'(t)\xi(t) + A(t)\eta(t).$$

The jumps of $\xi(t)$ are defined by the relations[6]

$$\xi(t_i) = \xi^-(t_i) + H(t_i)R^{-1}(t_i)H'(t_i)\eta(t_i),$$

$$\eta(t_i) = \eta^-(t_i), \qquad i = 0, \ldots, N,$$

$$\xi^-(t_i) = \lim_{t \to t_i - 0} \xi(t), \qquad \eta^-(t_i) = \lim_{t \to t_i - 0} \eta(t), \qquad i = 1, \ldots, N.$$

The boundary conditions take the form

$$\xi(T) = \xi(t_N) = a, \tag{4.33}$$

$$\eta(t_0) = P(t_0|0)\xi(t_0) = \big(E_n - K(t_0)H'(t_0)\big)P_0\xi(t_0)$$

[6]the jump at the point t_0 is not required for the definition of $\xi(t)$ and is inserted to simplify the formula for φ_- and the left boundary condition.

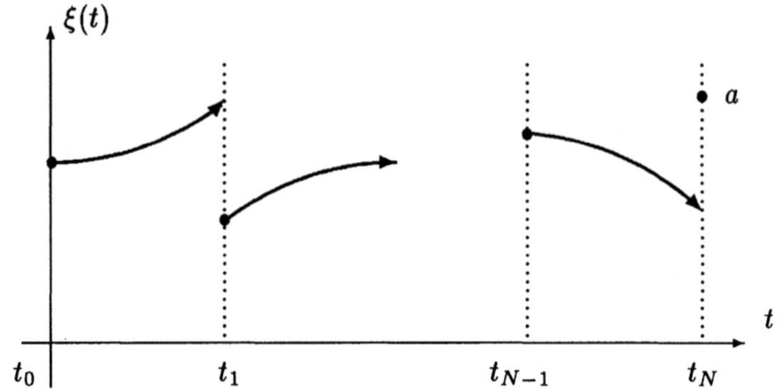

Fig. 2. Function $\xi(t)$

$$= \big(E_n + P_0 H(t_0) R^{-1}(t_0) H'(t_0)\big)^{-1} P_0 \xi(t_0)$$

$$= P_0\big(E_n + H(t_0) R^{-1}(t_0) H'(t_0) P_0\big)^{-1} \xi(t_0) = P_0 \xi^-(t_0).$$

Moreover,

$$\eta(t_i) = P(t_i|i)\xi(t_i), \qquad i = 0, \dots, N,$$

and, using (4.32),

$$\dot{\xi}(t) = -A'(t)\xi(t), \qquad t_i \le t < t_{i+1}, \qquad i = 0, \dots, N-1, \tag{4.34}$$

$$\xi^-(t_i) = \big(E_n - K(t_i)H'(t_i)\big)'\xi(t_i), \qquad \xi(t_N) = a, \qquad i = 0, \dots, N,$$

where $K(t)$, $P(t_i|i)$, and $P(t_i|i-1)$ are defined by (4.29) and (4.30).

It follows from (4.32) and (4.34) that

$$\xi(t) = \Upsilon'(t_{i+1}, t)\xi^-(t_{i+1}), \qquad t_i \le t < t_{i+1}, \qquad i = 0, \dots, N-1, \tag{4.35}$$

$$\xi^-(t_i) = \xi(t_i) - H(t_i)\varphi(t_i), \qquad \xi(t_N) = a, \qquad i = 0, \dots, N,$$

or with the initial value $\xi^-(t_0) = \varphi_-$. The function $\xi(t)$ is shown in Figure 2.

The only modification in Theorem 4.2 is that the quantity $\chi(t)$ is now computed by the formula

$$\chi_j(t) = \beta_2 \gamma_j(t) B_j'(t)\xi(t), \qquad B(t) = (B_1(t), \dots, B_r(t)), \quad j = 1, \dots, r,$$

and $\xi(t)$ is determined by (4.34) or (4.35).

4.1. *Computational Relations*

In order to determine the level of nonoptimality Δ_D^0 we have to compute the pair (φ_-, φ) and the function $\xi(t)$. The collection of formulas (4.26), (4.29) – (4.32), (4.34), and (4.35) generates different versions of computational relations. For example, {(4.26), (4.29), (4.30), (4.31), (4.34)}, {(4.26), (4.29), (4.30), (4.31), (4.35)}, or {(4.26), (4.29), (4.30), (4.34), (4.32)}.

If we have to compute Δ_D^0 given current values of T, then it is convenient to use the following iterative formulas. Let t_L be a current instant. We denote by

$$\left(\varphi_-^{(L)},\ \{\varphi^{(L)}(t_i)\}_{i=0}^{i=L}\right)$$

the pair that corresponds to the estimation of the quantity $l_*^{(L)} = a'x(t_L)$ by the measurements at the instants $t_0, \ldots, t_L$. Let us represent this pair in the form

$$\varphi_-^{(L)} = \left(a'\varpi_-^{(L)}\right)', \qquad \varpi_-^{(L)} = \Theta(t_L, 0)\big(E_n - K(t_0)H'(t_0)\big), \tag{4.36}$$

$$\varphi^{(L)}(t_i) = \left(a'\varpi^{(L)}(t_i)\right)', \qquad \varpi^{(L)}(t_i) = \Theta(t_L, t_i)K(t_i), \quad i = 0, \ldots, L.$$

For $L = 0, \ldots, N-1$, we have the following iterative procedure from (4.36):

$$\begin{cases} \varpi_-^{(0)} = E_n - K(t_0)H'(t_0), \\ \cdots \\ \varpi_-^{(L+1)} = \Theta(t_{L+1}, t_L)\varpi_-^{(L)}, \end{cases} \tag{4.37}$$

$$\begin{cases} \varpi^{(0)}(t_0) = K(t_0), \\ \cdots \\ \varpi^{(L+1)}(t_i) = \Theta(t_{L+1}, t_L)\varpi^{(L)}(t_i), \qquad i = 0, \ldots, L, \\ \varpi^{(L+1)}(t_{L+1}) = K(t_{L+1}). \end{cases}$$

Then $\varphi_-^{(N)} = \varphi_-$, $\varphi^{(N)}(t_i) = \varphi(t_i)$, $i = 0, \ldots, N$.

Formulas {(4.30), (4.26), (4.35), (4.37)} were used for computing the examples that will be discussed below.

REMARK 4.1. If the linear homogeneous equation $\dot{x}(t) = Ax(t)$ is stable, then the computation in (4.35) in backward time is more preferable than in the forward one.

REMARK 4.2. The function $\xi(t)$ can be also computed by iterative formulas similar to these of (4.37).

REMARK 4.3. In order to implement the estimate Δ_D a researcher must specify the scale multipliers β_1 and β_2. Let us focus upon the choice of these multipliers. We suggest an informal way that allows us to localize the exponents of these multipliers.

For definiteness, we assume that B, γ, H, and σ are time-invariant and the measurements are carried out at regular intervals $\Delta t = t_{i+1} - t_i$, $i = 0, \dots, N-1$. We also assume that the entries of $X(t)$ are varied "slowly" on the intervals of length Δt (that is typical for practice). Then the integral term in the formula for $I_D(\Phi_-, \Phi)$ from (4.10) may be approximated by the integral sum:

$$I_D(\Phi_-, \Phi) \approx I_D^{(1)},$$

$$I_D^{(1)} = \sum_{d=1}^{n} \sigma_{-d}|\Phi_{-d}| + \sum_{l=1}^{m}\sum_{i=0}^{N} \sigma_l |\Phi_l(t_i)| + \sum_{j=1}^{r}\sum_{i=0}^{N-1} \gamma_j |w_j(t_i)| \Delta t,$$

$$w(t_i) = (w_1(t_i) \dots, w_r(t_i))' = b'(t_i)\left(\sum_{s=0}^{i} h(t_s)\Phi(t_s) + \Phi_-\right).$$

The expression for $I_D^{(1)}$ defines a nonsmooth norm in $\mathbf{R}^{n+(N+1)m+Nr}$. We replace $I_D^{(1)}$ by the quadratic norm

$$I_D^{(2)} = \left\{\sum_{d=1}^{n}\left(\sigma_{-d}|\Phi_{-d}|\right)^2 + \sum_{l=1}^{m}\sum_{i=0}^{N}\left(\sigma_l|\Phi_l(t_i)|\right)^2 + \sum_{j=1}^{r}\sum_{i=0}^{N-1}\left(\gamma_j|w_j(t_i)|\right)^2 \Delta t^2\right\}^{\frac{1}{2}}.$$

The quantity $I_D^{(2)}$ approximates, in its turn, the expression (4.17) for $J_D(\Phi_-,\Phi)$, where P_0, $Q(t)$, and $R(t_i)$ are defined by (4.26) with

$$\beta_1 = 1, \qquad \beta_2 = \Delta t. \tag{4.38}$$

Though our argument is quite fuzzy, the numerical experience indicates that the initial point (4.38) yields a successfull neighborhood for more detailed search. Usually, several iterations near this point provide the minimal value for Δ_D (see the examples below).

4.2. *Prediction Problem in Discrete Case*

Let us consider a prediction problem for the system (4.1) – (4.5), in which we have to evaluate the quantity [139]

$$l_* = a'x(T+\kappa), \qquad \kappa > 0, \qquad a \in \mathbf{R}^n.$$

Denote by $\Delta_{D\kappa}$ the level of nonoptimality of the discrete Kalman filter in the prediction problem (this quantity is analogous to Δ_κ from Theorem 2.3).

THEOREM 4.3. *The following inequalities hold:*

$$1 \le \Delta_{D\kappa} \le \Delta^0_{D\kappa}, \qquad \Delta^0_{D\kappa} = \frac{\mathcal{N}_{D1} + F^{\kappa}}{\mathcal{N}^2_{D2}\mathcal{N}^{-1}_{D\infty} + F^{\kappa}},$$

where $\mathcal{N}_{D1}$, $\mathcal{N}_{D2}$, $\mathcal{N}_{D\infty}$, *and* F^{κ} *are defined by* (4.28) *and* (2.24), *respectively.*

Theorem 4.3 can be proved similarly to Theorem 4.2.

5. Proofs for the Case of Discrete Measurements

The idea for the derivation of the estimate of nonoptimality level for discrete case is the same as for the continuous case. Therefore only technical details will be emphasized in this section.

5.1. *Proof of Theorem 4.1*

Assertion 1 of Theorem 4.1 follows directly from (4.15), (4.18), and (4.23). Certainly, the direct computation of the variance for the estimation error results in the functional (4.17) similar to the continuous case.

Let us now prove Assertion 2 of Theorem 4.1. Introduce two Hilbert spaces: the space $\mathcal{H} = \mathbf{R}^n \times \mathbf{R}^{(N+1)\times m} \times L_2^r$ with the elements $\Gamma = (\Phi_-, \Phi, w)$ and the inner product

$$\langle \Gamma_{(1)}, \Gamma_{(2)} \rangle_{\mathcal{H}} = \Phi'_{-(1)}\Phi_{-(2)} + \sum_{i=0}^{N} \Phi'_{(1)}(t_i)\Phi_{(2)}(t_i) + \int_0^T w'_{(1)}(t)w_{(2)}(t)\,dt,$$

$$\Gamma_{(i)} \in \mathcal{H}, \qquad \Phi_{-(i)} \in \mathbf{R}^n, \qquad \Phi_{(i)} = (\Phi(t_0), \ldots, \Phi(t_N))'_{(i)} \in \mathbf{R}^{(N+1)\times m},$$

$$w_{(i)} \in L_2^r, \qquad i = 1, 2,$$

and the space $\tilde{\mathcal{H}} = L_2^r \times \mathbf{R}^n$ with the elements $\tilde{\Gamma} = (G, g)$ and the inner product

$$\langle \tilde{\Gamma}_{(1)}, \tilde{\Gamma}_{(2)} \rangle_{\tilde{\mathcal{H}}} = \int_0^T G'_{(1)}(t)G_{(2)}(t)\,dt + g'_{(1)}g_{(2)}.$$

$$\tilde{\Gamma}_{(i)} \in \tilde{\mathcal{H}}, \qquad G_{(i)} \in L_2^r, \qquad g_{(i)} \in \mathbf{R}^n, \qquad i = 1, 2.$$

Let us represent the problem (4.16) in the form

$$J_{D0} = \inf_{\Phi_- \in \mathbf{R}^n,\, \Phi \in \mathbf{R}^{(N+1)\times m},\, w \in L_2^r} J_D(\Phi_-, \Phi, w), \tag{5.1}$$

where

$$J_D(\Phi_-, \Phi, w) \tag{5.2}$$

$$= \left\{ \Phi'_- P_0 \Phi_- + \sum_{i=0}^{N} \Phi'(t_i) R(t_i) \Phi(t_i) + \int_0^T w'(t) Q(t) w(t)\, dt \right\}^{\frac{1}{2}},$$

under the constraints

$$b'(t)\left(\sum_{\{i \mid t_i \le t\}} h(t_i)\Phi(t_i) + \Phi_- \right) - w(t) = 0, \tag{5.3}$$

$$\tilde{a} - \sum_{i=0}^{N} h(t_i)\Phi(t_i) - \Phi_- = 0.$$

Thus the variational problem (5.1) – (5.3) takes the form

$$J_{D0} = \inf_{\mathcal{H}} J_D(\Phi_-, \Phi, w) \quad \text{under the constraint} \quad \Lambda_D(\Phi_-, \Phi, w) = 0. \tag{5.4}$$

The functional $J_D(\Phi_-, \Phi, w) : \mathcal{H} \to \mathbf{R}^1$ defined by (5.2) is a convex continuous functional, the operator $\Lambda_D : \mathcal{H} \to \tilde{\mathcal{H}}$ defined by (5.3) is a continuous affine operator. So, the problem (5.4) is a variational problem of the type $\mathcal{P}_1$ from Section 1 of Chapter 2.

Similarly to Theorem 2.1 it can easily be checked that the problem (5.4) has a unique solution $(\varphi_-, \varphi, w_\varphi)$.

Let us turn to the dual problem. The Lagrange function for (5.4) has the form

$$\mathcal{L}_D(\Phi_-, \Phi, w; p, \lambda) = J_D(\Phi_-, \Phi, w) \tag{5.5}$$

$$+ \int_0^T p'(t) \left[b'(t)\left(\sum_{\{i \mid t_i \le t\}} h(t_i)\Phi(t_i) + \Phi_- \right) - w(t) \right] dt$$

$$+ \lambda' \left(\tilde{a} - \sum_{i=0}^{N} h(t_i)\Phi(t_i) - \Phi_- \right).$$

Then, in accordance with the duality theory (the details are analogous to these of the proof of Theorem 2.1), the dual problem has the form

$$J_D^0 = \sup_{p \in L_2^r,\, \lambda \in \mathbf{R}^n} \left\{ \inf_{\Phi_- \in \mathbf{R}^n, \Phi \in \mathbf{R}^{(N+1)\times m}, w \in L_2^r} \mathcal{L}_D(\Phi_-, \Phi, w; p, \lambda) \right\}. \tag{5.6}$$

Moreover, the dual problem has a solution (p^0, λ^0), the Lagrange function (5.5) has a saddle point:

$$\mathcal{L}_D(\varphi_-, \varphi, w_\varphi; p, \lambda) \leq \mathcal{L}_D(\varphi_-, \varphi, w_\varphi; p^0, \lambda^0) \leq \mathcal{L}_D(\Phi_-, \Phi, w; p^0, \lambda^0),$$

$$\forall\ \Phi_-, \lambda \in \mathbf{R}^n, \quad \Phi \in \mathbf{R}^{(N+1)\times m}, \quad w, p \in L_2^r,$$

and the duality relation holds:

$$J_{D0} = J_D^0. \tag{5.7}$$

In order to compute the infimum in (5.6) let us represent the Lagrange function in a convenient form, using the structure of Hilbert space. First note that

$$\int_0^T p'(t)b'(t) \sum_{\{i|t_i\leq t\}} h(t_i)\Phi(t_i)\,dt \tag{5.8}$$

$$= \sum_{s=0}^{N-1} \int_{t_s}^{t_{s+1}} p'(t)b'(t) \sum_{\{i|t_i\leq t\}} h(t_i)\Phi(t_i)\,dt$$

$$= \sum_{s=0}^{N-1} \int_{t_s}^{t_{s+1}} p'(t)b'(t)\,dt \left(\sum_{i=0}^{s} h(t_i)\Phi(t_i)\right)$$

$$= \sum_{s=0}^{N-1} \sum_{i=0}^{s} \Phi'(t_i)h'(t_i) \left(\int_{t_s}^{t_{s+1}} b(t)p(t)\,dt\right)$$

$$= \sum_{i=0}^{N-1} \sum_{s=i}^{N-1} \Phi'(t_i)h'(t_i) \left(\int_{t_s}^{t_{s+1}} b(t)p(t)\,dt\right)$$

$$= \sum_{i=0}^{N} \Phi'(t_i)h'(t_i) \left(\int_{t_i}^{t_N} b(t)p(t)\,dt\right) = \sum_{i=0}^{N} \Phi'(t_i)h'(t_i) \left(\int_{t_i}^{T} b(t)p(t)\,dt\right).$$

In accordance with (5.5) and (5.8),

$$\mathcal{L}_D(\Phi_-, \Phi, w; p, \lambda) = \bar{\mathcal{L}}_D(\Gamma; \Xi) \triangleq \|\Gamma\|_{\mathcal{H}} - \langle \Xi, \Gamma \rangle_{\mathcal{H}} + \lambda'\tilde{a},$$

where

$$\Gamma = \Gamma(\Phi_-, \Phi, w)$$

$$= \left(P_0^{\frac{1}{2}}\Phi_-,\ \left(R^{\frac{1}{2}}(t_0)\Phi(t_0), \ldots, R^{\frac{1}{2}}(t_N)\Phi(t_N)\right)',\ Q^{\frac{1}{2}}(t)w(t)\right),$$

$$\Xi = \Xi(p, \lambda)$$

$$= \Bigg(P_0^{-\frac{1}{2}} \Big(\lambda - \int_0^T b(t)p(t)\,dt \Big),\ \Big(R^{-\frac{1}{2}}(t_0)h'(t_0)\Big(\lambda - \int_{t_0}^T b(s)p(s)\,ds\Big), \ldots,\ R^{-\frac{1}{2}}(t_N)h'(t_N)\Big(\lambda - \int_{t_N}^T b(s)p(s)\,ds\Big)\Big)',\ Q^{-\frac{1}{2}}(t)p(t) \Bigg).$$

In these new variables, the dual problem is represented in the form

$$J_D^0 = \sup_{\Xi(p,\lambda)} \Big(\inf_{\Gamma \in \mathcal{H}} \bar{\mathcal{L}}_D(\Gamma; \Xi) \Big) \tag{5.9}$$

and has a solution $\Xi^0 = \Xi(p^0, \lambda^0)$. The saddle point inequalities take the form

$$\bar{\mathcal{L}}_D(\Gamma^0; \Xi) \le \bar{\mathcal{L}}_D(\Gamma^0; \Xi^0) \le \bar{\mathcal{L}}_D(\Gamma; \Xi^0), \qquad \Gamma \in \mathcal{H}, \quad \Xi = \Xi(p, \lambda). \tag{5.10}$$

Computing the infimum in (5.9), we get

$$\inf_{\Gamma \in \mathcal{H}} \bar{\mathcal{L}}_D(\Gamma, \Xi) = \begin{cases} \tilde{a}'\lambda & \text{if } \|\Xi\|_{\mathcal{H}} \le 1 \\ -\infty & \text{otherwise.} \end{cases}$$

Hence the dual problem is determined by the relation

$$\sup_{\lambda \in \mathbf{R}^n,\, p \in L_2^r} \tilde{a}'\lambda \quad \text{under the constraint} \quad \|\Xi(p, \lambda)\|_{\mathcal{H}} \le 1. \tag{5.11}$$

It can easily be checked (see the proof of Theorem 2.1) that the second inequality of (5.10) implies

$$\Gamma^0 = k\,\Xi^0 \qquad \text{for a certain} \quad k \ge 0. \tag{5.12}$$

Returning to the initial variables, we obtain from (5.11) and (5.12) that the dual problem has the form

$$\sup_{p \in L_2^r,\, \lambda \in \mathbf{R}^n} \tilde{a}'\lambda \quad \text{under the constraint} \quad S_D(p, \lambda) \le 1, \tag{5.13}$$

where

$$S_D(p, \lambda) = \Big(\lambda - \int_0^T b(s)p(s)\,ds \Big)' P_0^{-1} \Big(\lambda - \int_0^T b(s)p(s)\,ds \Big) \tag{5.14}$$

$$+ \sum_{i=0}^N \Big[h'(t_i) \Big(\lambda - \int_{t_i}^T b(s)p(s)\,ds \Big) \Big]' R^{-1}(t_i) \Big[h'(t_i) \Big(\lambda - \int_{t_i}^T b(s)p(s)\,ds \Big) \Big]$$

$$+ \int_0^T p'(t) Q^{-1}(t) p(t)\,dt.$$

Moreover, for a certain $k \geq 0$,

$$\varphi_- = kP_0^{-1}\left(\lambda^0 - \int_0^T b(t)p^0(t)\,dt\right), \tag{5.15}$$

$$\varphi(t_i) = kR^{-1}(t_i)h'(t_i)\left(\lambda^0 - \int_{t_i}^T b(s)p^0(s)\,ds\right),$$

$$w_\varphi(t) = k\,Q^{-1}(t)p^0(t),$$

where (p^0, λ^0) is the solution of the dual problem (5.13).

Let us solve the dual problem (5.13). Since $\sup \tilde{a}'\lambda = -\inf(-\tilde{a}'\lambda)$, the problem (5.13) has the type $\mathcal{P}_2$ from Section 1 of Chapter 2 with a linear functional and a convex scalar constraint.

It is not hard to prove that (5.13) has a solution (p^0, λ^0) (cf. the proof of Theorem 2.1).

By virtue of the Kuhn-Tucker theorem (Theorem 1.5 of Chapter 2), the solution (p^0, λ^0) yields an unconstrained minimum to the Lagrange function

$$\mathcal{L}_D^0(p, \lambda; \kappa) = -\tilde{a}'\lambda + \kappa\,(S_D(p,\lambda) - 1) \qquad \text{for a certain} \quad \kappa \geq 0.$$

Moreover, $\kappa\,(S_D(p,\lambda) - 1) = 0$. It can be proved that if the problem is not trivial, i.e., $\tilde{a} \neq 0$, then $\kappa > 0$. Obviously, for $\tilde{a} = 0$ Theorem 4.1 is valid. Hence

$$S_D(p^0, \lambda^0) = 1. \tag{5.16}$$

Since $\mathcal{L}_D^0(p, \lambda)$ is a smooth convex function, the necessary and sufficient condition for its minimum are determined by the relations

$$\frac{\partial \mathcal{L}_D^0}{\partial p} = 0, \qquad \frac{\partial \mathcal{L}_D^0}{\partial \lambda} = 0. \tag{5.17}$$

Relations (5.17) result in the equalities

$$p(t) = Q(t)b'(t)\left\{ P_0^{-1}\left(\lambda - \int_0^T b(s)p(s)\,ds\right)\right.$$

$$\left. + \sum_{\{i|t_i \leq t\}} h(t_i)R^{-1}(t_i)h'(t_i)\left(\lambda - \int_{t_i}^T b(s)p(s)\,ds\right)\right\}$$

and

$$P_0^{-1}\left(\lambda - \int_0^T b(s)p(s)\,ds\right)$$

$$+\sum_{i=0}^{N} h(t_i)R^{-1}(t_i)h'(t_i)\left(\lambda - \int_{t_i}^{T} b(s)p(s)\,ds\right) = \frac{\tilde{a}}{2\kappa}$$

The last relations imply that

$$p(t) = Q(t)b'(t)\bar{\xi}(t), \tag{5.18}$$

where

$$\bar{\xi}(t) = P_0^{-1}\left(\lambda - \int_0^T b(s)Q(s)b'(s)\bar{\xi}(s)\,ds\right)$$

$$+ \sum_{\{i|t_i\le t\}} h(t_i)R^{-1}(t_i)h'(t_i)\left(\lambda - \int_{t_i}^{T} b(s)Q(s)b'(s)\bar{\xi}(s)\,ds\right),$$

and

$$P_0^{-1}\left(\lambda - \int_0^T b(s)Q(s)b'(s)\bar{\xi}(s)\,ds\right) \tag{5.19}$$

$$+\sum_{i=0}^{N} h(t_i)R^{-1}(t_i)h'(t_i)\left(\lambda - \int_{t_i}^{T} b(s)Q(s)b'(s)\bar{\xi}(s)\,ds\right) = \frac{\tilde{a}}{2\kappa}.$$

Let us introduce an auxiliary variable

$$\bar{\eta}(t) = \bar{\eta}(T) - \int_t^T b(s)Q(s)b'(s)\bar{\xi}(s)\,ds, \qquad \bar{\eta}(T) = \lambda. \tag{5.20}$$

It follows from (5.18) – (5.20) that

$$\bar{\xi}(t_{i+1}) = \bar{\xi}(t_i) + h(t_{i+1})R^{-1}(t_{i+1})h'(t_{i+1})\bar{\eta}(t_{i+1}), \tag{5.21}$$

$$\bar{\eta}(t_{i+1}) = \bar{\eta}(t_i) + Q_i\bar{\xi}(t_i), \qquad Q_i = \int_{t_i}^{t_{i+1}} b(s)Q(s)b'(s)\,ds,$$

$$\bar{\xi}(t) = \bar{\xi}(t_i), \quad \bar{\eta}(t) = \bar{\eta}(t_i) + \int_{t_i}^{t} b(s)Q(s)b'(s)\,ds\,\bar{\xi}(t_i), \qquad t_i \le t < t_{i+1},$$

$$i = 0, \ldots, N-1,$$

and

$$\bar{\xi}(T) = \bar{\xi}(t_N) = \frac{\tilde{a}}{2\kappa}, \tag{5.22}$$

$$\bar{\xi}(0) = \bar{\xi}(t_0) = \left(P_0^{-1} + h(t_0)R^{-1}(t_0)h'(t_0)\right)\bar{\eta}(t_0).$$

Resolving the last relation with respect to $\bar{\eta}(t_0)$, we get

$$\bar{\eta}(t_0) = P_0\big(E_n + h(t_0)R^{-1}(t_0)h'(t_0)P_0\big)^{-1}\bar{\xi}(t_0). \tag{5.23}$$

Using the well-known matrix identities [79]

$$\big(E_n - Ph(h'Ph + R)^{-1}h'\big)^{-1} = E_n + PhR^{-1}h', \tag{5.24}$$

$$Ph(h'Ph + R)^{-1} = \big(E_n + PhR^{-1}h'\big)^{-1}PhR^{-1},$$

the left boundary condition may be rewritten as follows:

$$\bar{\eta}(t_0) = \big(E_n + P_0h(t_0)R^{-1}(t_0)h'(t_0)\big)^{-1}P_0\bar{\xi}(t_0) \tag{5.25}$$

$$= \big(E_n - P_0h(t_0)\big(h'(t_0)P_0h(t_0) + R(t_0)\big)^{-1}h'(t_0)\big)P_0\bar{\xi}(t_0)$$

$$= \big(E_n - K(t_0)h'(t_0)\big)P_0\bar{\xi}(t_0) = P(t_0|0)\bar{\xi}(t_0),$$

where $P(t_i|i)$ and $K(t_i)$ are defined by (4.13) and (4.14).

Since the problem (5.13) has a solution, the boundary value problem (5.21) must have a solution. Moreover, this solution is unique. In fact, if there are two solutions $(\bar{\xi}_{(1)}(t), \bar{\eta}_{(1)}(t))$ and $(\bar{\xi}_{(2)}(t), \bar{\eta}_{(2)}(t))$, then the non-zero functions

$$\delta\xi(t) = \bar{\xi}_{(2)}(t) - \bar{\xi}_{(1)}(t), \qquad \delta\eta(t) = \bar{\eta}_{(2)}(t) - \bar{\eta}_{(1)}(t)$$

satisfy the equations

$$\delta\xi(t_{i+1}) = \delta\xi(t_i) + h(t_{i+1})R^{-1}(t_{i+1})h'(t_{i+1})\,\delta\eta(t_{i+1}), \tag{5.26}$$

$$\delta\eta(t_{i+1}) = \delta\eta(t_i) + Q_i\delta\xi(t_i),$$

$$\delta\xi(t) = \delta\xi(t_i), \quad \delta\eta(t) = \delta\eta(t_i) + \int_{t_i}^{t} b(s)Q(s)b'(s)\,ds\,\delta\xi(t_i), \quad t_i \le t < t_{i+1},$$

$$i = 0, \dots, N-1,$$

and

$$\delta\xi(t_N) = 0,$$

$$\delta\eta(t_0) = \big(E_n + P_0h(t_0)R^{-1}(t_0)h'(t_0)\big)^{-1}P_0\delta\xi(t_0)$$

$$= P_0\big(E_n + h(t_0)R^{-1}(t_0)h'(t_0)P_0\big)^{-1}\delta\xi(t_0).$$

It follows from (5.26) that

$$\delta\xi'(t_{i+1})\delta\eta(t_{i+1}) - \delta\xi'(t_i)\delta\eta(t_i)$$

$$= \left(\delta\xi'(t_i) + \delta\eta'(t_{i+1})h(t_{i+1})R^{-1}(t_{i+1})h'(t_{i+1})\right)\delta\eta(t_{i+1}) - \delta\xi'(t_i)\delta\eta(t_i)$$

$$= \delta\eta'(t_{i+1})h(t_{i+1})R^{-1}(t_{i+1})h'(t_{i+1})\delta\eta(t_{i+1}) + \delta\xi'(t_i)Q_i\delta\xi(t_i),$$

$$i = 0, \ldots, N-1,$$

whence,

$$\sum_{i=1}^{N} \delta\eta'(t_i)h(t_i)R^{-1}(t_i)h'(t_i)\delta\eta(t_i) + \delta\xi'(t_0)\delta\eta(t_0) \tag{5.27}$$

$$+ \sum_{j=0}^{N-1} \delta\xi'(t_j)Q_j\delta\xi(t_j) = 0.$$

Using the first matrix equality from (5.24), we have

$$E_n - hR^{-1}h'\left(P_0 - P_0h\left(h'P_0h + R\right)^{-1}h'P_0\right)$$

$$= E_n - hR^{-1}h'P_0\left(E_n - h\left(h'P_0h + R\right)^{-1}h'P_0\right)$$

$$= E_n - hR^{-1}h'P_0\left(E_n + hR^{-1}h'P_0\right)^{-1} = \left(E_n + hR^{-1}h'P_0\right)^{-1}.$$

Consequently, taking into account the boundary conditions, we get

$$\delta\eta'(t_0)\delta\xi(t_0) - \delta\eta'(t_0)h(t_0)R^{-1}(t_0)h'(t_0)\delta\eta(t_0)$$

$$= \delta\eta'(t_0)\left(E_n - h(t_0)R^{-1}(t_0)h'(t_0)P(t_0|0)\right)\delta\xi(t_0)$$

$$= \delta\xi'(t_0)\left(\left(E_n + h(t_0)R^{-1}(t_0)h'(t_0)P_0\right)^{-1}\right)'P_0$$
$$\times\left(E_n + h(t_0)R^{-1}(t_0)h'(t_0)P_0\right)^{-1}\delta\xi(t_0).$$

Then, taking into account (5.27), we obtain

$$\sum_{i=0}^{N} \delta\eta'(t_i)h(t_i)R^{-1}(t_i)h'(t_i)\delta\eta(t_i) + \sum_{i=0}^{N-1} \delta\xi'(t_i)Q_i\delta\xi(t_i)$$

$$+ \delta\xi'(t_0)\left(\left(E_n + h(t_0)R^{-1}(t_0)h'(t_0)P_0\right)^{-1}\right)'P_0$$
$$\times\left(E_n + h(t_0)R^{-1}(t_0)h'(t_0)P_0\right)^{-1}\delta\xi(t_0) = 0.$$

Thus,

$$P_0^{\frac{1}{2}}\left(E_n + h(t_0)R^{-1}(t_0)h'(t_0)P_0\right)^{-1}\delta\xi(t_0) = 0,$$

$$R^{-\frac{1}{2}}(t_i)h'(t_i)\delta\eta(t_i) = 0, \qquad i = 0, \ldots, N,$$

$$Q_j^{\frac{1}{2}}\delta\xi(t_j) = 0, \qquad j = 0, \ldots, N-1.$$

Then we get from (5.26) that $\delta\xi(t) \equiv 0$ and $\delta\eta(t) \equiv 0$.

Thus the boundary value problem has the unique solution $(\bar{\xi}(t), \bar{\eta}(t))$ and the solution of (5.13) has the form

$$p^0(t) = Q(t)b'(t)\bar{\xi}(t), \qquad \lambda^0 = \bar{\eta}(T). \tag{5.28}$$

Moreover, using the boundary value problem, the solution (5.15) of (5.1) – (5.3) may be rewritten in the form

$$\varphi_- = kP_0^{-1}\bar{\eta}(t_0), \tag{5.29}$$

$$\varphi(t_i) = kR^{-1}(t_i)h'(t_i)\bar{\eta}(t_i), \qquad i = 0, \ldots, N,$$

$$w_\varphi(t) = k\, b'(t)\bar{\xi}(t), \qquad t \in [0, T].$$

Substituting (5.29) into (4.11), we have from (5.21) and (5.22)

$$k\left(P_0^{-1}\bar{\eta}(t_0) + \sum_{i=0}^{N} h(t_i)R^{-1}(t_i)h'(t)\bar{\eta}(t_i)\right)$$

$$= k\left(\left(P_0^{-1} + h(t_0)R^{-1}(t_0)h'(t_0)\right)\bar{\eta}(t_0) + \sum_{i=1}^{N} h(t_i)R^{-1}(t_0)h'(t_0)\bar{\eta}(t_i)\right)$$

$$= k\left(\bar{\xi}(t_0) + \sum_{i=1}^{N} h(t_i)R^{-1}(t_0)h'(t_0)\bar{\eta}(t_i)\right) = k\bar{\xi}(t_N) = k\bar{\xi}(T) = k\frac{\tilde{a}}{2\kappa} = \tilde{a},$$

whence,

$$k = 2\kappa.$$

The boundary value problem (4.20), (4.21), differs from (5.21) – (5.23), and (5.25) by the scale 2κ only. Consequently the problem (4.20), (4.21) has the unique solution $(\tilde{\xi}(t), \tilde{\eta}(t))$ and the solution of (5.13) has the form

$$p^0(t) = (2\kappa)^{-1}\, Q(t)b'(t)\tilde{\xi}(t), \qquad \lambda^0 = (2\kappa)^{-1}\, \tilde{\eta}(T). \tag{5.30}$$

Moreover, the solution (5.15) of (5.1) – (5.3) takes the form

$$\varphi_- = P_0^{-1}\tilde{\eta}(t_0) = \left(E_n + h(t_0)R^{-1}(t_0)h'(t_0)P_0\right)^{-1}\tilde{\xi}(t_0), \tag{5.31}$$

$$\varphi(t_i) = R^{-1}(t_i)h'(t_i)\tilde{\eta}(t_i), \qquad i = 0, \ldots, N,$$

$$w_\varphi(t) = b'(t)\tilde{\xi}(t), \qquad t \in [0, T].$$

Substituting (5.30) into constraint (5.16), we get

$$\tilde{\eta}'(t_0)P_0^{-1}\tilde{\eta}(t_0) + \sum_{i=0}^{N} \tilde{\eta}'(t_i)h(t_i)R^{-1}(t_i)h'(t_i)\tilde{\eta}(t_i)$$

$$+ \int_0^T \tilde{\xi}'(t)b(t)Q(t)b'(t)\tilde{\xi}(t)\, dt = 4\kappa^2.$$

In accordance with (5.31) and (5.1) – (5.3), the last equality takes the form $J_{D0}^2 = 4\kappa^2$, whence,

$$2\kappa = J_{D0}.$$

Finally, the solution of the dual problem (5.13) has the form

$$p^0(t) = J_{D0}^{-1}\, Q(t)b'(t)\tilde{\xi}(t), \qquad \lambda^0 = J_{D0}^{-1}\, \tilde{\eta}(T). \tag{5.32}$$

Let us directly check that the functions

$$\tilde{\xi}(t_i) = \tilde{\Theta}'(T, t_i)\tilde{a}, \qquad \tilde{\eta}(t_i) = \tilde{P}(t_i|i)\tilde{\xi}(t_i), \qquad i = 0, \ldots, N, \tag{5.33}$$

where $\tilde{\Theta}(T, t_i)$ determines the Kalman estimator in (4.15), satisfy the boundary value problem (4.20), (4.21). In fact, it follows from (4.14) and (5.33) that

$$\tilde{\eta}'(t_i) = \tilde{\xi}'(t_i)\tilde{P}(t_i|i) = \tilde{\xi}'(t_i)\big(E_n - \tilde{K}(t_i)h'(t_i)\big)\tilde{P}(t_i|i-1) \tag{5.34}$$

$$= \tilde{\xi}'(t_{i-1})\tilde{P}(t_i|i-1).$$

Then the second equation of the boundary value problem follows from (5.34) and (4.14):

$$\tilde{\eta}'(t_{i+1}) = \tilde{\xi}'(t_i)\tilde{P}(t_{i+1}|i) = \tilde{\xi}'(t_i)\big(\tilde{P}(t_i|i) + Q_i\big) = \tilde{\eta}'(t_i) + \tilde{\xi}'(t_i)Q_i.$$

Moreover, we have from the definition of $\tilde{\Theta}(T, t_i)$, the first matrix identity of (5.24), and (5.34) that

$$\tilde{\xi}'(t_{i+1}) = \tilde{\xi}'(t_i)\big(E_n - \tilde{K}(t_{i+1})h'(t_{i+1})\big)^{-1}$$

$$= \tilde{\xi}'(t_i) + \tilde{\xi}'(t_i)\tilde{P}(t_{i+1}|i)h(t_{i+1})R^{-1}(t_{i+1})h'(t_{i+1})$$

$$= \tilde{\xi}'(t_i) + \tilde{\eta}'(t_{i+1})h(t_{i+1})R^{-1}(t_{i+1})h'(t_{i+1}).$$

So, the first equation of the boundary value problem also holds.

By virtue of the uniquness of a solution of the boundary value problem, (4.22) and (4.23) hold. In particular,

$$\varphi(t_i) = R^{-1}(t_i)h'(t_i)\tilde{\eta}(t_i) = R^{-1}(t_i)h'(t_i)\tilde{P}(t_i|i)\tilde{\xi}(t_i). \tag{5.35}$$

Let us prove that

$$\tilde{P}(t_i|i)h(t_i)R^{-1}(t_i) = \tilde{K}(t_i),$$

where $\tilde{K}(t_i)$ is defined by (4.13). Denoting by P^-, P^+, h, and R the matrices $\tilde{P}(t_i|i-1)$, $\tilde{P}(t_i|i)$, $h(t_i)$, and $R(t_i)$, respectively, we obtain from the matrix identities (5.24)

$$P^+hR^{-1} = \left(E_n - P^-h\left(h'P^-h + R\right)^{-1}h'\right)P^-hR^{-1}$$

$$= P^-h\left(h'P^-h + R\right)^{-1} = \tilde{K}(t_i).$$

Then it follows from (5.31) and (5.35) that

$$\varphi(t_i) = \tilde{K}'(t_i)\tilde{\xi}(t_i).$$

In accordance with the first matrix identity of (5.24),

$$\varphi_- = \left(E_n + h(t_i)R^{-1}(t_i)h'(t_i)P_0\right)^{-1}\tilde{\xi}(t_0) = \left(E_n - \tilde{K}(t_i)h'(t_i)\right)'\tilde{\xi}(t_0).$$

Thus all of equalities (4.18) are valid.

It follows directly from (4.21), (4.22), (5.7), and (5.32) that

$$J_{D0} = J_D^0 = \tilde{a}'\lambda^0 = J_{D0}^{-1}\,\tilde{a}'\tilde{\eta}(t_N) = J_{D0}^{-1}\,\tilde{a}'\tilde{P}(t_N|N)\tilde{a}.$$

In our proof we used the inverse matrices P_0^{-1} and $Q^{-1}(t)$. Nevertheless, Theorem 4.1 is valid for singular matrices P_0 and $Q(t)$ as well. This case can be treated by the same way as in the proof of Theorem 2.1. □

5.2. *Proof of Theorem 4.2*

In accordance with formulas (4.10) – (4.12),

$$\Delta_D = \frac{I_D(\varphi_-, \varphi)}{I_{D0}}, \tag{5.36}$$

where (φ_-, φ) are subjected to (4.11) and φ is the Kalman estimator. The numerator of (5.36) can be easily computed. The denominator is unknown; so, Δ_D is unknown. In order to obtain the upper bound for Δ_D we have to evaluate I_{D0} from below.

The problem (4.12) can be represented in the form (cf. (5.1) – (5.3))

$$I_{D0} = \inf_{\Phi_- \in \mathbf{R}^n,\, \Phi \in \mathbf{R}^{(N+1)\times m},\, w \in L_1^r} I_D(\Phi_-, \Phi, w), \tag{5.37}$$

$$I_D(\Phi_-, \Phi, w)$$

$$= \sum_{d=1}^{n} \sigma_{-d}|\Phi_{-d}| + \sum_{l=1}^{m}\sum_{i=0}^{N} \sigma_l(t_i)|\Phi_l(t_i)| + \sum_{j=1}^{r} \int_0^T \gamma_j(t)|w_j(t)|\, dt$$

under constraints (5.3).

By virtue of Theorem 4.1, the pair (φ_-, φ) is the solution of the problem (4.16); this problem can be also represented in the form (5.1) – (5.3).

Let us construct the dual problems for the variational problems (5.37), (5.3) and (5.1), (5.3).

THEOREM 5.1. *The problem dual to* (5.37), (5.3) *has the form*

$$I_D^0 = \sup_{p \in L_\infty^r,\, \lambda \in \mathbf{R}^n} \lambda' \tilde{a} \tag{5.38}$$

under the constraints

$$\sigma_l^{-1}(t_i) \left| h_l'(t_i) \left(\lambda - \int_{t_i}^{T} b(s)p(s)\, ds \right) \right| \le 1, \tag{5.39}$$

$$i = 0, \ldots, N, \qquad l = 1, \ldots, m,$$

$$\gamma_j^{-1}(t)\, |p_j(t)| \le 1 \quad \textit{a.e. on } [0,T], \qquad j = 1, \ldots, r,$$

$$\sigma_{-d}^{-1} \left| e_d' \left(\lambda - \int_0^T b(t)p(t)\, dt \right) \right| \le 1, \qquad d = 1, \ldots, n,$$

$$e_d = (\underbrace{0, \ldots, 1}_{d}, 0, \ldots, 0)'.$$

Moreover,

$$I_{D0} = I_D^0. \tag{5.40}$$

PROOF. Theorem 5.1 can be proved in perfect analogy to the proof of Theorem 5.2 of Chapter 2.

THEOREM 5.2. *The problem dual to* (5.1), (5.3) *has the form*

$$J_D^0 = \sup_{p \in L_2^r,\, \lambda \in \mathbf{R}^n} \tilde{a}' \lambda \tag{5.41}$$

under the constraint

$$S_D(p,\lambda) \le 1, \tag{5.42}$$

where

$$S_D(p,\lambda) = \left(\lambda - \int_0^T b(s)p(s)\,ds\right)' P_0^{-1}\left(\lambda - \int_0^T b(s)p(s)\,ds\right)$$

$$+ \sum_{i=0}^{N}\left[h'(t_i)\left(\lambda - \int_{t_i}^T b(s)p(s)\,ds\right)\right]' R^{-1}(t_i)\left[h'(t_i)\left(\lambda - \int_{t_i}^T b(s)p(s)\,ds\right)\right]$$

$$+ \int_0^T p'(t)Q^{-1}(t)p(t)\,dt.$$

The solution of the dual problem is determined by the formulas

$$p^0(t) = J_{D0}^{-1}\, Q(t)b'(t)\tilde{\xi}(t), \qquad \lambda^0 = J_{D0}^{-1}\,\tilde{\eta}(T). \tag{5.43}$$

Moreover,

$$J_{D0} = J_D^0.$$

PROOF. Theorem 5.2 follows directly from (5.13), (5.14), (5.28), and (5.7).

In accordance with our approach, we take as the "proper" pair the pair that is a solution of a restricted problem (5.38), (5.39) along the direction determined by the solution (p^0, λ^0) of the problem (5.41), (5.42).

In accordance with (5.39), (5.43), the pair $\nu\,(p^0,\lambda^0)$, $\nu \in \mathbf{R}^1$, is feasible if and only if

$$\mu\sigma_l^{-1}(t_i)\left|h_l'(t_i)\left(\tilde{\eta}(T) - \int_{t_i}^T b(s)Q(s)b'(s)\tilde{\xi}(s)\,ds\right)\right| \le 1, \tag{5.44}$$

$$i = 0,\ldots,N, \qquad l = 1,\ldots,m,$$

$$\mu\gamma_j^{-1}(t)\left|\left(Q(t)b'(t)\tilde{\xi}(t)\right)_j\right| \le 1 \quad \text{a.e. on } [0,T], \qquad j = 1,\ldots,r,$$

$$\mu\sigma_{-d}^{-1}\left|e_d'\left(\tilde{\eta}(T) - \int_0^T b(s)Q(s)b'(s)\tilde{\xi}(s)\,ds\right)\right| \le 1, \qquad d = 1,\ldots,n,$$

$$\mu = |\nu| J_{D0}^{-1}.$$

In accordance with (4.18), (4.20), (4.21), (4.26), and (5.20) (cf. (4.32), (4.33)), the conditions (5.44) are equivalent to the system of inequalities

$$\mu\left|\beta_1\sigma_l(t_i)\varphi_l(t_i)\right| \leq 1, \qquad i = 0,\ldots,N, \qquad l = 1,\ldots,m,$$

$$\mu\left|\beta_2\gamma_j(t)b_j'(t)\tilde{\xi}(t)\right| \leq 1 \text{ a.e. on } [0,T], \qquad j = 1,\ldots,r,$$

$$\mu\left|\sigma_{-d}\varphi_{-d}\right| \leq 1, \qquad d = 1,\ldots,n.$$

Consequently the pair $\nu(p^0(t),\lambda^0)$ is admissible if and only if

$$|\nu| \leq \nu^0 = J_{D0}\,\mathcal{N}_{\max}^{-1},$$

where

$$\mathcal{N}_{\max} = \max\left\{\max_{1\leq d\leq n}|\Psi_{-d}|, \sup_{1\leq l\leq m,\ 0\leq i\leq N}|\Psi_l(t_i)|, \operatorname*{ess\,sup}_{1\leq j\leq r,\ t\in[0,T]}|\chi_j(t)|\right\}.$$

By virtue of (4.21), (4.22), (4.24), and (5.43), the value of the objective function $\tilde{a}'\lambda$ along the chosen direction is equal to

$$\nu\tilde{a}'\lambda^0 = \nu J_{D0}^{-1}\tilde{a}'\tilde{\eta}(T) = \nu J_{D0}^{-1}\tilde{a}'\tilde{P}(t_N|N)\tilde{a} = \nu J_{D0}.$$

Therefore the pair $\nu(p^0,\lambda^0)$ provides the maximal value to the objective function at $\nu = \nu^0$ and we should take $(\hat{p},\hat{\lambda}) = \nu^0(p^0,\lambda^0)$. By virtue of (4.25),

$$\varphi_- + \sum_{\{i|t_i\leq t\}} h(t_i)\varphi(t_i) = \tilde{\xi}(t).$$

Hence, taking into account (4.26), (4.10), and (4.17), we see that

$$I_D(\varphi_-,\varphi) = \mathcal{N}_{D1} \quad \text{and} \quad J_{D0} = J_D(\varphi_-,\varphi) = \mathcal{N}_{D2},$$

where $\mathcal{N}_{D1}$ and $\mathcal{N}_{D2}$ are defined by (4.28). Thus, by virtue of (5.40), the desired estimate from below for I_{D0} has the form

$$I_{D0} = I_D^0 \geq \nu^0 J_{D0} = J_{D0}^2\,\mathcal{N}_{\max}^{-1} = \mathcal{N}_{D2}^2\,\mathcal{N}_{\max}^{-1}. \tag{5.45}$$

In accordance with the initial assumption, the piecewise continuous functions $b_j(t)$ and $\gamma_j(t)$ are continuous from the right at every point $t \in [0,T)$ and are continuous from the left at T. The piecewise continuous function $\tilde{\xi}(t)$ is continuous from the right except for the point T, at which it may have an isolated jump. Therefore, in contrast to the case of continuous measurements, the function $\gamma_j(t)b_j'(t)\tilde{\xi}(t)$ may have an isolated jump at $t = T$ and

$$\operatorname*{ess\,sup}_{1\leq j\leq r,\ t\in[0,T]}|\gamma_j(t)b_j'(t)\tilde{\xi}(t)| = \sup_{1\leq j\leq r,\ t\in[0,T)}|\gamma_j(t)b_j'(t)\tilde{\xi}(t)|.$$

Hence,

$$\mathcal{N}_{\max} = \mathcal{N}_{D\infty}. \tag{5.46}$$

Finally, the estimate (4.27) follows from (5.36), (5.45), and (5.46). Theorem 4.2 is proved. □

6. Examples for the Bounds of Nonoptimality Levels

In this section, we shall obtain the numerical values of the upper bound for the nonoptimality level of the Kalman filter in various dynamic guaranteed estimation problems. A general summary is presented at the end of the section.

EXAMPLE 6.1. Let us consider the system with continuous measurements (1.1) – (1.5) with $n = m = 1$, $A(t) = 0$, $B(t) = 1$, $H(t) = 1$, $a = 1$, $\gamma(t) = \gamma$, $\sigma(t) = \sigma$. In other words, the system and the measurements have the form:

$$\dot{x}(t) = u(t), \qquad |u(t)| \leq \gamma, \qquad t \in [0, T],$$

$$|x(0)| \leq \sigma_{-1}, \qquad \sigma_{-1} \to \infty,$$

$$z(t) = x(t) + \varrho(t), \qquad |\varrho(t)| \leq \sigma, \qquad l_* = x(T).$$

We put $\beta_1 = \beta_2 = 1$, i.e. the auxiliary weight coefficients are set by the equalities

$$P_0 = \sigma_{-1}^2, \qquad R(t) = \sigma^2, \qquad Q(t) = \gamma^2.$$

The boundary value problem (2.20) is determined by the formulas

$$\dot{\xi}(t) = \sigma^{-2}\eta(t), \qquad \xi(T) = 1,$$

$$\dot{\eta}(t) = \gamma^2 \xi(t), \qquad \eta(0) = \sigma_{-1}^2 \xi(0).$$

The solution of the boundary value problem is expressed via the hyperbolic functions:

$$\xi(t) = \cosh \kappa t \; \xi(0) + \frac{\sinh \kappa t}{\sigma^2 \kappa} \; \eta(0),$$

$$\eta(t) = \sigma^2 \kappa \sinh \kappa t \; \xi(0) + \cosh \kappa t \; \eta(0),$$

where

$$\xi(0) = \left(\cosh \kappa T + \frac{\sinh \kappa T}{\sigma^2 \kappa} \sigma_{-1}^2 \right)^{-1},$$

$$\eta(0) = \sigma_{-1}^2 \left(\cosh \kappa T + \frac{\sinh \kappa T}{\sigma^2 \kappa} \sigma_{-1}^2\right)^{-1}, \qquad \kappa = \frac{\gamma}{\sigma}.$$

In accordance with (2.19),

$$\varphi_- = \xi(0), \qquad \varphi(t) = \sigma^{-2} \eta(t).$$

Then, as $\sigma_{-1} \to \infty$,

$$\xi(t) \to \xi_\infty(t) = \frac{\sinh \kappa t}{\sinh \kappa T}, \qquad \eta(t) \to \eta_\infty(t) = \sigma^2 \kappa \frac{\cosh \kappa t}{\sinh \kappa T},$$

$$\sigma_{-1} \varphi_- \to 0, \qquad \varphi(t) \to \varphi_\infty(t) = \kappa \frac{\cosh \kappa t}{\sinh \kappa T},$$

$$\Psi(t) \to \Psi_\infty(t) = \sigma \varphi_\infty(t), \qquad \chi(t) \to \chi_\infty(t) = \gamma \xi_\infty(t).$$

By virtue of Theorem 2.2 (cf. (2.15), (2.16)), the limiting value of the level of nonoptimality takes the form

$$\Delta^0 = \frac{\left(\int_0^T \cosh \kappa t \, dt + \int_0^T \sinh \kappa t \, dt\right) \max\left(\max_{t \in [0,T]} \sinh \kappa t, \ \max_{t \in [0,T]} \cosh \kappa t,\right)}{\int_0^T \cosh^2 \kappa t \, dt + \int_0^T \sinh^2 \kappa t \, dt}$$

$$= \frac{e^{\kappa T} - 1}{\sinh \kappa T} = \frac{2}{1 + e^{-\kappa T}}.$$

Obviously, for $\Delta^0 = \Delta^0(\omega)$, $\omega = \kappa T$,

$$\frac{d\Delta^0(\omega)}{d\omega} = \frac{2e^{-\omega}}{(1 + e^{-\omega})^2} > 0, \qquad \frac{d^2 \Delta^0(\omega)}{d\omega^2} = -\frac{2e^{-\omega}(1 - e^{-\omega})}{(1 + e^{-\omega})^3} < 0.$$

Hence $\Delta^0(\omega)$ is a monotonically increasing concave function of ω; moreover,

$$\Delta^0(0) = 1, \qquad \Delta^0(\infty) = 2, \qquad \left.\frac{d\Delta^0(\omega)}{d\omega}\right|_{\omega=0} = \frac{1}{2}.$$

In our simplest example we can compute the value of $\Delta = I(\varphi_-, \varphi) I_0^{-1}$. To do this, let us consider the dual problem (3.48), (3.49):

$$I_0 = I^0 = \sup \lambda \quad \text{under the constraint} \quad \left| \int_t^T \gamma p(s) \, ds + \lambda \right| \leq \sigma, \quad t \in [0, T].$$

On the one hand, it follows from the constraint at $t = T$ that $|\lambda| \leq \sigma$; on the other hand, the constraint is valid for $p(s) = 0$ and $\lambda = \sigma$. Consequently,

$$I_0 = I^0 = \sigma.$$

Then,

$$\Delta = \sigma^{-1}\left(\int_0^T \gamma\frac{\cosh\kappa t}{\sinh\kappa T}\,dt + \int_0^T \gamma\frac{\sinh\kappa t}{\sinh\kappa T}\,dt\right) = \frac{e^{\kappa T}-1}{\sinh\kappa T} = \frac{2}{1+e^{-\kappa T}}.$$

So, $\Delta^0 = \Delta$ and the estimate of the nonoptimality level is sharp.

EXAMPLE 6.2. Let a material point of unit mass moves along the straight line under the action of an unknown but bounded force $u(t)$ (cf. Example 4.1 of Chapter 2):

$$\dot{x}_1(t) = x_2(t), \quad \dot{x}_2(t) = u(t), \qquad |u(t)| \le \gamma, \qquad t \in [0,T],$$

where $x_1(t)$ is the position of the point, $x_2(t)$ is the velocity of the point.

The measurements have the form

$$z(t) = x_1(t) + \varrho(t), \qquad |\varrho(t)| \le \sigma, \qquad t \in [0,T].$$

It is required to evaluate the terminal velocity $l_* = x_2(T)$.

This problem is a special case of (1.1) – (1.5) with $n = 2,\ m = r = 1$,

$$A(t) = \begin{pmatrix} 0 & 1 \\ 0 & 0 \end{pmatrix}, \quad B(t) = \begin{pmatrix} 0 \\ 1 \end{pmatrix}, \quad H(t) = \begin{pmatrix} 1 \\ 0 \end{pmatrix}, \quad a = \begin{pmatrix} 0 \\ 1 \end{pmatrix},$$

$$\gamma(t) = \gamma, \quad \sigma(t) = \sigma, \quad \sigma_{-1} = \sigma_{-2} = \varepsilon^{-1} \to \infty \text{ as } \varepsilon \to 0^+. \tag{6.1}$$

We set $\beta_1 = \beta_2 = 1$, i.e. the weight matrices in the auxiliary Kalman filter are defined by the formulas

$$P_0 = \varepsilon^{-2}E_2, \qquad R(t) = \sigma^2, \qquad Q(t) = \gamma^2. \tag{6.2}$$

The boundary value problem (2.20) takes the form

$$\dot{v}(t) = \mathcal{A}v(t), \qquad v(t) = (\xi'(t), \eta'(t))' \in \mathbf{R}^4, \qquad t \in [0,T], \tag{6.3}$$

$$\mathcal{A} = \begin{pmatrix} 0 & 0 & \sigma^{-2} & 0 \\ -1 & 0 & 0 & 0 \\ 0 & 0 & 0 & 1 \\ 0 & \gamma^2 & 0 & 0 \end{pmatrix},$$

with the boundary conditions

$$\xi(T) = a, \qquad \eta(0) = P_0\xi(0).$$

Solving (6.3), we get

$$v(t) = e^{\mathcal{A}t}v(0), \qquad e^{\mathcal{A}t} = \begin{pmatrix} \Pi_{11}(t) & \Pi_{12}(t) \\ \Pi_{21}(t) & \Pi_{22}(t) \end{pmatrix},$$

$$\Pi_{11}(t) = \begin{pmatrix} \cosh \alpha t \cos \alpha t & \alpha (\cosh \alpha t \sin \alpha t - \sinh \alpha t \cos \alpha t) \\ -\frac{1}{2\alpha} (\cosh \alpha t \sin \alpha t + \sinh \alpha t \cos \alpha t) & \cosh \alpha t \cos \alpha t \end{pmatrix},$$

$$\Pi_{12}(t) = \begin{pmatrix} \frac{1}{2\alpha\sigma^2} (\cosh \alpha t \sin \alpha t + \sinh \alpha t \cos \alpha t) & \frac{1}{2\alpha^2\sigma^2} \sinh \alpha t \sin \alpha t \\ -\frac{1}{2\alpha^2\sigma^2} \sinh \alpha t \sin \alpha t & -\frac{1}{4\alpha^3\sigma^2} (\cosh \alpha t \sin \alpha t - \sinh \alpha t \cos \alpha t) \end{pmatrix},$$

$$\Pi_{21}(t) = \begin{pmatrix} -\frac{\gamma^2}{4\alpha^3} (\cosh \alpha t \sin \alpha t - \sinh \alpha t \cos \alpha t) & \frac{\gamma^2}{2\alpha^2} \sinh \alpha t \sin \alpha t \\ -\frac{\gamma^2}{2\alpha^2} \sinh \alpha t \sin \alpha t & \frac{\gamma^2}{2\alpha} (\cosh \alpha t \sin \alpha t + \sinh \alpha t \cos \alpha t) \end{pmatrix},$$

$$\Pi_{22}(t) = \begin{pmatrix} \cosh \alpha t \cos \alpha t & \frac{1}{2\alpha} (\cosh \alpha t \sin \alpha t + \sinh \alpha t \cos \alpha t) \\ -\alpha (\cosh \alpha t \sin \alpha t - \sinh \alpha t \cos \alpha t) & \cosh \alpha t \cos \alpha t \end{pmatrix},$$

$$\alpha = \sqrt{\frac{\kappa}{2}}, \qquad \kappa = \frac{\gamma}{\sigma}.$$

Thus,

$$\xi(t) = \Pi_{11}(t)\xi(0) + \Pi_{12}(t)\eta(0), \qquad \eta(t) = \Pi_{21}(t)\xi(0) + \Pi_{22}(t)\eta(0).$$

It follows from the boundary conditions that

$$\big(\Pi_{11}(T) + \Pi_{12}(T)P_0\big)\xi(0) = a,$$

whence

$$\xi(0) = P_0^{-1}\left(\Pi_{11}(T)P_0^{-1} + \Pi_{12}(T)\right)^{-1} a,$$

$$\eta(0) = \left(\Pi_{11}(T)P_0^{-1} + \Pi_{12}(T)\right)^{-1} a$$

(it is easy to verify that the inverse matrix above exists).

Using (6.1), (6.2), and (2.19), we see that if $\varepsilon \to 0^+$, then

$$\sigma_{-1}\varphi_{-1} \to 0, \qquad \sigma_{-2}\varphi_{-2} \to 0, \qquad \varphi(t) \to \varphi_\infty(t), \qquad B'\xi(t) \to B'\xi_\infty(t),$$

where

$$\varphi_\infty(t) = \kappa\Big[(\cosh \alpha t \sin \alpha t + \sinh \alpha t \cos \alpha t)(\cosh \alpha T \sin \alpha T + \sinh \alpha T \cos \alpha T)$$

$$- 2\cosh \alpha t \cos \alpha t \sinh \alpha T \sin \alpha T\Big] \left(\sinh^2 \alpha T - \sin^2 \alpha T\right)^{-1},$$

$$B'\xi_\infty(t) = 2\Big[\sinh \alpha t \sin \alpha t \sinh \alpha T \sin \alpha T$$

$$- (\cosh \alpha t \sin \alpha t - \sinh \alpha t \cos \alpha t)(\cosh \alpha T \sin \alpha T + \sinh \alpha T \cos \alpha T)\Big]$$

TABLE 1. Values of Δ and Δ^0

κT^2	0	2	4	6	8	10	50	100	700
Δ	1.50	1.33	1.24	1.19	1.19	1.19	1.27	1.28	1.28
Δ^0	1.50	1.80	1.86	1.80	1.74	1.71	1.80	1.81	1.81

$$\times \left(\sinh^2 \alpha T - \sin^2 \alpha T\right)^{-1}.$$

It can easily be checked that $\Delta^0 = \Delta^0(\kappa T^2)$. It was shown in [133] (cf. Example 4.1 of Chapter 2) that

$$I_0 = \begin{cases} \sigma T^{-1}\left(2 + \frac{1}{2}\kappa T^2\right) & \text{for } 0 \le \kappa T^2 \le 4 \\ 2\sigma T^{-1}\left(\kappa T^2\right)^{\frac{1}{2}} & \text{for } \kappa T^2 > 4. \end{cases}$$

The expression for I_0 allows us to compute the quantity $\Delta = \Delta(\kappa T^2) = I(\varphi_-, \varphi)/I_0$. The nonoptimality level of the Kalman filter is obtained in accordance with Theorem 2.2. The values of Δ and Δ^0 are shown in Table 1.

It follows from Table 1 that for a quite wide range of σ, γ, and T the value of $\Delta^0 < 2$ and $(\Delta^0 - \Delta)/\Delta \sim 1/2$.

EXAMPLE 6.3. The correction problem for the inertial navigation system with the aid of a discrete position information reduces to the estimation problem that is described by the system (4.1) – (4.5) with following parameters:

$$n = 4, \qquad m = r = 1, \qquad \gamma(t) = \gamma \qquad \sigma(t) = \sigma,$$

$$A(t) = \begin{pmatrix} 0 & 0 & 1 & 0 \\ 0 & 0 & 1 & -1 \\ 0 & -1 & 0 & 0 \\ 0 & 0 & 0 & 0 \end{pmatrix}, \qquad B(t) = \begin{pmatrix} 0 \\ -1 \\ 0 \\ 0 \end{pmatrix}, \qquad H(t_i) = \begin{pmatrix} 1 \\ 0 \\ 0 \\ 0 \end{pmatrix},$$

$$t_i = \frac{\pi i}{2N}, \qquad i = 0, \ldots, N, \qquad t_0 = 0, \qquad t_N = T = \frac{\pi}{2}.$$

It is required to evaluate the quantities $l_* = x_d(\pi/2)$, $d = 1, \ldots, 4$.

For our computation we set

$$\sigma = 0.75 \cdot 10^{-2}, \qquad \sigma_{-1} = 1.2, \quad \sigma_{-2} = 1.2, \quad \sigma_{-3} = 0.6, \quad \sigma_{-4} = 0.8.$$

The nonoptimality level of the Kalman filter in the guaranteed estimation problem is computed in accordance with Theorem 4.2. The values of Δ_D^0 are displayed in Table 2 for $\beta_1 = 1$, $\beta_2 = \Delta t = (2N)^{-1}\pi$ (cf. (4.38)), and various γ. The influence of the parameter β_2 on Δ_D^0 for $\gamma = 0.08$ ($\beta_1 = 1$) is shown

TABLE 2. Values of Δ_D^0 for $\beta_1 = 1$, $\beta_2 = \Delta t$

γ	0.40	0.08	0.00
x_1	3.02	2.68	2.57
x_2	2.79	2.74	2.45
x_3	3.45	2.98	2.77
x_4	2.15	2.52	2.26

TABLE 3. Values of Δ_D^0 for $\beta_1 = 1$ and various β_2 ($\gamma = 0.08$)

β_2	$0.1\Delta t$	$0.5\Delta t$	Δt	$5\Delta t$	$7\Delta t$	$10\Delta t$	$20\Delta t$	$25\Delta t$
x_1	2.68	2.68	2.68	2.66	2.65	2.64	2.60	3.58
x_2	2.80	2.77	2.74	2.52	2.42	2.80	4.54	5.18
x_3	2.99	2.98	2.98	2.91	2.89	2.84	2.72	3.10
x_4	2.60	2.56	2.52	2.21	2.13	2.78	4.31	4.83

in Table 3. The optimal values of β_2 and minimal values of Δ_D^0 are presented in Table 4.

Thus the values of Δ_D^0 can be improved somewhat by varying the parameter β_2. It can be checked that an additional optimization in β_1 does not improve Δ_D^0 significantly. On the whole, the recommendation (4.38) is successful: it provides a good reference point.

EXAMPLE 6.4. Consider the oscillating system of four perturbed oscillators on the interval $t \in [0, 2\pi]$:

$$\dot{x}_1(t) = \omega_1 x_2(t), \qquad \dot{x}_3(t) = \omega_2 x_4(t),$$
$$\dot{x}_2(t) = -\omega_1 x_1(t) + u_1(t), \qquad \dot{x}_4(t) = -\omega_2 x_3(t) + u_2(t),$$

TABLE 4. Optimal values of β_2 (Example 6.3)

	$\gamma = 0.40$		$\gamma = 0.08$	
	Δ_D^0	β_2	Δ_D^0	β_2
x_1	2.45	$10\Delta t$	2.60	$20\Delta t$
x_2	2.21	$2\Delta t$	2.42	$7\Delta t$
x_3	2.51	$7\Delta t$	2.72	$20\Delta t$
x_4	1.97	$2\Delta t$	2.13	$7\Delta t$

$$\dot{x}_5(t) = \omega_3 x_6(t), \qquad \dot{x}_7(t) = \omega_4 x_8(t),$$
$$\dot{x}_6(t) = -\omega_3 x_5(t) + u_3(t), \qquad \dot{x}_8(t) = -\omega_4 x_7(t) + u_4(t),$$

$$|u_j(t)| \le \gamma, \qquad j = 1, \dots, 4.$$

Assume that these four subsystems are combined by the measurements

$$z(t_i) = x_1(t_i) + x_3(t_i) + x_5(t_i) + x_7(t_i) + \varrho(t_i),$$

$$|\varrho(t_i)| \le \sigma, \qquad t_i = \frac{2\pi i}{N}, \quad i = 0, \dots, N, \qquad t_0 = 0, \quad t_N = T = 2\pi.$$

Suppose also that

$$|x_d(0)| \le \sigma_-, \qquad d = 1, \dots, 8.$$

Obviously, this estimation problem is described by the system (4.1) – (4.5) with

$$n = 8, \qquad r = 4, \qquad m = 1, \qquad \gamma_j(t) = \gamma, \qquad \sigma(t_i) = \sigma,$$

$$A = \begin{pmatrix} \begin{matrix} 0 & \omega_1 \\ -\omega_1 & 0 \end{matrix} & 0_{2\times2} & 0_{2\times2} & 0_{2\times2} \\ 0_{2\times2} & \begin{matrix} 0 & \omega_2 \\ -\omega_2 & 0 \end{matrix} & 0_{2\times2} & 0_{2\times2} \\ 0_{2\times2} & 0_{2\times2} & \begin{matrix} 0 & \omega_3 \\ -\omega_3 & 0 \end{matrix} & 0_{2\times2} \\ 0_{2\times2} & 0_{2\times2} & 0_{2\times2} & \begin{matrix} 0 & \omega_4 \\ -\omega_4 & 0 \end{matrix} \end{pmatrix},$$

$$B = \begin{pmatrix} \begin{matrix} 0 \\ 1 \end{matrix} & 0_{2\times2} & 0_{2\times2} & 0_{2\times2} \\ 0_{2\times2} & \begin{matrix} 0 \\ 1 \end{matrix} & 0_{2\times2} & 0_{2\times2} \\ 0_{2\times2} & 0_{2\times2} & \begin{matrix} 0 \\ 1 \end{matrix} & 0_{2\times2} \\ 0_{2\times2} & 0_{2\times2} & 0_{2\times2} & \begin{matrix} 0 \\ 1 \end{matrix} \end{pmatrix}, \quad u(t) = \begin{pmatrix} u_1(t) \\ u_2(t) \\ u_3(t) \\ u_4(t) \end{pmatrix}, \quad H = \begin{pmatrix} 1 \\ 0 \\ 1 \\ 0 \\ 1 \\ 0 \\ 1 \\ 0 \end{pmatrix}.$$

It is required to evaluate the quantities $l_* = x_d(2\pi)$, i.e.,

$$l_* = a'x(2\pi), \qquad a = e_d = (\underbrace{0, \dots, 0, 1}_{d}, \dots, 0)', \qquad d = 1, \dots, 8.$$

We set $\omega_1 = 1$, $\omega_2 = 2$, $\omega_3 = 3$, $\omega_4 = 4$. Note that if $\gamma = 0$, then

$$z(t_i) = H'(t_i)x(2\pi) + \varrho(t_i),$$

TABLE 5. Values of Δ_D^0 for $\beta_1 = 1$, $\beta_2 = 2\pi N^{-1}$ (Example 6.4)

γ	10^{-3}	10^{-2}	10^{-1}	1	10	10^2
x_1	1.32	1.49	2.62	2.77	1.99	2.87
x_2	1.29	1.45	2.78	2.82	2.77	2.77
x_3	1.31	1.48	2.52	2.57	2.42	2.39
x_4	1.29	1.46	2.42	3.09	3.10	3.10
x_5	1.32	1.48	2.36	2.59	2.46	2.46
x_6	1.29	1.46	2.43	3.21	3.25	3.29
x_7	1.30	1.45	2.17	3.53	3.70	3.72
x_8	1.29	1.45	2.39	3.92	4.50	4.52

where

$$H'(t_i) = H'e^{A(t_i-2\pi)} = (\cos t_i, \sin t_i, \ldots, \cos 4t_i, \sin 4t_i)\,,$$

and the problem reduces to the estimation of the coefficients of the Fourier series (cf. Section 7 of Chapter 1).

We also put

$$\sigma_- = 100, \qquad \sigma = 0.2, \qquad N = 100.$$

Theorem 4.2 determines the algorithm for the computation of Δ_D^0. The values of Δ_D^0 for $\beta_1 = 1$, $\beta_2 = \Delta t = \frac{2\pi}{N}$, and various γ are displayed in Table 5. It follows from Table 5 that, for small values of $\gamma \sim 10^{-3}, 10^{-2}$, the values of Δ_D^0 fall in the range $1.3 \div 1.5$ and are close to these of Example 7.3 of Chapter 1. For $\gamma \sim 10^{-1}$ the values of Δ_D^0 is about 2.5. If $1 \le \gamma \le 10^2$, then $2.57 \le \Delta_D^0 \le 4.52$.

We can try to improve the guaranteed level of nonoptimality by varying β_2. The influence of the parameter β_2 on Δ_D^0 for $\gamma = 1$ $(\beta_1 = 1)$ is shown in Table 6. The optimal values of β_2 and the minimal values of Δ_D^0 $(\beta_1 = 1)$ are presented in Table 7.

It follows from Tables 5 – 7 that the optimal in β_2 values of Δ_D^0 are close to the values of Δ_D^0 for $\beta_2 = \Delta t$. So, the choice of $\beta_2 = \Delta t$ is justified. Only for certain states $x_d(2\pi)$ the values of Δ_D^0 can be slightly improved. The additional varying of β_1 does not result in the significant decrease of Δ_D^0.

EXAMLE 6.5. Let us consider the problem of parameter estimation for the motion of a satellite that moves in a near-circular orbit. The additional information is provided with the aid of the measurements of the directions to the radio beacons.

TABLE 6. Dependence on β_2 (Example 6.4)

β_2	$0.05\Delta t$	$0.1\Delta t$	$0.5\Delta t$	Δt	$2\Delta t$	$5\Delta t$	$10\Delta t$	$50\Delta t$
x_1	11.13	8.47	3.96	2.77	2.20	2.22	2.25	2.35
x_2	3.82	2.52	2.77	2.82	2.87	2.93	2.98	3.12
x_3	10.18	7.48	3.14	2.57	2.62	2.67	2.69	2.73
x_4	3.94	2.90	3.02	3.09	3.17	3.27	3.35	3.58
x_5	8.70	6.02	2.56	2.59	2.61	2.61	2.62	2.65
x_6	5.77	5.15	3.11	3.21	3.31	3.45	3.55	3.88
x_7	6.53	4.34	3.31	3.53	3.73	3.97	4.23	4.72
x_8	8.41	7.17	3.86	3.92	4.29	4.67	4.89	5.17

TABLE 7. Optimal values of β_2 and Δ_D^0 (Example 6.4)

	$\gamma = 10^{-2}$		$\gamma = 10^{-1}$		$\gamma = 1$		$\gamma = 10$		$\gamma = 10^2$	
	Δ_D^0	β_2	Δ_D^0	β_2	Δ_D^0	β_2	Δ_D^0	β_2	Δ_D^0	β_2
x_1	1.40	$7\Delta t$	2.41	$2\Delta t$	2.20	$2\Delta t$	1.99	Δt	2.75	$0.5\Delta t$
x_2	1.38	$6\Delta t$	2.14	$0.5\Delta t$	2.52	$0.1\Delta t$	2.75	$0.05\Delta t$	2.75	$0.05\Delta t$
x_3	1.39	$7\Delta t$	2.52	Δt	2.57	Δt	2.38	$0.1\Delta t$	2.38	$0.1\Delta t$
x_4	1.39	$6\Delta t$	2.19	$0.5\Delta t$	2.90	$0.1\Delta t$	3.08	$0.1\Delta t$	3.08	$0.1\Delta t$
x_5	1.40	$7\Delta t$	2.36	Δt	2.56	$0.5\Delta t$	2.45	$0.05\Delta t$	2.45	$0.05\Delta t$
x_6	1.40	$6\Delta t$	2.32	$0.5\Delta t$	3.11	$0.5\Delta t$	3.25	Δt	3.25	$0.1\Delta t$
x_7	1.40	$6\Delta t$	2.17	Δt	3.31	$0.5\Delta t$	3.40	$0.2\Delta t$	3.50	$0.1\Delta t$
x_8	1.40	$6\Delta t$	2.32	$0.5\Delta t$	3.86	$0.5\Delta t$	4.49	$0.05\Delta t$	4.49	$0.05\Delta t$

The navigational errors are described by the dynamic system

$$
\begin{aligned}
&\dot{x}_1(t) = x_2(t), && \dot{x}_2(t) = -x_1(t), \\
&\dot{x}_3(t) = x_4(t) - x_5(t), && \dot{x}_4(t) = -x_3(t) - x_6(t), \\
&\dot{x}_5(t) = x_3(t) + x_6(t), && \dot{x}_6(t) = x_4(t) + 2x_5(t), \\
&\dot{x}_7(t) = -x_8(t) + u_1(t), && \dot{x}_8(t) = x_7(t) + u_2(t), \qquad \dot{x}_9(t) = u_3(t),
\end{aligned}
$$

$t \in [0, 2\pi]$, where the dimensionless time 2π corresponds to a one turn of the satellite.

In matrix form these equations can be rewritten as follows:

$$
\frac{dx(t)}{dt} = Ax(t) + Bu(t), \qquad x(t) \in \mathbf{R}^9, \quad u(t) \in \mathbf{R}^3,
$$

where

$$A = \begin{pmatrix} \begin{matrix} 0 & 1 \\ -1 & 0 \end{matrix} & 0_{2\times 4} & 0_{2\times 3} \\ 0_{4\times 2} & \begin{matrix} 0 & 1 & -1 & 0 \\ -1 & 0 & 0 & -1 \\ 1 & 0 & 0 & 1 \\ 0 & 1 & 2 & 0 \end{matrix} & 0_{4\times 3} \\ 0_{3\times 2} & 0_{3\times 4} & \begin{matrix} 0 & -1 & 0 \\ 1 & 0 & 0 \\ 0 & 0 & 0 \end{matrix} \end{pmatrix}, \qquad B = \begin{pmatrix} 0_{6\times 3} \\ \begin{matrix} 1 & 0 & 0 \\ 0 & 1 & 0 \\ 0 & 0 & 1 \end{matrix} \end{pmatrix}.$$

We suppose that

$$|u_j(t)| \le \gamma, \qquad j = 1, \ldots, 3, \qquad\qquad |x_d(0)| \le \sigma_{-d}, \qquad d = 1, \ldots, 9.$$

Let the measurements are carried out on the interval $[0, 2\pi]$ at the instants $t_i = 2\pi i N^{-1}$, $i = 0, \ldots, N$. Assume that at t_i (for fixed i) the information from one beacon only is used. Then the measurement equations take the form

$$z(t_i) = H'(t_i)x(t_i) + \varrho(t_i), \qquad z(t_i), \varrho(t_i) \in \mathbf{R}^3,$$

$$|\varrho_l(t_i)| \le \sigma, \qquad l = 1, 2, 3,$$

$$H(t_i) = \begin{pmatrix} -\frac{a}{2l_i}\sin^2\theta_i \sin 2\phi_i & \frac{a}{l_i}\left(1 - \sin^2\theta_i \sin^2\phi_i\right) & \frac{a}{2l_i}\sin 2\theta_i \sin\phi_i \\ 0 & 0 & 0 \\ \frac{a}{l_i}\left(1 - \sin^2\theta_i \cos^2\phi_i\right) & -\frac{a}{2l_i}\sin^2\theta_i \sin 2\phi_i & \frac{a}{l_i}\sin 2\theta_i \cos\phi_i \\ 0 & 0 & 0 \\ \frac{a}{2l_i}\sin 2\theta_i \cos\phi_i & \frac{a}{2l_i}\sin 2\theta_i \sin\phi_i & \frac{a}{l_i}\left(1 - \cos^2\theta_i\right) \\ 0 & 0 & 0 \\ 0 & \cos\theta_i & \sin\theta_i \sin\phi_i \\ -\sin\theta_i \sin\phi_i & \sin\theta_i \cos\phi_i & 0 \\ -\cos\theta_i & 0 & -\sin\theta_i \cos\phi_i \end{pmatrix},$$

$$a = \frac{r}{r_E}, \qquad l_i = a\cos\theta_i - \sqrt{1 - a^2 \sin^2\theta_i}, \qquad i = 0, \ldots, N,$$

where r is the radius of the orbit, r_E is the Earth radius, and θ_i, ϕ_i are determined by the orbit inclination, design velocity of the satellite, and the location of the beacons. Since the explicit formulas for θ_i and ϕ_i (or digital representation of $H(t_i)$) are very cumbersome, we do not cite them here.

The estimation problem is to evaluate the terminal values of the components of the state at the end of the first turn of the satellite, i.e.,

$$l_* = a'x(2\pi), \qquad a = e_d = (\underbrace{0, \ldots, 0, 1}_{d}, \ldots, 0)', \qquad d = 1, \ldots, 9.$$

TABLE 8. Values of Δ_D^0 for $\beta_1 = 1$, $\beta_2 = \Delta t = \frac{\pi}{50}$

	$\sigma_{-1} = \sigma_{-3} = \sigma_{-5} = 1.2$, $\sigma_{-2} = \sigma_{-4} = \sigma_{-6} = 0.7$, $\sigma_{-7} = \sigma_{-8} = \sigma_{-9} = 0.8$		$\sigma_{-d} = 100$, $d=1,\ldots,100$
	$\sigma = 0.05$	$\sigma = 0.3$	$\sigma = 0.3$
x_1	1.23	1.71	1.29
x_2	1.30	1.75	1.29
x_3	1.57	2.24	1.14
x_4	2.52	2.12	2.25
x_5	2.58	2.21	2.30
x_6	1.56	2.20	1.14
x_7	1.36	1.47	1.29
x_8	1.45	1.51	1.29
x_9	1.44	1.44	1.44

So, our estimation problem is described by the system (4.1) – (4.5) with

$$n = 9, \qquad r = 3, \qquad m = 3, \qquad \gamma_j(t) = \gamma, \qquad \sigma(t_i) = \sigma,$$

For our computation we put $\gamma = 0.04$, $N = 100$. Theorem 4.2 determines the algorithm for obtaining Δ_D^0. The values of Δ_D^0 for $\beta_1 = 1$, $\beta_2 = \Delta t$ and certain values of σ, σ_{-d} are shown in Table 8.

The influence of the parameter β_2 on Δ_D^0 for $\sigma_{-1} = \sigma_{-3} = \sigma_{-5} = 1.2$, $\sigma_{-2} = \sigma_{-4} = \sigma_{-6} = 0.7$, $\sigma_{-7} = \sigma_{-8} = \sigma_{-9} = 0.8$, $\gamma = 0.04$, and two values of σ is characterized by Tables 9 and 10. The optimal values of β_2 and minimal values of Δ_D^0 compared with Δ_D^0 for $\beta_2 = \Delta t$ are presented in Table 11. It follows from Table 11 that, though the optimal values of β_2 may differ significantly from the initial recommendation $\beta_2 = \Delta t$, the minimal values of Δ_D^0 are close to the relevant values of Δ_D^0 for $\beta_2 = \Delta t$.

Thus the use of the Kalman filter does not lead to a substantial loss of optimality: for almost all state components $\Delta_D^0 \sim 1.5$. Only for two components, x_4 and x_5, the optimal values of Δ_D^0 lie in the range 2.0 – 2.3.

EXAMPLE 6.6. Let us consider an example of the prediction problem that was discussed in Sections 2 and 4. The problem is to evaluate the transversal bias of a satellite in a near-circular orbit at the end of the second turn by the measurements of the transversal bias on the previous turn [139]. The dynamic equations and the measurement equation have the form (4.1) – (4.5) with

$$n = 4, \qquad m = r = 1, \qquad \gamma(t) = \gamma \qquad \sigma(t) = \sigma,$$

TABLE 9. Influence of β_2 on Δ_D^0 for $\beta_1 = 1$ and $\sigma = 0.3$ $(\gamma = 0.04)$

β_2	$0.01\Delta t$	$0.1\Delta t$	Δt	$5\Delta t$	$10\Delta t$	$100\Delta t$	$200\Delta t$	$500\Delta t$
x_1	1.72	1.72	1.71	1.70	1.67	1.56	1.53	1.87
x_2	1.75	1.75	1.75	1.73	1.71	1.66	1.66	2.60
x_3	2.25	2.25	2.24	2.23	2.21	1.94	1.73	2.36
x_4	2.15	2.14	2.13	2.09	2.03	7.00	9.82	13.16
x_5	2.23	2.23	2.21	2.15	2.09	7.21	9.84	12.75
x_6	2.20	2.20	2.20	2.18	2.16	1.91	1.71	2.69
x_7	1.45	1.45	1.47	1.57	1.66	1.94	1.94	2.37
x_8	1.48	1.48	1.51	1.61	1.71	2.12	2.18	3.37
x_9	1.44	1.44	1.44	1.44	1.43	1.31	1.20	1.53

TABLE 10. Influence of β_2 on Δ_D^0 for $\beta_1 = 1$ and $\sigma = 0.05$ $(\gamma = 0.04)$

β_2	$0.01\Delta t$	$0.1\Delta t$	Δt	$5\Delta t$	$10\Delta t$	$100\Delta t$	$200\Delta t$	$500\Delta t$
x_1	1.33	1.31	1.23	1.17	1.16	1.13	1.12	1.11
x_2	1.36	1.35	1.30	1.27	1.28	1.29	1.28	1.94
x_3	1.58	1.58	1.57	1.54	1.52	1.37	1.25	1.61
x_4	3.02	2.97	2.52	2.20	2.82	4.00	4.15	4.29
x_5	3.16	3.10	2.58	2.24	2.78	3.72	3.83	3.92
x_6	1.57	1.57	1.56	1.54	1.52	1.39	1.28	1.88
x_7	1.44	1.43	1.36	1.29	1.27	1.21	1.19	1.18
x_8	1.47	1.47	1.45	1.46	1.49	1.49	1.46	2.16
x_9	1.44	1.44	1.44	1.43	1.42	1.31	1.20	1.52

$$A(t) = \begin{pmatrix} 0 & 1 & 2 & 0 \\ 1 & 0 & 0 & 1 \\ 1 & 0 & 0 & 1 \\ 0 & 1 & 1 & 0 \end{pmatrix}, \qquad B(t) = \begin{pmatrix} 0 \\ 1 \\ 0 \\ 0 \end{pmatrix}, \qquad H(t_i) = \begin{pmatrix} 0 \\ 0 \\ 0 \\ 1 \end{pmatrix},$$

$$t_i = \frac{2\pi i}{N}, \qquad i = 0, \dots, N, \qquad t_0 = 0, \qquad t_N = T = 2\pi.$$

We assume also that

$$|x_d(0)| \leq \sigma_-, \qquad d = 1, 2, 3, 4.$$

The problem is to evaluate the quantity $l_* = x_4(4\pi)$, i.e., $a = (0, 0.0, 1)'$ and $\kappa = 2\pi$. The quadrature F^κ and the auxiliary vector a_κ can easily be computed from (2.24) and (2.25):

$$F^\kappa = \gamma\left\{6\pi^2 - 2\tau_0^2 + 8(1 - \cos\tau_0)\right\}, \qquad a_\kappa = (0, -6\pi, -6\pi, 1)',$$

TABLE 11. Optimal values of β_2 and Δ_D^0 ($\beta_1 = 1$)

	$\sigma = 0.3$			$\sigma = 0.05$		
	β_2	Δ_D^0 *optimal*	Δ_D^0 $\beta_2{=}\Delta t$	β_2	Δ_D^0 *optimal*	Δ_D^0 $\beta_2{=}\Delta t$
x_1	$200\Delta t$	1.53	1.71	$500\Delta t$	1.11	1.23
x_2	$200\Delta t$	1.66	1.75	$5\Delta t$	1.27	1.30
x_3	$200\Delta t$	1.73	2.24	$200\Delta t$	1.26	1.57
x_4	$10\Delta t$	2.03	2.13	$5\Delta t$	2.20	2.52
x_5	$10\Delta t$	2.09	2.21	$5\Delta t$	2.24	2.58
x_6	$200\Delta t$	1.71	2.20	$200\Delta t$	1.28	1.56
x_7	$0.1\Delta t$	1.45	1.47	$500\Delta t$	1.18	1.36
x_8	$0.1\Delta t$	1.48	1.51	Δt	1.45	1.45
x_9	Δt	1.44	1.44	$200\Delta t$	1.20	1.44

TABLE 12. Values of $\Delta_{D\kappa}^0$ and $\Delta_{D\kappa}$

	$\Delta_{D\kappa}^0$		$\Delta_{D\kappa}$
γ	$N = 100$	$N = 12$	$N = 12$
10^{-3}	2.01	1.76	1.54
10	1.44	1.14	1.02
40	1.14	1.13	1.03
100	1.14	1.12	1.05

where τ_0 is a unique root of the equation $4\sin\tau - 3\tau = 0$.

We put $\sigma_- = 10^4$, $\sigma = 1$. The values of $\Delta_{D\kappa}^0$ (computed in accordance with Theorem 4.2) for $\beta_1 = 1$, $\beta_2 = \Delta t = 2\pi N^{-1}$, various γ, and two values of N are presented in Table 12. The optimal estimators for the case of $N = 12$ were obtained in [139]. So, the true values of $\Delta_{D\kappa}$ for $N = 12$ can be computed; they are cited in the same table.

We see from Table 12 that $\Delta_{D\kappa}^0$ does not exceed 2.01 and, in certain cases, Δ_D^0 is close to unity. Moreover, for $N = 12$ it yields a good approximation for $\Delta_{D\kappa}$. For large γ, the term F^κ contributes significantly to the numerator and the denominator of $\Delta_{D\kappa}^0$ and $\Delta_{D\kappa}$. Partially by this reason $\Delta_{D\kappa}^0$ and $\Delta_{D\kappa}$ are close to unity for large γ.

Summary

The ranges of Δ_D^0 (Δ^0) for Examples 6.1 – 6.6 are presented in Table 13.

TABLE 13. Summarized results

Ex. 6.1	Ex. 6.2	Ex. 6.3	Ex. 6.4	Ex. 6.5	Ex. 6.6
$1.01 \div 2.00$	$1.50 \div 1.81$	$1.97 \div 2.72$	$1.38 \div 4.49$	$1.11 \div 2.24$	$1.12 \div 2.01$

If β_2 was varied, then the minimal values of Δ_D^0 are taken. The additional minimization in β_1 does not improve Δ_D^0 significantly. Formula (4.38) gives a good reference point for the choice of the free parameters β_1, β_2. In many cases the level of nonoptimality does not exceed 1.50.

The least successful example is Example 6.4. We see from Table 7 that a quarter of all values of Δ_D is greater than 3.0. On the average, Δ_D^0 is about 2.5. Note that the value 2.5 is an acceptable quantity for the loss of optimality in much of applied problems, in which the stability with respect to various noise model is of primal importance.

We hardly can expect that in the majority of guaranteed estimation problems the Kalman filter is almost optimal. However, with the help of our guaranteed level of nonoptimality we can select the problems in which the Kalman filter can be used without critical loss of optimality.

In the vast majority of the discussed examples the loss of optimality is not critical. A possible presence of white noises in perturbations makes the use of the Kalman filter all the more natural. This case is considered in Chapter 4 in detail.

Thus the obtained bound for the nonoptimality level is a useful tool for the analysis and design of estimation algorithms.

REMARK 6.1. It should be noted that we did not use all potentialities of our estimate. In fact, we used the same Kalman filter for the estimation itself (in the numerator of Δ_D (or Δ)) and for the evaluating the unknown optimal value I_{D0} (or I_0) (in the denominator of Δ_D (or Δ)). Obviously, we can use two separate Kalman filters for these different purposes (with parameters $\beta_1^{(1)}$, $\beta_2^{(1)}$ and $\beta_1^{(2)}$, $\beta_2^{(2)}$, respectively). In this case, the corresponding optimization of Δ_D^0 (or Δ^0) in $\beta_i^{(j)}$ is decomposed into two independent two-dimensional problems. Clearly, the nonoptimality estimate can be somewhat improved.

7. Comments to Chapter 3

7.1. *Section 1*

The theory of Kalman filter is outlined in a lot of papers and books. We cite here [84], [85], [41], [80], [81], [79], [126], [170], [143], [47], [187], [83], [82],

[1]. An overview of the least-squares estimation is given in [201], where an important work by P. Swerling [202] is also cited.

The measure of nonoptimality of the Kalman filter was introduced in [158], [66]. The term "level of nonoptimality" was suggested to the author by F.L. Chernousko.

The solution to an *a posteriori* guaranteed estimation problem with the help of a multi-parameter family of Kalman filters was obtained in [124] for discrete uncertain deterministic dynamic systems. By a variation of the covariance matrices in the auxiliary Kalman filters it turned out to be possible to approximate the solution of the primary problem with any degree of accuracy. Though the number of adjusted parameters is large and the numerical algorithms for adjusting the parameters were not discussed, the result of [124] is of great theoretical importance.

7.2. *Section 2*

Theorem 2.1 was proved in fact, e.g., in [79], [187]. The explanation for the link between $I(\Phi_-, \Phi)$ and $J(\Phi_-, \Phi)$ with the help of the Cauchy-Schwarz-Bunyakovskii inequality was proposed by F.L. Chernousko. Theorem 2.2 was first stated and proved in [158] and, in a more compact form, in [66]. Details are presented in the technical report [152]. Guaranteed prediction problem was discussed in [139].

Obviously, the optimal guaranteed estimation problem, which is considered in this Chapter, is equivalent to a corresponding control problem for a relative linear system with a nonsmooth cost functional. The same is true for the estimation problems in subsequent chapters. So, our results can be easily reformulated for the equivalent control problems as well: simplified control algorithms can be constructed and the level of their nonoptimality can be evaluated.

7.3. *Section 3*

To prove Theorem 2.1 and Theorem 2.2 we follow the method presented briefly in [66]. More detailed proofs (but in another style) were given in [152] and [158]. The proofs of these theorems are based on the duality theory for convex variational problems [185], [75], [51], [123], [5], [186].

In the proof of Theorem 2.1 for singular matrices P_0 and $Q(t)$ and in the proof of Lemma 2.1 we substantially used the notion of the strict convexity that allowed us to evaluate the proximity of an admissible point to the optimal one with the help of the corresponding values of the functional. We took this approach from a very clear book [87]. General features of convex minimization problems are given, e.g., in [51].

7.4. *Section 4*

The case of discrete measurements was studied after the case of continuous measurements in [32] and in [36]. A lot of technical details is contained in the reports [165] and [35].

7.5. *Section 5*

Though the idea of the construction of the upper bound for the nonoptimality level in the discrete case is the same as in the continuous one, the derivation contains some additional crucial points that we discuss in detail.

7.6. *Section 6*

Examples 6.1 and 6.2 were considered in [158]. Examples 6.3, 6.4, and 6.6 were studied in [35]. Example 6.5 is taken from [34].

8. Exercises to Chapter 3

1. Let $\tilde{\theta}(t)$ be the solution of the matrix equation $\dot{\tilde{\theta}}(t) = -\tilde{K}(t)h'(t)\tilde{\theta}(t)$, $\tilde{\theta}(0) = E_n$. Prove directly that if a pair (φ_-, φ) satisfies the condition $\varphi_- + \int_0^T h(t)\varphi(t)\,dt = \tilde{a}$ and φ is the Kalman estimator, i.e., $\varphi(t) = \tilde{K}'(t)\left(\tilde{\theta}^{-1}(t)\right)'\tilde{\theta}'(T)\tilde{a}$, then $\varphi_- = \tilde{\theta}'(T)\tilde{a}$.
2. Consider the problem (3.7) from Section 3. Prove that if P_0, $R(t)$, and $Q(t)$ are strictly positive definite (are subjected to (1.6)), then the functional $J(\Phi_-, \Phi, w)$ is coercive and strictly convex. Prove also that the manifold $\Lambda(\Phi_-, \Phi, w) = 0$ is affine and closed.
3. Prove Theorem 3.3.
4. Verify that in the proof of Theorem 2.1 for singular matrices P_0, $Q(t)$ and in Lemma 2.1
$$\lim_{\varepsilon \to 0+} \max_{t\in[0,T]} \|\varphi(t,\varepsilon) - \varphi(t)\|_{\mathbf{R}^n} = 0$$
and
$$\lim_{\varepsilon \to 0+} \max_{t\in[0,T]} \|\varphi(t,\varepsilon) - \varphi_\infty(t)\|_{\mathbf{R}^n} = 0.$$
5. Consider the problem (1.15), (2.14), and (2.26) with $A(t) = 0$, $B(t) = 0$. Obtain the explicit formulas for the solution of this problem and the solution of the boudary value problem. Verify directly the assertions of Lemma 2.1.
6. Formulate and prove the analog of Lemma 2.1 for the case of discrete measurements.
7. Prove that the value of Δ^0 for $\sigma_{-d} = \sigma_{-d}^0$, $\sigma_l(t) = \sigma_l^0(t)$, and $\gamma_j(t) = \gamma_j^0(t)$ is equal to the value of Δ^0 for $\sigma_{-d} = \nu\sigma_{-d}^0$, $\sigma_l(t) = \nu\sigma_l^0(t)$, and $\gamma_j(t) =$

$\nu\gamma_j^0(t)$, where $\nu \in \mathbf{R}^1$, $\nu > 0$.

8. Consider the three-parametric manifold of the auxiliary Kalman filters with $P_0 = \beta_0 \text{diag}(\sigma_{-1}^2, \ldots, \sigma_{-n}^2)$, $R(t) = \beta_1 \text{diag}(\sigma_1^2(t), \ldots, \sigma_n^2(t))$, $Q(t) = \beta_2 \text{diag}(\gamma_1^2(t), \ldots, \gamma_r^2(t))$, $\beta_0, \beta_1, \beta_2 > 0$. Prove that the corresponding estimate of nonoptimality coincides with that of Theorem 2.2. That is why we put $\beta_0 = 1$ in Theorem 2.2.
9. As mentioned in Section 2, the prediction problem for the quantity $l_* = a'x(T+\kappa)$ by the measurements on the interval $[0,T]$ is equivalent to a relative filtering problem on the interval $[0, T+\kappa]$ if we put $H(t) = 0$ for $t \in (T, T+\kappa]$. For this filtering problem on the interval $[0, T+\kappa]$ the estimate Δ^0 can be computed. Prove that $\Delta^0 \geq \Delta_\kappa^0$, where Δ_κ^0 is given by Theorem 2.3.
10. Consider the problem of Example 6.1. In accordance with Remark 6.1, obtain the modified guaranteed level of nonoptimality for the Kalman filter with $R = \alpha_1\sigma^2$, $Q = \alpha_2\gamma^2$, $\alpha_1, \alpha_2 > 0$. To evaluate I_0 from below use the Kalman filter with $R = \beta_1\sigma^2$, $Q = \beta_2\gamma^2$, $\beta_1, \beta_2 > 0$. Prove that, for fixed σ, γ, and T, this modified guaranteed level of nonoptimality depends on the ratios $\frac{\alpha_1}{\alpha_2}$ and $\frac{\beta_1}{\beta_2}$ only.
11. In accordance with our initial assumptions (see Section 1), the entries of the matrices $A(t)$, $B(t)$, $H(t)$ and the functions $u_j(t)$, $\varrho_l(t)$, $\gamma_j(t)$, $\sigma_l(t)$ are piecewise continuous and have no isolated jumps. Let the second restriction be removed, i.e., these quantities may have isolated jumps. What form Δ^0 will take then?

Chapter 4

Stochastic Guaranteed Estimation Problems

The classical noise model is the white noise. In deterministic guaranteed estimation problems the noise is modeled by uncertain bounded disturbances. In this chapter, we consider a combined noise model of the form $\varrho(t) = \varrho_u(t) + \varrho_w(t)$, where $\varrho_u(t)$ is an uncertain bounded disturbance and $\varrho_w(t)$ is the white noise process with uncertain but bounded intensity. This combined disturbance can be also treated as a white noise process with uncertain mathematical expectation and intensity.

The uncertainty in system statistics results in minimax approach to estimation. The minimax estimation problems reduce to difficult variational problems (particularly hard for dynamic systems). Our approach is the same as in Chapter 3. We use an appropriate Kalman filter for approximate solving the estimation problem. Thereafter we construct the level of nonoptimality for the Kalman filter in the stochastic guaranteed estimation problem without exact solving the original variational problem.

This approach is quite general. We also apply it to the classical problem on the sensitivity of the Kalman filter to white noise intensities.

1. Optimal Stochastic Guaranteed Estimation Problem

Let us consider a linear stochastic dynamic system

$$\dot{x}(t) = A(t)x(t) + B(t)u(t), \qquad t \in [0,T], \tag{1.1}$$

where $x(t) = (x_1(t), \ldots, x_n(t))' \in \mathbf{R}^n$ is the system state vector, $u(t) = (u_1(t), \ldots, u_r(t))' \in \mathbf{R}^r$ is the perturbation vector, $A(t) \in \mathbf{R}^{n\times n}$, $B(t) \in \mathbf{R}^{n\times r}$. A prime denotes the transposition sign as before.

Suppose the measurements $z(t)$ are made on the interval $[0,T]$:

$$z(t) = H'(t)x(t) + \varrho(t), \qquad z(t), \varrho(t) \in \mathbf{R}^m, \quad H(t) \in \mathbf{R}^{n\times m}, \tag{1.2}$$

where $\varrho(t) = (\varrho_1(t), \ldots, \varrho_m(t))'$ is a measurement noise.

Assume the following stochastic hypotheses:

(a) the initial state $x(0)$ is a random variable and the perturbations $u(t)$, $\varrho(t)$ are white noise processes:

$$\mathbf{E}\big(x(0) - \mathbf{E}x(0)\big)\big(x(0) - \mathbf{E}x(0)\big)' = \operatorname{diag}(c_1^x, \dots, c_n^x),$$

$$\mathbf{E}\big(u(t) - \mathbf{E}u(t)\big)\big(u(s) - \mathbf{E}u(s)\big)' = \operatorname{diag}(q_1^u(t), \dots, q_r^u(t))\,\delta(t-s),$$

$$\mathbf{E}\big(\varrho(t) - \mathbf{E}\varrho(t)\big)\big(\varrho(s) - \mathbf{E}\varrho(s)\big)' = \operatorname{diag}(r_1^\varrho(t), \dots, r_m^\varrho(t))\,\delta(t-s),$$

$\delta(t)$ is the Dirac delta-function;
(b) the first moments and the intensities of the perturbations are unknown but bounded:

$$|\mathbf{E}x_d(0)| \le \sigma_{-d}, \qquad |\mathbf{E}u_j(t)| \le \gamma_j(t), \qquad |\mathbf{E}\varrho_l(t)| \le \sigma_l(t),$$

$$c_d^x \le c_d, \qquad q_j^u(t) \le q_j(t), \qquad r_l^\varrho(t) \le r_l(t),$$

$$d = 1, \dots, n, \qquad j = 1, \dots, r, \qquad l = 1, \dots, m;$$

(c) the initial state $x(0)$ and the perturbations $u(t)$, $\varrho(t)$ are mutually independent.

Here σ_{-d}, c_d are specified positive numbers, $\gamma_j(t) \ge \gamma^0 > 0$, $q_j(t) \ge q^0 > 0$, $\sigma_l(t) \ge \sigma^0 > 0$, $r_l(t) \ge r^0 > 0$ are specified positive functions. The entries of the matrices $A(t)$, $B(t)$, $H(t)$, the quantities $\mathbf{E}u_j(t)$, $\gamma_j(t)$, $q_j^u(t)$, $q_j(t)$, $\mathbf{E}\varrho_l(t)$, $\sigma_l(t)$, $r_l^\varrho(t)$, $r_l(t)$ are assumed to be piecewise continuous functions on $[0, T]$. For stochastic elements $x(0)$, $u(t)$, and $\varrho(t)$ that satisfy hypotheses (a) – (c) we write

$$(x(0), u, \varrho) \in \mathcal{S}.$$

The estimation problem is to evaluate a scalar quantity $l_* = a'x(T)$, where $a \in \mathbf{R}^n$ is a given vector. Obviously, we may consider $a \ne 0$; otherwise, the problem is trivial.

Let us use a class of linear estimators of the form

$$\hat{l} = \int_0^T \Phi'(t) z(t)\, dt, \qquad \Phi(t) = (\Phi_1(t), \dots, \Phi_m(t))' \in \mathbf{R}^m, \qquad \Phi \in L_2^m,$$

where L_2^m is the Hilbert space of square integrable functions with the inner product

$$\langle \Phi, \Psi \rangle_{L_2^m} = \int_0^T \Phi'(t)\Psi(t)\, dt, \qquad \Phi, \Psi \in L_2^m.$$

We define the estimation quality by the guaranteed value of the second moment of the estimation error:

$$\mathcal{D}_{\text{guar}}(\Phi) = \sup_{(x(0), u, \varrho) \in \mathcal{S}} \mathbf{E}(\hat{l} - l_*)^2.$$

The optimal stochastic guaranteed estimation problem is to find an optimal estimator Φ^0 that minimizes the guaranteed value of the second moment:

$$\mathcal{D}_{\text{guar}}(\Phi^0) = \mathcal{J}_0 \triangleq \inf_{\Phi \in L_2^m} \mathcal{D}_{\text{guar}}(\Phi).$$

REMARK 1.1. Since the white noise is a generalized process, the form (1.1), (1.2) is conventional. In the pure mathematical literature the stochastic system (1.1) and the measurements (1.2) are usually written with the help of stochastic differentials. However, this form is also conventional. For linear systems the two representations are equivalent. We use the more simple form (1.1), (1.2), which is traditional for "engineering" literature.

Similar to Chapter 3, in order to reduce the computation we introduce the new variables

$$y(t) = X^{-1}(t)x(t) \in \mathbf{R}^n, \qquad h(t) = X'(t)H(t) \in \mathbf{R}^{n\times m}, \tag{1.3}$$

$$b(t) = X^{-1}(t)B(t) \in \mathbf{R}^{n\times r}, \qquad \tilde{a} = X'(T)a,$$

$$h(t) = (h_1(t), \dots, h_m(t)), \qquad h_l(t) \in \mathbf{R}^n, \quad l = 1, \dots, m,$$

$$b(t) = (b_1(t), \dots, b_r(t)), \qquad b_j(t) \in \mathbf{R}^n, \quad j = 1, \dots, r.$$

where $X(t) \in \mathbf{R}^{n\times n}$ is a fundamental matrix of the homogeneous equation

$$\dot{x}(t) = A(t)x(t), \qquad X(0) = E_n, \quad E_n \text{ is identity matrix.}$$

Then,

$$\dot{y}(t) = b(t)u(t), \quad y(t) \in \mathbf{R}^n, \quad y(0) = x(0), \qquad t \in [0, T], \tag{1.4}$$

$$z(t) = h'(t)y(t) + \varrho(t), \qquad l_* = \tilde{a}'y(T). \tag{1.5}$$

Introduce also the notation

$$\sigma_- = (\sigma_{-1}, \dots, \sigma_{-n})' \in \mathbf{R}^n, \tag{1.6}$$

$$\gamma(t) = (\gamma_1(t), \dots, \gamma_r(t))' \in \mathbf{R}^r,$$

$$\sigma(t) = (\sigma_1(t), \dots, \sigma_m(t))' \in \mathbf{R}^m,$$

$$c = \operatorname{diag}(c_1, \dots, c_n) \in \mathbf{R}^{n\times n}, \tag{1.7}$$

$$q(t) = \operatorname{diag}(q_1(t), \dots, q_r(t)) \in \mathbf{R}^{r\times r},$$

$$r(t) = \operatorname{diag}(r_1(t), \dots, r_m(t)) \in \mathbf{R}^{m\times m}.$$

For a vector $\omega = (\omega_1, \dots, \omega_k)' \in \mathbf{R}^k$, we define $|\omega| = (|\omega_1|, \dots, |\omega_k|)'$.

LEMMA 1.1. *The guaranteed value of the second moment of the estimation error is determined by the formula*

$$\mathcal{D}_{\text{guar}}(\Phi) = \mathcal{J}(\Phi_-(\Phi), \Phi, w(\Phi)), \tag{1.8}$$

where

$$\mathcal{J}(\Phi_-, \Phi, w) = \left\{\sigma_-' |\Phi_-| + \int_0^T \sigma'(t)|\Phi(t)|\, dt + \int_0^T \gamma'(t)|w(t)|\, dt\right\}^2 \tag{1.9}$$

$$+ \Phi_-' c\, \Phi_- + \int_0^T \Phi'(t) r(t) \Phi(t)\, dt + \int_0^T w'(t) q(t) w(t)\, dt,$$

$$\Phi_- = \tilde{a} - \int_0^T h(t)\Phi(t)\, dt, \tag{1.10}$$

$$w(t) = b'(t)\left(\Phi_- + \int_0^t h(s)\Phi(s)\, ds\right),$$

and the quantities σ_-, $\sigma(t)$, $\gamma(t)$, c, $r(t)$, *and* $q(t)$ *are defined by* (1.6) *and* (1.7).

PROOF. Using new variables (1.3) and equations (1.4), (1.5), we have

$$\hat{l} - l_* = \left(\int_0^T \Phi'(t) z(t)\, dt - \tilde{a}\right)' y(0) + \int_0^T \Phi'(t) \varrho(t)\, dt$$

$$+ \left(\int_0^T h(t)\Phi(t)\, dt - \tilde{a}\right)' \int_0^T b(\tau) u(\tau)\, d\tau$$

$$- \int_0^T \Phi'(t) h'(t) \int_t^T b(\tau) u(\tau)\, d\tau\, dt.$$

The order of integration in the last summand may be interchanged like for deterministic integrals (see, e.g., [47])

$$\int_0^T \Phi'(t) h'(t) \int_t^T b(\tau) u(\tau)\, d\tau\, dt = \int_0^T \left(\int_0^\tau h(t)\Phi(t)\, dt\right)' b(\tau) u(\tau)\, d\tau.$$

Then

$$\hat{l} - l_* = -\Phi_-' x(0) + \int_0^T \Phi'(t)\varrho(t)\,dt - \int_0^T w'(t)u(t)\,dt. \qquad (1.11)$$

It follows from hypothesis (a), (c), and (1.11) that

$$\mathbf{E}(\hat{l} - l_*)^2 = \left(-\Phi_-' \mathbf{E}x(0) + \int_0^T \Phi'(t)\mathbf{E}\varrho(t)\,dt - \int_0^T w'(t)\mathbf{E}u(t)\,dt\right)^2$$

$$+ \Phi_-' c^x \Phi_- + \int_0^T \Phi'(t)r^\varrho(t)\Phi(t)\,dt + \int_0^T w'(t)q^u(t)w(t)\,dt,$$

where

$$c^x = \mathrm{diag}\,(c_1^x, \ldots, c_n^x),$$

$$q^u(t) = \mathrm{diag}\,(q_1^u(t), \ldots, q_r^u(t)), \qquad r^\varrho(t) = \mathrm{diag}\,(r_1^\varrho(t), \ldots, r_m^\varrho(t)).$$

In accordance with assumption (b) it can easily be verified (see Section 2 of Chapter 2 for details) that the guaranteed value of the second moment is determined by (1.8), (1.9). □

Thus the optimal stochastic guaranteed estimation problem reduces to the following nonsmooth variational problem

$$\mathcal{J}_0 = \inf_{\Phi_- \in \mathbf{R}^n,\, \Phi \in L_2^m,\, w \in L_2^r} \mathcal{J}(\Phi_-, \Phi, w) \quad \text{under constraints (1.10)}, \qquad (1.12)$$

where $\mathcal{J}(\Phi_-, \Phi, w)$ is given by (1.9).

If (Φ_-^0, Φ^0, w^0) is the solution of (1.12), then Φ^0 is the optimal estimator. The numerical solving of (1.12) is a very difficult problem.

Similar to the deterministic case, which was studied in Chapter 3, we shall use the classical Kalman filter [79], [47] for approximate solving (1.12).

Here the term *Kalman filter* means the algorithm for measurement data processing regardless of the actual noise models.

The level of the nonoptimality of the Kalman filter in the stochastic guaranteed estimation problem we define by the ratio

$$\Delta_{\mathcal{S}} = \left\{ \frac{\sup_{(x(0),u,\varrho)\in\mathcal{S}} \mathbf{E}\left(\hat{l}_\varphi - l_*\right)^2}{\sup_{(x(0),u,\varrho)\in\mathcal{S}} \mathbf{E}\left(\hat{l}_0 - l_*\right)^2} \right\}^{\frac{1}{2}},$$

where

$$\hat{l}_0 = \int_0^T \Phi^{0\prime}(t)z(t)\,dt \quad \text{and} \quad \hat{l}_\varphi = \int_0^T \varphi'(t)z(t)\,dt$$

are the unknown optimal estimate and the Kalman filter estimate, respectively. Obviously, $\Delta_{\mathcal{S}} \geq 1$.

Using Lemma 1.1, we see that

$$1 \leq \Delta_{\mathcal{S}} = \left(\frac{\mathcal{D}_{\text{guar}}(\varphi)}{\mathcal{D}_{\text{guar}}(\Phi^0)}\right)^{\frac{1}{2}} = \left(\frac{\mathcal{D}_{\text{guar}}(\varphi)}{\mathcal{J}_0}\right)^{\frac{1}{2}} = \left(\frac{\mathcal{J}(\varphi_-, \varphi, w_\varphi)}{\mathcal{J}(\Phi^0_-, \Phi^0, w^0)}\right)^{\frac{1}{2}}. \tag{1.13}$$

Here (Φ^0_-, Φ^0, w^0) is the solution of the original problem (1.12) and the triple $(\varphi_-, \varphi, w_\varphi)$, where

$$\varphi_- = \tilde{a} - \int_0^T h(s)\varphi(s)\, ds, \qquad w_\varphi(t) = b'(t)\left(\varphi_- + \int_0^t h(s)\varphi(s)\, ds\right),$$

corresponds to the Kalman estimator φ. With the help of the same argument that we used in Subsection 3.2 of Chapter 3 it can be proved that the solution (Φ^0_-, Φ^0, w^0) exists. Howerever, this is not essential for our approach.

The Kalman filter estimator φ can be easily computed and the numerator $\mathcal{D}_{\text{guar}}(\varphi)$ is a known quantity. Since the optimal estimator is unknown, the denominator $\mathcal{D}_{\text{guar}}(\Phi^0)$ is unknown and $\Delta_{\mathcal{S}}$ is unknown.

We shall construct an upper bound $\Delta^0_{\mathcal{S}}$ for $\Delta_{\mathcal{S}}$ that can be efficiently computed *without solving the original nonsmooth problem* (1.12). If the value of $\Delta^0_{\mathcal{S}}$ is not large, then the use of the Kalman filter would be justified. The results of Section 7 of Chapter 1 for the simplest static systems and of Chapter 3 for deterministic dynamic systems give us the hope that our approach will be successful for stochastic dynamic systems as well.

2. Approximating Problem. Bound for the Level of Nonoptimality

Let us simplify the problem (1.12) replacing the first nonsmooth term by a quadratic one (cf. Chapter 3)

$$\beta_0\, \Phi'_- \sigma_{-\mathcal{S}} \Phi_- + \beta_1 \int_0^T \Phi'(t)\sigma_{\mathcal{S}}(t)\Phi(t)\, dt + \beta_2 \int_0^T w'(t)\gamma_{\mathcal{S}}(t)w(t)\, dt,$$

where

$$\sigma_{-\mathcal{S}} = \operatorname{diag}(\sigma^2_{-1}, \ldots, \sigma^2_{-n}),$$

$$\sigma_{\mathcal{S}}(t) = \operatorname{diag}(\sigma^2_1(t), \ldots, \sigma^2_m(t)), \qquad \gamma_{\mathcal{S}}(t) = \operatorname{diag}(\gamma^2_1(t), \ldots, \gamma^2_r(t)),$$

and β_0, β_1, β_2 are specified parameters. Then the whole functional $\mathcal{J}$ is approximated by the quadratic functional

$$\mathcal{J}_a(\Phi_-, \Phi, w) \tag{2.1}$$

$$= \Phi'_- P_0 \Phi_- + \int_0^T \Phi'(t) R(t) \Phi(t)\, dt + \int_0^T w'(t) Q(t) w(t)\, dt,$$

where

$$P_0 = \operatorname{diag}(P_{01}, \ldots, P_{0n}) = \beta_0\, \sigma_{-\mathcal{S}} + c, \tag{2.2}$$

$$R(t) = \operatorname{diag}(R_1(t), \ldots, R_m(t)) = \beta_1\, \sigma_{\mathcal{S}}(t) + r(t),$$

$$Q(t) = \operatorname{diag}(Q_1(t), \ldots, Q_r(t)) = \beta_2\, \gamma_{\mathcal{S}}(t) + q(t).$$

Thus we get the following approximating variational problem

$$\mathcal{J}_{a0} = \inf_{\Phi_- \in \mathbf{R}^n,\, \Phi \in L_2^m,\, w \in L_2^r} \mathcal{J}_a(\Phi_-, \Phi, w) \quad \text{under constraints (1.10).} \tag{2.3}$$

The Kalman filter for the system (1.1), (1.2) is described in detail in Section 2 of Chapter 3.

THEOREM 2.1. 1^0. *The Kalman estimator φ is determined by the solution $(\varphi_-, \varphi, w_\varphi)$ of the problem (2.3).*
2^0. *The solution of the problem (2.3) is determined by the formulas:*

$$\varphi_- = \tilde{\xi}(0), \tag{2.4}$$

$$\varphi(t) = R^{-1}(t) h'(t) \tilde{\eta}(t) = \tilde{K}'(t) \tilde{\xi}(t),$$

$$w_\varphi(t) = b'(t) \tilde{\xi}(t),$$

where $\tilde{\xi}(t)$, $\tilde{\eta}(t)$ are the solutions of the boundary value problem:

$$\dot{\tilde{\xi}}(t) = h(t) R^{-1}(t) h'(t) \tilde{\eta}(t), \qquad \tilde{\xi}(T) = \tilde{a}, \tag{2.5}$$

$$\dot{\tilde{\eta}}(t) = b(t) Q(t) b'(t) \tilde{\xi}(t), \qquad \tilde{\eta}(0) = P_0 \tilde{\xi}(0).$$

Moreover,

$$\tilde{\eta}(t) = \tilde{P}(t) \tilde{\xi}(t), \qquad t \in [0, T], \tag{2.6}$$

$$\dot{\tilde{\xi}}(t) = -\left(-\tilde{K}(t) h'(t)\right)' \tilde{\xi}(t), \qquad \tilde{\xi}(T) = \tilde{a}, \tag{2.7}$$

and

$$\mathcal{J}_{a0} = \tilde{a}' \tilde{P}(T) \tilde{a}, \tag{2.8}$$

where $\tilde{P}(t)$ and $\tilde{K}(t)$ are the solution of the Riccati equation and the Kalman gain, respectively[1].

Theorem 2.1 follows directly from Theorem 2.1 of Chapter 3. In Section 3 we shall derive the dual problem to (2.3). In derivation we shall obtain another proof of (2.4), (2.5), and (2.8).

It follows from (2.4) and (2.5) that

$$\tilde{\xi}(t) = \int_0^t h(\tau)\varphi(\tau)\,d\tau + \varphi_- = \tilde{a} - \int_t^T h(\tau)\varphi(\tau)\,d\tau.$$

In contrast to the initial problem, the solution of the linear-quadratic problem (2.3) can be implemented easily.

Let us obtain the upper bound for the level of nonoptimality for the Kalman filter. First, introduce the notation

$$s_d = \frac{\sigma_{-d}^2}{c_d}, \qquad S_d = \frac{P_{0d}\varphi_{-d}}{\sigma_{-d}}, \qquad d = 1, \ldots, n,$$

$$f_l(t) = \frac{\sigma_l^2(t)}{r_l(t)}, \qquad F_l(t) = \frac{R_l(t)\varphi_l(t)}{\sigma_l(t)}, \qquad l = 1, \ldots, m,$$

$$g_j(t) = \frac{\gamma_j^2(t)}{q_j(t)}, \qquad G_j(t) = \frac{Q_j(t)w_{\varphi j}(t)}{\gamma_j(t)}, \qquad j = 1, \ldots, r,$$

$$\varphi_- = (\varphi_{-1}, \ldots, \varphi_{-n})', \qquad \varphi(t) = (\varphi_1(t), \ldots, \varphi_m(t))',$$

$$w_\varphi(t) = (w_{\varphi 1}(t), \ldots, w_{\varphi r}(t))'.$$

$$\zeta_{\max} = \max\left\{ \max_{d=1,\ldots,n} |S_d|, \sup_{l=1,\ldots,m,\ t\in[0,T]} |F_l(t)|, \sup_{j=1,\ldots,r,\ t\in[0,T]} |G_j(t)| \right\}.$$

These quantities are uniquely determined by the solution of (2.2), (2.3).

Let the function $\mathcal{E}(\zeta)\colon [0, \zeta_{\max}] \to \mathbf{R}^1$ be defined by the formula

$$\mathcal{E}(\zeta) = \sum_{\{d:\,|S_d|>\zeta\}} s_d\Big(|S_d| - \zeta\Big) + \sum_{l=1}^{m} \int_{\{t:\,|F_l(t)|>\zeta\}} f_l(t)\Big(|F_l(t)| - \zeta\Big)\,dt$$

$$+ \sum_{j=1}^{r} \int_{\{t:\,|G_j(t)|>\zeta\}} g_j(t)\Big(|G_j(t)| - \zeta\Big)\,dt - \zeta.$$

[1] see Section 2 of Chapter 3 for details.

LEMMA 2.1. *The function $\mathcal{E}(\zeta)$ is a strictly monotonically decreasing function on the interval $[0, \zeta_{\max}]$. The equation*

$$\mathcal{E}(\zeta) = 0 \tag{2.9}$$

has a unique root on $[0, \zeta_{\max}]$.

The proof of Lemma 2.1 is given in Section 3.

By virtue of Lemma 2.1 the root $\tilde{\zeta}$ of (2.9) can easily be found by sequential halving the interval $[0, \zeta_{\max}]$.

Define the quantity

$$\begin{aligned}\Omega(\varphi_-, \varphi, w_\varphi) = &\sum_{\{d:\,|S_d| > \tilde{\zeta}\}} s_d \left(|S_d| - \tilde{\zeta}\right)^2 \\ &+ \sum_{l=1}^{m} \int_{\{t:\,|F_l(t)| > \tilde{\zeta}\}} f_l(t) \left(|F_l(t)| - \tilde{\zeta}\right)^2 dt \\ &+ \sum_{j=1}^{r} \int_{\{t:\,|G_j(t)| > \tilde{\zeta}\}} g_j(t) \left(|G_j(t)| - \tilde{\zeta}\right)^2 dt + \tilde{\zeta}^2,\end{aligned} \tag{2.10}$$

where $\tilde{\zeta}$ is the root of equation (2.9).

THEOREM 2.2. *Let $\{\varphi_-, \varphi, w_\varphi\}$ be the solution of the linear-quadratic problem (2.3). Then the level of nonoptimality Δ_S is bounded with the quantity Δ_S^0 that is determined by the following relations:*

$$1 \le \Delta_S \le \Delta_S^0 = \frac{\sqrt{\mathcal{J}(\varphi_-, \varphi, w_\varphi)\ \Omega(\varphi_-, \varphi, w_\varphi)}}{\mathcal{J}_a(\varphi_-, \varphi, w_\varphi)}, \tag{2.11}$$

where $\mathcal{J}(\Phi_-, \Phi, w)$ is defined by (1.9), $\mathcal{J}_a(\Phi_-, \Phi, w)$ is defined by (2.1), and $\Omega(\varphi_-, \varphi, w_\varphi)$ is defined by (2.10).

The proof of Theorem 2.2 is based on the duality theory for convex variational problems and is given in Section 3.

We see from Theorem 2.2 that the value of Δ_S^0 is determined by the characteristics of the Kalman filter only. These characteristics can be computed quite easily. Therefore the estimate for the nonoptimality level Δ_S^0 can be efficiently computed without solving the original nonsmooth variational problem (2.3).

If Δ_S^0 is not large, then there is no need to resolve the difficult nonsmooth problem (2.3) and the Kalman filter can be used in the optimal stochastic guaranteed estimation problem as well.

Let us return to the initial variables. Consider the system (1.1), (1.2) again. Introduce the new variables

$$K(t) = X(t)\tilde{K}(t), \qquad P(t) = X(t)\tilde{P}(t)X'(t),$$

$$\xi(t) = \left(X^{-1}(t)\right)'\tilde{\xi}(t), \qquad \eta(t) = X(t)\tilde{\eta}(t).$$

Then the Kalman filter for the system (1.1), (1.2) takes the standard form

$$\hat{l}_\varphi = a'\hat{x}(T),$$

$$\dot{\hat{x}}(t) = A(t)\hat{x}(t) + K(t)\big(z(t) - H'(t)\hat{x}(t)\big), \qquad \hat{x}(0) = 0,$$

$$K(t) = P(t)H(t)R^{-1}(t), \tag{2.12}$$

$$\dot{P}(t) = A(t)P(t) + P(t)A'(t) - P(t)H(t)R^{-1}(t)H'(t)P(t) \tag{2.13}$$

$$+ B(t)Q(t)B'(t), \qquad P(0) = P_0,$$

$$\hat{x}(t) \in \mathbf{R}^n, \quad K(t) \in \mathbf{R}^{n\times m}, \quad P(t) \in \mathbf{R}^{n\times n}, \qquad t \in [0,T].$$

It can easily be checked that

$$\varphi_- = \xi(0), \tag{2.14}$$

$$\varphi(t) = R^{-1}(t)H'(t)\eta(t) = K'(t)\xi(t),$$

$$w_\varphi(t) = B'(t)\xi(t),$$

and the corresponding boundary value problem turns into the system

$$\dot{\xi}(t) = -A'(t)\xi(t) + H(t)R^{-1}(t)H'(t)\eta(t), \qquad \xi(T) = a, \tag{2.15}$$

$$\dot{\eta}(t) = B(t)Q(t)B'(t)\xi(t) + A(t)\eta(t), \qquad \eta(0) = P_0\xi(0).$$

Moreover,

$$\eta(t) = P(t)\xi(t), \qquad t \in [0,T], \tag{2.16}$$

and

$$\dot{\xi}(t) = -\left(A(t) - K(t)H'(t)\right)'\xi(t), \qquad \xi(T) = a,$$

where $K(t)$ and $P(t)$ are defined by (2.12) and (2.13).

It follows from (2.14) and (2.15) that

$$\dot{\xi}(t) = -A'(t)\xi(t) + H(t)\varphi(t) \tag{2.17}$$

with the initial value $\xi(0) = \varphi_-$ or with the terminal value $\xi(T) = a$.

The only modification in Theorem 2.2 is that the quantity $w_\varphi(t)$ is now computed by the formula

$$w_{\varphi j}(t) = B_j'(t)\xi(t), \qquad B(t) = (B_1(t), \ldots, B_r(t)), \quad j = 1, \ldots, r,$$

and $\xi(t)$ is determined by the second equation of (2.16) or by relation (2.17).

2.1. *Polar Form for* Δ_S^0

V.N. Soloviov proposed an elegant form for the estimates of type (2.11) using the notion of the *caliber* and the *polar* [185], [195], [196].

Let us introduce a Hilbert space $\mathcal{H} = \mathbf{R}^n \times L_2^m \times L_2^r$ with the elements $\chi = (\Phi_-, \Phi, w)$ and the inner product

$$\langle \chi_{(1)}, \chi_{(2)} \rangle_{\mathcal{H}} = \Phi'_{-(1)}\Phi_{-(2)} + \int_0^T \Phi'_{(1)}(t)\Phi_{(2)}(t)\,dt \tag{2.18}$$
$$+ \int_0^T w'_{(1)}(t)w_{(2)}(t)\,dt.$$

Definition 2.1 [185]. A function $k(\chi): \mathcal{H} \to \mathbf{R}^1$ such that $k(\alpha\chi) = \alpha k(\chi)$ for any $0 < \alpha < \infty$ is called positively homogeneous.

Obviously, $k(0) = k(2 \cdot 0) = 2k(0)$ and $k(0) = 0$.

Definition 2.2 [185]. A nonnegative positively homogeneous convex function $k(\chi): \mathcal{H} \to \mathbf{R}^1$ is called the caliber[2].

Suppose that

$$k(\chi) > 0 \quad \text{if} \quad \chi \neq 0. \tag{2.19}$$

Definition 2.3 [185]. The function $k^\circ: \mathcal{H} \to \mathbf{R}^1$

$$k^\circ(\chi^*) = \sup_{k(\chi)=1} \langle \chi^*, \chi \rangle_{\mathcal{H}}, \qquad \chi^* \in \mathcal{H}$$

[2] Another conventional name of the caliber is the Minkowski function.

is called the polar of k.

Consider the caliber

$$k(\chi) = \mathcal{J}^{\frac{1}{2}}\left(P_0^{-\frac{1}{2}}\Phi_-,\ R^{-\frac{1}{2}}\Phi,\ Q^{-\frac{1}{2}}w\right), \qquad \chi = (\Phi_-, \Phi, w) \in \mathcal{H}, \tag{2.20}$$

where P_0, R, and Q are specified by (2.2). Note that (2.19) holds.

Construct the element

$$\chi^\varphi = \left(P_0^{\frac{1}{2}}\varphi_-,\ R^{\frac{1}{2}}\varphi,\ Q^{\frac{1}{2}}w_\varphi\right) \in \mathcal{H}, \tag{2.21}$$

where the triple $(\varphi_-, \varphi, w_\varphi)$ is defined by (2.14).

LEMMA 2.2. *The quantity Δ_S^0 is represented in the form*

$$\Delta_S^0 = \frac{k(\chi^\varphi)\ k^\circ(\chi^\varphi)}{\|\chi^\varphi\|_{\mathcal{H}}^2}, \tag{2.22}$$

where k° is the polar of the caliber k from (2.20) and $\|\chi^\varphi\|_{\mathcal{H}}^2 = \langle\chi^\varphi, \chi^\varphi\rangle_{\mathcal{H}}$.

The proof of Lemma 2.2 is contained in Section 3.

3. Derivation of Main Result for Stochastic Problem

3.1. *Proof of Lemma 2.1*

Consider the term

$$\mathcal{E}_1(\zeta) = \int\limits_{\{t:\,|F_1(t)|>\zeta\}} f_1(t)\left(|F_1(t)| - \zeta\right) dt.$$

Suppose $\zeta_2 > \zeta_1$; then

$$\mathcal{E}_1(\zeta_2) - \mathcal{E}_1(\zeta_1) = \int\limits_{\{t:\,|F_1(t)|>\zeta_2\}} f_1(t)\left(|F_1(t)| - \zeta_2\right) dt \tag{3.1}$$

$$- \int\limits_{\{t:\,\zeta_2 \geq |F_1(t)|>\zeta_1\}} f_1(t)\left(|F_1(t)| - \zeta_1\right) dt \ - \int\limits_{\{t:\,|F_1(t)|>\zeta_2\}} f_1(t)\left(|F_1(t)| - \zeta_1\right) dt$$

$$= \int\limits_{\{t:\,|F_1(t)|>\zeta_2\}} f_1(t)(\zeta_1 - \zeta_2)\, dt \ - \int\limits_{\{t:\,\zeta_2 \geq |F_1(t)|>\zeta_1\}} f_1(t)\left(|F_1(t)| - \zeta_1\right) dt.$$

The continuity and monotone decrease of $\mathcal{E}_1(\zeta)$ follow directly from (3.1). The continuity and the monotone decrease of other summands can be verified similarly. Since $-\zeta$ is continuous and strictly monotonically decreasing function, $\mathcal{E}(\zeta)$ is a continuous and strictly monotonically decreasing function.

Obviously, $\mathcal{E}(0) \geq 0$ and $\mathcal{E}(\zeta_{\max}) = -\zeta_{\max} \leq 0$. Hence equation (2.9) has a unique root on $[0, \zeta_{\max}]$. □

3.2. *Proof of Theorem 2.2*

In accordance with (1.12) and (1.13),

$$\Delta_{\mathcal{S}} = \left(\frac{\mathcal{J}(\varphi_-, \varphi, w_\varphi)}{\mathcal{J}_0} \right)^{\frac{1}{2}}. \tag{3.2}$$

Therefore in order to find the upper bound for $\Delta_{\mathcal{S}}$ we shall construct a lower bound for $\mathcal{J}_0$. As in Chapter 3, the duality theory will be used for this purpose.

In addition to the space $\mathcal{H}$, let us introduce the Hilbert space $\tilde{\mathcal{H}} = L_2^r \times \mathbf{R}^n$ with the elements $\tilde{\Gamma} = (G, g)$ and the inner product

$$\langle \tilde{\Gamma}_{(1)}, \tilde{\Gamma}_{(2)} \rangle_{\tilde{\mathcal{H}}} = \int_0^T G'_{(1)}(t) G_{(2)}(t)\, dt + g'_{(1)} g_{(2)},$$

$$\tilde{\Gamma}_{(i)} \in \tilde{\mathcal{H}}, \qquad G_{(i)} \in L_2^r, \qquad g_{(i)} \in \mathbf{R}^n, \qquad i = 1, 2.$$

We define the affine operator $\Lambda\colon \mathcal{H} \to \tilde{\mathcal{H}}$ by the relations:

$$\Lambda(\Phi_-, \Phi, w) = (v, d), \qquad \Phi_-, d \in \mathbf{R}^n, \quad \Phi \in L_2^m, \quad w, v \in L_2^r, \tag{3.3}$$

$$v(t) = b'(t) \left(\Phi_- + \int_0^t h(s)\Phi(s)\, ds \right) - w(t), \qquad t \in [0, T],$$

$$d = -\Phi_- - \int_0^T h(t)\Phi(t)\, dt + \tilde{a}.$$

Thus the extremal problem (1.12) can be rewritten in the form

$$\mathcal{J}_0 = \inf_{\mathcal{H}} \mathcal{J}(\Phi_-, \Phi, w) \quad \text{under the constraint} \quad \Lambda(\Phi_-, \Phi, w) = 0, \tag{3.4}$$

where the functional $\mathcal{J}\colon \mathcal{H} \to \mathbf{R}^1$ is continuous and convex and the operator $\Lambda\colon \mathcal{H} \to \tilde{\mathcal{H}}$ is continuous and affine.

So, the problem (3.4) is a special case of the variational problem $\mathcal{P}_1$ posed in Section 1 of Chapter 2.

The dual space $\tilde{\mathcal{H}}^*$ can be identified with $\tilde{\mathcal{H}}$. Then the Lagrange function for the problem (3.4) takes the form

$$\mathcal{L}_{\mathcal{S}}(\Phi_-, \Phi, w; p, \lambda) = \mathcal{J}(\Phi_-, \Phi, w)$$

$$+\int_0^T p'(t)\left[b'(t)\left(\int_0^t h(s)\Phi(s)\,ds+\Phi_-\right)-w(t)\right]dt$$

$$+\lambda'\left(\tilde{a}-\int_0^T h(t)\Phi(t)\,dt-\Phi_-\right),$$

where $p \in L_2^r$ and $\lambda \in \mathbf{R}^n$ are Lagrange multipliers.

By definition, the dual problem to (3.4) has the form

$$\mathcal{J}^0 = \sup_{p\in L_2^r,\,\lambda\in\mathbf{R}^n} \Pi(p,\lambda),$$

where

$$\Pi(p,\lambda) \triangleq \inf_{\Phi_-\in\mathbf{R}^n,\Phi\in L_2^m,\,w\in L_2^r} \mathcal{L}_S(\Phi_-,\Phi,\,w;p,\lambda).$$

THEOREM 3.1. *The duality relation holds:*

$$\mathcal{J}_0 = \mathcal{J}^0 = \sup_{p\in L_2^r,\,\lambda\in\mathbf{R}^n} \Pi(p,\lambda). \tag{3.5}$$

PROOF. Let us define the perturbation of the extremal problem (3.4) by the variation of the constraint:

$$\mathcal{J}_0(\delta w,\delta\tilde{a}) = \inf_{\mathcal{H}} \mathcal{J}(\Phi_-,\Phi,\,w)$$

under the constraint

$$\Lambda(\Phi_-,\Phi,\,w) = \begin{pmatrix}\delta w\\ \delta\tilde{a}\end{pmatrix}, \qquad \delta w\in L_2^r, \quad \delta\tilde{a}\in\mathbf{R}^n.$$

If $\delta w = 0$ and $\delta\tilde{a} = 0$, then we get the original problem (3.4).

Similar to Proposition 5.1 of Chapter 2 we can prove that for any neighbourhood of zero $\mathcal{U}\subset\tilde{\mathcal{H}}$ the function $\mathcal{J}_0(\delta w,\delta\tilde{a})$ is bounded:

$$0 \le \mathcal{J}_0(\delta w,\delta\tilde{a}) \le C \quad \forall\,(\delta w,\delta\tilde{a})\in\mathcal{U}, \qquad C = C(\mathcal{U}) = \text{const}.$$

Consequently, by virtue of Lemma 1.4 and Lemma 1.5 of Chapter 2, $\mathcal{J}_0(\delta w,\delta\tilde{a})$ is convex and continuous at the point $(0,0)$. Then, in accordance with Theorem 1.7 of Chapter 2, the duality relation holds:

$$\mathcal{J}_0 = \mathcal{J}^0 = \sup_{p\in L_2^r,\,\lambda\in\mathbf{R}^n}\left\{\inf_{\Phi_-\in\mathbf{R}^n,\Phi\in L_2^m,\,w\in L_2^r} \mathcal{L}_S(\Phi_-,\Phi,\,w;p,\lambda)\right\}. \tag{3.6}$$

Then (3.5) follows from (3.6). □

Let us obtain the dual problem to the approximating variational problem (2.3).

THEOREM 3.2. 1^0. *The problem dual to the problem* (2.3) *has the form*

$$\mathcal{J}^{a0} = \sup_{p\in L_2^r,\, \lambda\in\mathbf{R}^n}\left\{-\frac{1}{4}\left[\left(\lambda-\int_0^T b(t)p(t)\,dt\right)' P_0^{-1}\left(\lambda-\int_0^T b(t)p(t)\,dt\right)\right.\right. \quad (3.7)$$

$$+\int_0^T\left(\lambda-\int_t^T b(s)p(s)\,ds\right)' h(t)R^{-1}(t)h'(t)\left(\lambda-\int_t^T b(s)p(s)\,ds\right)dt$$

$$\left.\left.+\int_0^T p'(t)Q^{-1}(t)p(t)\,dt\right]+a'\lambda\right\}.$$

2^0. *The duality relation holds:*

$$\mathcal{J}_{a0} = \mathcal{J}^{a0}.$$

3^0. *The solution of* (3.7) *is determined by the equalities*

$$p^0(t) = 2Q(t)b'(t)\tilde{\xi}(t), \qquad \lambda^0 = 2\tilde{\eta}(T). \quad (3.8)$$

where $\tilde{\xi}(t)$ *and* $\tilde{\eta}(t)$ *are determined by the boundary value problem*

$$\dot{\tilde{\xi}}(t) = h(t)R^{-1}(t)h'(t)\tilde{\eta}(t), \qquad \tilde{\xi}(T) = \tilde{a}, \quad (3.9)$$

$$\dot{\tilde{\eta}}(t) = b(t)Q(t)b'(t)\tilde{\xi}(t), \qquad \tilde{\eta}(0) = P_0\tilde{\xi}(0).$$

PROOF. The approximating problem (2.3) can be represented in the form

$$\mathcal{J}_{a0} = \inf_{\mathcal{H}} \mathcal{J}_a(\Phi_-, \Phi, w) \quad \text{under the constraint} \quad \Lambda(\Phi_-, \Phi, w) = 0, \quad (3.10)$$

where $\mathcal{J}_a(\Phi_-, \Phi, w)\colon \mathcal{H} \to \mathbf{R}^1$ is defined by (2.1), (2.2), and the affine operator $\Lambda\colon \mathcal{H} \to \tilde{\mathcal{H}} = L_2^r \times \mathbf{R}^n$ is defined by (3.3).

The functional $\mathcal{J}_a(\Phi_-, \Phi, w)$ is a convex continuous functional, the operator Λ is a continuous affine operator. So, the problem (3.10) is a variational problem of type $\mathcal{P}_1$ from Section 1 of Chapter 2.

Let us prove the existence and uniqueness of the solution for the problem (3.10)[3]. The space $\mathcal{H}$ is a Hilbert space; consequently it is a reflexive Banach space. By virtue of the initial assumptions, the bounds for the uncertainties are positive. Consequently the matrices P_0, $R(t)$, and $Q(t)$ are nonsingular.

[3]These facts follow directly from Theorem 2.1 of Chapter 3. Nevertheless, we repeat our argument for completeness.

Then it can easily be checked that the functional $\mathcal{J}_a(\Phi_-, \Phi, w)$ is coercive and strictly convex. Since the operator Λ is continuous and affine, the set $\Lambda(\Phi_-, \Phi, w) = 0$ is an affine closed manifold. Then, by virtue of Lemma 1.2 of Chapter 2, the problem (3.10) has a unique solution $(\varphi_-, \varphi, w_\varphi)$.

Let us turn to the dual problem. Since the space $\tilde{\mathcal{H}}$ is Hilbert, the dual space can be identified with the original space: $\tilde{\mathcal{H}}^* = \tilde{\mathcal{H}}$. So, the Lagrange function has the form

$$\mathcal{L}_a(\Phi_-, \Phi, w; p, \lambda) = \mathcal{J}_a(\Phi_-, \Phi, w) \tag{3.11}$$

$$+ \int_0^T p'(t)\left[b'(t)\left(\int_0^t h(s)\Phi(s)\, ds + \Phi_-\right) - w(t)\right] dt$$

$$+ \lambda'\left(\tilde{a} - \int_0^T h(t)\Phi(t)\, dt - \Phi_-\right).$$

The perturbation of problem (3.10) has the form (cf. Section 1 of Chapter 2)

$$\mathcal{J}_{a0}(\delta w, \delta\tilde{a}) = \inf_{\mathcal{H}} \mathcal{J}_a(\Phi_-, \Phi, w)$$

under the constraint

$$\Lambda(\Phi_-, \Phi, w) = \begin{pmatrix} \delta w \\ \delta\tilde{a} \end{pmatrix}, \qquad \delta w \in L_2^r, \quad \delta\tilde{a} \in \mathbf{R}^n.$$

With $\delta w = 0$ and $\delta\tilde{a} = 0$ we get the initial problem (3.10).

Similar to the proof of Theorem 5.2 of Chapter 2 we can verify that $\mathcal{J}_{a0}(\delta w, \delta\tilde{a})$ is bounded in a neighborhood of the point $(0,0)$ (cf. Proposition 5.1 of Chapter 2). Then, using Lemmas 1.4 – 1.6 of Chapter 2, we can establish that $\mathcal{J}_{a0}(\delta w, \delta\tilde{a})$ is convex, and at the point $(0,0)$ $\mathcal{J}_{a0}(\delta w, \delta\tilde{a})$ is continuous and has a non-empty subdifferential. Then, by virtue of Theorems 1.8 and 1.9 of Chapter 2, the dual problem

$$\mathcal{J}^{a0} = \sup_{p \in L_2^r, \lambda \in \mathbf{R}^n} \left\{ \inf_{\Phi_- \in \mathbf{R}^n, \Phi \in L_2^m, w \in L_2^r} \mathcal{L}_a(\Phi_-, \Phi, w; p, \lambda) \right\} \tag{3.12}$$

has a solution (p^0, λ^0), the Lagrange function has a saddle point:

$$\mathcal{L}_a(\varphi_-, \varphi, w_\varphi; p, \lambda) \le \mathcal{L}_a(\varphi_-, \varphi, w_\varphi; p^0, \lambda^0) \le \mathcal{L}_a(\Phi_-, \Phi, w; p^0, \lambda^0),$$

$$\forall\, \Phi_-, \lambda \in \mathbf{R}^n, \quad \Phi \in L_2^m, \quad w, p \in L_2^r,$$

and the duality relation holds:

$$\mathcal{J}_{a0} = \mathcal{J}^{a0}. \tag{3.13}$$

Since $\mathcal{L}_a(\Phi_-, \Phi, w; p, \lambda)$ is a linear-quadratic convex function, the necessary and sufficient condition for its minimum are determined by the relations (see Theorem 1.4 of Chapter 2)

$$\frac{\partial \mathcal{L}_a}{\partial \Phi_-} = 0, \qquad \frac{\partial \mathcal{L}_a}{\partial \Phi} = 0, \qquad \frac{\partial \mathcal{L}_a}{\partial w} = 0, \tag{3.14}$$

where $\frac{\partial \mathcal{L}_a}{\partial \Phi}$ and $\frac{\partial \mathcal{L}_a}{\partial w}$ are the Frechét derivatives. Differentiating $\mathcal{L}_a$ with resprect to Φ_-, Φ, and w, we obtain from (3.11), (3.14) the following equalities

$$\varphi_- = \frac{1}{2} P_0^{-1}\left(\lambda - \int_0^T b(s)p(s)\, ds\right), \tag{3.15}$$

$$\varphi(t) = \frac{1}{2} R^{-1}(t)h'(t)\left(\lambda - \int_t^T b(s)p(s)\, ds\right),$$

$$w_\varphi(t) = \frac{1}{2} Q^{-1}(t)p(t).$$

Substituting (3.15) into (3.11), (3.12), we obtain the dual problem (3.7).

The function in curly brackets in (3.7) is a linear-quadratic function; moreover, the function in square brackets is convex. Since $\sup\{-f\} = -\inf f$, the maximum for (3.7) is determined by the equations

$$\frac{\partial \mathcal{L}_a(\varphi_-(p,\lambda), \varphi(p,\lambda), w_\varphi(p,\lambda)); p, \lambda)}{\partial p} = 0, \tag{3.16}$$

$$\frac{\partial \mathcal{L}_a(\varphi_-(p,\lambda), \varphi(p,\lambda), w_\varphi(p,\lambda)); p, \lambda)}{\partial \lambda} = 0.$$

Formulas (3.16) imply the equalities

$$p(t) = Q(t)b'(t)\left\{ P_0^{-1}\left(\lambda - \int_0^T b(s)p(s)\, ds\right)\right.$$

$$\left. + \int_0^t h(s)R^{-1}(s)h'(s)\left(\lambda - \int_s^T b(\tau)p(\tau)\, d\tau\right) ds\right\}$$

and

$$P_0^{-1}\left(\lambda - \int_0^T b(s)p(s)\, ds\right)$$

$$+ \int_0^T h(s)R^{-1}h'(s)\left(\lambda - \int_s^T b(\tau)p(\tau)\, d\tau\right) ds = 2\tilde{a}.$$

The last relations imply that

$$p(t) = 2Q(t)b'(t)\tilde{\xi}(t),$$

where

$$\tilde{\xi}(t) = P_0^{-1}\left(\frac{\lambda}{2} - \int_0^T b(s)Q(s)b'(s)\tilde{\xi}(s)\,ds\right)$$

$$+ \int_0^t h(s)R^{-1}(s)h'(s)\left(\frac{\lambda}{2} - \int_s^T b(\tau)Q(\tau)b'(\tau)\tilde{\xi}(\tau)\,d\tau\right)ds,$$

and

$$P_0^{-1}\left(\frac{\lambda}{2} - \int_0^T b(s)Q(s)b'(s)\tilde{\xi}(s)\,ds\right) \tag{3.17}$$

$$+ \int_0^T h(s)R^{-1}(s)h'(s)\left(\frac{\lambda}{2} - \int_s^T b(\tau)Q(\tau)b'(\tau)\tilde{\xi}(\tau)\,d\tau\right)ds = \tilde{a}.$$

Let us introduce an auxiliary variable

$$\tilde{\eta}(s) = \tilde{\eta}(T) - \int_s^T b(\tau)Q(\tau)b'(\tau)\tilde{\xi}(\tau)\,d\tau, \qquad \tilde{\eta}(T) = \frac{\lambda}{2}. \tag{3.18}$$

It follows from (3.17), (3.18) that

$$\dot{\tilde{\xi}}(t) = h(t)R^{-1}(t)h'(t)\tilde{\eta}(t), \qquad \tilde{\xi}(T) = \tilde{a}, \tag{3.19}$$

$$\dot{\tilde{\eta}}(t) = b(t)Q(t)b'(t)\tilde{\xi}(t), \qquad \tilde{\eta}(0) = P_0\tilde{\xi}(0).$$

Since the problem (3.7) has a solution, the boundary value problem (3.19) must have a solution. It was already proved in Theorem 2.1 of Chapter 3 that this solution is unique.

Thus the solution of (3.7) has the form

$$p^0(t) = 2Q(t)b'(t)\tilde{\xi}(t), \qquad \lambda^0 = 2\tilde{\eta}(T). \tag{3.20}$$

Note that, correct to a scale, the solution (3.20) coincides with the solution of the similar dual problem from Section 3 of Chapter 3.

Using the boundary value problem, the solution (3.15) of (3.10) takes the form

$$\varphi_- = \tilde{\xi}(0), \qquad \varphi(t) = R^{-1}(t)h'(t)\tilde{\eta}(t), \qquad w_\varphi(t) = b'(t)\tilde{\xi}(t).$$

So, (2.4) and (2.5) hold.

It can be easily checked by direct differentiation that the function $\tilde{P}(t)\tilde{\xi}(t)$ satisfies the second equation of (2.5). Then, by virtue of the uniquness of a solution, (2.6) holds. In particular,

$$\varphi(t) = R^{-1}(t)h'(t)\tilde{\eta}(t) = \tilde{K}'(t)\tilde{\xi}(t)$$

and the last equality in the second relation from (2.4) is valid.

Substituting (3.20) into the dual problem (3.7), we have

$$\mathcal{J}^{a0} = -\frac{1}{4}\Bigg[4\,\tilde{\eta}'(0)P_0^{-1}\tilde{\eta}(0) + 4\int_0^T \tilde{\xi}'(t)b(t)Q(t)b'(t)\tilde{\xi}(t)\,dt$$

$$+4\int_0^T \varphi'(t)R(t)\varphi(t)\,dt\Bigg] + 2\tilde{a}'\tilde{\eta}(T)$$

or, taking into account (3.13) and (2.6),

$$\mathcal{J}_{a0} = -\mathcal{J}_{a0} + 2\tilde{a}'\tilde{P}(T)\tilde{a},$$

whence,

$$\mathcal{J}_{a0} = \tilde{a}'\tilde{P}(T)\tilde{a}.$$

Hence (2.8) is also valid. □

The case of singular P_0 and $Q(t)$ can be studied in the same way as in Chapter 3.

Following the approach, which we already applied in Chapter 3, we use the pair (p^0, λ^0) to obtain the lower estimate for $\mathcal{J}_0$. Let us find an approximate solution of the dual problem (3.7) along the direction defined by (p^0, λ^0). In other words, let us replace the dual problem (3.7) by a more rough one-dimensional extremal problem

$$\mathcal{D}_0 = \sup_{\nu\in\mathbf{R}^1} \Pi(\nu\lambda^0, \nu p^0), \tag{3.21}$$

where (p^0, λ^0) is defined by (3.8). In accordance with Theorem 3.1,

$$\mathcal{J}_0 = \mathcal{J}^0 \geq \mathcal{D}_0. \tag{3.22}$$

Then it follows from (3.2) that the desired lower estimate takes the form

$$1 \leq \Delta_S \leq \Delta_S^0 = \left(\frac{\mathcal{D}_{\text{guar}}(\varphi)}{\mathcal{D}_0}\right)^{\frac{1}{2}} = \left(\frac{\mathcal{J}(\varphi_-, \varphi, w_\varphi)}{\mathcal{D}_0}\right)^{\frac{1}{2}}. \tag{3.23}$$

Let us show that Δ_S^0 can eaisly be computed via the characteristics of the Kalman filter.

The objective function of the dual problem can be represented in another form

$$\Pi(p,\lambda) = a'\lambda - \Gamma(p,\lambda), \tag{3.24}$$

where

$$\Gamma(p,\lambda) = -\inf_{\Phi_-\in\mathbf{R}^n,\Phi\in L_2^m,\, w\in L_2^r} \tilde{\mathcal{L}}_{\mathcal{S}}(\Phi_-,\Phi,\, w; p,\lambda), \tag{3.25}$$

$$\tilde{\mathcal{L}}_{\mathcal{S}}(\Phi_-,\Phi,\, w; p,\lambda) = \mathcal{J}(\Phi_-,\Phi,\, w)$$

$$-\Psi'\Phi_- - \int_0^T \Theta'(t)\Phi(t)\,dt - \int_0^T p'(t)w(t)\,dt,$$

$$\Psi = (\Psi_1,\dots,\Psi_n)' = \Psi(p,\lambda) = \lambda - \int_0^T b(t)p(t)\,dt \in \mathbf{R}^n,$$

$$\Theta(t) = (\Theta_1(t),\dots,\Theta_m(t))' = \Theta(t;p,\lambda)$$

$$= h'(t)\left(\lambda - \int_t^T b(s)p(s)\,ds\right) \in \mathbf{R}^m, \qquad t\in[0,T].$$

LEMMA 3.1. *The optimal value of the problem* (3.21) *is determined by the equalities:*

$$\mathcal{D}_0 = \frac{\mathcal{J}_{a0}^2}{\Gamma(p^0,\lambda^0)} = \frac{\mathcal{J}^2(\varphi_-,\varphi,\, w_\varphi)}{\Gamma(p^0,\lambda^0)} = \frac{\left(\tilde{a}'\tilde{P}(T)\tilde{a}\right)^2}{\Gamma(p^0,\lambda^0)}. \tag{3.26}$$

PROOF. Obviously,

$$\tilde{\mathcal{L}}_{\mathcal{S}}(\nu\Phi_-,\nu\Phi,\, \nu w; \nu p,\nu\lambda) \equiv \nu^2\tilde{\mathcal{L}}_{\mathcal{S}}(\Phi_-,\Phi,\, w; p,\lambda) \qquad \forall\,\nu\in\mathbf{R}^1.$$

Therefore,

$$\Gamma(\nu p,\nu\lambda) = \nu^2\Gamma(p,\lambda) \qquad \forall\,\nu\in\mathbf{R}^1. \tag{3.27}$$

Moreover, if $\tilde{a}\neq 0$, then

$$\Gamma(p^0,\lambda^0) > 0. \tag{3.28}$$

In fact, assume that $\Gamma(p^0,\lambda^0) < 0$. Then, by virtue of (3.24) and (3.27),

$$\Pi(\nu p^0,\nu\lambda^0) \to \infty \qquad \text{as} \qquad \nu\to\infty$$

that contradicts to (3.21) and (3.22). Assume now that $\Gamma(p^0, \lambda^0) = 0$. Then $\tilde{a}'\lambda^0 = 0$ by the same reason. In accordance with (3.8), (2.6), (3.9), and (2.8),

$$\tilde{a}'\lambda^0 = 2\tilde{\eta}'(T)\tilde{a} = 2\tilde{a}'\tilde{P}(T)\tilde{a} = 2\mathcal{J}_{a0} \tag{3.29}$$

and $\mathcal{J}_{a0} = 0$. By our initial assumption, $P_0 > 0$ and $R(t) \geq R_0 > 0$. Therefore we have from (2.1) and (1.10) that $\varphi_- = 0$, $\varphi = 0$, and $\tilde{a} = a = 0$.

Thus it follows from (3.21), (3.24), (3.27), and (3.28) that

$$\mathcal{D}_0 = \sup_{\nu \in \mathbf{R}^1} \Big(\nu \tilde{a}'\lambda^0 - \nu^2 \Gamma(p^0, \lambda^0) \Big) = \frac{\left(\tilde{a}'\lambda^0\right)^2}{4\Gamma(p^0, \lambda^0)}. \tag{3.30}$$

Formulas (3.29) and (3.30) imply (3.26). Lemma 3.1 is proved.

Let us compute the quantity $\Gamma(p^0, \lambda^0)$. It follows from (2.4) – (2.6), (3.8), (3.9), and (3.25) that, with $(p, \lambda) = (p^0, \lambda^0)$,

$$\Psi = 2\tilde{\eta}(T) - 2\int_0^T b(t)Q(t)b'(t)\tilde{\xi}(t)\, dt = 2\tilde{\eta}(0) = 2P_0\tilde{\xi}(0) = 2P_0\varphi_- \tag{3.31}$$

and

$$\Theta(t) = h'(t)\left(2\tilde{\eta}(T) - 2\int_t^T b(s)Q(s)b'(s)\tilde{\xi}(s)\, ds\right) = 2h'(t)\tilde{\eta}(t) = 2R(t)\varphi(t).$$

Let us compute the infimum in (3.25). The function $\tilde{\mathcal{L}}_S(\,\cdot\,; p^0, \lambda^0)$ is a proper convex function that maps $\mathcal{H}$ into $\mathbf{R}^1$. In accordance with Theorem 1.4 of Chapter 2, the necessary and sufficient condition for a minimum are determined by the inclusion

$$\partial\tilde{\mathcal{L}}_S \ni 0, \tag{3.32}$$

where $\partial\tilde{\mathcal{L}}_S$ is the subdifferential of $\tilde{\mathcal{L}}_S$ with respect to $(\Phi_-, \Phi, w) \in \mathcal{H}$ at the minimum point ((p^0, λ^0) are fixed).

Using the formula for the subdifferential of a sum of convex functions (Theorem 1.2 of Chapter 2), the Dubovitskii-Milyutin formula for the subdifferential of a composite function (Theorem 1.3 of Chapter 2), and the formula for the subdifferential of a norm type function (Lemma 1.1 of Chapter 2), we can represent (3.32) in the form (taking into account (3.31)):

$$\begin{aligned}
&c_d\Phi_{-d} + \zeta\sigma_{-d}\operatorname{sign}^*\Phi_{-d} - P_{0d}\varphi_{-d} \ni 0, && d = 1, \dots, n, \\
&r_l(t)\Phi_l(t) + \zeta\sigma_l(t)\operatorname{sign}^*\Phi_l(t) - R_l(t)\varphi_l(t) \ni 0, && l = 1, \dots, m, \\
&q_j(t)w_j(t) + \zeta\gamma_j(t)\operatorname{sign}^* w_j(t) - Q_j(t)w_{\varphi j}(t) \ni 0, && j = 1, \dots, r,
\end{aligned} \tag{3.33}$$

where

$$\zeta = \sigma'_-|\Phi_-| + \int_0^T \sigma'(t)|\Phi(t)|\, dt + \int_0^T \gamma'(t)|w(t)|\, dt, \tag{3.34}$$

$$\operatorname{sign}^*\Phi_{-d} = \begin{cases} +1 & \text{for } \Phi_{-d} > 0 \\ [-1,+1] & \text{for } \Phi_{-d} = 0 \\ -1 & \text{for } \Phi_{-d} < 0, \end{cases}$$

$$\operatorname{sign}^*\Phi_l(t) = \begin{cases} +1 & \text{for } t \in [0,T]\colon \Phi_l(t) > 0 \\ e(t) \in L_2,\ |e(t)| \le 1 & \text{for } t \in [0,T]\colon \Phi_l(t) = 0 \\ -1 & \text{for } t \in [0,T]\colon \Phi_l(t) < 0, \end{cases}$$

$$\operatorname{sign}^* w_j(t) = \begin{cases} +1 & \text{for } t \in [0,T]\colon w_j(t) > 0 \\ v(t) \in L_2,\ |v(t)| \le 1 & \text{for } t \in [0,T]\colon w_j(t) = 0 \\ -1 & \text{for } t \in [0,T]\colon w_j(t) < 0. \end{cases}$$

Then we have from (3.33)

$$\Phi_{-d} = \begin{cases} c_d^{-1}\big(P_{0d}\varphi_{-d} - \zeta\sigma_{-d}\big) & \text{for } S_d > \zeta \\ 0 & \text{for } |S_d| \le \zeta \\ c_d^{-1}\big(P_{0d}\varphi_{-d} + \zeta\sigma_{-d}\big) & \text{for } S_d < -\zeta, \end{cases} \tag{3.35}$$

$d = 1, \ldots, n,$

$$\Phi_l(t) = \begin{cases} r_l^{-1}(t)\big(R_l(t)\varphi_l(t) - \zeta\sigma_l(t)\big) & \text{for } t \in [0,T]\colon F_l(t) > \zeta \\ 0 & \text{for } t \in [0,T]\colon |F_l(t)| \le \zeta \\ r_l^{-1}(t)\big(R_l(t)\varphi_l(t) + \zeta\sigma_l(t)\big) & \text{for } t \in [0,T]\colon F_l(t) < -\zeta, \end{cases}$$

$l = 1, \ldots, m,$

$$w_j(t) = \begin{cases} q_j^{-1}(t)\big(Q_j(t)w_{\varphi j}(t) - \zeta\gamma_j(t)\big) & \text{for } t \in [0,T]\colon G_j(t) > \zeta \\ 0 & \text{for } t \in [0,T]\colon |G_j(t)| \le \zeta \\ q_j^{-1}(t)\big(Q_j(t)w_{\varphi j}(t) + \zeta\gamma_j(t)\big) & \text{for } t \in [0,T]\colon G_j(t) < -\zeta, \end{cases}$$

$j = 1, \ldots, r.$

Substituting (3.35) into (3.34) we obtain that ζ satisfies equation (2.9).

In accordance with Lemma 2.1 equation (2.9) has a unique root $\tilde{\zeta}$. Consequently the formulas (3.35) determine the minimum point for $\tilde{\mathcal{L}}_S$.

Let us now transform the expression for $\Gamma(p^0, \lambda^0)$ by substituting the formulas (3.35) into (3.25) with (3.31). For brevity, consider the summand

$$\int_0^T \Theta_1(t)\Phi_1(t)\,dt - \int_0^T r_1(t)\Phi_1^2(t)\,dt$$

$$= \int\limits_{\{t:\, F_1(t) > \tilde{\zeta}\}} \frac{2R_1(t)\varphi_1(t)\sigma_1(t)}{r_1(t)} \left(\frac{R_1(t)\varphi_1(t)}{\sigma_1(t)} - \tilde{\zeta}\right)$$

$$- \frac{\sigma_1^2(t)}{r_1(t)} \left(\frac{R_1(t)\varphi_1(t)}{\sigma_1(t)} - \tilde{\zeta}\right)^2 dt$$

$$= \int\limits_{\{t:\, F_1(t) < -\tilde{\zeta}\}} \frac{2R_1(t)\varphi_1(t)\sigma_1(t)}{r_1(t)} \left(\frac{R_1(t)\varphi_1(t)}{\sigma_1(t)} + \tilde{\zeta}\right)$$

$$- \frac{\sigma_1^2(t)}{r_1(t)} \left(\frac{R_1(t)\varphi_1(t)}{\sigma_1(t)} + \tilde{\zeta}\right)^2 dt$$

$$= \int\limits_{\{t:\, F_1(t) > \tilde{\zeta}\}} 2f_1(t)F_1(t)\left(F_1(t) - \tilde{\zeta}\right) - f_1(t)\left(F_1(t) - \tilde{\zeta}\right)^2 dt$$

$$+ \int\limits_{\{t:\, F_1(t) < -\tilde{\zeta}\}} 2f_1(t)F_1(t)\left(F_1(t) + \tilde{\zeta}\right) - f_1(t)\left(F_1(t) + \tilde{\zeta}\right)^2 dt$$

$$= \int\limits_{\{t:\, F_1(t) > \tilde{\zeta}\}} f_1(t)\left(F_1^2(t) - \tilde{\zeta}^2\right) dt + \int\limits_{\{t:\, F_1(t) < -\tilde{\zeta}\}} f_1(t)\left(F_1^2(t) - \tilde{\zeta}^2\right) dt$$

$$= \int\limits_{t:\, |F_1(t)| > \tilde{\zeta}\,\}} f_1(t)\left(F_1^2(t) - \tilde{\zeta}^2\right) dt.$$

Other summands from (3.25) can be transformed similarly. Then we get

$$\Gamma(p^0, \lambda^0) = \sum_{\{d:\, |S_d| > \tilde{\zeta}\}} s_d\left(S_d^2 - \tilde{\zeta}^2\right) \tag{3.36}$$

$$+ \sum_{l=1}^{m} \int\limits_{\{t:\, |F_l(t)| > \tilde{\zeta}\}} f_l(t)\left(F_l^2(t) - \tilde{\zeta}^2\right) dt$$

$$+ \sum_{j=1}^{r} \int\limits_{\{t:\, |G_j(t)| > \tilde{\zeta}\}} g_j(t)\left(G_j^2(t) - \tilde{\zeta}^2\right) dt \; - \tilde{\zeta}^2$$

$$= \sum_{\{d:\, |S_d| > \tilde{\zeta}\}} s_d\left(\left(|S_d| - \tilde{\zeta}\right)^2 + 2\tilde{\zeta}\left(|S_d| - \tilde{\zeta}\right)\right)$$

$$+\sum_{l=1}^{m} \int\limits_{\{t:|F_l(t)|>\tilde{\zeta}\}} f_l(t)\Big(\big(|F_l(t)|-\tilde{\zeta}\big)^2 + 2\tilde{\zeta}\big(|F_l(t)|-\tilde{\zeta}\big)\Big)\,dt$$

$$+\sum_{j=1}^{r} \int\limits_{\{t:|G_j(t)|>\tilde{\zeta}\}} g_j(t)\Big(\big(|G_j(t)|-\tilde{\zeta}\big)^2 + 2\tilde{\zeta}\big(|G_j(t)|-\tilde{\zeta}\big)\Big)\,dt - \tilde{\zeta}^2.$$

Then, using (2.9), we obtain from (3.36) that

$$\Gamma(p^0,\lambda^0) = \Omega(\varphi_-,\varphi,\, w_\varphi), \tag{3.37}$$

where $\Omega(\varphi_-,\varphi,\, w_\varphi)$ is defined by (2.10). Finally, the inequality (2.11) from Theorem 2.2 follows from (3.23), (3.26), and (3.37). □

The case of singular P_0 and $Q(t)$ can be studied in the same way as in Chapter 3.

3.3. *Proof of Lemma 2.2*

In accordance with (3.25),

$$\Gamma(p,\lambda) = 2 \sup_{\Phi_-\in\mathbf{R}^n,\Phi\in L_2^m,\, w\in L_2^r} \left\{ \frac{1}{2}\Psi'\Phi_- + \int_0^T \frac{1}{2}\Theta'(t)\Phi(t)\,dt \right. \tag{3.38}$$

$$\left. + \int_0^T \frac{1}{2}p'(t)w(t)\,dt - \frac{1}{2}\Big(\mathcal{J}^{\frac{1}{2}}(\Phi_-,\Phi,\, w)\Big)^2 \right\}.$$

The function $\mathcal{J}^{\frac{1}{2}}(\Phi_-,\Phi,\, w)$ is a positively homogeneous function.

Assertion 3.1. *If $k(\chi)$ is a caliber and $k^\circ(\chi^*)$ is a polar of $k(\chi)$, then*

$$\sup_{\chi\in\mathcal{H}}\Big\{\langle\chi^*,\chi\rangle_{\mathcal{H}} - \frac{1}{2}k^2(\chi)\Big\} = \frac{1}{2}\big(k^\circ(\chi^*)\big)^2. \tag{3.39}$$

Proof. In accordance with the definition of a polar, we have

$$\sup_{\chi\in\mathcal{H}}\Big\{\langle\chi^*,\chi\rangle_{\mathcal{H}} - \frac{1}{2}k^2(\chi)\Big\} = \sup_{\alpha\geq 0}\ \sup_{k(\chi)=\alpha}\Big\{\langle\chi^*,\chi\rangle_{\mathcal{H}} - \frac{1}{2}k^2(\chi)\Big\}$$

$$= \sup_{\alpha\geq 0}\Big\{\alpha k^\circ(\chi^*) - \frac{1}{2}\alpha^2\Big\} = \frac{1}{2}\big(k^\circ(\chi^*)\big)^2.$$

The assertion is proved.

By virtue of Assertion 3.1,

$$\Gamma(p, \lambda) = \left((\mathcal{J}^{\frac{1}{2}})^{\circ} (\tfrac{1}{2}\Psi, \tfrac{1}{2}\Theta, \tfrac{1}{2}p) \right)^2.$$

Since $(\mathcal{J}^{\frac{1}{2}})^{\circ}(\chi^*) > 0$, it follows from (3.31), (3.38), and (3.39) that

$$\Gamma^{\frac{1}{2}}(p^0, \lambda^0) = (\mathcal{J}^{\frac{1}{2}})^{\circ}(P_0\varphi_-, R\varphi, Qw_\varphi). \tag{3.40}$$

Introducing the function $k(\chi)$ by formula (2.20) and the element χ^φ by formula (2.21), it can easily be checked that

$$\mathcal{J}^{\frac{1}{2}}(\varphi_-, \varphi, w_\varphi) = k(\chi^\varphi) \tag{3.41}$$

and

$$(\mathcal{J}^{\frac{1}{2}})^{\circ}(P_0\varphi_-, R\varphi, Qw_\varphi) = k^{\circ}(\chi^\varphi).$$

It is obviously from (2.1) and (2.18) that

$$\mathcal{J}_a(\varphi_-, \varphi, w_\varphi) = \|\chi^\varphi\|^2_{\mathcal{H}}. \tag{3.42}$$

Then the desired formula (2.22) follows from (2.11), (3.37), and (3.40) – (3.42).

□

3.4. *Limiting Cases*

Let us discuss the asymptotic behavior of the upper bound Δ^0_S in the limiting cases such that: (a) the perturbations are arbitrarily close to the white noise; (b) the perturbations are almost deterministic.

First consider case (a). Suppose

$$\sigma_{-d} = \varepsilon\sigma^0_{-d}, \qquad \sigma_l(t) = \varepsilon\sigma^0_l(t), \qquad \gamma_j(t) = \varepsilon\gamma^0_j(t), \qquad \varepsilon \ll 1, \tag{3.43}$$

$$d = 1, \ldots, n, \qquad l = 1, \ldots, m, \qquad j = 1, \ldots, r.$$

If $\varepsilon \to 0^+$, then

$$P_0 \to c, \qquad R(t) \to r(t), \qquad Q(t) \to q(t). \tag{3.44}$$

By virtue of the continuous dependence of the solutions of ordinary differential equations on a parameter,

$$\varphi_- \to \varphi^0_-, \qquad \varphi(t) \to \varphi^0(t), \qquad w_\varphi(t) \to w^0_\varphi(t), \tag{3.45}$$

where $\{\varphi^0_-, \varphi^0, w^0_\varphi\}$ are the characteristics of the Kalman filter for $P_0 = c$, $R(t) = r(t)$, and $Q(t) = q(t)$.

ASSERTION 3.2. *If $\varepsilon \to 0^+$ in (3.43), then the root of equation (2.9) tends to zero:*

$$\tilde{\zeta} \to 0. \tag{3.46}$$

PROOF. Introduce the function

$$\mathcal{E}^0(\zeta) = \sum_{\{d:\,|S_d|>\zeta\}} s_d\Big(|S_d| - \zeta\Big) \tag{3.47}$$

$$+\sum_{l=1}^{m} \int_{\{t:\,|F_l(t)|>\zeta\}} f_l(t)\Big(|F_l(t)| - \zeta\Big)\,dt$$

$$+\sum_{j=1}^{r} \int_{\{t:\,|G_j(t)|>\zeta\}} g_j(t)\Big(|G_j(t)| - \zeta\Big)\,dt \;\equiv \mathcal{E}(\zeta) + \zeta.$$

In the proof of Lemma 2.2 we showed that $\mathcal{E}^0(\zeta)$ is a monotonically decreasing function. Obviously, $\mathcal{E}^0(0) = \mathcal{E}(0) > 0$. Consequently, if $\zeta > \mathcal{E}(0)$, then

$$\mathcal{E}(\zeta) = \mathcal{E}^0(\zeta) - \zeta \le \mathcal{E}^0(0) - \zeta = \mathcal{E}(0) - \zeta < 0.$$

Then, due to Lemma 2.2, the root of (2.9) $\tilde{\zeta}$ satisfies the inequalities $0 < \tilde{\zeta} < \mathcal{E}(0)$. By virtue of (3.43) – (3.45),

$$\mathcal{E}(0) = \sum_{d=1}^{n} s_d|S_d| + \sum_{l=1}^{m} \int_0^T f_l(t)|F_l(t)|\,dt + \sum_{j=1}^{r} \int_0^T g_j(t)|G_j(t)|\,dt$$

$$= \varepsilon\left\{\sum_{d=1}^{n} \frac{\sigma^0_{-d}}{c_d} P_{0d}\,|\varphi_{-d}| + \sum_{l=1}^{m} \int_0^T \frac{\sigma^0_l(t)}{r_l(t)} R_l(t)\,|\varphi_l(t)|\,dt\right.$$

$$\left.+\sum_{j=1}^{r} \int_0^T \frac{\gamma^0_j(t)}{q_j(t)} Q_j(t)\,|w_{\varphi j}(t)|\,dt\right\} \to 0 \qquad \text{as } \varepsilon \to 0^+.$$

Hence (3.46) is valid. Assertion 3.2 is proved.

It follows from (3.43) – (3.46) that the three quantities $\mathcal{J}(\varphi_-, \varphi, w_\varphi)$, $\Omega(\varphi_-, \varphi, w_\varphi)$, and $\mathcal{J}_a(\varphi_-, \varphi, w_\varphi)$ tend to

$$\mathcal{J}_a(\varphi^0_-, \varphi^0, w^0_\varphi)\Big|_{P_0=c,\; R(t)=r(t),\; Q(t)=q(t)} \qquad \text{as } \varepsilon \to 0^+.$$

Thus, in accordance with (2.11),

$$\Delta^0_S \to 1 \quad \text{as} \quad \varepsilon \to 0^+. \tag{3.48}$$

Relation (3.48) has a clear meaning: the Kalman filter is optimal if the perturbations are white noises.

Let us now consider case (b):

$$c = \varepsilon c^{\infty}, \qquad r(t) = \varepsilon r^{\infty}(t), \qquad q(t) = \varepsilon q^{\infty}(t), \qquad \varepsilon \ll 1. \tag{3.49}$$

If $\varepsilon \to 0^+$, then

$$P_0 \to \beta_0 \sigma_{-S}, \qquad R(t) \to \beta_1 \sigma_S(t), \qquad Q(t) \to \beta_2 \gamma_S(t), \tag{3.50}$$

$$\varphi_- \to \varphi_-^{\infty}, \qquad \varphi(t) \to \varphi^{\infty}(t), \qquad w_\varphi(t) \to w_\varphi^{\infty}(t),$$

$$s_d \to \infty, \qquad S_d \to S_d^{\infty}, \qquad d = 1, \dots, n,$$

$$f_l(t) \to \infty, \qquad F_l(t) \to F_l^{\infty}(t), \qquad l = 1, \dots, m,$$

$$g_j(t) \to \infty, \qquad G_j(t) \to G_j^{\infty}(t), \qquad j = 1, \dots, r,$$

$$\zeta_{\max} \to \zeta_{\max}^{\infty}, \tag{3.51}$$

where

$$\zeta_{\max}^{\infty} = \max\left\{ \max_{d=1,\dots,n} |S_d^{\infty}|, \sup_{l=1,\dots,m,\ t\in[0,T]} |F_l^{\infty}(t)|, \sup_{j=1,\dots,r,\ t\in[0,T]} |G_j^{\infty}(t)| \right\}$$

$$= \max\left\{ \max_{d=1,\dots,n} |\beta_0 \sigma_{-d} \varphi_{-d}^{\infty}|, \sup_{l=1,\dots,m,\ t\in[0,T]} |\beta_1 \sigma_l(t) \varphi_l^{\infty}(t)|, \sup_{j=1,\dots,r,\ t\in[0,T]} |\beta_2 \gamma_j(t) w_{\varphi j}^{\infty}(t)| \right\}.$$

In formulas (3.50) and (3.51), the triple $\{\varphi_-^{\infty}, \varphi^{\infty}, w_\varphi^{\infty}\}$ consists of the Kalman filter characteristics for $P_0 = \beta_0 \sigma_{-S}$, $R(t) = \beta_1 \sigma_S(t)$, and $Q(t) = \beta_2 \gamma_S(t)$; the quantities S_d^{∞}, $F_l^{\infty}(t)$, $G_j^{\infty}(t)$, $\zeta_{\max}^{\infty}$ are the corresponding limiting values.

ASSERTION 3.3. *If* $\varepsilon \to 0^+$ *in* (3.49), *then the root of equation* (2.9)

$$\tilde{\zeta} \to \zeta_{\max}^{\infty}. \tag{3.52}$$

PROOF. Assume that (3.52) does not hold, i.e., there exists a sequence ε_k such that, for a $\delta > 0$,

$$|\tilde{\zeta}(\varepsilon_k) - \zeta_{\max}^{\infty}| > \delta, \qquad \mathcal{E}(\tilde{\zeta}(\varepsilon_k)) = 0, \qquad k = 1, 2, \dots . \tag{3.53}$$

Since

$$0 \le \tilde{\zeta}(\varepsilon_k) \le \zeta_{\max}(\varepsilon_k), \quad \text{and} \quad \zeta_{\max}(\varepsilon_k) \to \zeta_{\max}^{\infty},$$

we get that $\tilde{\zeta}(\varepsilon_k)$ is bounded and, therefore, there exists a subsequence $\{\varepsilon_{k'}\} \subset \{\varepsilon_k\}$ such that $\tilde{\zeta}(\varepsilon_{k'}) \to \zeta^{\infty}$. Without loss of generality we may consider $\{\varepsilon_{k'}\} = \{\varepsilon_k\}$. So,

$$\tilde{\zeta}(\varepsilon_k) \to \zeta^{\infty}. \tag{3.54}$$

It follows from (3.53) and (3.54) that

$$|\zeta^{\infty} - \zeta_{\max}^{\infty}| > \delta. \tag{3.55}$$

Obviously, $\mathcal{E}(\zeta) = \varepsilon^{-1}\bar{\mathcal{E}}(\zeta) - \zeta$, where

$$\begin{aligned} \bar{\mathcal{E}}(\zeta) = & \sum_{\{d:\,|S_d| > \zeta\}} \frac{\sigma_{-d}^2}{c_d^{\infty}} \Big(|S_d| - \zeta\Big) \\ & + \sum_{l=1}^{m} \int\limits_{\{t:\,|F_l(t)| > \zeta\}} \frac{\sigma_l^2(t)}{r_l^{\infty}(t)} \Big(|F_l(t)| - \zeta\Big)\, dt \\ & + \sum_{j=1}^{r} \int\limits_{\{t:\,|G_j(t)| > \zeta\}} \frac{\gamma_j^2(t)}{q^{\infty}(t)} \Big(|G_j(t)| - \zeta\Big)\, dt. \end{aligned} \tag{3.56}$$

Let us treat the limiting behaviour of $\bar{\mathcal{E}}(\tilde{\zeta}(\varepsilon_k);\varepsilon_k)$ as $k \to \infty$. Consider the first integral term in (3.56). We have

$$\begin{aligned} & \left| \int\limits_{\{t:\,|F_1(t)| > \zeta\}} \frac{\sigma_1^2(t)}{r_1^{\infty}(t)} \Big(|F_1(t)| - \zeta\Big)\, dt \; - \int\limits_{\{t:\,|F_1^{\infty}(t)| > \zeta^{\infty}\}} \frac{\sigma_1^2(t)}{r_1^{\infty}(t)} \Big(|F_1^{\infty}(t)| - \zeta^{\infty}\Big)\, dt \right| \\ & = \left| \int_0^T \frac{\sigma_1^2(t)}{r_1^{\infty}(t)} \Big(\max\Big(|F_1(t)| - \zeta, 0\Big) - \max\Big(|F_1^{\infty}(t)| - \zeta^{\infty}, 0\Big)\Big)\, dt \right| \\ & \le \int_0^T \frac{\sigma_1^2(t)}{r_1^{\infty}(t)} \left| \max\Big(|F_1(t)| - \zeta, 0\Big) - \max\Big(|F_1^{\infty}(t)| - \zeta^{\infty}, 0\Big) \right| dt \\ & \le \int_0^T \frac{\sigma_1^2(t)}{r_1^{\infty}(t)} \Big| |F_1(t)| - |F_1^{\infty}(t)| + \zeta^{\infty} - \zeta \Big|\, dt. \end{aligned}$$

Similar inequalities can be obtained for the other terms in (3.56). Hence, in accordance with (3.50) and (3.54),

$$\bar{\mathcal{E}}(\tilde{\zeta}(\varepsilon_k);\varepsilon_k) \to \mathcal{E}^{\infty} \qquad \text{as } k \to \infty, \tag{3.57}$$

where

$$\mathcal{E}^{\infty} = \sum_{\{d:\,|S_d^{\infty}|>\zeta^{\infty}\}} \frac{\sigma_{-d}^2}{c_d^{\infty}}\left(|S_d^{\infty}|-\zeta^{\infty}\right)$$

$$+\sum_{l=1}^{m} \int_{\{t:\,|F_l^{\infty}(t)|>\zeta^{\infty}\}} \frac{\sigma_l^2(t)}{r_l^{\infty}(t)}\left(|F_l^{\infty}(t)|-\zeta^{\infty}\right) dt$$

$$+\sum_{j=1}^{r} \int_{\{t:\,|G_j^{\infty}(t)|>\zeta^{\infty}\}} \frac{\gamma_j^2(t)}{q^{\infty}(t)}\left(|G_j^{\infty}(t)|-\zeta^{\infty}\right) dt.$$

By virtue of (3.55),

$$\mathcal{E}^{\infty} > 0. \tag{3.58}$$

It follows from (3.54), (3.57), and (3.58) that

$$\mathcal{E}(\tilde{\zeta}(\varepsilon_k);\varepsilon_k) = \varepsilon^{-1}\bar{\mathcal{E}}(\tilde{\zeta}(\varepsilon_k);\varepsilon_k) - \tilde{\zeta}(\varepsilon_k) \to \infty \qquad \text{as } k \to \infty.$$

The last relation contradicts to the equalities $\mathcal{E}(\tilde{\zeta}(\varepsilon_k)) = 0\ \forall k$. Thus our initial assumption that (3.52) does not hold is false. Assertion 3.3 is proved.

Let us now show that if (3.49), (3.50), and (3.52) hold, then

$$\Omega(\varphi_-, \varphi, w_\varphi) \to \left(\zeta_{\max}^{\infty}\right)^2 \qquad \text{as } \varepsilon \to 0^+, \tag{3.59}$$

where $\Omega(\varphi_-, \varphi, w_\varphi)$ is defined by (2.10).
In fact,

$$\Omega(\varphi_-, \varphi, w_\varphi) = \Omega^0(\varphi_-, \varphi, w_\varphi) + \tilde{\zeta}^2,$$

$$\Omega^0(\varphi_-, \varphi, w_\varphi) \triangleq \sum_{\{d:\,|S_d|>\tilde{\zeta}\}} s_d\left(|S_d|-\tilde{\zeta}\right)^2$$

$$+\sum_{l=1}^{m} \int_{\{t:\,|F_l(t)|>\tilde{\zeta}\}} f_l(t)\left(|F_l(t)|-\tilde{\zeta}\right)^2 dt$$

$$+\sum_{j=1}^{r} \int_{\{t:\,|G_j(t)|>\tilde{\zeta}\}} g_j(t)\left(|G_j(t)|-\tilde{\zeta}\right)^2 dt.$$

Obviously[4],

$$|\Omega^0(\varphi_-, \varphi, w_\varphi)| \le \mathcal{E}^0(\tilde{\zeta})\, M(\tilde{\zeta}),$$

where $\mathcal{E}^0(\tilde{\zeta})$ is defined by (3.47) and $M(\tilde{\zeta})$ is defined by the expression

$$\max\left\{ \max_{d:\,|S_d|>\tilde{\zeta}} \Big(|S_d| - \tilde{\zeta}\Big),\ \sup_{\substack{l=1,\dots,m\\ t:\,|F_l(t)|>\tilde{\zeta}}} \Big(|F_l(t)| - \tilde{\zeta}\Big),\ \sup_{\substack{j=1,\dots,r\\ t:\,|G_j(t)|>\tilde{\zeta}}} \Big(|G_j(t)| - \tilde{\zeta}\Big) \right\}.$$

The quantity $\tilde{\zeta}$ is the root of equation (2.9): $\mathcal{E}^0(\tilde{\zeta}) = \tilde{\zeta}$; hence,

$$|\Omega^0(\varphi_-, \varphi, w_\varphi)| \le \tilde{\zeta}\, M(\tilde{\zeta}).$$

By definition of $\zeta_{\max}$, we have

$$|S_d| \le \zeta_{\max}, \qquad |F_l(t)| \le \zeta_{\max}, \qquad |G_j(t)| \le \zeta_{\max},$$

$$d = 1, \dots, n, \qquad l = 1, \dots, m, \qquad j = 1, \dots, r.$$

By virtue of (3.51) and (3.52),

$$\zeta_{\max} \to \zeta_{\max}^\infty \qquad \text{and} \qquad \tilde{\zeta} \to \zeta_{\max}^\infty.$$

Consequently,

$$M(\tilde{\zeta}) \to 0, \qquad \Omega^0(\varphi_-, \varphi, w_\varphi) \to 0 \qquad \text{as } \varepsilon \to 0^+$$

and (3.59) is valid.

Then, in accordance with (1.9), (2.1), (2.11), (3.50), and (3.59),

$$\Delta_S^0 \to \Delta_S^\infty \qquad \text{as } \varepsilon \to 0^+.$$

The limiting value is determined by the formula

$$\Delta_S^\infty = \frac{(\mathcal{J}^\infty)^{\frac{1}{2}}\, \zeta_{\max}^\infty}{\mathcal{J}_a^\infty}, \tag{3.60}$$

where

$$\mathcal{J}^\infty = \sigma'_- |\varphi_-^\infty| + \int_0^T \sigma'(t) |\varphi^\infty(t)|\, dt + \int_0^T \gamma'(t) |w_\varphi^\infty(t)|\, dt,$$

$$\mathcal{J}_a^\infty = \beta_0 \varphi_-^{\infty\prime} \sigma_{-S} \varphi_-^\infty + \beta_1 \int_0^T \varphi^{\infty\prime}(t) \sigma_S(t) \varphi^\infty(t)\, dt$$
$$+ \beta_2 \int_0^T w_\varphi^{\infty\prime}(t) \gamma_S(t) w_\varphi^\infty(t)\, dt,$$

[4]by definition, $\max\limits_{\emptyset} |\cdots| = 0$.

and $\zeta_{\max}^{\infty}$ is defined by (3.51).

Formula (3.60) with $\beta_0 = 1$ was obtained in Chapter 3 for the case of deterministic bounded uncertain perturbations.

Thus we have proved that for the main formula (2.11) the natural limiting relations are valid.

4. Discrete Measurements in Stochastic Estimation Problem

Consider a continuous linear stochastic dynamic system

$$\dot{x}(t) = A(t)x(t) + B(t)u(t), \qquad t \in [0,T], \tag{4.1}$$

where $x(t) = (x_1(t), \dots, x_n(t))' \in \mathbf{R}^n$ is the system state vector, $u(t) = (u_1(t), \dots, u_r(t))' \in \mathbf{R}^r$ is the perturbation vector, $A(t) \in \mathbf{R}^{n\times n}$, $B(t) \in \mathbf{R}^{n\times r}$. As before, a prime denotes the transposition sign.

Suppose the discrete measurements $z(t_i)$ are made on the interval $[0,T]$:

$$z(t_i) = H'(t_i)x(t_i) + \varrho(t_i), \qquad z(t_i), \varrho(t_i) \in \mathbf{R}^m, \quad H(t_i) \in \mathbf{R}^{n\times m}, \tag{4.2}$$

$$i = 0, \dots, N, \qquad t_0 = 0, \quad t_N = T,$$

where $z(t_i)$ are the measurements, $\varrho(t_i) = (\varrho_1(t_i), \dots, \varrho_m(t_i))'$ is a measurement noise.

Assume that the following stochastic hypotheses are valid:

(a_D) the initial state $x(0)$ is a random variable and the perturbations $u(t)$, $\varrho(t_i)$ are white noise processes:

$$\mathbf{E}\big(x(0) - \mathbf{E}x(0)\big)\big(x(0) - \mathbf{E}x(0)\big)' = \operatorname{diag}(c_1^x, \dots, c_n^x),$$

$$\mathbf{E}\big(u(t) - \mathbf{E}u(t)\big)\big(u(s) - \mathbf{E}u(s)\big)' = \operatorname{diag}(q_1^u(t), \dots, q_r^u(t))\,\delta(t-s),$$

$$\mathbf{E}\big(\varrho(t_i) - \mathbf{E}\varrho(t_i)\big)\big(\varrho(t_p) - \mathbf{E}\varrho(t_p)\big)' = \operatorname{diag}(r_1^\varrho(t_i), \dots, r_m^\varrho(t_i))\,\delta_{ip},$$

$\delta(t)$ is the Dirac delta-function, δ_{ij} is the Kronecker delta ($t, s \in [0,T]$, $i, p = 0, \dots, N$);

(b_D) the mathematical expectations and intensities of the perturbations are unknown but bounded:

$$|\mathbf{E}x_d(0)| \le \sigma_{-d}, \qquad |\mathbf{E}u_j(t)| \le \gamma_j(t), \qquad |\mathbf{E}\varrho_l(t_i)| \le \sigma_l(t_i),$$

$$c_d^x \le c_d, \qquad q_j^u(t) \le q_j(t), \qquad r_l^\varrho(t_i) \le r_l(t_i),$$

$$d = 1, \dots, n, \qquad j = 1, \dots, r, \qquad l = 1, \dots, m, \qquad i = 0, \dots, N;$$

(c_D) the initial state $x(0)$ and the perturbations $u(t)$, $\varrho(t_i)$ are mutually independent for all $t \in [0,T]$ and $i = 0,\dots,N$.

Here σ_{-d}, c_d, $\sigma_l(t_i)$, $r_l(t_i)$ are specified positive numbers, $\gamma_j(t) \geq \gamma^0 > 0$, $q_j(t) \geq q^0 > 0$ are specified positive functions. The entries of the matrices $A(t)$, $B(t)$, the quantities $\mathbf{E}u_j(t)$, $\gamma_j(t)$, $q_j^u(t)$, $q_j(t)$ are assumed to be piecewise continuous functions on $[0,T]$. For stochastic elements $x(0)$, $u(t)$, and $\varrho(t_i)$ that satisfy hypotheses (a_D) – (c_D), we write

$$(x(0), u, \varrho) \in \mathcal{S}_D.$$

The problem is to evaluate a scalar quantity $l_* = a'x(T)$ with the help of linear functionals

$$\hat{l} = \sum_{i=0}^{N} \Phi'(t_i)z(t_i), \qquad \Phi(t_i) \in \mathbf{R}^m,$$

where $a \in \mathbf{R}^n$ is a given vector.

We define the estimation quality by the guaranteed value of the second moment of the estimation error:

$$\mathcal{D}_{\text{guar}\,D}(\Phi) = \sup_{(x(0),u,\varrho) \in \mathcal{S}_D} \mathbf{E}(\hat{l} - l_*)^2. \tag{4.3}$$

The optimal stochastic guaranteed estimation problem is to find an optimal estimator

$$\Phi^0 = \{\Phi_1^0(t_i), \dots, \Phi_m^0(t_i)\}_{i=0}^{i=N}$$

that minimizes the guaranteed value of the second moment

$$\mathcal{D}_{\text{guar}\,D}(\Phi^0) = \mathcal{J}_{D0} \triangleq \inf_{\Phi(t_i),\, i=0,\dots,N} \mathcal{D}_{\text{guar}\,D}(\Phi).$$

Similar to Section 1, let us introduce the notation

$$\sigma_- = (\sigma_{-1}, \dots, \sigma_{-n})' \in \mathbf{R}^n,$$

$$\gamma(t) = (\gamma_1(t), \dots, \gamma_r(t))' \in \mathbf{R}^r,$$

$$\sigma(t_i) = (\sigma_1(t_i), \dots, \sigma_m(t_i))' \in \mathbf{R}^m,$$

$$c = \operatorname{diag}(c_1, \dots, c_n) \in \mathbf{R}^{n\times n},$$

$$q(t) = \operatorname{diag}(q_1(t), \dots, q_r(t)) \in \mathbf{R}^{r\times r},$$

$$r(t_i) = \operatorname{diag}(r_1(t_i), \dots, r_m(t_i)) \in \mathbf{R}^{m\times m}.$$

For a vector $\omega = (\omega_1, \ldots, \omega_k)' \in \mathbf{R}^k$, we define $|\omega| = (|\omega_1|, \ldots, |\omega_k|)'$.

LEMMA 4.1. *The guaranteed value of the second moment of the estimation error defined by* (4.3) *is represented by the formula*

$$\mathcal{D}_{\text{guar}\,D}(\Phi) = \mathcal{J}_D(\Phi_-(\Phi), \Phi, w(\Phi)),$$

where

$$\mathcal{J}_D(\Phi_-, \Phi, w) = \tag{4.4}$$

$$= \left\{ \sigma'_- |\Phi_-| + \sum_{i=0}^{N} \sigma'(t_i)|\Phi(t_i)| + \int_0^T \gamma'(t)|w'(t)|\, dt \right\}^2$$

$$+ \Phi'_- c\, \Phi_- + \sum_{i=0}^{N} \Phi'(t_i) r(t_i) \Phi(t_i) + \int_0^T w'(t) q(t) w(t)\, dt,$$

$$\Phi_- = \xi_{\Phi}^-(t_0), \qquad w(t) = B'(t)\xi_{\Phi}(t), \tag{4.5}$$

and $\xi_{\Phi}^-(t_0)$, $\xi_{\Phi}(t)$ *are defined by the differential equation*

$$\dot{\xi}_{\Phi}(t) = -A'(t)\xi_{\Phi}(t), \qquad t_i \le t < t_{i+1}, \qquad i = 0, \ldots, N-1, \tag{4.6}$$

with the jumps at the points t_i

$$\xi_{\Phi}^-(t_i) = \xi_{\Phi}(t_i) - H(t_i)\Phi(t_i), \qquad \xi_{\Phi}(t_N) = \xi_{\Phi}(T) = a, \qquad i = 0, \ldots, N.$$

$$\xi_{\Phi}^-(t_i) = \lim_{t \to t_i - 0} \xi_{\Phi}(t), \qquad i = 1, \ldots, N.$$

PROOF. Using the auxiliary variables (see Section 1 and Section 4 of Chapter 3)

$$h(t) = X'(t)H(t), \qquad b(t) = X^{-1}(t)B(t), \qquad \tilde{a} = X'(T)a,$$

where $X(t)$ is the solution of the equation $\dot{X}(t) = A(t)X(t)$, $X(0) = E_n$, it can easily be checked that

$$\mathcal{D}_{\text{guar}\,D}(\Phi) = \mathcal{J}_D(\Phi_-(\Phi), \Phi, w(\Phi)),$$

where

$$\Phi_- = \tilde{a} - \sum_{i=0}^{N} h(t_i)\Phi(t_i),$$

$$w(t) = b'(t)\Big(\Phi_- + \sum_{\{i|t_i\le t\}} h(t_i)\Phi(t_i)\Big).$$

Obviously, Φ_- and $w(t)$ can be represented in the form

$$\Phi_- = \tilde{\xi}_{\Phi}^-(t_0), \qquad w(t) = b'(t)\tilde{\xi}_{\Phi}(t),$$

where

$$\dot{\tilde{\xi}}_{\Phi}(t) = 0, \qquad t_i \le t < t_{i+1}, \qquad i = 0,\dots,N-1, \tag{4.7}$$

$$\tilde{\xi}_{\Phi}^-(t_i) = \tilde{\xi}_{\Phi}(t_i) - h(t_i)\Phi(t_i), \qquad \tilde{\xi}_{\Phi}(t_N) = \tilde{\xi}_{\Phi}(T) = \tilde{a}, \qquad i = 0,\dots,N.$$

$$\tilde{\xi}_{\Phi}^-(t_i) = \lim_{t\to t_i-0} \tilde{\xi}_{\Phi}(t), \qquad i = 1,\dots,N.$$

With the help of the notation $\xi_{\Phi}(t) = \big(X^{-1}(t)\big)'\tilde{\xi}_{\Phi}(t)$, $\xi_{\Phi}^-(t_0) = \tilde{\xi}_{\Phi}^-(t_0)$, relations (4.7) can be rewritten in the form (4.6); moreover,

$$\Phi_- = \xi_{\Phi}^-(t_0), \qquad w(t) = B'(t)\xi_{\Phi}(t).$$

□

Thus the optimal stochastic guaranteed estimation problem reduces to the following nonsmooth variational problem

$$\mathcal{J}_{D0} = \inf_{\Phi_-\in\mathbf{R}^n,\ \Phi(t_i)\in\mathbf{R}^m,\ w\in L_2^r} \mathcal{J}_D(\Phi_-,\Phi,w) \tag{4.8}$$

under constraints (4.5), (4.6),

where $\mathcal{J}(\Phi_-,\Phi,w)$ is given by (4.4). The solution of (4.8) $\{\Phi_-^0,\Phi^0,w^0\}$ determines the optimal guaranteed estimator Φ^0.

The solving of (4.8) is a difficult numerical problem. Therefore, similar to the continuous case, we shall use the discrete Kalman filter (see, e.g., [79], [47]) to obtain an approximate solution for (4.8).

The Kalman filter for the system (4.1) with discrete measurements (4.2) has the form (cf. Chapter 3)

$$\hat{l}_{\varphi} = a'\hat{x}(T|N) = a'\hat{x}(t_N|N),$$

$$\hat{x}(t_i|i) = \hat{x}(t_i|i-1) + K(t_i)\big(z(t_i) - H'(t_i)\hat{x}(t_i|i-1)\big), \qquad i = 0,\dots,N,$$

$$\hat{x}(t_i|i-1) = \Upsilon(t_i,t_{i-1})\hat{x}(t_{i-1}|i-1), \qquad i = 1,\dots,N,$$

$$\hat{x}(t_0|-1) = 0,$$

where $\Upsilon(t_i|t_{i-1}) = X(t_i)X^{-1}(t_{i-1})$ is the transition matrix of the homogeneous equation $\dot{x}(t) = A(t)x(t)$,

$$K(t_i) = P(t_i|i-1)H(t_i)\big(H'(t_i)P(t_i|i-1)H(t_i) + R(t_i)\big)^{-1} \tag{4.9}$$

$$= P(t_i|i)H(t_i)R^{-1}(t_i),$$

$$P(t_i|i) = P(t_i|i-1) - K(t_i)H'(t_i)P(t_i|i-1), \tag{4.10}$$

$$P(t_0|-1) = P_0, \qquad i = 0, \dots, N,$$

$$P(t_i|i-1) = \Upsilon(t_i, t_{i-1})P(t_{i-1}|i-1)\Upsilon'(t_i, t_{i-1})$$

$$+ \int_{t_{i-1}}^{t_i} \Upsilon(t_i, \tau)B(\tau)Q(\tau)B'(\tau)\Upsilon'(t_i, \tau)\, d\tau, \qquad i = 1, \dots, N,$$

$P_0 \geq 0$, $R(t_i) > 0$, and $Q(t) \geq 0$ are specified weight matrices.

We define the level of the nonoptimality of the Kalman filter in the stochastic guaranteed estimation problem by the ratio

$$\Delta_{SD} = \left\{ \frac{\sup_{(x(0),u,\varrho)\in \mathcal{S}_D} \mathbf{E}\left(\hat{l}_\varphi - l_*\right)^2}{\sup_{(x(0),u,\varrho)\in \mathcal{S}_D} \mathbf{E}\left(\hat{l}_0 - l_*\right)^2} \right\}^{\frac{1}{2}},$$

where

$$\hat{l}_0 = \sum_{i=0}^{N} \Phi^{0\prime}(t_i)z(t_i) \quad \text{and} \quad \hat{l}_\varphi = \sum_{i=0}^{N} \varphi'(t_i)z(t_i)$$

are the unknown optimal estimator and the discrete Kalman filter estimator, respectively. Obviously, $\Delta_{SD} \geq 1$.

Using Lemma 4.1, we see that

$$1 \leq \Delta_{SD} = \left(\frac{\mathcal{D}_{\text{guar}\,D}(\varphi)}{\mathcal{D}_{\text{guar}\,D}(\Phi^0)}\right)^{\frac{1}{2}} = \left(\frac{\mathcal{D}_{\text{guar}\,D}(\varphi)}{\mathcal{J}_{D0}}\right)^{\frac{1}{2}} = \left(\frac{\mathcal{J}_D(\varphi_-, \varphi, w_\varphi)}{\mathcal{J}_D(\Phi^0_-, \Phi^0, w^0)}\right)^{\frac{1}{2}}.$$

Here $\{\Phi^0_-, \Phi^0, w^0\}$ is the solution of the original problem (4.8), φ is the Kalman estimator, and, in accordance with (4.5), the quantities φ_-, w_φ are defined by the formulas

$$\varphi_- = \xi^-_\varphi(t_0), \qquad w_\varphi(t) = B'(t)\xi_\varphi(t).$$

The Kalman filter estimator φ can be computed quite easily and the numerator $\mathcal{D}_{\text{guar}\,D}(\varphi)$ is a known quantity. Since the optimal estimator is unknown, the denominator $\mathcal{D}_{\text{guar}\,D}(\Phi^0)$ is unknown and Δ_{SD} is unknown.

We shall construct an upper bound Δ^0_{SD} for Δ_{SD} that can be efficiently computed *without solving the original nonsmooth problem* (4.8). If the value of Δ^0_{SD} is not large, then the use of the Kalman filter will be justified.

Similar to the continuous case, we define the weight matrices of the Kalman filter by the formulas

$$P_0 = \operatorname{diag}(P_{01}, \ldots, P_{0n}) = \beta_0\, \sigma_{-S} + c,$$

$$R(t_i) = \operatorname{diag}(R_1(t_i), \ldots, R_m(t_i)) = \beta_1\, \sigma_S(t_i) + r(t_i),$$

$$Q(t) = \operatorname{diag}(Q_1(t), \ldots, Q_r(t)) = \beta_2\, \gamma_S(t) + q(t),$$

where

$$\sigma_{-S} = \operatorname{diag}(\sigma_{-1}^2, \ldots, \sigma_{-n}^2),$$

$$\sigma_S(t_i) = \operatorname{diag}(\sigma_1^2(t_i), \ldots, \sigma_m^2(t_i)), \qquad \gamma_S(t) = \operatorname{diag}(\gamma_1^2(t), \ldots, \gamma_r^2(t)),$$

and β_0, β_1, β_2 are specified parameters.

THEOREM 4.1. 1^0. *The Kalman estimator φ is determined by the solution $\{\varphi_-, \varphi, w_\varphi\}$ of the following linear-quadratic problem*

$$\mathcal{J}_{aD0} = \inf_{\Phi_- \in \mathbf{R}^n,\, \Phi(t_i) \in \mathbf{R}^m,\, w \in L_2^r} \mathcal{J}_{aD}(\Phi_-, \Phi, w) \tag{4.11}$$

under constraints (4.5), (4.6),

where

$$\mathcal{J}_{aD}(\Phi_-, \Phi, w) \tag{4.12}$$

$$= \Phi'_- P_0\, \Phi_- + \sum_{i=0}^{N} \Phi'(t_i) R(t_i) \Phi(t_i) + \int_0^T w'(t) Q(t) w(t)\, dt.$$

2^0. *The solution of the problem* (4.11) *has the form*

$$\varphi_- = \xi^-(t_0), \qquad \varphi(t_i) = K'(t_i)\xi(t_i), \qquad w_\varphi(t) = B'(t)\xi(t), \tag{4.13}$$

where the piecewise continuous (continuous from the right) function $\xi(t)$ satisfies the differential equation

$$\dot{\xi}(t) = -A'(t)\xi(t), \qquad t_i \le t < t_{i+1}, \qquad i = 0, \ldots, N-1, \tag{4.14}$$

with the jumps

$$\xi^-(t_i) = (E_n - K(t_i)H'(t_i))'\xi(t_i), \qquad \xi(t_N) = a, \qquad i = 0, \ldots, N,$$

$$\xi^-(t_i) = \lim_{t \to t_i - 0} \xi(t), \qquad i = 1, \ldots, N.$$

Moreover,

$$\mathcal{J}_{aD0} = a'P(t_N|N)a.$$

In these formulas, the Kalman gain $K(t_i)$ are the terminal value of the solution for the Riccati equation $P(t_N|N)$ are defined by (4.9) *and* (4.10).

PROOF. Using the variables from the proof of Lemma 4.1, it can easily be checked that Theorem 4.1 is an another form of Theorem 4.1 of Chapter 3. □

It follows from (4.13) and (4.14) that

$$\dot{\xi}(t) = A(t)\xi(t), \qquad t_i \le t < t_{i+1}, \qquad i = 0, \ldots, N-1, \tag{4.15}$$

$$\xi^-(t_i) = \xi(t_i) - H(t_i)\varphi(t_i), \qquad \xi(t_N) = a, \qquad i = 0, \ldots, N.$$

Thus we have from (4.6), (4.15) that

$$\xi(t) \equiv \xi_\varphi(t), \qquad \xi^-(t_i) \equiv \xi_\varphi^-(t_i)$$

and

$$\varphi_- = \xi^-(t_0) = \xi_\varphi^-(t_0), \qquad w_\varphi(t) = B'(t)\xi(t) = B'(t)\xi_\varphi(t).$$

Note that if we approximate the first nonsmooth term in (4.4) by a quadratic one (cf. Section 2)

$$\beta_0\, \Phi'_-\sigma_{-S}\Phi_- + \beta_1 \sum_{i=0}^{N} \Phi'(t_i)\sigma_S(t_i)\Phi(t_i) + \beta_2 \int_0^T w'(t)\gamma_S(t)w(t)\, dt,$$

then (4.4) is approximated by the quadratic functional (4.12).

Various computational formulas for the implementation of the Kalman estimator φ and the quantities φ_-, w_φ are described in detail in Section 4 of Chapter 3.

Let us obtain the upper bound for the level of nonoptimality for the Kalman filter. First, introduce the notation

$$s_d = \frac{\sigma_{-d}^2}{c_d}, \qquad S_d = \frac{P_{0d}\varphi_{-d}}{\sigma_{-d}}, \qquad d = 1, \ldots, n,$$

$$f_l(t_i) = \frac{\sigma_l^2(t_i)}{r_l(t_i)}, \qquad F_l(t_i) = \frac{R_l(t_i)\varphi_l(t_i)}{\sigma_l(t_i)}, \qquad l = 1, \ldots, m,$$

$$g_j(t) = \frac{\gamma_j^2(t)}{q_j(t)}, \qquad G_j(t) = \frac{Q_j(t)w_{\varphi j}(t)}{\gamma_j(t)}, \qquad j = 1, \ldots, r,$$

$$\varphi_- = (\varphi_{-1}, \ldots, \varphi_{-n})', \qquad \varphi(t_i) = (\varphi_1(t_i), \ldots, \varphi_m(t_i))',$$

$$w_\varphi(t) = (w_{\varphi 1}(t), \ldots, w_{\varphi r}(t))',$$

$$\zeta_{\max D} = \max \left\{ \max_{d=1,\ldots,n} |S_d|, \sup_{\substack{l=1,\ldots,m \\ i=0,\ldots,N}} |F_l(t_i)|, \sup_{\substack{j=1,\ldots,r \\ t\in[0,T)}} |G_j(t)| \right\}.$$

These quantities are uniquely determined by the characteristics of the Kalman filter (4.13), (4.14).

Let the function $\mathcal{E}_D(\zeta)\colon [0, \zeta_{\max D}] \to \mathbf{R}^1$ be defined by the formula

$$\mathcal{E}_D(\zeta) = \sum_{\{d:\, |S_d| > \zeta\}} s_d\Big(|S_d| - \zeta\Big) + \sum_{l=1}^{m} \sum_{\{i:\, |F_l(t_i)| > \zeta\}} f_l(t_i)\Big(|F_l(t_i)| - \zeta\Big)$$

$$+ \sum_{j=1}^{r} \int\limits_{\{t:\, |G_j(t)| > \zeta\}} g_j(t)\Big(|G_j(t)| - \zeta\Big)\, dt - \zeta.$$

LEMMA 4.2. *The function $\mathcal{E}_D(\zeta)$ is a strictly monotonically decreasing function on the interval $[0, \zeta_{\max D}]$. The equation*

$$\mathcal{E}_D(\zeta) = 0 \tag{4.16}$$

has a unique root on $[0, \zeta_{\max D}]$.

The proof of Lemma 4.2 is analogous to the proof of Lemma 2.1.

By virtue of Lemma 4.2, the root ζ_D of (4.16) can easily be found by sequential halving the interval $[0, \zeta_{\max D}]$.

Define the quantity

$$\Omega_D(\varphi_-, \varphi, w_\varphi) = \sum_{\{d:\, |S_d| > \zeta_D\}} s_d\Big(|S_d| - \zeta_D\Big)^2 \tag{4.17}$$

$$+ \sum_{l=1}^{m} \sum_{\{i:\, |F_l(t_i)| > \zeta_D\}} f_l(t_i)\Big(|F_l(t_i)| - \zeta_D\Big)^2 dt$$

$$+ \sum_{j=1}^{r} \int\limits_{\{t:\, |G_j(t)| > \zeta_D\}} g_j(t)\Big(|G_j(t)| - \zeta_D\Big)^2 dt + \zeta_D^2,$$

where ζ_D is the root of equation (4.16).

THEOREM 4.2. *Let $\{\varphi_-, \varphi, w_\varphi\}$ be defined by* (4.13), (4.14). *Then the level of nonoptimality Δ_{SD} is bounded with the quantity Δ_{SD}^0, which is determined by the following relations:*

$$1 \le \Delta_{SD} \le \Delta_{SD}^0 = \frac{\sqrt{\mathcal{J}_D(\varphi_-, \varphi, w_\varphi)\ \Omega_D(\varphi_-, \varphi, w_\varphi)}}{\mathcal{J}_{aD}(\varphi_-, \varphi, w_\varphi)}, \tag{4.18}$$

where $\mathcal{J}_D$, $\mathcal{J}_{aD}$, and Ω_D are defined by (4.4), (4.12), and (4.17), *respectively.*

PROOF. The proof of Theorem 4.2 is based on the duality theory for convex variational problems and is completely analogous to the proof of Theorem 2.2. □

If Δ_{SD}^0 is not large, then we do not need to resolve the difficult nonsmooth problem (4.8) and the Kalman filter can be used in the optimal stochastic guaranteed estimation probem as well.

5. Examples for Stochastic Problems

First we consider two simple examples, in which all calculations can be carried out in explicit form.

EXAMPLE 5.1. Assume that a scalar system and the measurements have the form

$$\dot{x} = 0, \qquad |\mathbf{E}x| \le \sigma_-, \quad \mathbf{E}\left(x - \mathbf{E}x\right)^2 \le c, \tag{5.1}$$

$$z(t) = x + \varrho(t), \qquad t \in [0, T], \tag{5.2}$$

$$|\mathbf{E}\varrho(t)| \le \sigma, \tag{5.3}$$

$$\mathbf{E}\left(\varrho(t) - \mathbf{E}\varrho(t)\right)\left(\varrho(s) - \mathbf{E}\varrho(s)\right) = r^\varrho \delta(t-s), \quad r^\varrho \le r.$$

For definiteness, we also assume that

$$(\sigma_- - \sigma)\frac{\sigma T}{r} > 1. \tag{5.4}$$

The estimation problem is to find an optimal guaranteed estimate for x with the help of measurements (5.2). In general notation of Section 1,

$$n = m = 1, \quad A(t) = B(t) = 0, \quad H(t) = 1, \quad a = 1, \quad \sigma(t) = \sigma, \quad r(t) = r.$$

Then the boundary value problem (2.15) takes the form

$$\dot{\xi}(t) = R^{-1}\eta(t), \qquad \xi(T) = 1, \tag{5.5}$$

$$\dot{\eta}(t) = 0, \qquad \eta(0) = P_0 \xi(0),$$

where

$$P_0 = \beta_0 \sigma_-^2 + c, \qquad R = \beta_1 \sigma^2 + r, \qquad \beta_0, \beta_1 > 0. \tag{5.6}$$

The solution of (5.5), (5.6) is determined by the relations

$$\xi(t) = \frac{1 + R^{-1} P_0 t}{1 + R^{-1} P_0 T}, \qquad \eta(t) = \frac{P_0}{1 + R^{-1} P_0 T}.$$

In accordance with (2.14),

$$\varphi_- = \frac{1}{1 + R^{-1} P_0 T}, \qquad \varphi(t) = \varphi = \frac{R^{-1} P_0}{1 + R^{-1} P_0 T}, \qquad w_\varphi(t) = 0.$$

Introducing the quantities

$$s = \frac{\sigma_-^2}{c}, \qquad S = \frac{P_0 \varphi_-}{\sigma_-}, \qquad f = \frac{\sigma^2}{r}, \qquad F = \frac{R\varphi}{\sigma},$$

we obtain the equation (2.9), where the continuous function $\mathcal{E}(\zeta)$ is determined by the expression

$$\mathcal{E}(\zeta) = \begin{cases} s(S - \zeta) + f(F - \zeta)T - \zeta, & 0 \le \zeta \le S \\ f(F - \zeta)T - \zeta, & S < \zeta \le F \\ -\zeta, & F < \zeta < \infty. \end{cases}$$

By virtue of (5.4), the root of the equation $\mathcal{E}(\zeta) = 0$ has the form

$$\tilde{\zeta} = (1 + fT)^{-1} fFT.$$

Then we have from (1.9), (2.1), and (2.10) that

$$\mathcal{J}(\varphi_-, \varphi, w_\varphi) = (\sigma_- \varphi_- + \sigma \varphi T)^2 + c\varphi_-^2 + r\varphi^2 T,$$

$$\mathcal{J}_a(\varphi_-, \varphi, w_\varphi) = P_0 \varphi_-^2 + R\varphi^2 T,$$

$$\Omega(\varphi_-, \varphi, w_\varphi) = (1 + fT)^{-1} F^2 fT.$$

Finally, the upper bound from (2.11) takes the form

$$\Delta_{\mathcal{S}}^0 = (1 + \alpha_0^{-1})^{-1} \left\{ 1 + (1 + \alpha_1)^{-1} \alpha_0^{-2} \left[\sigma_-^2 + c + 2\alpha_0 \sigma_- \sigma \right] r^{-1} T \right\}^{\frac{1}{2}},$$

where $\alpha_0 = R^{-1} P_0 T$, $\alpha_1 = r^{-1} \sigma^2 T$, and P_0, R are defined by (5.6).

Let the *a priori* information about the state be poor, i.e.,

$$\sigma_- = \varepsilon^{-1} \bar{\sigma}_-, \quad c = \varepsilon^{-2} \bar{c}, \qquad \bar{\sigma}_-, \bar{c} > 0, \qquad \text{and } \varepsilon \ll 1. \tag{5.7}$$

Then it can easily be verified that

$$\Delta_{\mathcal{S}}^{0} \to 1 \qquad \text{as } \varepsilon \to 0^{+}.$$

Thus, in the case of poor *a priori* information, the Kalman filter for our example is asymptotically optimal estimation algorithm.

REMARK 5.1. A more detailed analysis of the problems (3.5) and (3.21) leads to the assertion that if $n = 1$ and $b(t) = B(t) = 0$, then $\mathcal{J}_0 = \mathcal{D}_0$ and $\Delta_{\mathcal{S}} = \Delta_{\mathcal{S}}^{0}$.

EXAMPLE 5.2. Consider the previous system (5.1). Let the measurement equation be time-variant:

$$z(t) = t\,x + \varrho(t), \tag{5.8}$$

so that $H(t) = t$. The hypotheses about measurement noise are still defined by (5.3). The estimation problem is to find an optimal guaranteed estimate for x with the help of measurements (5.8).

Then the boundary value problem (2.15) takes the form

$$\begin{aligned} &\dot{\xi}(t) = R^{-1}t^{2}\eta(t), \qquad \xi(T) = 1, \\ &\dot{\eta}(t) = 0, \qquad \eta(0) = P_0\xi(0), \end{aligned} \tag{5.9}$$

where the auxiliary weight matrices P_0 and R are defined by (5.6).

The solution of (5.9), (5.6) is determined by the formulas

$$\xi(t) = \frac{1 + \frac{1}{3}R^{-1}P_0\,t^{3}}{1 + \frac{1}{3}R^{-1}P_0T^{3}}, \qquad \eta(t) = \frac{P_0}{1 + \frac{1}{3}R^{-1}P_0T^{3}}.$$

In accordance with (2.14),

$$\varphi_{-} = \frac{1}{1 + \frac{1}{3}R^{-1}P_0T^{3}}, \qquad \varphi(t) = \frac{R^{-1}P_0\,t}{1 + \frac{1}{3}R^{-1}P_0T^{3}}, \qquad w_{\varphi}(t) = 0.$$

In addition, we assume that the *a priori* information about the state x is poor, i.e., (5.7) holds and $\varepsilon \to 0^{+}$.

Then

$$P_0^{\frac{1}{2}}\varphi_{-} \to 0, \qquad \varphi(t) \to \frac{3\,t}{T^{3}}, \qquad s = \frac{\bar{\sigma}_{-}^{2}}{\bar{c}}, \qquad S = \frac{P_0\varphi_{-}}{\sigma_{-}} \to 0, \tag{5.10}$$

and the transcendental equation (2.9) for the system (5.1), (5.8), (5.3) (taking into account relations (5.7), (5.10)) turns into the equation

$$\frac{\sigma^{2}}{r}\int_{t^{0}(\zeta)}^{T}\left(\frac{3Rt}{\sigma} - \zeta\right)dt - \zeta = 0, \qquad t^{0}(\zeta) = \frac{\sigma T^{3}}{3R}\zeta, \tag{5.11}$$

where, in accordance with Lemma 2.1,

$$0 \leq \zeta \leq \zeta_{\max} = \frac{3R}{\sigma T^2}. \tag{5.12}$$

Calculating the integral in (5.11), we get

$$\zeta^2 - \frac{6R}{\sigma T^2}\left(1 + \frac{r}{\sigma^2 T}\right)\zeta + \frac{9R^2}{\sigma^2 T^4} = 0,$$

whence, taking into account (5.12),

$$\tilde{\zeta} = \frac{3R}{\sigma T^2}\left\{1 + \theta - \sqrt{(1+\theta)^2 - 1}\right\}, \qquad \theta = \frac{r}{\sigma^2 T}. \tag{5.13}$$

If $\varepsilon \to 0^+$, then, by virtue of (5.10), formula (2.10) for $\Omega(\varphi_-, \varphi, w_\varphi)$ turns into the equality

$$\Omega(\varphi_-, \varphi, w_\varphi) = \frac{\sigma^2}{r}\int_{t^0(\tilde{\zeta})}^{T}\left(\frac{3Rt}{\sigma T^3} - \tilde{\zeta}\right)^2 dt + \tilde{\zeta}^2. \tag{5.14}$$

Substituting (5.13) into (5.14), we have

$$\Omega(\varphi_-, \varphi, w_\varphi) = \frac{3R^2}{\sigma^2 T^4}\left\{\left(2 + \theta - \sqrt{(1+\theta)^2 - 1}\right)^2 - 1\right\}. \tag{5.15}$$

It can easily be checked that if $\varepsilon \to 0^+$, then, due to (1.9), (2.1), and (5.10), the expressions for $\mathcal{J}(\varphi_-, \varphi, w_\varphi)$ and $\mathcal{J}_a(\varphi_-, \varphi, w_\varphi)$ turn into the formulas

$$\mathcal{J}(\varphi_-, \varphi, w_\varphi) = \frac{3\sigma^2}{T^2}\left(\theta + \tfrac{3}{4}\right), \qquad \mathcal{J}_a(\varphi_-, \varphi, w_\varphi) = \frac{3R}{T^3}. \tag{5.16}$$

Using (5.15) and (5.16), the upper bound from (2.11) can be written in the form

$$\Delta_S^0 = \frac{1}{2}\sqrt{(4\theta + 3)\left[2\theta^2 + 6\theta + 3 - 2(\theta + 2)\sqrt{(1+\theta)^2 - 1}\right]}, \tag{5.17}$$

where θ is defined in (5.13).

The values of Δ_S^0 from (5.17) are presented in Table 1.

It follows from Table 1 that for our simplest example the estimate (2.11) provides a good result: if $\theta > 0.25$, then the Kalman filter is a virtually optimal estimator; for other values of θ, the Kalman filter accuracy is also acceptable.

Let us treat two examples with discrete measurements discussed in Section 6 of Chapter 3. Here we consider the case that white noises are also present in

TABLE 1. Values of Δ_S^0 for Example 5.2

θ	0.0	0.25	1.0	∞
Δ_S^0	1.50	1.13	1.03	1.0

TABLE 2. Values of Δ_{SD}^0 for $\beta_0 = 1$, $\beta_1 = 1$, $\beta_2 = \Delta t$

	$\gamma = 0.08$ $q^{\frac{1}{2}} = 7 \cdot 10^{-3}$	$\gamma = 0.08$ Deterministic disturbances	$\gamma = 0.4$ $q^{\frac{1}{2}} = 4 \cdot 10^{-2}$	$\gamma = 0.4$ Deterministic disturbances
x_1	2.40	2.68	2.65	3.02
x_2	2.39	2.74	2.18	2.79
x_3	2.62	2.98	2.91	3.45
x_4	2.28	2.52	1.71	2.15

disturbances. The corresponding general theory is expounded in Section 4. We shall restrict our consideration to the final numerical results.

EXAMPLE 5.3. Consider the correction problem for the inertial navigation system with the aid of a discrete position information (see Example 6.3 of Chapter 3).

In accordance with Example 6.3 of Chapter 3, we put

$$\sigma = 0.75 \cdot 10^{-2}, \qquad \sigma_{-1} = 1.2, \quad \sigma_{-2} = 1.2, \quad \sigma_{-3} = 0.6, \quad \sigma_{-4} = 0.8,$$

$$c_d^{\frac{1}{2}} = \sigma_{-d}, \quad d = 1, \ldots, 4, \qquad r^{\frac{1}{2}} = \sigma,$$

$$T = \frac{\pi}{2}, \qquad N = 420,$$

(a) $\gamma = 0.08$, $q^{\frac{1}{2}} = 7 \cdot 10^{-3}$; (b) $\gamma = 0.4$, $q^{\frac{1}{2}} = 4 \cdot 10^{-2}$;

$$\beta_0 = 1, \qquad \beta_1 = 1, \qquad \beta_2 = \Delta t = \frac{T}{N}. \tag{5.18}$$

The nonoptimality level of the Kalman filter in the stochastic guaranteed estimation problem is computed in accordance with Theorem 4.2. The values of Δ_{SD}^0 are displayed in Table 2 for two values of γ.

Except for the set (5.18), the computations were carried out for the following values of parameters:

$$\beta_0 = 10^{-2}, 10^{-1}, 10, \ 10^2,$$

TABLE 3. The influence of white noise in Example 5.3.

	x_1	x_2	x_3	x_4
1:2	2.34	1.90	2.45	1.72
1:10	1.34	1.28	1.28	1.28

$$\beta_0^{-1}\beta_1 = 10^{-2}, 10^{-1}, 10,$$

$$\beta_0^{-1}\beta_2 = 10^{-2}\Delta t,\ 10^{-1}\Delta t,\ 10\Delta t,\ 10^2\Delta t.$$

Nevertheless, the results were not improved significantly.

Note that the influence of the uncertain deterministic components of disturbances (for example, for measurement noise) is greater than the influence of the white noise components (with the same weight coefficients: $\sigma \approx r^{\frac{1}{2}}$). In other words, the nonsmooth part of the functional $\mathcal{J}_D$ is greater that the quadratic smooth part. This is due to the fact that the spectrum of the uncertain deterministic components has the "worst" nature. Therefore the presence of white noise does not influence strongly in the level of nonoptimality in Table 2.

The values of Δ_{SD}^0 decrease substantially when the influence of the white noise components is comparable with the influence of the uncertain deterministic components. Consider the modifications of case (b). Let the bounds of all uncertain deterministic components be two and ten times less than these in case (b). The corresponding values of Δ_{SD}^0 are shown in Table 3.

EXAMPLE 5.4. Consider the oscillating system from Example 6.4 of Chapter 3. Let us treat the following cases:

(a) $\sigma_- = 100, \quad \gamma = 1, \quad \sigma = 0.2;$

(b) $\sigma_- = 50, \quad \gamma = 0.5, \quad \sigma = 0.1;$

(c) $\sigma_- = 10, \quad \gamma = 0.1, \quad \sigma = 0.02.$

For cases (a) – (c) we put

$$c_{-d} = c = 100, \quad d = 1, \ldots, 8, \qquad q_j^{\frac{1}{2}} = q^{\frac{1}{2}} = 0.35, \quad j = 1, \ldots, 4,$$

$$r^{\frac{1}{2}} = 0.2, \qquad T = 2\pi, \qquad N = 100,$$

$$\beta_0 = 1, \qquad \beta_1 = 1, \qquad \beta_2 = \Delta t = \frac{T}{N}.$$

TABLE 4. The influence of white noise in Example 5.4

	x_1	x_2	x_3	x_4	x_5	x_6	x_7	x_8
Ex.6.4 of Ch.3	2.77	2.82	2.57	3.09	2.59	3.21	3.53	3.92
(a)	2.05	2.49	2.25	2.42	2.24	2.33	2.87	2.78
(b)	1.81	2.12	1.90	1.99	1.93	1.92	2.27	2.18
(c)	1.13	1.14	1.12	1.13	1.13	1.13	1.14	1.13

Theorem 4.2 determines the algorithm for computing Δ^0_{SD}. The values of Δ^0_{SD} are displayed in Table 4. In the first row of the table, the results for the pure deterministic disturbances from Example 6.4 of Chapter 3 are cited (the corresponding values of the bounds are the same ones as the values of deterministic bounds for case (a)).

In order to improve the estimate Δ^0_{SD} for case (a), the values of the free parameters were varied:

$$\beta_0 = 10^{-2}, 10^{-1}, 10,$$

$$\beta_0^{-1}\beta_1 = 10^{-1}, 10,$$

$$\beta_0^{-1}\beta_2 = 10^{-6}\Delta t,\ 10^{-1}\Delta t,\ \Delta t,\ 10\Delta t.$$

The values of Δ^0_{SD} were not reduced decidedly.

Let us consider two cases of "large" and "small" dynamic perturbations:

$$\sigma_- = c^{\frac{1}{2}} = 100, \qquad \sigma = r^{\frac{1}{2}} = 0.2, \qquad \begin{array}{ll}\text{(d)}\ \gamma = 10^2, & q^{\frac{1}{2}} = 35,\\ \text{(e)}\ \gamma = 10^{-2}, & q^{\frac{1}{2}} = 3.5 \cdot 10^{-3},\end{array}$$

$$\beta_0 = 1, \qquad \beta_1 = 1, \qquad \beta_2 = \Delta t.$$

The corresponding results are presented in Table 5. For comparison, the values of Δ^0_D for the pure deterministic disturbances from Example 6.4 of Chapter 3 (Table 5) are also cited.

It follows from Examples 5.3 and 5.4 that the values of Δ^0_{SD} are not critical and, in number of cases, the presence of white noise decreases Δ^0_{SD} in comparison with the corresponding deterministic versions of Chapter 3. If the levels of white noises are relatively large, then the values of Δ^0_{SD} may come close to unity (the Kalman filter is an optimal estimation algorithm for the case of white noises). Moreover, the choice of free parameters $\beta_0 = 1$, $\beta_1 = 1$, $\beta_2 = \Delta t$ turned out to be appropriate that coincides with the analysis of Chapter 3. By analogy with Remark 6.1 of Chapter 3, the estimate for the level of nonoptimality can be improved somewhat by a separation of

TABLE 5. "Large" and "small" disturbances in Example 5.4

	x_1	x_2	x_3	x_4	x_5	x_6	x_7	x_8
Ex.6.4 of Ch.3 ($\gamma = 10^2$)	2.87	2.77	2.39	3.10	2.46	3.29	3.72	4.52
(d)	1.90	2.40	2.10	2.12	2.13	2.05	2.71	2.61
Ex.6.4 of Ch.3 ($\gamma = 10^{-2}$)	1.49	1.45	1.48	1.46	1.48	1.46	1.45	1.45
(e)	1.40	1.39	1.41	1.41	1.40	1.41	1.37	1.38

the procedures for finding the best estimator and for the obtaining the lower bound for the unknown optimal estimation accuracy.

Let us now consider a real complicated guaranteed estimation problem.

EXAMPLE 5.5: *Topographical Referencing Problem.*

One way for the topographical referencing of objects in undeveloped districts is to use a helicopter with an on-board navigation equipment. A helicopter takes off a point with well-known coordinates and is headed towards the object that is to be located. The coordinates of the object are indicated by an inertial navigation system at a terminal instant when the helicopter lands at the object.

However, the inertial navigation system has growing in time errors. In particular, the uncertain bounded components of accelerometer noises (besides white noise components) considerably affect the accuracy of the inertial navigation system (the contribution of the accelerometer noises to final error can reach 70 percent). Therefore the navigation system must be corrected externally.

To obtain the correcting velocity information, the helicopter may make short landings along its route. In these instants, the relative velocity of the helicopter is equal to zero that forms a correcting signal. The advantage of this correction method is a full autonomy, the drawback is that the frequent landinds are difficult to follow by a pilot.

To obtain the correcting position information, the navigation satellites may be used, which orbits are known with high precision. On the one hand, the distances to the satellites can be computed by the readings of the navigation system; on the other hand, these distances can be measured directly by a radio-navigation hardware. The difference between these signals characterizes the errors of the inertial navigation system. The invoking of the satellite information enchances the accuracy of correction.

In both cases, the main aim is to find optimal guaranteed estimates for the errors of the helicopter location at a terminal instant.

The system of equations for navigation errors (in dimensionless variables) has the form [177]

$$\frac{d\Gamma_1}{dt} = \omega_3\Gamma_2 + \pi_1 + \omega_2\beta_3, \qquad \frac{d\Gamma_2}{dt} = -\omega_3\Gamma_1 + \pi_2 - \omega_1\beta_3, \tag{5.19}$$

$$\frac{d\pi_1}{dt} = -\Gamma_1 - \beta_1 + \omega_3\pi_2 - \epsilon_1(t), \qquad \frac{d\pi_2}{dt} = -\Gamma_2 - \beta_2 - \omega_3\pi_1 + \epsilon_2(t),$$

$$\frac{d\beta_1}{dt} = \omega_3\beta_2 - \omega_2\beta_3 + \nu_1, \qquad \frac{d\nu_1}{dt} = 0,$$

$$\frac{d\beta_2}{dt} = \omega_1\beta_3 - \omega_3\beta_1 + \nu_2, \qquad \frac{d\nu_2}{dt} = 0,$$

$$\frac{d\beta_3}{dt} = \omega_2\beta_1 - \omega_1\beta_2 + \nu_3, \qquad \frac{d\nu_3}{dt} = 0,$$

$$\frac{d\Delta H}{dt} = \pi_3, \qquad \frac{d\pi_3}{dt} = 2\omega_1\pi_1 + 2\omega_2\pi_2 + \epsilon_{30} + \epsilon_3(t), \qquad \frac{d\epsilon_{30}}{dt} = 0,$$

$$t \in [0, T].$$

Here Γ_1, Γ_2 are the angle positioning errors ($a_E\Gamma_1$ and $a_E\Gamma_2$ are the lengthwise positioning errors, where a_E is the semimajor axis of the Earth ellipsoid); $\beta_1 = \tilde{\beta}_1 - \frac{\tilde{\epsilon}_{10}}{g}$, $\beta_2 = \tilde{\beta}_2 + \frac{\tilde{\epsilon}_{20}}{g}$, $\beta_3 = \tilde{\beta}_3$, where $\tilde{\beta}_s$ $(s = 1, 2, 3)$ are the components of the kinematic error of the inertial navigation system; $\tilde{\epsilon}_{10}$, $\tilde{\epsilon}_{20}$ are the constant errors of the horizontal accelerometers; g is the acceleration of gravity; $\pi_1 = -\frac{\delta v_2}{a_E\omega_0}$, $\pi_2 = \frac{\delta v_1}{a_E\omega_0}$, where δv_1, δv_2 are the dynamic errors in velocity; $\omega_0 = \sqrt{\frac{g}{a_E}}$ is the Schuler frequency; $\omega_s = \frac{\tilde{\omega}_s}{\omega_0}$, where $\tilde{\omega}_s$ $(s = 1, 2, 3)$ are the projections of the absolute angular velocity of the moving trihedron on the axes of the moving trihedron (these angular velocities are determined by a specified reference trajectory);

$$\nu_1 = \frac{\tilde{\nu}_1 - \tilde{\omega}_3^0\tilde{\epsilon}_{20}/g}{\omega_0}, \quad \nu_2 = \frac{\tilde{\nu}_2 + \tilde{\omega}_3^0\tilde{\epsilon}_{10}/g}{\omega_0}, \quad \nu_3 = \frac{\tilde{\nu}_3 - \left(\tilde{\omega}_2^0\tilde{\epsilon}_{10} + \tilde{\omega}_1^0\tilde{\epsilon}_{20}\right)/g}{\omega_0},$$

where $\tilde{\omega}_i^0$ are the values of $\tilde{\omega}_s(t)$ $(s = 1, 2, 3)$ at the start point; $\tilde{\nu}_s$ $(s = 1, 2, 3)$ are the constant components of the gyroscopic drift; $\Delta H = \frac{\Delta\tilde{H}}{a_E}$, where $\Delta\tilde{H}$ is the altitude error; $\pi_3 = \frac{\Delta\tilde{V}_3}{a_E\omega_0}$, where $\Delta\tilde{V}_3$ is the error in vertical velocity; $\epsilon_{30} = \frac{\tilde{\epsilon}_{30}}{g}$, where $\tilde{\epsilon}_{30}$ is the constant error of vertical accelerometer; $t = \omega_0\tilde{t}$, where $\tilde{t}$ is the dimensional time; $T = \omega_0\tilde{T}$, where $\tilde{T}$ is the dimensional time of correction; $\epsilon_s(t) = \frac{\tilde{\epsilon}_s(t)}{g}$, where $\tilde{\epsilon}_s(t)$ $(s = 1, 2, 3)$ describe the instability of accelerometers' errors [177].

So, the equations (5.19) can be represented in the form

$$\frac{dx}{dt} = Ax + Bu(t), \qquad t \in [0,T], \tag{5.20}$$

where

$$x = \left(\Gamma_1, \Gamma_2, \pi_1, \pi_2, \beta_1, \beta_2, \beta_3, \nu_1, \nu_2, \nu_3, \Delta H, \pi_3, \epsilon_{30}\right)' \in \mathbf{R}^{13},$$

$$u(t) = \left(\epsilon_1(t), \epsilon_2(t), \epsilon_3(t)\right)' \in \mathbf{R}^3,$$

$$A = \left(\begin{array}{cccc:ccc:ccc:ccc}
0 & \omega_3 & 1 & 0 & 0 & 0 & \omega_2 & 0 & 0 & 0 & 0 & 0 & 0 \\
-\omega_3 & 0 & 0 & 1 & 0 & 0 & -\omega_1 & 0 & 0 & 0 & 0 & 0 & 0 \\
-1 & 0 & 0 & \omega_3 & -1 & 0 & 0 & 0 & 0 & 0 & 0 & 0 & 0 \\
0 & -1 & -\omega_3 & 0 & 0 & -1 & 0 & 0 & 0 & 0 & 0 & 0 & 0 \\
\hdashline
0 & 0 & 0 & 0 & 0 & \omega_3 & -\omega_2 & 1 & 0 & 0 & 0 & 0 & 0 \\
0 & 0 & 0 & 0 & \omega_3 & 0 & \omega_1 & 0 & 1 & 0 & 0 & 0 & 0 \\
0 & 0 & 0 & 0 & \omega_2 & -\omega_1 & 0 & 0 & 0 & 1 & 0 & 0 & 0 \\
\hdashline
0 & 0 & 0 & 0 & 0 & 0 & 0 & 0 & 0 & 0 & 0 & 0 & 0 \\
0 & 0 & 0 & 0 & 0 & 0 & 0 & 0 & 0 & 0 & 0 & 0 & 0 \\
0 & 0 & 0 & 0 & 0 & 0 & 0 & 0 & 0 & 0 & 0 & 0 & 0 \\
\hdashline
0 & 0 & 0 & 0 & 0 & 0 & 0 & 0 & 0 & 0 & 0 & 1 & 0 \\
0 & 0 & 2\omega_1 & 2\omega_2 & 0 & 0 & 0 & 0 & 0 & 0 & 0 & 0 & 1 \\
0 & 0 & 0 & 0 & 0 & 0 & 0 & 0 & 0 & 0 & 0 & 0 & 0
\end{array}\right),$$

and

$$B = \begin{pmatrix}
0 & 0 & -1 & 0 & 0 & 0 & 0 & 0 & 0 & 0 & 0 & 0 & 0 \\
0 & 0 & 0 & 1 & 0 & 0 & 0 & 0 & 0 & 0 & 0 & 0 & 0 \\
0 & 0 & 0 & 0 & 0 & 0 & 0 & 0 & 0 & 0 & 0 & 1 & 0
\end{pmatrix}'.$$

Note that the first ten equations of (5.19) ((5.20)), which are called the *equations of horizontal errors,* are separated from the last three ones, which are called the *equations of altitude channel*.

Our aim is to evaluate $\Gamma_1(T) = x_1(T)$, $\Gamma_2(T) = x_2(T)$, and $\Delta H(T) = x_{11}(T)$ with the help of an external information about the phase vector.

Let us write the equations for the external correcting information.

• *Velocity information*

At the moment of a landing, the relative velocity of the helicopter (relative to the Earth) is equal to zero. The errors in the determining of the relative velocity $\Delta\tilde{V}_1$, $\Delta\tilde{V}_2$, $\Delta\tilde{V}_3$ at the moment of landing are expressed by the formulas

$$\Delta\tilde{V}_s = a_E\,\omega_0\,\Delta V_s, \qquad s = 1,2,3, \tag{5.21}$$

$$\Delta V_1 = -u_3\Gamma_1 + \pi_2 - u_1\beta_3,$$

$$\Delta V_2 = -u_3\Gamma_2 - \pi_1 - u_2\beta_3,$$

$$\Delta V_3 = \pi_3,$$

where $u = (u_1, u_2, u_3)' = \frac{\tilde{u}}{\omega_0}$ and $\tilde{u} = (\tilde{u}_1, \tilde{u}_2, \tilde{u}_3)'$ is the angular velocity of the Earth in the projection on the moving trihedron. Thus, at the i-th moment of landing, the velocity measurements have the form

$$z_s^V(t_i) = \Delta V_s(t_i) + \varrho_s^V(t_i), \qquad i = 0, \dots, N_V, \qquad s = 1,2,3, \tag{5.22}$$

where $\Delta V_s(t_i)$ are defined by (5.21), $\varrho_s^V(t_i)$ are the measurement errors (adjusted to scale).

Suppose the landings are carried out through the equal time moments, i.e.,

$$t_i = i\,\Delta t_V, \qquad i = 0, \dots, N_V, \qquad t_0 = 0, \qquad t_{N_V} = T.$$

The velocity measurements (5.22) can be presented in a standard form

$$z^V(t_i) = H_V' x(t_i) + \varrho^V(t_i), \qquad i = 0, \dots, N_V, \tag{5.23}$$

$$H_V = \begin{pmatrix} -u_3 & 0 & 0 & 1 & 0 & 0 & -u_1 & 0 & 0 & 0 & 0 & 0 & 0 \\ 0 & -u_3 & -1 & 0 & 0 & 0 & -u_2 & 0 & 0 & 0 & 0 & 0 & 0 \\ 0 & 0 & 0 & 0 & 0 & 0 & 0 & 0 & 0 & 0 & 0 & 1 & 0 \end{pmatrix}',$$

$$\varrho^V(t_i) = \left(\varrho_1^V(t_i),\, \varrho_2^V(t_i),\, \varrho_3^V(t_i)\right)', \qquad z^V(t_i) = \left(z_1^V(t_i),\, z_2^V(t_i),\, z_3^V(t_i)\right)'.$$

• *Satellite information*

Suppose three navigational satellites move in pricisely known orbits. Denote by M and M_0 the center of mass of the helicopter and the location of this center of mass computed by inertial navigation system, respectively. Let $S_{(j)}$ be the exact location of the j-th satellite ($j = 1,2,3$). Then $\overrightarrow{\Delta r} = \overrightarrow{MM_0}$ is the vector of positioning error, $\overrightarrow{R_{(j)}} = \overrightarrow{MS_{(j)}}$ is the vector "helicopter –

j-th satellite", $\overrightarrow{O_{(j)}} = \overrightarrow{M_0 S_{(j)}}$ is an on-board numerical image of the vector "helicopter – j-th satellite". Obviously,

$$\overrightarrow{R_{(j)}} = \overrightarrow{O_{(j)}} + \overrightarrow{MM_0}.$$

Taking the square of the last vector equality and linearizing the right-hand side in $\overrightarrow{\Delta r}$, we get

$$\text{dist}^R_{(j)} = \text{dist}^O_{(j)} + \left(\text{dist}^O_{(j)}\right)^{-1} \left(O_{(j)1}\Delta r_1 + O_{(j)2}\Delta r_2 + O_{(j)3}\Delta r_3\right), \quad (5.24)$$

where

$$\text{dist}^R_{(j)} = \sqrt{R^2_{(j)1} + R^2_{(j)2} + R^2_{(j)3}}, \qquad \text{dist}^O_{(j)} = \sqrt{O^2_{(j)1} + O^2_{(j)2} + O^2_{(j)3}}$$

and the vectors

$$\overrightarrow{R_{(j)}} = \left(R_{(j)1}, R_{(j)2}, R_{(j)3}\right)', \qquad \overrightarrow{O_{(j)}} = \left(O_{(j)1}, O_{(j)2}, O_{(j)3}\right)',$$

$$\overrightarrow{\Delta r} = \left(\Delta r_1, \Delta r_2, \Delta r_3\right)'$$

are given by their projections onto the axes of the moving trihedron.

The values of $O_{(j)1}$, $O_{(j)2}$, $O_{(j)3}$, and $\text{dist}^O_{(j)}$ are computed directly by the readings of the inertial navigation system and by the exact orbits of the satellites. The values of $\text{dist}^R_{(j)}$ are measured by a specialized receiver. This measurements yield the values $\text{dist}_{(j)}$. A specific nature of these measurements is such that $\text{dist}_{(j)}$ contain a substantial additive noise component which is common for all the satellites. Therefore, in order to form the position satellite correcting measurement $\tilde{z}^P(t_k)$ at the instant t_k, we use the measurements for a pair of satellites (for example, for the first and second one) that are subtracted then:

$$\tilde{z}^P_{1,2}(t_k) = \left[\text{dist}_{(2)}(t_k) - \text{dist}^O_{(2)}(t_k)\right] - \left[\text{dist}_{(1)}(t_k) - \text{dist}^O_{(1)}(t_k)\right]. \quad (5.25)$$

We assume further that the satellite measurements are equidistant, i.e.,

$$t_k = k\,\Delta t_P, \qquad k = 0, \ldots, N_P, \qquad t_0 = 0, \qquad t_{N_P} = T.$$

Taking into account that, in the projection on the axes of the moving trihedron,

$$\overrightarrow{\Delta r} = (a_E\Gamma_2, -a_E\Gamma_1, a_E\Delta H)',$$

we get from (5.24), (5.25) the position measurement in the form

$$z^P_{1,2}(t_k) = H'_{1,2}(t_k)x(t_k) + \varrho^P_{1,2}(t_k), \qquad k = 0, \ldots, N_P, \quad (5.26)$$

where

$$z_{1,2}^P(t_k) = a_E^{-1} \tilde{z}_{1,2}^P(t_k),$$

$$H_{1,2}(t_k) = \left(\mathcal{H}_1^{1,2}(t_k), \mathcal{H}_2^{1,2}(t_k), 0, 0, 0, 0, 0, 0, 0, 0, \mathcal{H}_3^{1,2}(t_k), 0, 0\right)',$$

$$\mathcal{H}_1^{1,2}(t_k) = -\frac{O_{(1)2}(t_k)}{\text{dist}_{(1)}^O(t_k)} + \frac{O_{(2)2}(t_k)}{\text{dist}_{(2)}^O(t_k)}, \tag{5.27}$$

$$\mathcal{H}_2^{1,2}(t_k) = \frac{O_{(1)1}(t_k)}{\text{dist}_{(1)}^O(t_k)} - \frac{O_{(2)1}(t_k)}{\text{dist}_{(2)}^O(t_k)},$$

$$\mathcal{H}_3^{1,2}(t_k) = \frac{O_{(1)3}(t_k)}{\text{dist}_{(1)}^O(t_k)} - \frac{O_{(2)3}(t_k)}{\text{dist}_{(2)}^O(t_k)};$$

$\varrho_{1,2}^P(t_k)$ is the resulting (adjusted to scale) error in determining the distance. Obviously, the summands in (5.27) are the direction cosines of the directions "helicopter – satellite".

For another pair of satellites we have an equation that is analogous to (5.26):

$$z_{1,3}^P(t_k) = H_{1,3}'(t_k)x(t_k) + \varrho_{1,3}^P(t_k), \tag{5.28}$$

where the matrix $H_{1,3}(t_k)$ is analogous to the matrix $H_{1,2}(t_k)$.

Further we shall assume that all system perturbations and measurement noises are either uncertain bounded deterministic functions:

$$|x_d(0)| \le \sigma_{-d}, \quad d = 1, \ldots, 13, \qquad |\epsilon_s(t)| \le \gamma, \quad t \in [0, T], \quad s = 1, 2, 3,$$

$$|\varrho_s^V(t_i)| \le \sigma^V, \quad i = 0, \ldots, N_V, \quad s = 1, 2, 3,$$

$$|\varrho_{1,2}^P(t_k)| \le \sigma^P, \quad |\varrho_{1,3}^P(t_k)| \le \sigma^P, \qquad k = 0, \ldots, N_P,$$

or combined disturbances. So, the equations for the optimal guaranteed estimation problems are defined by (5.20), (5.23) or (5.20), (5.26), (5.28). For our purposes, the coefficients of matrices A and H_V may be taken constant (they depend on specific reference trajectory). In accordance with our approach we shall use the Kalman filter for approximate solving the guaranteed estimation problems; moreover we shall find the upper bound for the nonoptimality level of the Kalman filter. In this book, we are mainly interested in this nonoptimality level. The topographical referencing problem itself, in particular, the accuracy of the correction of the inertial navigation system (the guaranteed values of the estimation errors of $\Gamma_1(T)$ and $\Gamma_2(T)$) is discussed in [37]. So, let us turn to the evaluation of the levels of nonoptimality for various kinds of correcting information.

- *Estimate for level of nonoptimality — velocity information*

Consider the system for the "horizontal errors" defined by the first ten equations of (5.19) or (5.20). Suppose that at an initial instant we have a sufficiently precise knowledge about a location of the helicopter defined by $\Gamma_1(0)$, $\Gamma_2(0)$ and, during the motion, the information about the relative velocity is available. Then the correcting equations are defined by (5.23), where we use the first two equations for the horizontal errors in velocity ΔV_1 and ΔV_2. So, in general notation, $n = 10$, $m = 2$, $r = 2$, $a = (1, 0, \ldots, 0)'$, and $a = (0, 1, 0, \ldots, 0)'$. For our computation, we take the following values of parameters.
System coefficients:

$$\omega_1 = u_1 = 4.928 \cdot 10^{-2}, \quad \omega_2 = u_2 = 0, \quad \omega_3 = u_3 = 3.179 \cdot 10^{-2}; \qquad (5.29)$$

Bounds for initial uncertainty $(\times 10^6)$:

$$\sigma_{-1} = \sigma_{-2} = 0.8, \qquad \sigma_{-3} = \sigma_{-4} = 100.0, \qquad (5.30)$$

$$\sigma_{-5} = \sigma_{-6} = 300.0, \qquad \sigma_{-7} = 1000.0, \qquad \sigma_{-8} = \sigma_{-9} = \sigma_{-10} = 80.0;$$

Bounds for system noises $(\times 10^6)$:

$$\gamma_1 = \gamma_2 = \gamma = 10.0; \qquad (5.31)$$

Bounds for measurement noises $(\times 10^6)$:

$$\sigma_1 = \sigma_2 = \sigma^V = 0.65; \qquad (5.32)$$

Time interval between measurements:

$$\Delta\tilde{t}_V = 5\,\text{min}, \qquad \Delta t_V = \omega_0 \Delta\tilde{t}_V = 0.372;$$

Maximal time of correction and number of measurements:

$$\max \tilde{T} = 80\,\text{min}, \qquad \max T = \omega_0 \tilde{T} = 5.952, \qquad \max N_V = 16.$$

First let us consider the deterministic case (see Section 4 of Chapter 3 for details). The values of the upper bound of the nonoptimality level Δ_D^0 for the component $x_1(T) = \Gamma_1(T)$ are presented in Table 6. In accordance with the recommendation of Section 4 of Chapter 3 we put $\beta_2 = 0.4 \approx \Delta t_V$. It follows from Table 6 that, with $\beta_1 = 3$ and $\beta_2 = 0.4$, the values of Δ_D^0 are close to unity. So, the use of the Kalman filter is justified. The variation of β_2 about $\beta_2 = 0.4$ does not lead to the considerable decrease of Δ_D^0.

TABLE 6. Velocity correction: Δ_D^0 for deterministic case

$\tilde{T}$	N_V	$\beta_1 = 1$	$\beta_1 = 2$	$\beta_1 = 3$	$\beta_1 = 4$
40 min	8	1.83	1.55	1.36	1.43
60 min	12	1.85	1.56	1.35	1.42
80 min	16	1.89	1.56	1.34	1.42

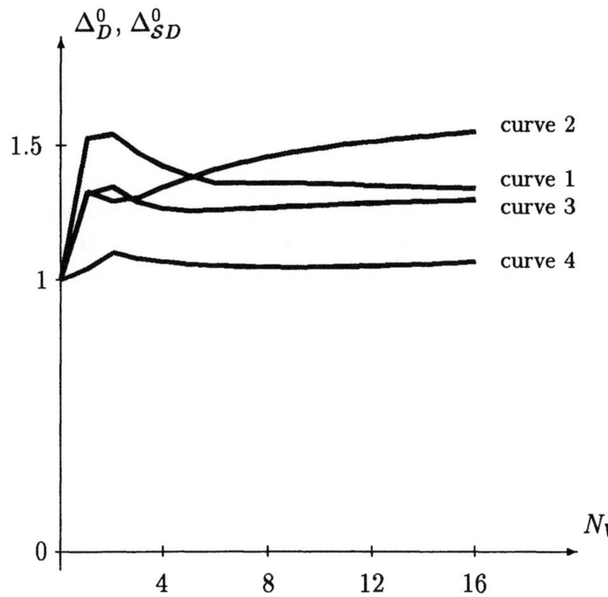

Fig. 1. Level of nonoptimality (velocity information)

The graph of Δ_D^0 as a function of N_V for $\beta_1 = 3$ and $\beta_2 = 0.4$ is depicted in Figure 1 (curve 1). We see from curve 1 that Δ_D^0 is quite stable (except for a small transient in amplitude and time).

The values of Δ_D^0 for the case of estimating $x_2(T) = \Gamma_2(T)$ coincide virtually with these for the case of estimating $x_1(T) = \Gamma_1(T)$.

Let us now consider the case of combined disturbances (see Section 4). In addition to the bounds of deterministic disturbances, we take the following values of the stochastic components of uncertainty.

Parameters of initial random values $(\times 10^6)$:

$$\sqrt{c_1} = \sqrt{c_2} = 0.8, \qquad \sqrt{c_3} = \sqrt{c_4} = 100.0, \tag{5.33}$$

$$\sqrt{c_5} = \sqrt{c_6} = 300.0, \qquad \sqrt{c_7} = 1000.0, \qquad \sqrt{c_8} = \sqrt{c_9} = \sqrt{c_{10}} = 80.0;$$

Bounds for intensities of system and measurement noises ($\times 10^6$):

$$\sqrt{q_1} = \sqrt{q_2} = 0.5, \qquad \sqrt{r_1} = \sqrt{r_2} = 0.65. \tag{5.34}$$

In other words, $\sqrt{c_d} = \sigma_{-d}$ $(d = 1, \ldots, 10)$ and $\sqrt{r_1} = \sqrt{r_2} = \sigma^V$.

In Figure 1, the values of Δ^0_{SD} as a function of N_V are shown for $\beta_0 = 1$, $\beta_1 = 3$, $\beta_2 = 0.4$ (curve 2)[5] and for $\beta_0 = 1$, $\beta_1 = 4$, $\beta_2 = 0.6$ (curve 3). It follows from curves 2 and 3 that, for $N_V \geq 3$, $\Delta^0_{SD} \leq 1.55$ and $\Delta^0_{SD} \leq 1.30$, respectively[6]. Thus the use of Kalman filter is justified here too.

Note that the contribution of the nonsmooth term in the functional $\mathcal{J}_D$ defined by (4.4) is considerably greater than the contribution of the quadratic one. Therefore the presence of white noise did not affect significantly the level of nonoptimality.

Let us study the case that all the bounds for the deterministic disturbances are lowered by ten times as compared with (5.30) – (5.32). The values of Δ^0_{SD} are shown in Figure 1 (curve 4). For almost all values of N_V ($N_V \geq 4$), $\Delta^0_{SD} \leq 1.07$. The guaranteed value of estimation error is reduced by six times, approximately. Even in this case of "small" deterministic noises, the contribution of the nonsmooth term is comparable with the contribution of the quadratic one.

As for the previous cases, the results for estimating $x_2(T) = \Gamma_2(T)$ are completely analogous to the results for $x_1(T) = \Gamma_1(T)$.

- *Estimate for level of nonoptimality — satellite information*

Consider the full system of the thirteenth order (5.19) (or (5.20)). Suppose now that the external position information is extracted from the measurements of the distances to navigation satellites. The corresponding correcting equations are defined by (5.26) – (5.28). The coefficients in (5.26) – (5.28) are determined by the specified orbits of satellites and the reference trajectory of the helicopter. We do not cite the corresponding explicit formulas (or numerical tables) because they are too cumbersome. We shall study two cases: a correction with the help of three and four satellites (a pair of satellites gives one measurement equation). In general notation, $n = 13$, $m = 2$ or $m = 3$, $r = 3$, $a = (1, 0, \ldots, 0)'$, $a = (0, 1, 0, \ldots, 0)'$ and $a = (0, \ldots, 1, 0, 0)'$. For our computation, the system coefficients are set by (5.29) and the bounds for deterministic uncertain disturbances are defined by (5.30), (5.31). Other bounds are set by the following equalities.

[5] For curve 2, β_1 and β_2 are the same as for the deterministic case considered above.

[6] The functions Δ^0_D and Δ^0_{SD} are defined on a discrete set of points but the graphs are drawn with continuous curves to be more visual.

Bounds for initial uncertainty in altitude channel $(\times 10^6)$:

$$\sigma_{-11} = 0.8, \qquad \sigma_{-12} = 0.65, \qquad \sigma_{-13} = 200.0; \tag{5.35}$$

Bound for system altitude noise $(\times 10^6)$:

$$\gamma_3 = 10.0; \tag{5.36}$$

Bounds for measurement noises $(\times 10^6)$:

$$\sigma_1 = \sigma_2 = \sigma_3 = \sigma^P = 0.78; \tag{5.37}$$

Time interval between measurements:

$$\Delta\tilde{t}_P = 3\,\text{min}, \qquad \Delta t_P = \omega_0 \Delta\tilde{t}_P = 0.223; \tag{5.38}$$

Maximal time of correction and number of measurements:

$$\max\tilde{T} = 81\,\text{min}, \qquad \max T = \omega_0\tilde{T} = 6.026, \qquad \max N_P = 27.$$

Similarly to the case of velocity correction, let us consider three versions for the characterization of disturbances:

d – all disturbances are deterministic with bounds of uncertainty (5.30),(5.31), (5.35) – (5.37);

c – the disturbances are combined; deterministic bounds are defined by (5.30), (5.31), (5.35) – (5.37); stochastic intensities are defined by the equalities: $\sqrt{c_d} = \sigma_{-d}$ $(d = 1, \ldots, 13)$, $\sqrt{q_j} = 0.5$ $(j = 1, 2, 3)$, $\sqrt{r_l} = \sigma^P$ $(l = 1, 2, 3)$ (cf. (5.33), (5.34));

cs – the disturbances are combined but the bounds for deterministic components are lowered by ten times as compared with version **c**.

Consider the case of estimating $x_1(T) = \Gamma_1(T)$. The values of Δ_D^0 and Δ_{SD}^0 were optimized in free parameters β_0, β_1, and β_2. They are shown in Figures 2 and 3 as a function of N_P. The quantity N_P is varied from 13 to 27 (that corresponds to the number of measurements from 14 to 28 or to the terminal time of observation from 40 min to 81 min). Curves 1, curves 2, and curves 3 correspond to version **d**, version **c**, and version **cs**, respectively. The optimal values of free parameters are presented in Table 7.

Note that the general recommendation on the choice of free parameters from Section 4 of Chapter 3 yielded a good reference point to start optimization. In all numerical experiments the sensitivity of Δ_D^0, Δ_{SD}^0 to the variation of β_0, β_1, β_2 in the neighborhood of a "successful" point was not sharp; moreover, the sensitivity of cost functional (which describes the guaranteed estimation error) was weak.

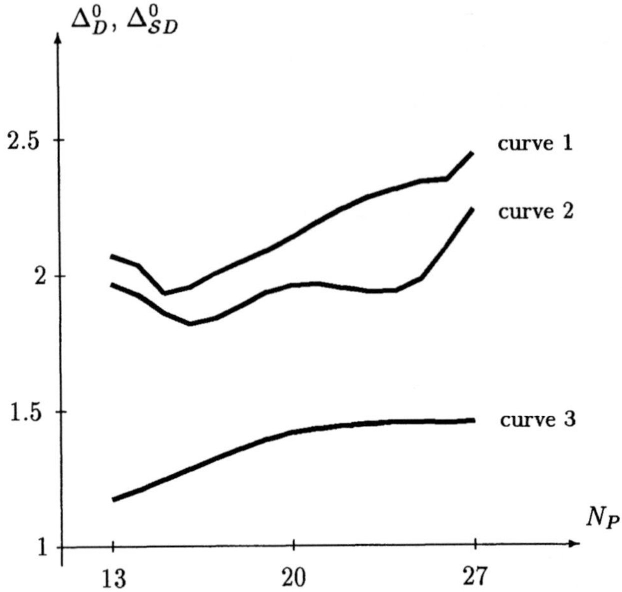

Fig. 2. Level of nonoptimality (satellite information; $m = 2$)

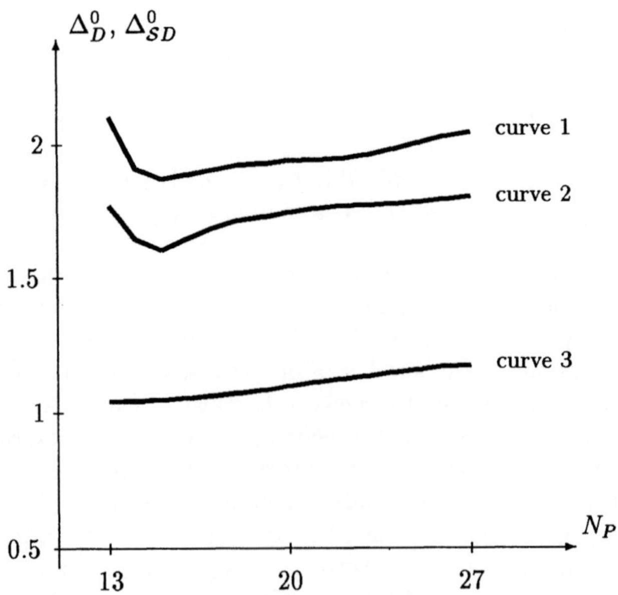

Fig. 3. Level of nonoptimality (satellite information; $m = 3$)

TABLE 7. Optimal values of free parameters

		d:	β_1	β_2	c:	β_0	β_1	β_2	cs:	β_0	β_1	β_2
$m=2$	Γ_1,Γ_2		1	0.4		1	0.5	0.5		1	1	5
	ΔH		1	0.8		1	1	1.5		1	1	5
$m=3$	Γ_1,Γ_2		1	3		1	2	4		1	1	5
	ΔH		1	2		1	2	4		1	1	5

For combined noises, the contribution of the nonsmooth term in the cost functional is essential, i.e., the influence of “small” uncertain deterministic disturbances may not be neglected. As for all previous cases, the values of Δ_D^0, Δ_{SD}^0 for $x_2(T) = \Gamma_2(T)$ and $x_{11}(T) = \Delta H(T)$ are close to these ones for $x_1(T) = \Gamma_1(T)$.

In summary, the optimal values of Δ_D^0 and Δ_{SD}^0 for $40\,\text{min} < T < 80\,\text{min}$ fall in the following intervals:
velocity correction — $1.07 \div 1.36$;
satellite correction ($m = 2$) — $1.17 \div 2.54$;
satellite correction ($m = 3$) — $1.03 \div 2.18$.
Thus the Kalman filter may be used for solving the topographical referencing problem. Optimal guaranteed estimation algorithms will not provide the striking improvement of estimation accuracy.

The examples discused above show that the estimate of nonoptimality level is useful for analyzing the estimation problems.

6. Kalman Filter under Uncertainty in Intensities of Noises

In previous sections we studied the level of nonoptimality for simplified estimation algorithms in guaranteed estimation problems. Our approach, which is based on the duality theory, can be successfully applied to other classes of estimation problems, in particular, to the classical mean-square filtering problem.

The optimal mean-square estimate for the state of linear dynamic system is provided by the Kalman filter [84], [85], [79], [143], [47], [187]. When using the Kalman filter, we have to set the exact values of intensities of white noises. In applied problems, an uncertainty in these intensities is always present. So, we have to use a designed Kalman filter with some designed noise intensities. The true level of nonoptimality of the designed filter can be easily calculated provided the true intensities are known. In the presence of the uncertainty the performance of the designed filter can not be evaluated because of very complicated dependence of this performance on true intensities (especially for time-varying intensities).

In this section an upper bound for the level of nonoptimality of the designed filter is obtained. This upper bound has a relatively simple dependence on the true intensities. Moreover, this bound can be efficiently computed basing on the designed filter parameters only. Therefore, this estimate can be useful for the analysis of the Kalman filter sensitivity. In particular, it can be sucessfully used for the optimization over uncertain elements for the evaluation of a guaranteed sensitivity of the designed filter. Let us now discuss the problem in detail.

Consider a linear stochastic dynamic system (cf. Remark 1.1)

$$\dot{x}(t) = A(t)x(t) + B(t)u(t), \quad x(t) \in \mathbf{R}^n, \quad u(t) \in \mathbf{R}^r, \quad t \in [0, T],$$

where $x(0)$ is a random variable such that

$$\mathbf{E}x(0) = 0, \qquad \mathbf{E}x(0)x'(0) = P_0 > 0,$$

and $u(t)$ is a white noise process:

$$\mathbf{E}u(t) = 0, \qquad \mathbf{E}u(t)u'(s) = Q_0(t)\delta(t-s), \quad Q_0(t) \geq Q_0 > 0.$$

Suppose that the measurements of the state vector are available on $[0, T]$:

$$z(t) = H'(t)x(t) + \varrho(t), \qquad z(t), \varrho(t) \in \mathbf{R}^m,$$

where $\varrho(t)$ is also a white noise processes:

$$\mathbf{E}\varrho(t) = 0, \qquad \mathbf{E}\varrho(t)\varrho'(s) = R_0(t)\delta(t-s), \quad R_0(t) \geq R_0 > 0.$$

Assume that $x(0)$, $u(t)$, and $\varrho(t)$ are mutually independent. The matrix P_0 is the true initial covariance; the matrices $R_0(t)$ and $Q_0(t)$ are the true white noise intensities. The elements of all time-varying matrices mentioned above are assumed to be piecewise-continuous. The notation $K > 0$ (≥ 0) means that K is positively definite (semi-definite). The problem is to find an estimate for a scalar quantity $l_* = a'x(T)$, where $a \in \mathbf{R}^n$ is a given vector.

Let us use a class of linear estimators

$$\hat{l} = \int_0^T \Phi'(t)z(t)\,dt, \qquad \Phi(t) \in \mathbf{R}^m, \qquad \Phi \in L_2^m.$$

The estimation quality is defined by the second moment of the estimation error

$$\mathcal{D}(\Phi) = \mathbf{E}(\hat{l} - l_*)^2.$$

The classical optimal filtering problem is to find an optimal estimator Φ^0 that minimizes the second moment of the estimation error:

$$\mathcal{D}(\Phi^0) = \inf_{\Phi \in L_2^m} \mathcal{D}(\Phi).$$

Similar to previous chapters let us introduce new variables

$$y(t) = X^{-1}(t)x(t), \qquad h(t) = X'(t)H(t),$$

$$b(t) = X^{-1}(t)B(t), \qquad \tilde{a} = X'(T)a,$$

where $X(t)$ is a fundamental matrix of the homogeneous equation

$$\dot{x}(t) = A(t)x(t), \qquad X(0) = E_n, \quad E_n \text{ is identity matrix.}$$

Then the dynamic system and the measurements take the form

$$\dot{y}(t) = b(t)u(t), \qquad y(t) \in \mathbf{R}^n, \quad y(0) = x(0), \qquad t \in [0, T],$$

$$z(t) = h'(t)y(t) + \varrho(t), \qquad\qquad l_* = \tilde{a}'y(T).$$

Arguing as in the proof of Lemma 1.1, it can easily be checked that the second moment of the estimation error it determined by the formula

$$\mathcal{D}(\Phi) = J^2(\Phi_-(\Phi), \Phi, w(\Phi); P_0, Q_0, R_0),$$

where

$$J(\Phi_-, \Phi, w; P_0, Q_0, R_0) \tag{6.1}$$

$$= \left\{ \Phi_-' P_0 \Phi_- + \int_0^T \Phi'(t) R_0(t) \Phi(t)\, dt + \int_0^T w'(t) Q_0(t) w(t)\, dt \right\}^{\frac{1}{2}}$$

and

$$\Phi_- = \tilde{a} - \int_0^T h(t)\Phi(t)\, dt, \tag{6.2}$$

$$w(t) = b'(t)\left(\Phi_- + \int_0^t h(s)\Phi(s)\, ds \right).$$

Thus the optimal filtering problem reduces to the following variational problem

$$J_0(P_0, Q_0, R_0) = \inf_{\Phi_- \in \mathbf{R}^n,\, \Phi \in L_2^m,\, w \in L_2^r} J(\Phi_-, \Phi, w; P_0, Q_0, R_0) \tag{6.3}$$

under constraints (6.2).

If (Φ_-^0, Φ^0, w^0) is the solution of (6.2), (6.3), then Φ^0 is the optimal estimator and

$$\hat{l}_0 = \int_0^T \Phi^{0\prime}(t) z(t)\, dt$$

is the optimal estimate. By virtue of Theorem 2.1 of Chapter 3, Φ^0 is the Kalman estimator and the Kalman filter is the optimal estimation algorithm:

$$\hat{l}_0 = a'\hat{x}(T),$$

$$\dot{\hat{x}}(t) = A(t)\hat{x}(t) + K(t)\big(z(t) - H'(t)\hat{x}(t)\big), \qquad \hat{x}(0) = 0,$$

$$K(t) = P(t)H(t)R_0^{-1}(t),$$

$$\dot{P}(t) = A(t)P(t) + P(t)A'(t) - P(t)H(t)R_0^{-1}(t)H'(t)P(t) + B(t)Q_0(t)B'(t), \qquad P(0) = P_0.$$

In contrast to the classical problem statement, let the true values of P_0, $Q_0(t)$, and $R_0(t)$ be unknown but belong to specified sets:

$$P_0 \in \mathcal{P}, \qquad Q_0(t) \in \mathcal{Q}, \qquad R_0(t) \in \mathcal{R}. \tag{6.4}$$

In this case, the Kalman filter can not be implemented. In order to construct the desired estimate by a designed Kalman filter, let us set some designed values P, $Q(t)$, and $R(t)$ for the intensities. Then, instead of the optimal estimate $\hat{l}_0$, we have a "suboptimal" estimate

$$\hat{l} = \int_0^T \varphi'(t) z(t)\, dt,$$

which is provided by the designed Kalman filter

$$\hat{l} = a'\hat{x}(T), \tag{6.5}$$

$$\dot{\hat{x}}(t) = A(t)\hat{x}(t) + K(t)\big(z(t) - H'(t)\hat{x}(t)\big), \qquad \hat{x}(0) = 0,$$

$$K(t) = P(t)H(t)R^{-1}(t),$$

$$\dot{P}(t) = A(t)P(t) + P(t)A'(t) - P(t)H(t)R^{-1}(t)H'(t)P(t) + B(t)Q(t)B'(t), \qquad P(0) = P.$$

Similarly to (6.1) – (6.3), the designed Kalman estimator is determined by the solution $\{\varphi_-, \varphi, w_\varphi\}$ of the variational problem

$$J_0(P, Q, R) = \inf_{\Phi_- \in \mathbf{R}^n,\, \Phi \in L_2^m,\, w \in L_2^r} J(\Phi_-, \Phi, w; P, Q, R) \tag{6.6}$$

under constraints (6.2).

We define the sensitivity of the designed Kalman filter by the ratio

$$\Delta_K = \left(\frac{\mathcal{D}(\varphi)}{\mathcal{D}(\Phi^0)}\right)^{\frac{1}{2}} = \frac{J(\varphi_-, \varphi, w_\varphi; P_0, Q_0, R_0)}{J(\Phi^0_-, \Phi^0, w^0; P_0, Q_0, R_0)}, \qquad \Delta_K \geq 1. \tag{6.7}$$

The values of $\mathcal{D}(\varphi)$ and $\mathcal{D}(\Phi^0)$ can be calculated given P, $Q(t)$, $R(t)$, P_0, $Q_0(t)$, and $R_0(t)$. Howerever, the dependence of Δ_K on P_0, $Q_0(t)$, and $R_0(t)$ is very complex. Therefore the analysis of the quantity Δ_K, in particular, the maximization in (6.7) over P_0, $Q_0(t)$, $R_0(t)$ under restrictions (6.4) is a very difficult problem, especially for time-varying $Q_0(t)$ and $R_0(t)$.

Our aim is to construct an upper bound Δ_K^0 for Δ_K such that the dependence of Δ_K^0 on P_0, $Q_0(t)$, $R_0(t)$ is relatively simple and suitable for the Kalman filter sensitivity analysis.

It follows from the results of Section 2 of Chapter 3, Theorem 2.1, and formulas (2.14) – (2.17) that the solution of (6.6), (6.2) has the form

$$\varphi_- = \xi(0), \tag{6.8}$$

$$\varphi(t) = K'(t)\xi(t) = R^{-1}(t)H'(t)\eta(t),$$

$$w_\varphi(t) = B'(t)\xi(t),$$

where $\xi(t)$ and $\eta(t)$ are determined by the boundary value problem

$$\dot{\xi}(t) = -A'(t)\xi(t) + H(t)R^{-1}(t)H'(t)\eta(t), \qquad \xi(T) = a,$$

$$\dot{\eta}(t) = B(t)Q(t)B'(t)\xi(t) + A(t)\eta(t), \qquad \eta(0) = P\xi(0).$$

Moreover,

$$\dot{\xi}(t) = -(A(t) - K(t)H'(t))'\xi(t), \qquad \xi(T) = a,$$

and

$$\dot{\xi}(t) = -A'(t)\xi(t) + H(t)\varphi(t), \qquad \xi(T) = a,$$

where $K(t)$ and $P(t)$ are defined by (6.5).

Two simple schemes for computing $\{\varphi_-, \varphi, w_\varphi\}$ are discussed in Section 2 of Chapter 3. The main result of this section is the following theorem.

THEOREM 6.1. *The level of nonoptimality for the designed Kalman filter* Δ_K *satisfies the inequalites*

$$1 \leq \Delta_K \leq \Delta_K^0, \tag{6.9}$$

$$\Delta_K^0 = \frac{\mathcal{N}^{1/2}(U, V(t), W(t)) \cdot \mathcal{N}^{1/2}(U^{-1}, V^{-1}(t), W^{-1}(t))}{\mathcal{N}_0},$$

where[7]

$$\mathcal{N}(U, V(t), W(t)) \tag{6.10}$$

$$= \Psi'_- U \Psi_- + \int_0^T \Psi'(t) V(t) \Psi(t)\, dt + \int_0^T \chi'(t) W(t) \chi(t)\, dt,$$

$$\mathcal{N}_0 = \Psi'_- \Psi + \int_0^T \Psi'(t) \Psi(t)\, dt + \int_0^T \chi'(t) \chi(t)\, dt;$$

$$U = P^{-1/2} P_0 P^{-1/2},$$

$$V(t) = R^{-1/2}(t) R_0(t) R^{-1/2}(t), \qquad W(t) = Q^{-1/2}(t) Q_0(t) Q^{-1/2}(t);$$

$$\Psi_- = P^{1/2} \xi(0),$$

$$\Psi(t) = R^{1/2}(t) \varphi(t) = R^{-1/2}(t) H'(t) \eta(t), \qquad \chi(t) = Q^{1/2}(t) B'(t) \xi(t).$$

The values of Ψ_-, $\Psi(t)$, and $\chi(t)$ are uniquely determined by the elements $\{\varphi_-, \varphi, w_\varphi\}$ of the designed Kalman filter. The uncertainty in Δ_K^0 is contained in the matrices U, $V(t)$, and $W(t)$ only (they depend on unknown P_0, $Q_0(t)$, $R_0(t)$). Thus the uncertainty is localized in the coefficients of the expression that can be efficiently computed.

PROOF: The value of sensitivity (6.7) can be represented in the form

$$\Delta_K = \frac{J(\varphi_-, \varphi, w_\varphi; P_0, Q_0, R_0)}{J_0(P_0, Q_0, R_0)}. \tag{6.11}$$

It follows from (2.4) – (2.6), (2.8) (see also Theorem 2.1 of Chapter 3) that the solution of (6.6), (6.2) is determined by the formulas (cf. (6.8))

$$\varphi_- = \tilde{\xi}(0), \tag{6.12}$$

$$\varphi(t) = R^{-1}(t) h'(t) \tilde{\eta}(t) = \tilde{K}'(t) \tilde{\xi}(t),$$

$$w_\varphi(t) = b'(t) \tilde{\xi}(t),$$

[7] For a symmetric matrix $A \geq 0$, $A^{\frac{1}{2}}$ is a symmetric matrix such that $A^{\frac{1}{2}} A^{\frac{1}{2}} = A$.

where

$$\dot{\tilde{\xi}}(t) = h(t)R^{-1}(t)h'(t)\tilde{\eta}(t), \qquad \tilde{\xi}(T) = \tilde{a}, \tag{6.13}$$

$$\dot{\tilde{\eta}}(t) = b(t)Q(t)b'(t)\tilde{\xi}(t), \qquad \tilde{\eta}(0) = P\tilde{\xi}(0).$$

Moreover,

$$\tilde{\eta}(t) = \tilde{P}(t)\tilde{\xi}(t), \qquad t \in [0, T], \tag{6.14}$$

and

$$J_0 = \left(\tilde{a}'\tilde{P}(T)\tilde{a}\right)^{\frac{1}{2}}, \tag{6.15}$$

where

$$\tilde{K}(t) = \tilde{P}(t)h(t)R^{-1}(t),$$

$$\dot{\tilde{P}}(t) = -\tilde{P}(t)h(t)R^{-1}(t)h'(t)\tilde{P}(t) + b'(t)Q(t)b(t), \qquad \tilde{P}(0) = P.$$

By virtue of Theorem 3.2 of Chapter 3 the denominator in (6.11) has the dual variational representation

$$J_0(P_0, Q_0, R_0) = \sup_{p \in L_2^r, \, \lambda \in \mathbf{R}^n} \tilde{a}'\lambda \tag{6.16}$$

under the constraint

$$S(p, \lambda; P_0, Q_0, R_0) \leq 1, \tag{6.17}$$

where

$$S(p, \lambda; P_0, Q_0, R_0) = \left(\lambda - \int_0^T b(s)p(s)\, ds\right)' P_0^{-1}\left(\lambda - \int_0^T b(s)p(s)\, ds\right)$$

$$+ \int_0^T \left[h'(t)\left(\lambda - \int_t^T b(s)p(s)\, ds\right)\right]' R_0^{-1}(t)\left[h'(t)\left(\lambda - \int_t^T b(s)p(s)\, ds\right)\right] dt$$

$$+ \int_0^T p'(t)Q_0^{-1}(t)p(t)\, dt.$$

Moreover, the solution of the dual problem has the form

$$p_0^0(t) = J_0^{-1}(P_0, Q_0, R_0)\, Q_0(t)b'(t)\tilde{\xi}_0(t), \tag{6.18}$$

$$\lambda_0^0 = J_0^{-1}(P_0, Q_0, R_0)\, \tilde{\eta}_0(T),$$

where $\tilde{\xi}_0(t)$, $\tilde{\eta}_0(t)$ are defined similary to relations (6.13) in which P, $Q(t)$, and $R(t)$ are replaced by P_0, $Q_0(t)$, and $R_0(t)$, respectively.

If we find an appropriate admissible element for the dual problem (6.16), (6.17), then we get the estimate from below for $J_0(P_0, Q_0, R_0)$. Thus we can obtain the upper bound Δ_K^0 for Δ_K.

In complete analogy with (6.16) – (6.18), we have the similar dual representation for the designed filter:

$$J_0(P, Q, R) = \sup_{p \in L_2^r,\, \lambda \in \mathbf{R}^n} \tilde{a}'\lambda \tag{6.19}$$

under the constraint

$$S(p, \lambda; P, Q, R) \leq 1. \tag{6.20}$$

The solution of (6.19), (6.20) has the form

$$p^0(t) = J_0^{-1}(P, Q, R)\, Q(t)b'(t)\tilde{\xi}(t), \tag{6.21}$$

$$\lambda^0 = J_0^{-1}(P, Q, R)\, \tilde{\eta}(T),$$

where $\tilde{\xi}(t)$, $\tilde{\eta}(t)$ are defined by (6.13).

In accordance with our approach we shall search an admissible element on the one-dimensional set

$$p(t) = \nu p^0(t), \qquad \lambda = \nu\lambda^0, \qquad \nu \in \mathcal{R}^1. \tag{6.22}$$

Thus we obtain a one-dimensional problem (6.16), (6.17), (6.22). As far as this problem has more narrow restrictions than (6.17), it provides the required estimate from below. The implementation of this idea leads to the desired estimate (6.9), (6.10).

In fact, the pair defined by (6.22), (6.21) is admissible if and only if $S(p, \lambda; P_0, Q_0, R_0) \leq 1$, i.e.,

$$\begin{aligned}
\nu^2 J_0^{-2}(P, Q, R)\Bigg\{ &\left(\tilde{\eta}(T) - \int_0^T b(s)Q(s)b'(s)\tilde{\xi}(s)\,ds\right)' P_0^{-1} \\
&\qquad \times \left(\tilde{\eta}(T) - \int_0^T b(s)Q(s)b'(s)\tilde{\xi}(s)\,ds\right) \\
&+ \int_0^T \left(\tilde{\eta}(T) - \int_t^T b(s)Q(s)b'(s)\tilde{\xi}(s)\,ds\right)' h(t)R_0^{-1}(t)h'(t) \\
&\qquad \times \left(\tilde{\eta}(T) - \int_t^T b(s)Q(s)b'(s)\tilde{\xi}(s)\,ds\right) dt \\
&+ \int_0^T \tilde{\xi}'(t)b(t)Q(t)Q_0^{-1}(t)Q(t)b'(t)\tilde{\xi}(t)\,dt \Bigg\} \leq 1.
\end{aligned}$$

Taking into account (6.13), the last inequality can be represented in the form

$$\nu^2 J_0^{-2}(P,Q,R)\left\{\tilde{\eta}'(0)P_0^{-1}\tilde{\eta}(0) + \int_0^T \tilde{\eta}'(t)h(t)R_0^{-1}(t)h'(t)\tilde{\eta}(t)\,dt\right.$$

$$\left. + \int_0^T \tilde{\xi}'(t)b(t)Q(t)Q_0^{-1}(t)Q(t)b'(t)\tilde{\xi}(t)\,dt\right\} \leq 1$$

or, by virtue of (6.12), (6.14),

$$\nu^2 J_0^{-2}(P,Q,R)\left\{\varphi_-'PP_0^{-1}P\varphi_- + \int_0^T \varphi'(t)R(t)R_0^{-1}(t)R(t)\varphi(t)\,dt\right.$$

$$\left. + \int_0^T \tilde{\xi}'(t)b(t)Q(t)Q_0^{-1}(t)Q(t)b'(t)\tilde{\xi}(t)\,dt\right\} \leq 1.$$

Thus the pair (6.22) is admissible if and only if $|\nu| \leq \nu_0$, where

$$\nu_0 = J_0(P,Q,R)\left\{\varphi_-'PP_0^{-1}P\varphi_- + \int_0^T \varphi'(t)R(t)R_0^{-1}(t)R(t)\varphi(t)\,dt\right.$$

$$\left. + \int_0^T \tilde{\xi}'(t)b(t)Q(t)Q_0^{-1}(t)Q(t)b'(t)\tilde{\xi}(t)\,dt\right\}^{-\frac{1}{2}}.$$

In accordance with (6.16), (6.14), (6.15), and (6.21),

$$J_0(P_0,Q_0,R_0) \geq \nu\,\tilde{a}'\lambda^0 = \nu J_0^{-1}(P,Q,R)\,\tilde{a}'\tilde{\eta}(T) \tag{6.23}$$

$$= \nu J_0^{-1}(P,Q,R)\,\tilde{a}'\tilde{P}(T)\tilde{a} = \nu J_0(P,Q,R), \qquad |\nu| \leq \nu_0.$$

Hence the best lower estimate for $J_0(P_0,Q_0,R_0)$ is attained with $\nu = \nu_0$ and (6.23) implies the following inequality

$$J_0(P_0,Q_0,R_0) \geq J_0^2(P,Q,R) \tag{6.24}$$

$$\times\left\{\varphi_-'PP_0^{-1}P\varphi_- + \int_0^T \varphi'(t)R(t)R_0^{-1}(t)R(t)\varphi(t)\,dt\right.$$

$$\left. + \int_0^T \tilde{\xi}'(t)b(t)Q(t)Q_0^{-1}(t)Q(t)b'(t)\tilde{\xi}(t)\,dt\right\}^{-\frac{1}{2}}.$$

By virtue of (6.8) and (6.12),

$$\tilde{\xi}(0) = \xi(0) = \varphi_-, \qquad b'(t)\tilde{\xi}(t) = B'(t)\xi(t) = w_\varphi(t).$$

Then, in accordance with (6.10),

$$J(\varphi_-, \varphi, w_\varphi; P_0, Q_0, R_0) = \mathcal{N}^{1/2}(U, V(t), W(t)), \qquad J_0^2(P, Q, R) = \mathcal{N}_0,$$

and the expression in curly brackets from (6.24) is equal to

$$\mathcal{N}(U^{-1}, V^{-1}(t), W^{-1}(t)).$$

Thus the desired inequality (6.9) follows from (6.11) and (6.24). □

Let us discuss some applications of the estimate obtained. We shall evaluate the guaranteed sensitivity of the designed Kalman filter for the restrictions on $\mathcal{P}$, $\mathcal{Q}$, and $\mathcal{R}$ of various types.

EXAMPLE 6.1. Let the nonempty sets $\mathcal{P}$, $\mathcal{Q}$, $\mathcal{R}$ are determined by the following matrix inequalities:

$$\mathcal{P} = \{P_0 > 0 \,|\, P_{01} \le P_0 \le P_{02}\}, \qquad P_{01}, P_{02} > 0,$$

$$\mathcal{Q} = \{Q_0(t) > 0 \,|\, Q_1(t) \le Q_0(t) \le Q_2(t)\}, \qquad Q_1(t), Q_2(t) \ge \tilde{Q} > 0,$$

$$\mathcal{R} = \{R_0(t) > 0 \,|\, R_1(t) \le R_0(t) \le R_2(t)\}, \qquad R_1(t), R_2(t) \ge \tilde{R} > 0.$$

Then it follows from Theorem 6.1 that

$$\Delta_K \le \sup_{\mathcal{P},\mathcal{Q},\mathcal{R}} \Delta_K \le \sup_{\mathcal{P},\mathcal{Q},\mathcal{R}} \Delta_K^0 \le \Delta_{(1)},$$

$$\Delta_{(1)} = \frac{\mathcal{N}^{1/2}(U_2, V_2(t), W_2(t)) \cdot \mathcal{N}^{1/2}(U_1, V_1(t), W_1(t))}{\mathcal{N}_0},$$

where

$$U_1 = P^{1/2} P_{01}^{-1} P^{1/2},$$

$$V_1(t) = R^{1/2}(t) R_1^{-1}(t) R^{1/2}(t), \qquad W_1(t) = Q^{1/2}(t) Q_1^{-1}(t) Q^{1/2}(t),$$

$$U_2 = P^{-1/2} P_{02} P^{-1/2},$$

$$V_2(t) = R^{-1/2}(t) R_2(t) R^{-1/2}(t), \qquad W_2(t) = Q^{-1/2}(t) Q_2(t) Q^{-1/2}(t),$$

and $\mathcal{N}$, $\mathcal{N}_0$ are determined by (6.10).

EXAMPLE 6.2. Let the diagonal entries of the matrices P_0, $Q_0(t)$, and $R_0(t)$ be known. At the same time we assume that the mutual correlation coefficients

$$k_{ij}^P = \frac{(P_0)_{ij}}{\sqrt{(P_0)_{ii}\,(P_0)_{jj}}}, \qquad i \neq j, \qquad i,j = 1,\ldots,n, \tag{6.25}$$

$$k_{kl}^Q(t) = \frac{(Q_0(t))_{kl}}{\sqrt{(Q_0(t))_{kk}\,(Q_0(t))_{ll}}}, \qquad k \neq l, \qquad k,l = 1,\ldots,r,$$

$$k_{pq}^R(t) = \frac{(R_0(t))_{pq}}{\sqrt{(R_0(t))_{pp}\,(R_0(t))_{qq}}}, \qquad p \neq q, \qquad p,q = 1,\ldots,m,$$

$(n,r,m > 1)$ are unknown but are bounded by specified constants k^P, k^Q, and k^R:

$$|k_{ij}^P| \leq k^P < (n-1)^{-1}, \tag{6.26}$$

$$|k_{kl}^Q(t)| \leq k^Q < (r-1)^{-1}, \qquad |k_{pq}^R(t)| \leq k^R < (m-1)^{-1}, \qquad t \in [0,T].$$

We put

$$P = \operatorname{diag}\{(P_0)_{11},\ldots,(P_0)_{nn}\} \tag{6.27}$$

$$Q(t) = \operatorname{diag}\{(Q_0(t))_{11},\ldots,(Q_0(t))_{rr}\},$$

$$R(t) = \operatorname{diag}\{(R_0(t))_{11},\ldots,(R_0(t))_{mm}\}.$$

Using formulas (6.25) – (6.27) and the Gershgorin theorem [59], [127], it can be easily checked that

$$\Psi'_- U \Psi_- \leq \lambda_{\max}(U)\,\kappa_1 \leq \left(1 + (n-1)k^P\right)\kappa_1, \tag{6.28}$$

$$\Psi'(t)V(t)\Psi(t) \leq \lambda_{\max}(V(t))\,\kappa_2 \leq \left(1 + (m-1)k^R\right)\kappa_2,$$

$$\chi'(t)W(t)\chi(t) \leq \lambda_{\max}(W(t))\,\kappa_3 \leq \left(1 + (r-1)k^Q\right)\kappa_3$$

and

$$\Psi'_- U^{-1} \Psi_- \leq \lambda_{\min}^{-1}(U)\,\kappa_1 \leq \left(1 - (n-1)k^P\right)^{-1}\kappa_1, \tag{6.29}$$

$$\Psi'(t)V^{-1}(t)\Psi(t) \leq \lambda_{\min}^{-1}(V(t))\,\kappa_2 \leq \left(1 - (m-1)k^R\right)^{-1}\kappa_2,$$

$$\chi'(t)W^{-1}(t)\chi(t) \leq \lambda_{\min}^{-1}(W(t))\,\kappa_3 \leq \left(1 - (r-1)k^Q\right)^{-1}\kappa_3,$$

where $\lambda_{\min}(A)$ and $\lambda_{\max}(A)$ are the minimal and maximal eigenvalues of a matrix A and

$$\kappa_1 = \|\Psi_-\|^2_{\mathbf{R}^n}, \qquad \kappa_2 = \int_0^T \|\Psi(t)\|^2_{\mathbf{R}^m}\, dt, \qquad \kappa_3 = \int_0^T \|\chi(t)\|^2_{\mathbf{R}^r}\, dt.$$

Then it follows from (6.9), (6.10) and (6.28), (6.29) that

$$\Delta_K \le \Delta_K^0 \le \Delta_{(2)}, \tag{6.30}$$

$$\Delta_{(2)} = \Big[\big(1+(n-1)k^P\big)\,\kappa_1 + \big(1+(m-1)k^R\big)\,\kappa_2 + \big(1+(r-1)k^Q\big)\,\kappa_3\Big]^{\frac{1}{2}}$$

$$\times \Big[\big(1-(n-1)k^P\big)^{-1}\kappa_1 + \big(1-(m-1)k^R\big)^{-1}\kappa_2 + \big(1-(r-1)k^Q\big)^{-1}\kappa_3\Big]^{\frac{1}{2}}$$

$$\times \big(\kappa_1+\kappa_2+\kappa_3\big)^{-1}.$$

Certain coefficients at κ_s can be quite large but, sometimes, the corresponding values of κ_s can be sufficiently small for the estimate $\Delta_{(2)}$ to be useful.

Obviously, the last inequality implies a more rough inequality

$$\Delta_{(2)} \le \left[\frac{1+c}{1-c}\right]^{\frac{1}{2}}, \qquad c = \max\left\{(n-1)k^P, (m-1)k^R, (r-1)k^Q\right\}.$$

In special cases, more accurate estimates than (6.30) can be obtained (see the next example).

EXAMPLE 6.3. Let the matrix P_0 and the diagonal elements of the matrices $Q_0(t)$, $R_0(t)$ are known and the noises are two-dimensional: $r = m = 2$. Assume that the mutual correlation intensities

$$k_Q(t) = \frac{(Q_0(t))_{12}}{\sqrt{(Q_0(t))_{11}\,(Q_0(t))_{22}}} \quad \text{and} \quad k_R(t) = \frac{(R_0(t))_{12}}{\sqrt{(R_0(t))_{11}\,(R_0(t))_{22}}}$$

are unknown but bounded by specified constants:

$$|k_Q(t)| \le k_Q, \qquad |k_R(t)| \le k_R, \qquad t \in [0,T].$$

We put $P = P_0$,

$$Q(t) = \operatorname{diag}\left\{(Q_0(t))_{11}, (Q_0(t))_{22}\right\}, \quad R(t) = \operatorname{diag}\left\{(R_0(t))_{11}, (R_0(t))_{22}\right\}.$$

Then it follows from (6.9), (6.10) that

$$\mathcal{N}_0\, \Delta_K \le \mathcal{N}_0 \Delta_K^0 \tag{6.31}$$

$$= \left[\Psi'_- \Psi_- + \int_0^T \left(\Psi_1^2(t) + \Psi_2^2(t) + 2k_R(t)\Psi_1(t)\Psi_2(t) \right) dt \right.$$

$$\left. + \int_0^T \left(\chi_1^2(t) + \chi_2^2(t) + 2k_Q(t)\chi_1(t)\chi_2(t) \right) dt \right]^{\frac{1}{2}}$$

$$\times \left[\Psi'_- \Psi_- + \int_0^T \frac{\Psi_1^2(t) + \Psi_2^2(t) - 2k_R(t)\Psi_1(t)\Psi_2(t)}{1 - k_R^2(t)} \, dt \right.$$

$$\left. + \int_0^T \frac{\chi_1^2(t) + \chi_2^2(t) - 2k_Q(t)\chi_1(t)\chi_2(t)}{1 - k_Q^2(t)} \, dt \right]^{\frac{1}{2}},$$

$$\Psi(t) = (\Psi_1(t), \Psi_2(t))', \qquad \chi(t) = (\chi_1(t), \chi_2(t))'.$$

The analysis of the dependence of Δ_K on the functions $k_Q(t), k_R(t)$ is difficult. At the same time, Δ_K^0 has an explicit and quite simple dependence on unknown correlations. In fact, we have from (6.31) that

$$\Delta_K \leq \Delta_K^0 \leq \left(\mathcal{N}_0^2 - \omega^2 \right)^{\frac{1}{2}} \mathcal{N}_0^{-1} \cdot \left(1 - \max\left\{ k_Q^2, k_R^2 \right\} \right)^{-\frac{1}{2}}, \tag{6.32}$$

where

$$\omega = 2 \int_0^T \left(k_R(t)\Psi_1(t)\Psi_2(t) + k_Q(t)\chi_1(t)\chi_2(t) \right) dt.$$

Finally, we have from (6.32) that

$$\Delta_K \leq \Delta_K^0 \leq \frac{1}{\sqrt{1 - \max\left\{ k_Q^2, k_R^2 \right\}}}. \tag{6.33}$$

The guaranteed bound (6.33) shows that the sensitivity of the Kalman filter to the uncertain correlation coefficients is quite weak. For example, with $\max\{k_Q, k_R\} = 0.5$, $\Delta_K \leq 1.15$, and, with $\max\{k_Q, k_R\} = 0.75$, $\Delta_K \leq 1.51$.

Let us assume (in addition to the conditions mentioned above) that the matrix P_0 is almost known and almost diagonal: it has only one unknown nondiagonal entry $k_{i^0 j^0}$ such that $|k_{i^0 j^0}| \leq k_P$. Then it is easy to verify that

$$\Delta_K \leq \Delta_K^0 \leq \frac{1}{\sqrt{1 - \max\left\{ k_P^2, k_Q^2, k_R^2 \right\}}}. \tag{6.34}$$

It should be noted that the estimates (6.33) and (6.34) were obtained for a quite broad class of problems based on the structure of the expression for Δ_K^0 only (and without any substantial calculations).

Let us further consider the case that *a priori* information about the initial state is poor:

$$P = P_0 = \varepsilon^{-2}\bar{P}, \quad \bar{P} > 0, \qquad \varepsilon \to 0^+. \tag{6.35}$$

LEMMA 6.1. *Let the system* $\dot{x}(t) = A(t)x(t)$, $z(t) = H'(t)x(t)$ *be completely observable on the segment* $[0,T]$. *Then, under condition* (6.35), *there exist the finite limits*:

$$\lim_{\varepsilon\to 0+} \Psi_- = 0, \qquad \lim_{\varepsilon\to 0+} \Psi(t) = \Psi_\infty(t), \qquad \lim_{\varepsilon\to 0+} \chi(t) = \chi_\infty(t).$$

PROOF. Lemma 6.1 follows directly from Lemma 2.1 of Chapter 3 (see also Exercise 4 of Chapter 3).

Assume also that there is no perturbations in the object, i.e.,

$$B(t) = 0, \qquad t \in [0,T]. \tag{6.36}$$

In this simple case, the exact solutions of the corresponding maximization problems can be obtained.

If (6.35) and (6.36) hold, then it follows from Theorem 6.1 and Lemma 6.1 that

$$\Delta_K \le \Delta_{K\infty}^0,$$

$$\Delta_{K\infty}^0 = \left[\int_0^T \left(\Psi_{\infty 1}^2(t) + \Psi_{\infty 2}^2(t) + 2k_R(t)\Psi_{\infty 1}(t)\Psi_{\infty 2}(t)\right) dt\right]^{\frac{1}{2}} \tag{6.37}$$

$$\times \left[\int_0^T \frac{\Psi_{\infty 1}^2(t) + \Psi_{\infty 2}^2(t) - 2k_R(t)\Psi_{\infty 1}(t)\Psi_{\infty 2}(t)}{1 - k_R^2(t)}\, dt\right]^{\frac{1}{2}} \mathcal{N}_{0\infty}^{-1},$$

where

$$\mathcal{N}_{0\infty} = \int_0^T \left(\Psi_{\infty 1}^2(t) + \Psi_{\infty 2}^2(t)\right) dt, \qquad \Psi_\infty(t) = (\Psi_{\infty 1}(t), \Psi_{\infty 2}(t))'.$$

ASSERTION 6.1: *The maximum in* (6.37) *is determined by the formula*

$$\max_{|k_R(t)| \le k_R} \Delta_{K\infty}^0 = \left(1 - k_R^2\right)^{-\frac{1}{2}},$$

and is attained at

$$k_R(t) = k_R^0(t) = \begin{cases} k_R \operatorname{sign}(\Psi_{\infty 1}(t)\Psi_{\infty 2}(t)) & \text{if } 0 \le t \le \alpha \\ -k_R \operatorname{sign}(\Psi_{\infty 1}(t)\Psi_{\infty 2}(t)) & \text{if } \alpha < t \le T, \end{cases}$$

where α *is the median of "the distribution"* $|\Psi_{\infty 1}(t)\Psi_{\infty 2}(t)|$:

$$\int_0^\alpha |\Psi_{\infty 1}(t)\Psi_{\infty 2}(t)|\, dt = \int_\alpha^T |\Psi_{\infty 1}(t)\Psi_{\infty 2}(t)|\, dt.$$

EXAMPLE 6.4. Assume that conditions (6.35) and (6.36) are still valid. Moreover, let the measurement noise be scalar ($m = 1$) and its intensity be unknown but bounded:

$$R_{\min} \le R_0(t) \le R_{\max}, \qquad t \in [0, T].$$

Recall that the system dimension n is arbitrary. We set the designed intensity of the measurement noise by a certain constant parameter:

$$R(t) = \bar{R} = \text{const} > 0.$$

Then, by virtue of Theorem 6.1 and Lemma 6.1,

$$\Delta_K \le \Delta_{K\infty}^0 = \frac{\left(\int_0^T R_0(t)\varphi_\infty^2(t)\, dt\right)^{\frac{1}{2}} \left(\int_0^T R_0^{-1}(t)\varphi_\infty^2(t)\, dt\right)^{\frac{1}{2}}}{\int_0^T \varphi_\infty^2(t)\, dt}, \tag{6.38}$$

where $\varphi_\infty(t) = \bar{R}^{-\frac{1}{2}}\Psi_\infty(t)$ is the designed estimator. For example, if $A(t) \equiv 0$, then it follows from Lemma 7.1 of Chapter 1 and Lemma 2.1 of Chapter 3 that

$$\varphi_\infty(t) = a' \left(\int_0^T H(t)H'(t)\, dt\right)^{-1} H(t),$$

i.e., $\varphi_\infty(t)$ is the weight function of the least-squares method.

Let us consider the optimal problem

$$\Delta_{\max} = \sup_{R_{\min} \le R_0(t) \le R_{\max}} \Delta_{K\infty}^0, \tag{6.39}$$

where $\Delta_{K\infty}^0$ is defined by (6.38).

LEMMA 6.2. *The maximum in* (6.39) *is determined by the formula*

$$\Delta_{\max} = \frac{\theta^2 + 1}{2\theta}, \qquad \theta = \left(\frac{R_{\max}}{R_{\min}}\right)^{\frac{1}{2}}, \tag{6.40}$$

and is attained at

$$R_0(t) = R_0^0(t) = \begin{cases} R_{\max} & \text{if } 0 \le t \le \beta \\ R_{\min} & \text{if } \beta < t \le T, \end{cases}$$

where β *is the median of "the distribution"* $\varphi_\infty^2(t)$:

$$\int_0^\beta \varphi_\infty^2(t)\,dt = \int_\beta^T \varphi_\infty^2(t)\,dt.$$

PROOF. Passing on to the integral sums in the expression for $\Delta_{K\infty}^0$ from (6.38), it is not hard to prove that the supremum in (6.39) can be taken over the functions that take the bounding values $R_{\min}$ and $R_{\max}$ only. By $\mathcal{T}^-$ and $\mathcal{T}^+$ denote the sets on which $R_0(t)$ takes the values $R_{\min}$ and $R_{\max}$, respectively. Introduce the quantities

$$g_1 = \frac{\int_{\mathcal{T}^-} \varphi_\infty^2(t)\,dt}{\int_0^T \varphi_\infty^2(t)\,dt}, \qquad g_2 = \frac{\int_{\mathcal{T}^+} \varphi_\infty^2(t)\,dt}{\int_0^T \varphi_\infty^2(t)\,dt}.$$

Then the optimal problem (6.39) takes the form

$$\Delta_{\max} = \sup_{g_1+g_2=1} \left(g_1 R_{\min} + g_2 R_{\max}\right)^{\frac{1}{2}} \left(g_1 R_{\min}^{-1} + g_2 R_{\max}^{-1}\right)^{\frac{1}{2}} \tag{6.41}$$

$$= \sup_{g_1+g_2=1} \left[\left(g_1+g_2\right)^2 + \left(R_{\max}R_{\min}^{-1} + R_{\min}R_{\max}^{-1} - 2\right) g_1 g_2\right]^{\frac{1}{2}}.$$

Since $\left(R_{\max}R_{\min}^{-1} + R_{\min}R_{\max}^{-1} - 2\right) > 0$, the solution of (6.41) is attained at

$$g_1 = g_2 = \frac{1}{2}. \tag{6.42}$$

Then Lemma 6.2 follows directly from (6.41) and (6.42). □

Finally, we have from (6.38) – (6.40) that

$$\Delta_K \le \frac{\theta^2+1}{2\theta}, \qquad \theta = \left(\frac{R_{\max}}{R_{\min}}\right)^{\frac{1}{2}}. \tag{6.43}$$

Obviously, for $\theta \ge 1$, $(\theta^2+1)/2\theta \ge \theta$; therefore the inequality (6.43) implies a more rough inequality:

$$\Delta_K \le \theta. \tag{6.44}$$

The latter inequality can be easily obtained from the explicit expression for Δ_K as well. Some numerical values of θ and $\Delta_{\max}$ are presented in Table 8.

TABLE 8. **Guaranteed sensitivity**

θ	2.0	3.0	4.0	5.0	10.0
$\Delta_{\max}$	1.25	1.67	2.13	2.60	5.05

Table 8 shows that the Kalman filter sensitivity is rather temperate. Besides, our nontrivial estimate $\Delta_{\max}$ is approximately two times better than the trivial rough one.

So, using the duality theory of convex variational problems, the upper bound for the sensitivity of the designed Kalman filter is obtained. This formula is considerably simpler for analysis than the initial exact relation. In number of cases, this formula allows us to obtain immediately useful nontrivial inequalities that describe the guaranteed sensitivity of the designed Kalman filter.

7. Comments to Chapter 4

7.1. *Section 1*

The minimax stochastic estimation problems were studied by many authors (see, e.g., [76], [2], [39], [170], [88], [207], [188], [180], [33], [150], [209], [210], [208], [132], [6], [134], [7], [120], [21], [193], [194], [26], [195], [196], [8], [197], [198], [199], [9], [200]; for nonparametric problems see, e.g., [73], [74]). We set up the stochastic guaranteed estimation problem following the papers [132], [133].

7.2. *Section 2*

The problem on the evaluation of the nonoptimality level of the Kalman filter was posed in [161], where Theorem 2.2 was first proved. The main result was then published in [163].

7.3. *Section 3*

When deriving Theorem 2.2 we essentially used the representation of the Kalman filter via a boundary-value problem developed in Chapter 3. Our main tool is the duality theory of convex variational problems [185], [75], [51], [123], [5], [182], [183], [186].

7.4. *Section 4*

The discrete case was studied in [65], [67]. The applications of the duality theory to guaranteed estimation problems in static stochastic systems were investigated in [21], [193], [194]. In [195], [196], [198] the dual algorithms for solving the guaranteed estimation problems in static systems are presented. In these papers, the estimate of nonoptimality is obtained in a polar form. The question of the optimality of linear algorithms in guaranteed estimation problems for static stochastic systems are studied in [155], [197], [199], [200] (the latter paper is a survey devoted to the duality in nonconvex extremal problems, new minimax theorems and their applications to minimax estimation). Nevertheless, the question on the optimality of linear algorithms in the problems discussed in Chapter 4 is still open.

7.5. *Section 5*

Examples 5.1 – 5.4 are taken from [65] and [67]. An important Example 5.5 is published in [37]. The topographical referencing problem itself was investigated by different methods. A significant advance has been made in the studying of this problem in the Laboratory of Navigation and Control (Faculty of Mechanics and Mathematics, M.V. Lomonosov Moscow State University) headed by Prof. N.A. Parusnikov [63]. Almost all software for computing the levels of nonoptimality in Chapters 3 and 4, particularly for the topographical referencing problem, was developed by Dr. A.A. Golovan.

7.6. *Section 6*

The construction of the guaranteed level of the Kalman filter sensitivity to the uncertainty in noise intensities was performed in [159] and, in a brief modified form, in [164].

8. Exercises to Chapter 4

1. Consider the Kalman filter for the original system from Section 1 with $A(t) = 0$ and $B(t) = 0$. In addition, assume that $\mathbf{E}(x(0)x'(0)) = P_0 \to \infty$ $(P_0 > 0)$. Prove directly that, in this case, the Kalman estimator tends to the estimator of the least-squares method (LSM-estimator), which is discussed in Section 7 of Chapter 1.
2. Consider the guaranteed estimation problem for a simplest static system
$$z(t) = H'(t)q + \varrho(t), \qquad q = \text{const}, \quad z(t), \varrho(t) \in \mathbf{R}^1, \quad t \in [0, T],$$
where $\varrho(t)$ is a combined noise. The parameter $q \in \mathbf{R}^n$ is supposed to be completely determinisitic and completely unknown. Construct the

corresponding primal variational problem of the type (1.9), (1.10):

$\inf \mathcal{J}$ under the unbiasness condition.

3. Consider the previous guaranteed estimation problem. Obtain the dual problem for the primal problem of the form

   $\inf \mathcal{J}^{\frac{1}{2}}$ under the unbiasness condition.
4. Consider the guaranteed estimation problem from Exercise 2. Construct the algorithm for the computation of the upper bound for the level of nonoptimality of the LSM-estimator (see Section 7 of Chapter 1).
5. Let us put for the problem from Exercise 2 that $n = 1$, $H(t) = \cos t$, $T = 2\pi$. Obtain the upper bound for the level of nonoptimality of the LSM-estimator (see Section 7 of Chapter 1).
6. Consider the guaranteed estimation problem for a special case of system (4.1), (4.2). Let a scalar system be described by the discrete measurements

   $$z(t_i) = q + \varrho(t_i), \qquad i = 0, \dots, N, \qquad z(t_i), q, \varrho(t_i) \in \mathbf{R}^1,$$

   where q is an unknown stochastic quantity. Construct the upper bound for the level of nonoptimality of the discrete Kalman filter Δ_{SD}^0 (see Theorem 4.2).
7. What form the expressions for $\mathcal{E}(\zeta)$, $\Omega(\varphi_-, \varphi, w_\varphi)$, and Δ_S^0 from Theorem 2.2 will take if some of the quantities σ_{-d}, c_d, $\gamma_j(t)$, $q_j(t)$ are equal to zero?
8. Prove inequality (6.34).
9. Prove Assertion 6.1.
10. Obtain the explicit expression for the measure of the sensitivity of the designed Kalman filter Δ_K defined by (6.7) for the special case discussed in Example 6.4. Using this explicit form for Δ_K, prove inequality (6.44).
11. Construct the upper bound for the sensitivity of the designed Kalman filter to the uncertainty in noise intensities in case of discrete measurements.

Chapter 5

Estimation in Dynamic Systems with Aftereffect

A lot of processes in nature, technologies, economics are described by dynamic systems with aftereffect. The theory of systems with aftereffect is developing at an intensive rate. Estimation and control problems occupy an important place in this theory. In particular, different optimal filtering problems can be formulated for dynamic systems with aftereffect. They can be reduced to nontrivial variational problems. In many cases, instead of searching for an optimal estimator, it is more preferable to use the existing simpler estimation algorithms. In such an approach, however, we must evaluate (without finding the optimal algorithm), the increment in estimation error due to the rejection of the optimal filter. In this chapter, we shall construct such estimates for dynamic systems with aftereffect. Generally, the analysis of systems with aftereffect is much more complex than for systems without delays. Therefore, the construction of simplified algorithms with guaranteed level of nonoptimality is valuable for a designer.

1. Pseudo-Fundamental Matrix and Cauchy Formula

Let us consider a linear dynamic system with aftereffect

$$\dot{x}(t) = A(t)x(t) + B(t)x(t-h) + \int_{-h}^{0} G(t,\tau)x(t+\tau)\,d\tau, \tag{1.1}$$

$$x(t) \in \mathbf{R}^n, \qquad 0 \le t \le T,$$

with the boundary condition

$$x(\tau) = x^0(\tau), \quad -h \le \tau \le 0, \tag{1.2}$$

where $x(t)$, $0 \le t \le T$, is an absolutely continuous function and $x^0(\tau)$, $-h \le \tau \le 0$, is a given piecewise continuous function, h is a constant delay. The matrix coefficients $A(t)$, $B(t)$, and $G(t,\tau)$ are square-integrable bounded functions on $[0,T]$, $[0,T]$, and $[0,T]\times[-h,0]$, respectively. We assume for simplicity that $A(t)$, $B(t)$, and $G(t,\tau)$ are piecewise continuous in each variable. Without loss of generality, we also assume that, in the points of discontinuity in t, they are continuous from the right, possibly, except for the point $t=T$, at which these functions are continuous from the left.

It is not hard to prove that, given $x(\tau)$, $-h\le\tau\le 0$, equation (1.1) has a unique solution on $[0,T]$ (see, e.g., [105], [70], [106]). Generally, given $x(\tau)$, $s-h\le\tau\le s$, equation (1.1) has a unique solution on $[s,T]$. The general solution of an ordinary differential equation without aftereffect is expressed via a fundamental matrix. Let us obtain an analogous formula for the system (1.1).

LEMMA 1.1. *If $x(t)$ is a solution of (1.1), then, for $0\le s<t\le T$, the Cauchy formula holds:*

$$x(t)=Y(t,s)x(s)+\int_{s-h}^{s}Y(t,\xi+h)B(\xi+h)x(\xi)\,d\xi \tag{1.3}$$

$$+\int_0^h\int_{s-\eta}^{s}Y(t,\xi+\eta)G(\xi+\eta,-\eta)x(\xi)\,d\xi d\eta,$$

where the $n\times n$ matrix function $Y(t,s)$, $(t,s)\in[-h,T]\times[0,T+h]$, is absolutely continuous in each variable in the domain $t\ge s$ and satisfies the differential equations

$$\frac{\partial Y(t,s)}{\partial t}=A(t)Y(t,s)+B(t)Y(t-h,s)+\int_{-h}^{0}G(t,\xi)Y(t+\xi,s)\,d\xi, \tag{1.4}$$

$$\frac{\partial Y(t,s)}{\partial s}=-Y(t,s)A(s)-Y(t,s+h)B(s+h) \tag{1.5}$$

$$-\int_0^h Y(t,s+\eta)G(s+\eta,-\eta)\,d\eta$$

with the boundary conditions[1]

$$Y(t,t)=E_n,\qquad Y(t,s)=0,\quad t<s. \tag{1.6}$$

Given $x(\tau)$, $s-h\le\tau\le s$, the Cauchy formula (1.3) determines the general form of the solution $x(t)$ for $t>s$. In particilar, if $s=0$, then (1.3) gives the general solution for the problem (1.1), (1.2). The matrix $Y(t,s)$ is called *pseudo-fundamental*. Note that $Y(t,s)$, generally, has no inverse one.

[1]The functions $B(T+\xi)$ and $G(T+\xi,-\xi)$ for $0<\xi\le h$ may be defined arbitrarily; these values do not affect the result since $Y(t,t+\xi)=0$ for $0<\xi\le h$.

PROOF. Let $Y(t,s) \in \mathbf{R}^{n\times n}$ be an absolutely continuous in s, $0 \le s \le t$, matrix function. From the one hand, integrating by parts,

$$\int_s^t Y(t,\xi)\dot{x}(\xi)\,d\xi = Y(t,t)x(t) - Y(t,s)x(s) - \int_s^t \frac{\partial Y(t,\xi)}{\partial \xi}x(\xi)\,d\xi; \tag{1.7}$$

from the other hand, by virtue of (1.1),

$$\int_s^t Y(t,\xi)\dot{x}(\xi)\,d\xi = \int_s^t Y(t,\xi)A(\xi)x(\xi)\,d\xi \tag{1.8}$$

$$+ \int_s^t Y(t,\xi+h)B(\xi+h)x(\xi)\,d\xi + \int_{s-h}^s Y(t,\xi+h)B(\xi+h)x(\xi)\,d\xi$$

$$- \int_{t-h}^t Y(t,\xi+h)B(\xi+h)x(\xi)\,d\xi + \int_s^t\int_{-h}^0 Y(t,\xi)G(\xi,\tau)x(\xi+\tau)\,d\tau d\xi.$$

Let us transform the last integral in (1.8). Interchanging the integrations and carrying out standard substitutions, we have

$$\mathcal{G} \triangleq \int_s^t\int_{-h}^0 Y(t,\xi)G(\xi,\tau)x(\xi+\tau)\,d\tau d\xi \tag{1.9}$$

$$= \int_{-h}^0\int_s^t Y(t,\xi)G(\xi,\tau)x(\xi+\tau)\,d\xi d\tau$$

$$= \int_0^h\int_{s-\eta}^{t-\eta} Y(t,\xi+\eta)G(\xi+\eta,-\eta)x(\xi)\,d\xi d\eta.$$

The last integral in (1.9) can be partitioned into three integrals (see Figure 1) such that

$$\mathcal{G} = \int_0^h\int_{s-\eta}^{s} Y(t,\xi+\eta)G(\xi+\eta,-\eta)x(\xi)\,d\xi d\eta \tag{1.10}$$

$$+ \int_0^h\int_s^t Y(t,\xi+\eta)G(\xi+\eta,-\eta)x(\xi)\,d\xi d\eta$$

$$- \int_0^h\int_{t-\eta}^t Y(t,\xi+\eta)G(\xi+\eta,-\eta)x(\xi)\,d\xi d\eta$$

$$= \int_0^h\int_{s-\eta}^{s} Y(t,\xi+\eta)G(\xi+\eta,-\eta)x(\xi)\,d\xi d\eta$$

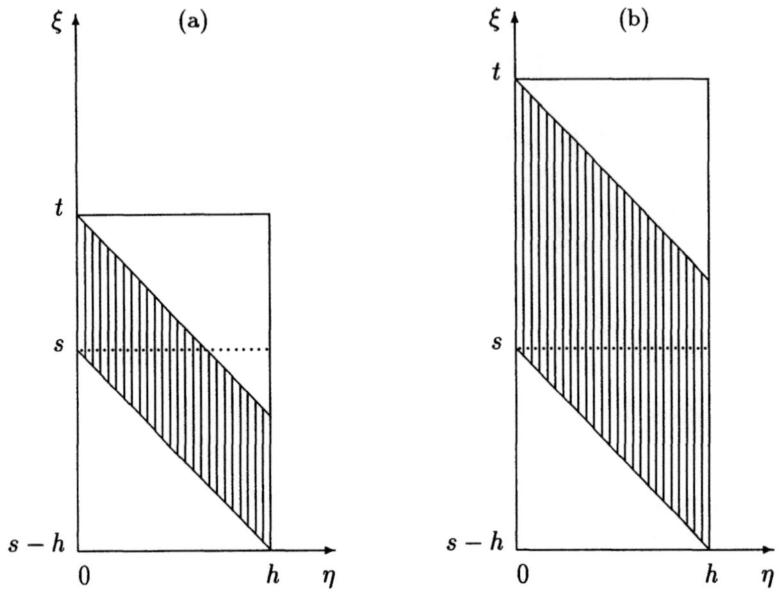

Fig. 1. Domain of integration: (a) $t - s < h$; (b) $t - s \geq h$

$$+ \int_s^t \int_0^h Y(t, \xi + \eta) G(\xi + \eta, -\eta) \, d\eta \, x(\xi) \, d\xi$$

$$- \int_0^h \int_{t-\eta}^t Y(t, \xi + \eta) G(\xi + \eta, -\eta) x(\xi) \, d\xi d\eta.$$

Thus, it follows from (1.7) – (1.10) that

$$Y(t,t)x(t) - Y(t,s)x(s) - \int_{s-h}^s Y(t, \xi + h) B(\xi + h) x(\xi) \, d\xi \tag{1.11}$$

$$- \int_0^h \int_{s-\eta}^s Y(t, \xi + \eta) G(\xi + \eta, -\eta) x(\xi) \, d\xi d\eta$$

$$= \int_s^t \left(\frac{\partial Y(t, \xi)}{\partial \xi} + Y(t, \xi) A(\xi) + Y(t, \xi + h) B(\xi + h) \right.$$

$$\left. + \int_0^h Y(t, \xi + \eta) G(\xi + \eta, -\eta) \, d\eta \right) x(\xi) \, d\xi$$

$$- \int_{t-h}^{t} Y(t, \xi + h) B(\xi + h) x(\xi)\, d\xi$$

$$- \int_{0}^{h} \int_{t-\eta}^{t} Y(t, \xi + \eta) G(\xi + \eta, -\eta) x(\xi)\, d\xi d\eta.$$

Suppose that $Y(t, \xi)$ satisfies (1.5) with right boundary condition (1.6)[2]. Then the right-hand side of (1.11) is equal to zero and the Cauchy formula (1.3) follows from (1.11).

The matrix $Y(t, s)$ is uniquely defined by (1.5) and (1.6). The equality (1.4) remains to be proved.

Consider a solution $\bar{x}(t)$ of (1.1) with the boundary condition

$$\bar{x}(\tau) = 0, \quad s - h \leq \tau < s, \qquad \bar{x}(s) = \alpha. \tag{1.12}$$

Then we have from (1.3) that

$$\bar{x}(t) = Y(t, s)\alpha, \qquad s \leq t \leq T. \tag{1.13}$$

It follows from (1.6) and (1.12) that (1.13) is valid for $s - h \leq t \leq T$ as well. The function $\bar{x}(t)$ satisfies (1.1) and α is arbitrary. Hence, $Y(t, s)$ as a function of t is absolutely continuous for $s \leq t \leq T$ and satisfies (1.4). □

COROLLARY 1.1. *The Cauchy formula can be represented in another forms:*

$$x(t) = Y(t, s) x(s) + \int_{s-h}^{s} Y(t, \xi + h) B(\xi + h) x(\xi)\, d\xi \tag{1.14}$$

$$+ \int_{s-h}^{s} Y(t, \xi + h) \int_{\xi}^{s} G(\xi + h, \eta - \xi - h) x(\eta)\, d\eta d\xi$$

or,

$$x(t) = Y(t, s) x(s) + \int_{s-h}^{s} Y(t, \xi + h) B(\xi + h) x(\xi)\, d\xi$$

$$+ \int_{s-h}^{s} \int_{s}^{\eta + h} Y(t, \xi) G(\xi, \eta - \xi)\, d\xi\, x(\eta)\, d\eta.$$

PROOF. Relations (1.14) can be obtained from (1.3) by the standard change of variables and the interchange of integrations. □

[2]Obviously, equation (1.5) with condition (1.6) can be reduced to an equation of type (1.1) with left boundary condition (1.2) by the reversing of time. Therefore the solution of (1.5), (1.6) exists and is unique.

Let us consider a non-homogeneous dynamic system

$$\dot{x}(t) = A(t)x(t) + B(t)x(t-h) + \int_{-h}^{0} G(t,\tau)x(t+\tau)\,d\tau + C(t)u(t), \quad (1.15)$$

$$x(\tau) = x^0(\tau), \qquad -h \le \tau \le 0,$$

where $u(t)$ is a piecewise continuous function and $C(t)$ is an $n \times r$ matrix function satisfying the same conditions as $A(t)$ and $B(t)$.

COROLLARY 1.2. *If $x(t)$ is a solution of (1.15), then, for $0 \le s < t \le T$, the Cauchy formula holds:*

$$x(t) = Y(t,s)x(s) + \int_{s-h}^{s} Y(t,\xi+h)B(\xi+h)x(\xi)\,d\xi \quad (1.16)$$

$$+ \int_0^h \int_{s-\eta}^{s} Y(t,\xi+\eta)G(\xi+\eta,-\eta)x(\xi)\,d\xi d\eta + \int_s^t Y(t,\eta)C(\eta)u(\eta)\,d\eta.$$

Certainly, the modifications of type (1.14) also take place.

PROOF. Let us prove that the function

$$\tilde{x}(t) = \int_s^t Y(t,\eta)C(\eta)u(\eta)\,d\eta, \qquad s \le t \le T, \quad (1.17)$$

$$\tilde{x}(t) = 0, \qquad s-h \le t < s, \quad (1.18)$$

satisfies (1.15).

For $s+h \le t$, we may use formula (1.17) only. Since $Y(t,s)$ satisfies (1.4), the function $\tilde{x}(t)$ satisfies (1.15). For $s \le t < s+h$, we have to prove that if $\tilde{x}(t)$ is defined by (1.17), (1.18), then

$$\dot{\tilde{x}}(t) = A(t)\tilde{x}(t) + \int_{s-t}^{0} G(t,\tau)x(t+\tau)\,d\tau + C(t)u(t). \quad (1.19)$$

For $s \le t < s+h$, equation (1.4) takes the form

$$\frac{\partial Y(t,s)}{\partial t} = A(t)Y(t,s) + \int_{s-t}^{0} G(t,\xi)Y(t+\xi,s)\,d\xi, \quad Y(t,t) = E_n. \quad (1.20)$$

Then (1.19) follows from (1.17) and (1.20).

If $\hat{x}(t)$ defined by (1.3) is a solution of (1.1) with the initial condition

$$\hat{x}(\tau) = x^0(\tau), \qquad s-h \le \tau \le s,$$

then

$$x(t) = \hat{x}(t) + \tilde{x}(t), \qquad s \le t \le T,$$

is a solution of (1.15) with the same initial condition. Obviously, if (1.1) has a unique solution, then (1.15) also has a unique solution. Thus formula (1.16) determines the general form for the solution of (1.15). □

2. Guaranteed Estimation in Dynamic Systems with Delay

The main results of optimal system theory are related to linear-quadratic estimation and control problems. Powerful analytical methods have been developed for studying these problems, and effective numerical algorithms are available for solving them. In engineering practice, however, nonsmooth problems are frequently encountered, such as, for example, minimax estimation under unknown bounded perturbations. Linear-quadratic theory can be applied fruitfully for these cases as well. In the section, we apply this theory to find the approximate solution of minimax filtering problem in dynamic systems with aftereffect.

2.1. *Problem Statement*

Consider a linear dynamic system with delay

$$\dot{x}(t) = A(t)x(t) + B(t)x(t-h) + C(t)u(t), \qquad t \in [0, T], \tag{2.1}$$

$$x(\tau) = 0, \qquad -h \le \tau < 0, \qquad h = \text{const}, \qquad h < T,$$

where $x(t) = (x_1(t), \dots, x_n(t))' \in \mathbf{R}^n$ is the system state vector; $A(t), B(t) \in \mathbf{R}^{n\times n}$, $C(t) \in \mathbf{R}^{n\times r}$, and $u(t) = (u_1(t), \dots, u_r(t))' \in \mathbf{R}^r$ is a perturbation vector. A prime denotes the transposition sign. Dynamic system (2.1) is a special case of system (1.15) discussed in Section 1 with $G(t, \tau) = 0$.

We assume that

$$|u_j(t)| \le \gamma_j(t), \qquad j = 1, \dots, r, \tag{2.2}$$

where $\gamma_j(t)$ are given positive functions.

Let measurements be made on the segment $[0, T]$[3]:

$$z(t) = H'(t)x(t-\bar{h}) + \varrho(t), \qquad z(t), \varrho(t) \in \mathbf{R}^m, \quad H(t) \in \mathbf{R}^{n\times m}, \tag{2.3}$$

$$\bar{h} = \text{const}, \quad \bar{h} < T.$$

[3]If $\bar{h} > h$, then we assume that $x(\tau) = 0$ for $-\bar{h} \le \tau < 0$.

We assume that the measurement errors $\varrho(t) = (\varrho_1(t), \dots, \varrho_m(t))'$ obey the constraints

$$|\varrho_l(t)| \le \sigma_l(t), \qquad t \in [0,T], \quad l = 1, \dots, m, \tag{2.4}$$

where $\sigma_l(t)$, $l = 1, \dots, m$, are known positive functions. Moreover, the initial state vector of the system is assumed to belong to the parallelepiped

$$|x_d(0)| \le \sigma_{-d}, \qquad d = 1, \dots, n, \tag{2.5}$$

where σ_{-d} , $d = 1, \dots, n$, are known positive numbers.

The elements of the matrices $A(t)$, $B(t)$, $C(t)$, $H(t)$, and the functions $u_j(t)$, $\gamma_j(t)$, $\varrho_l(t)$, $\sigma_l(t)$ are assumed to be piecewise continuous. For simplicity we assume also that $A(t)$, $B(t)$, $C(t)$, $\gamma_j(t)$, and $\sigma_l(t)$ have no isolated jumps (see Section 1 or Section 1 of Chapter 3 for details). Furthermore, we use the notation $x(0) \in \Pi$, $u \in U$, $\varrho \in \sigma$ for the sets mentioned above.

Our problem now is to evaluate the scalar quantity $l_* = a'x(T)$ (where $a \in \mathbf{R}^n$ is a given vector) with the help of linear functionals

$$\hat{l} = \int_0^T \Phi'(t) z(t)\, dt, \qquad \Phi(t) = (\Phi_1(t), \dots, \Phi_m(t))' \in \mathbf{R}^m, \tag{2.6}$$

$$\Phi(t) = \Phi_{pc}(t) + \sum_{s=1}^{M} \Phi_{(s)} \delta(t - t_s). \tag{2.7}$$

Here $\Phi_{pc}(t)$ is a piecewise continuous function; $\Phi_{(s)} \in \mathbf{R}^m$, $t_s \in [0,T]$, $s = 1, \dots M$, M is an arbitrary integer, and $\delta(t)$ is the Dirac delta-function. We denote set (2.7) by $\mathcal{F}^m$. The estimate $\hat{l}$ includes both averaged and individual values of $z(t)$.

The optimal *a priori* guaranteed estimation problem is to find an estimator Φ^0 that minimizes the guaranteed estimation error

$$d(\Phi^0) = d_0 = \inf_{\Phi \in \mathcal{F}^m} d(\Phi), \qquad d(\Phi) = \sup_{x(0) \in \Pi,\, u \in U,\, \varrho \in \sigma} |\hat{l} - l_*|. \tag{2.8}$$

Thus we must solve a minimax problem. For $h = \bar{h} = 0$ this problem was treated in Chapters 2 and 3.

REMARK 2.1. Similar to the case of systems without delay, it can be shown that the use of nonlinear estimators does not improve an estimation accuracy (see Section 5 of Chapter 2).

For the system (2.1), the pseudo-fundamental matrix $Y(t,s)$, which was introduced in the previous section, takes the form

$$\frac{\partial Y(t,s)}{\partial t} = A(t) Y(t,s) + B(t) Y(t-h, s), \qquad t \ge s, \tag{2.9}$$

$$\frac{\partial Y(t,s)}{\partial s} = -Y(t,s)A(s) - Y(t,s+h)B(s+h), \qquad s \le t,$$

$$Y(t,t) = E_n, \qquad Y(t,s) = 0, \quad s > t,$$

where E_n is the identity matrix.

Now, from (2.1), (2.9), and Corollary 1.2, we find that

$$x(t) = Y(t,0)x(0) + \int_0^t Y(t,\tau)C(\tau)u(\tau)\,d\tau. \tag{2.10}$$

By virtue of (2.1), (2.3), (2.6), and (2.10), the estimation error is given by the expression

$$\begin{aligned}\hat{l} - l_* = {} & \left\{\int_{\bar{h}}^T \Phi'(t)H'(t)Y(t-\bar{h},0)\,dt - a'Y(T,0)\right\}x(0) \\ & + \int_0^T \Phi'(t)\varrho(t)\,dt \\ & + \int_0^{T-\bar{h}} \left[\int_{\tau+\bar{h}}^T \Phi'(t)H'(t)Y(t-\bar{h},\tau)\,dt - a'Y(T,\tau)\right] C(\tau)u(\tau)\,d\tau \\ & - a'\int_{T-\bar{h}}^T Y(T,\tau)C(\tau)u(\tau)\,d\tau. \end{aligned} \tag{2.11}$$

Note that the Dirichlet formula $\int_{\bar{h}}^T \int_0^{t-\bar{h}} f(t,\tau)\,d\tau dt = \int_0^{T-\bar{h}} \int_{\tau+\bar{h}}^T f(t,\tau)\,dtd\tau$ is used in deriving (2.11) (cf. Lemma 2.1 from Chapter 2).

Let us also introduce the quantities:

(1) a function $\xi_\Phi(t)$, which for given $\Phi(t)$, $\Phi(\tau) = 0$, $T < \tau \le T + \bar{h}$, is such that[4]

$$\begin{aligned}&\dot{\xi}_\Phi(t) = -A'(t)\xi_\Phi(t) - B'(t+h)\xi_\Phi(t+h) + H(t+\bar{h})\Phi(t+\bar{h}), \\ &0 \le t \le T, \qquad \xi_\Phi(T) = a, \\ &\xi_\Phi(\tau) = 0, \qquad B(\tau) = 0, \quad H(\tau) = 0, \qquad T < \tau \le T + \max(h,\bar{h});\end{aligned} \tag{2.12}$$

(2) a functional $I(\Phi_-, \Phi, w)$: $\mathbf{R}^n \times \mathcal{F}^m \times L_1^r \to \mathbf{R}^1$:

$$I(\Phi_-, \Phi, w) = \sum_{d=1}^n \sigma_{-d}|\Phi_{-d}| + \sum_{l=1}^m \int_0^T \sigma_l(t)|\Phi_l(t)|\,dt \tag{2.13}$$

[4] As is usually done in linear systems, for $\Phi(t) = \Phi_{pc}(t) + \sum_{s=1}^M \Phi_{(s)}\delta(t - t_s) \in \mathcal{F}^m$, $\xi_\Phi(t) \triangleq \xi_{\Phi_{pc}}(t) + \sum_{s=1}^M \Phi_{(s)}\xi_{\delta_s}(t)$, where $\xi_{\delta_s}(t) \triangleq -Y'(t_s - \bar{h}, t)H(t_s)$.

$$+\sum_{j=1}^{r}\int_{0}^{T-\bar{h}} \gamma_j(t)|w_j(t)|\,dt$$

and the number

$$\kappa = \sum_{j=1}^{r}\int_{T-\bar{h}}^{T} \gamma_j(t)|C_j'(t)\xi_0(t)|\,dt, \qquad C(t) = (C_1(t),\ldots,C_r(t))'. \tag{2.14}$$

Using (2.9) and (2.12) we can easily show that

$$Y'(T,\tau)a - \int_{\tau+\bar{h}}^{T} Y'(t-\bar{h},\tau)H(t)\Phi(t)\,dt = \xi_\Phi(\tau), \qquad 0 \le \tau \le T. \tag{2.15}$$

Therefore, by virtue of (2.11), (2.15), we have

$$\hat{l} - l_* = -\xi_\Phi'(0)x(0) + \int_0^T \Phi'(t)\varrho(t)\,dt \tag{2.16}$$

$$-\int_0^{T-\bar{h}} \xi_\Phi'(\tau)C(\tau)u(\tau)\,d\tau - \int_{T-\bar{h}}^{T} \xi_0'(\tau)C(\tau)u(\tau)\,d\tau.$$

Hence, from (2.2), (2.4), (2.5), (2.8), (2.13), (2.14), and (2.16), we obtain

$$d(\Phi) = I(\Phi_-,\Phi,w) + \kappa, \tag{2.17}$$

where $\Phi_-(\Phi)$ and $w(\Phi)$ are defined by the constraints

$$0 = \Phi_- - \xi_\Phi(0), \qquad 0 = w(t) - C'(t)\xi_\Phi(t), \quad 0 \le t \le T-\bar{h}. \tag{2.18}$$

Thus, the optimal guaranteed estimation problem is reduced to the following nonsmooth variational problem:

$$I_0 = \inf_{\Phi_-\in\mathbf{R}^n,\,\Phi\in\mathcal{F}^m,\,w\in L_1^r} I(\Phi_-,\Phi,w) \tag{2.19}$$

under constraints (2.18).

Furthermore, $d_0 = I_0 + \kappa$.

Problem (2.19), (2.18) is not easy to solve numerically.

If $x(0)$ is a Gaussian random variable, and $u(t)$ and $\varrho(t)$ are Gaussian white noises, then the corresponding linear-quadratic filtering problem has a constructive solution algorithm [96], [97], [105], [10], [108], [106]. The optimal estimator φ of this problem extends the Kalman filter to systems with aftereffect. Moreover, the corresponding variational problem can be shown to be similar to problem (2.19), (2.18), in which quadratic terms are substituted for the nonsmooth terms in (2.13) (see Subsection 2.2 below).

We shall derive an approximate solution of problem (2.19), (2.18) by substituting a quadratic functional for the functional I and using the estimator φ instead of the optimal algorithm Φ^0.

In this case, the approximation quality is defined by the ratio

$$\Delta_A = \frac{d(\varphi)}{d(\Phi^0)} = \frac{d(\varphi)}{d_0}. \tag{2.20}$$

Here Δ_A is unknown, since d_0 is unknown.

Our aim here is to determine the upper bound Δ_A^0 of Δ_A, which can be computed without knowing the exact solution of problem (2.19), (2.18). The use of estimator φ is justified if Δ_A^0 is close to unity.

2.2. *Approximating Problem and Main Result*

Instead of the nonsmooth problem (2.19), (2.18), we consider a much simpler linear-quadratic problem

$$J_0 = \inf_{\Phi_- \in \mathbf{R}^n,\, \Phi \in L_2^m,\, w \in L_2^r} J(\Phi_-, \Phi, w), \tag{2.21}$$

$$J(\Phi_-, \Phi, w) \tag{2.22}$$

$$= \left\{ \Phi'_- P_0 \Phi_- + \int_0^T \Phi'(t) R(t) \Phi(t)\, dt + \int_0^{T-\bar{h}} w'(t) Q(t) w(t)\, dt \right\}^{\frac{1}{2}}$$

under constraints (2.18).

It is clear that the problem (2.21), (2.22), (2.18) can be obtained from (2.16) provided that we take the classical stochastic hypotheses

$$\mathbf{E}x(0) = 0, \qquad \mathbf{E}x(0)x'(0) = P_0,$$

$$\mathbf{E}u(t) = 0, \qquad \mathbf{E}u(t)u'(s) = Q(t)\delta(t-s),$$

$$\mathbf{E}\varrho(t) = 0, \qquad \mathbf{E}\varrho(t)\varrho'(s) = R(t)\delta(t-s)$$

and the second moment of estimation error is to be minimized.

For our deterministic guaranteed estimation problem, we formally define the matrices for (2.22) as

$$P_0 = \operatorname{diag}\{\sigma_{-1}^2, \ldots, \sigma_{-n}^2\}, \tag{2.23}$$

$$R(t) = \beta_1 \operatorname{diag}\{\sigma_1^2(t), \ldots, \sigma_m^2(t)\}, \qquad Q(t) = \beta_2 \operatorname{diag}\{\gamma_1^2(t), \ldots, \gamma_r^2(t)\},$$

where β_1, $\beta_2 > 0$ are given constants.

THEOREM 2.1. 1^0. *The solution* $\{\varphi_-, \varphi, w_\varphi\}$ *of the linear-quadratic problem* (2.18), (2.21), (2.22) *is given by the relations*

$$\varphi_- = \xi(0), \tag{2.24}$$

$$\varphi(t) = \begin{cases} R^{-1}(t)H'(t)\eta(t-\bar{h}) & \text{if } \bar{h} \le t \le T \\ 0 & \text{if } 0 \le t < \bar{h}, \end{cases}$$

$$w_\varphi(t) = C'(t)\xi(t), \qquad 0 \le t \le T - \bar{h},$$

where $\xi(t)$, $\eta(t)$ *satisfy the boundary value problem*

$$\dot{\xi}(t) = -A'(t)\xi(t) - B'(t+h)\xi(t+h) + H(t+\bar{h})R^{-1}(t+\bar{h})H'(t+\bar{h})\eta(t), \tag{2.25}$$

$$\dot{\eta}(t) = A(t)\eta(t) + B(t)\eta(t-h) + C(t)Q(t)C'(t)\xi(t), \qquad 0 \le t \le T,$$

with the boundary conditions[5]

$$\xi(T) = a, \qquad \xi(\tau) = 0, \qquad T < \tau \le T + h,$$

$$\eta(0) = P_0\xi(0), \qquad \eta(s) = 0, \qquad -h \le s < 0.$$

2^0. *Boundary value problem* (2.25) *has a unique solution. Moreover,*

$$\eta(t) = P(t)\xi(t) + \int_{-h}^{0} P_1(t,\tau)\xi(t-\tau)\,d\tau, \qquad 0 \le t \le T, \tag{2.26}$$

where the matrices $P(t)$, $P_1(t,\tau)$ *obey the system of functional differential equations:*

$$\dot{P}(t) = A(t)P(t) + P(t)A'(t) - P(t)H(t+\bar{h})R^{-1}(t+\bar{h})H'(t+\bar{h})P(t) + C(t)Q(t)C'(t) + P_1(t,0) + P_1'(t,0), \tag{2.27}$$

$$\left(\frac{\partial}{\partial t} + \frac{\partial}{\partial \tau}\right) P_1(t,\tau) = A(t)P_1(t,\tau) - P(t)H(t+\bar{h})R^{-1}(t+\bar{h})H'(t+\bar{h})P_1(t,\tau) + P_2(t,0,\tau),$$

[5] By definition, we put $H(\tau) = 0$ and $R^{-1}(\tau) = 0$ for $T < \tau \le T + \bar{h}$. The values of $B(\tau)$ for $T < \tau \le T + h$ can be defined arbitrarily; for definiteness, we also set them zero.

$$\left(\frac{\partial}{\partial t}+\frac{\partial}{\partial \tau}+\frac{\partial}{\partial s}\right) P_2(t,\tau,s)$$
$$= - P_1'(t,\tau)H(t+\bar{h})R^{-1}(t+\bar{h})H'(t+\bar{h})P_1(t,s),$$

$$0 \le t \le T, \qquad -h \le \tau, s \le 0$$

with the boundary conditions

$$P(0) = P_0, \qquad P_1(0,\tau) = P_2(0,\tau,s) = 0, \quad -h < \tau, s \le 0, \tag{2.28}$$

$$P_1'(t,-h) = B(t+h)P(t),$$

$$P_2(t,-h,\tau) = P_2'(t,\tau,-h) = B(t+h)P_1(t,\tau),$$

$$0 \le t \le T, \qquad -h \le \tau \le 0.$$

3^0. *The following equality holds:*

$$\xi(t) = \xi_\varphi(t), \qquad 0 \le t \le T, \tag{2.29}$$

where $\xi_\varphi(t)$ *is determined by* (2.12), *and*

$$\dot{\xi}(t) = -A'(t)\xi(t) - B'(t+h)\xi(t+h) \tag{2.30}$$
$$+ H(t+\bar{h})R^{-1}(t+\bar{h})H'(t+\bar{h})\left[P(t)\xi(t) + \int_{-h}^{0} P_1(t,\tau)\xi(t-\tau)\,d\tau\right],$$

$$0 \le t \le T, \qquad \xi(T) = a, \qquad \xi(\tau) = 0, \quad T < \tau \le T+h.$$

The proof of the theorem is contained in Subsection 2.5. For $h = \bar{h} = 0$, relations (2.24) – (2.30) are transformed into the well-known formulas of the Kalman filtering theory.

There exist constructive numerical algorithms for solving problem (2.18), (2.21), (2.22). Let us discuss briefly basic ideas of these algorithms.

The only nontrival part of the solution of problem (2.18), (2.21), (2.22) is the solution of the system of functional differential equations (2.27), (2.28). A constructive iterative numerical algorithm for solving system (2.27), (2.28) was developed in [96], [91], [97]. This method is based on the Bellman method of successive approximations. A k-th iteration consists of solving the linearized system

$$\dot{P}^{(k)}(t) = A(t)P^{(k)}(t) + P^{(k)}(t)A'(t)$$
$$- P^{(k-1)}(t)H(t+\bar{h})R^{-1}(t+\bar{h})H'(t+\bar{h})P^{(k)}(t)$$

$$- P^{(k)}(t)H(t+\bar{h})R^{-1}(t+\bar{h})H'(t+\bar{h})P^{(k-1)}(t)$$

$$+ P^{(k-1)}(t)H(t+\bar{h})R^{-1}(t+\bar{h})H'(t+\bar{h})P^{(k-1)}(t)$$

$$+ C(t)Q(t)C'(t) + P_1^{(k)}(t,0) + P_1^{(k)'}(t,0),$$

$$\left(\frac{\partial}{\partial t} + \frac{\partial}{\partial \tau}\right) P_1^{(k)}(t,\tau) = A(t)P_1^{(k)}(t,\tau)$$

$$- P^{(k-1)}(t)H(t+\bar{h})R^{-1}(t+\bar{h})H'(t+\bar{h})P_1^{(k)}(t,\tau)$$

$$- P^{(k)}(t)H(t+\bar{h})R^{-1}(t+\bar{h})H'(t+\bar{h})P_1^{(k-1)}(t,\tau)$$

$$+ P^{(k-1)}(t)H(t+\bar{h})R^{-1}(t+\bar{h})H'(t+\bar{h})P_1^{(k-1)}(t,\tau) + P_2^{(k)}(t,0,\tau),$$

$$\left(\frac{\partial}{\partial t} + \frac{\partial}{\partial \tau} + \frac{\partial}{\partial s}\right) P_2^{(k)}(t,\tau,s)$$

$$= - P_1^{(k-1)'}(t,\tau)H(t+\bar{h})R^{-1}(t+\bar{h})H'(t+\bar{h})P_1^{(k)}(t,s)$$

$$- P_1^{(k)'}(t,\tau)H(t+\bar{h})R^{-1}(t+\bar{h})H'(t+\bar{h})P_1^{(k-1)}(t,s),$$

$$+ P_1^{(k-1)'}(t,\tau)H(t+\bar{h})R^{-1}(t+\bar{h})H'(t+\bar{h})P_1^{(k-1)}(t,s),$$

$$0 \le t \le T, \qquad -h \le \tau, s \le 0$$

with the boundary conditions

$$P^{(k)}(0) = P_0, \qquad P_1^{(k)}(0,\tau) = P_2^{(k)}(0,\tau,s) = 0, \quad -h < \tau, s \le 0,$$

$$P_1^{(k)'}(t,-h) = B(t+h)P^{(k)}(t),$$

$$P_2^{(k)}(t,-h,\tau) = P_2^{(k)'}(t,\tau,-h) = B(t+h)P_1^{(k)}(t,\tau),$$

$$0 \le t \le T, \qquad -h \le \tau \le 0;$$

$$k = 1, \dots .$$

The starting matrices $P^{(0)}(t)$, $P_1^{(0)}(t,\tau)$, $P_2^{(0)}(t,\tau,s)$ may be taken arbitrary (they must satisfy the boundary conditions only). It is proved in [91] that

$$\max_{0 \le t \le T} \|P(t) - P^{(k)}(t)\| \le \frac{C}{k!}, \qquad k \ge 1,$$

where $\|K\|$ is the Euclidean norm of a matrix $K \in \mathbf{R}^{n\times n}$ and $C = \text{const}$. The norms of the differences $P_1 - P_1^{(k)}$ and $P_2 - P_2^{(k)}$ can be evaluated similarly:

$$\max_{t,\tau} \|P_1(t,\tau) - P_1^{(k)}(t,\tau)\| \le \frac{C_1}{k!}, \qquad \max_{t,\tau,s} \|P_2(t,\tau,s) - P_2^{(k)}(t,\tau,s)\| \le \frac{C_2}{k!},$$

$0 \le t \le T$, $-h \le \tau, s \le 0$. The crucial point of this method is that, for any k, the solution of this linearized system can be expressed explicitly via $P^{(k-1)}(t)$, $P_1^{(k-1)}(t,\tau)$, $P_2^{(k-1)}(t,\tau,s)$ with the help of certain quadratures (see also [105], [10], [108], [106]).

Another numerical method for solving system (2.27), (2.28) is as follows. Suppose that, for fixed t, we have computed a t-fiber $P(t)$, $P_1(t,\tau)$, $P_2(t,\tau,s)$ $(-h \le \tau, s \le 0)$. Then, for $\Delta t \ll 1$, we compute the $(t+\Delta t)$-fiber using the derivatives along the direction $e^0 = (1,1)'$ in the equation for $P_1(t,\tau)$ and along the direction $n^0 = (1,1,1)'$ in the equation for $P_2(t,\tau,s)$. The missing points are computed with the help of boundary conditions.

Thus, the estimator $\varphi(t)$ and the function $\xi_\varphi(t)$ (2.12) can be computed. Our main result is the following theorem.

THEOREM 2.2. *The nonoptimality level Δ_A satisfies the inequalities*

$$1 \le \Delta_A \le \Delta_A^0,$$

$$\Delta_A^0 = \frac{\mathcal{N}_1(\psi_-, \psi, \chi) + \kappa}{\mathcal{N}_2^2(\psi_-, \psi, \chi)\mathcal{N}_\infty^{-1}(\psi_-, \psi, \chi) + \kappa}. \tag{2.31}$$

Here,

$$\mathcal{N}_1 = \sum_{d=1}^{n} |\psi_{-d}| + \sum_{l=1}^{m} \beta_1^{-1} \int_{\bar{h}}^{T} |\psi_l(t)|\,dt + \sum_{j=1}^{r} \beta_2^{-1} \int_0^{T-\bar{h}} |\chi_j(t)|\,dt, \tag{2.32}$$

$$\mathcal{N}_2 = \left\{ \sum_{d=1}^{n} \psi_{-d}^2 + \sum_{l=1}^{m} \beta_1^{-1} \int_{\bar{h}}^{T} \psi_l^2(t)\,dt + \sum_{j=1}^{r} \beta_2^{-1} \int_0^{T-\bar{h}} \chi_j^2(t)\,dt \right\}^{\frac{1}{2}},$$

$$\mathcal{N}_\infty = \max \left\{ \max_{1\le d\le n} |\psi_{-d}|, \quad \sup_{\substack{1\le l\le m,\\ t\in[\bar{h},T]}} |\psi_l(t)|, \quad \sup_{\substack{1\le j\le r,\\ t\in[0,T-\bar{h}]}} |\chi_j(t)| \right\},$$

where

$$\psi_{-d} = \sigma_{-d}\varphi_{-d}, \quad \psi_l(t) = \beta_1\sigma_l(t)\varphi_l(t), \quad \chi_j(t) = \beta_2\gamma_j(t)C_j'(t)\xi_\varphi(t), \tag{2.33}$$

$$\varphi_- = (\varphi_{-1}, \dots, \varphi_{-n})', \qquad \varphi(t) = (\varphi_1(t), \dots, \varphi_m(t))'.$$

The proof of this theorem is based on the duality theory of convex variational problems and is given in Subsection 2.3. The estimate (2.31) extends the corresponding estimate from Section 2 of Chapter 3 to systems with delay.

The algorithm for calculating (2.31) consists of the following steps.

Step 1. Solution of differential equations (2.27) under boundary conditions (2.28), where the matrices P_0, $Q(t)$, and $R(t)$ are given by (2.23); constructive numerical procedures exist to realize this step.

Step 2. Solution of differential equation (2.30) to determine the function $\xi(t) = \xi_\varphi(t)$ and the vector $\xi(0) = \xi_\varphi(0) = \varphi_-$.

Step 3. Computation of the estimator for the linear-quadratic problem φ by formulas (2.24), (2.26).

Step 4. Computation of the auxiliary functions (2.33) and quadratures (2.32).

Step 5. Computation of the nonoptimality level (2.31).

Steps 1 – 3 consist in solving the corresponding linear-quadratic problem, while steps 4 and 5 consist in computing the quadratures and elementary operations over them. Hence, the linear-quadratic theory can be fruitfully applied to analyze minimax filtering problems in systems with delay.

REMARK 2.2. The result can be generalized to several and distributed delays (cf. Section 5). Moreover, it can be extended to the case in which, along with uncertain bounded perturbations, white noises are also present (see Section 3).

REMARK 2.3. Simpler algorithms for approximately solving the mean-square filtering problem (2.21), (2.18) can be constructed from the structure of the optimal estimator φ. An estimate similar to (2.31) can also be obtained for these simplified algorithms (see Sections 4 and 5).

2.3. *Proof of Main Result*

First, let us construct the corresponding dual problems.

THEOREM 2.3. 1^0. *The problem dual to* (2.19), (2.18) *has the form*

$$I^0 = \sup_{p \in L_\infty^r, \lambda \in \mathbf{R}^n} a' \left[\int_0^{T-\bar{h}} Y(T,t)C(t)p(t)\,dt + Y(T,0)\lambda \right] \tag{2.34}$$

under constraints

$$\sigma_l^{-1}(t) \left| H_l'(t) \left[Y(t-\bar{h},0)\lambda + \int_0^{t-\bar{h}} Y(t-\bar{h},s)C(s)p(s)\,ds \right] \right| \leq 1, \tag{2.35}$$

$$l = 1, \ldots, m, \qquad \bar{h} \leq t \leq T,$$

$$\gamma_j^{-1}(\tau)|p_j(\tau)| \le 1, \qquad p(\tau) = (p_1(\tau), \ldots, p_r(\tau))',$$

$$j = 1, \ldots, r, \qquad 0 \le \tau \le T - \bar{h},$$

$$\sigma_{-d}^{-1} |\lambda_d| \le 1, \qquad \lambda = (\lambda_1, \ldots, \lambda_n)', \qquad d = 1, \ldots, n.$$

2^0. *The following equality holds:*

$$I_0 = I^0. \tag{2.36}$$

The proof of the theorem is given in Subsection 2.6.

THEOREM 2.4. 1^0. *The problem dual to* (2.18), (2.21) *has the form*

$$J^0 = \sup_{p \in L_2^r, \lambda \in \mathbf{R}^n} a' \left[\int_0^{T-\bar{h}} Y(T,t)C(t)p(t)\,dt + Y(T,0)\lambda \right] \tag{2.37}$$

under the constraint

$$\left\{ \lambda' P_0^{-1} \lambda + \int_{\bar{h}}^T \left[Y(t-\bar{h},0)\lambda + \int_0^{t-\bar{h}} Y(t-\bar{h},s)C(s)p(s)\,ds \right]' \right. \tag{2.38}$$

$$\times H(t)R^{-1}(t)H'(t) \left[Y(t-\bar{h},0)\lambda + \int_0^{t-\bar{h}} Y(t-\bar{h},s)C(s)p(s)\,ds \right] dt$$

$$\left. + \int_0^{T-\bar{h}} p'(t)Q^{-1}(t)p(t)\,dt \right\}^{\frac{1}{2}} \le 1.$$

2^0. *The following equalities hold:*

$$J^0 = J_0 = \left\{ a'P(T)a - \int_{T-\bar{h}}^T \xi'(t)C(t)Q(t)C'(t)\xi(t)\,dt \right\}^{\frac{1}{2}}; \tag{2.39}$$

3^0. *The solution of* (2.37), (2.38) *is given by the formulas*

$$p^0(t) = J_0^{-1}Q(t)C'(t)\xi(t), \qquad 0 \le t \le T - \bar{h}, \tag{2.40}$$

$$\lambda^0 = J_0^{-1}\eta(0) = J_0^{-1}P_0\xi(0),$$

where $\xi(t)$, $\eta(t)$ *are defined by the boundary value problem* (2.25).

The proof is given in Subsection 2.5.

The dual problem (2.34), (2.35) is as complicated as the primal one. But, using the dual problem, the optimal value of the primal problem can be estimated from below.

In accordance with our general approach let us search the solution of problem (2.34), (2.35) along the ray $\nu(p^0, \lambda^0)$, $\nu \in \mathcal{R}^1$, $\nu > 0$.

According to (2.23), (2.35), (2.40), the points on the ray are admissible if and only if

$$\nu\sigma_l^{-1}(t)J_0^{-1}\Big|H_l'(t)\Big[Y(t-\bar{h},0)\eta(0) + \int_0^{t-\bar{h}} Y(t-\bar{h},s)C(s)Q(s)C'(s)\xi(s)\,ds\Big]\Big| \le 1, \tag{2.41}$$

$$l = 1,\dots,m, \qquad \bar{h} \le t \le T.$$

$$\nu J_0^{-1}|\beta_2\gamma_j(\tau)C_j'(\tau)\xi(\tau)| \le 1, \qquad j = 1,\dots,r, \qquad 0 \le \tau \le T-\bar{h},$$

$$\nu J_0^{-1}|\sigma_{-d}\xi_d(0)| \le 1, \qquad d = 1,\dots,n, \qquad \xi(0) = (\xi_1(0),\dots,\xi_n(0))'.$$

By virtue of (2.24) and the second equation of (2.25), we find from (2.41) that

$$\nu J_0^{-1}|\beta_1\sigma_l(t)\varphi_l(t)| \le 1, \qquad l = 1,\dots,m, \qquad \bar{h} \le t \le T, \tag{2.42}$$

$$\nu J_0^{-1}|\beta_2\gamma_j(\tau)C_j'(\tau)\xi(\tau)| \le 1, \qquad j = 1,\dots,r, \qquad 0 \le \tau \le T-\bar{h},$$

$$\nu J_0^{-1}|\sigma_{-d}\varphi_{-d}| \le 1, \qquad d = 1,\dots,n, \qquad \varphi_- = (\varphi_{-1},\dots,\varphi_{-n})'.$$

Hence, with regard for (2.29) and (2.32), from (2.42) we find that

$$0 < \nu \le \nu^0 = J_0\mathcal{N}_\infty^{-1}. \tag{2.43}$$

Under condition (2.43), the optimal value of the cost functional satisfies the inequality

$$I_0 = I^0 \ge \nu^0 a'\Big[\int_0^{T-\bar{h}} Y(T,t)C(t)p^0(t)\,dt + Y(T,0)\lambda^0\Big] = \nu^0 J_0 \tag{2.44}$$

$$= \mathcal{N}_\infty^{-1}J_0^2 = \mathcal{N}_\infty^{-1}J^2(\varphi_-,\varphi,w_\varphi) = \mathcal{N}_\infty^{-1}\mathcal{N}_2^2, \qquad w_\varphi(t) = C'(t)\xi_\varphi(t).$$

Obviously, from (2.17), (2.18), (2.29), we obtain

$$\mathcal{D}(\varphi) = \mathcal{N}_1 + \kappa. \tag{2.45}$$

Then the upper estimate for (2.20) follows from (2.44) and (2.45). □

Finally, let us formulate a useful asymptotic result for the case of poor *a priori* information about the initial state.

The system (2.1), (2.3) is completely observable on $[0, T]$ if the matrix

$$\mathcal{O} = \int_{\bar{h}}^{T} Y'(t-\bar{h}, 0)H(t)R^{-1}(t)H'(t)Y(t-\bar{h}, 0)\,dt \quad \text{is nonsingular,} \tag{2.46}$$

where $Y(t, 0)$ and $R(t)$ are defined by (2.9) and (2.23), respectively.

LEMMA 2.1. *Let condition (2.46) be satisfied. If*

$$\sigma_{-d} = \sigma_{-d}(\varepsilon) \equiv \varepsilon^{-1}\bar{\sigma}_{-d}, \qquad \bar{\sigma}_{-d} = \text{const}, \quad d = 1, \ldots, n,$$

and $\varepsilon \to 0^+$, *then the following asymptotic relations hold:*

$$\lim_{\varepsilon \to 0+} \varphi_-(\varepsilon) = 0, \qquad \lim_{\varepsilon \to 0+} \sigma_{-d}(\varepsilon)\varphi_{-d}(\varepsilon) = 0, \qquad d = 1, \ldots, n,$$

$$\lim_{\varepsilon \to 0+} \|\varphi(\cdot, \varepsilon) - \varphi_\infty(\cdot)\|_{L_2^m} = 0,$$

$$\lim_{\varepsilon \to 0+} \max_{t \in [0,T]} \|\xi(t, \varepsilon) - \xi_\infty(t)\|_{\mathbf{R}^n} = 0, \quad \lim_{\varepsilon \to 0+} \max_{t \in [0,T]} \|\eta(t, \varepsilon) - \eta_\infty(t)\|_{\mathbf{R}^n} = 0,$$

where $\varphi_\infty(t)$ *is a unique solution of the variational problem*

$$\inf_{\Phi \in L_2^m} \left(\int_0^T \Phi'(t)R(t)\Phi(t)\,dt + \int_0^{T-\bar{h}} \xi'_\Phi(t)C(t)Q(t)C'(t)\xi_\Phi(t)\,dt \right)$$

under the constraint $\xi_\Phi(0) = 0$.

The proof of Lemma 2.1 is completely analogous to the proof of Lemma 2.1 in Section 2 of Chapter 3.

2.4. *Simplest Example*

EXAMPLE 2.1. Let $n = m = 1$, $A(t) = 0$, $B(t) = B = \text{const}$, $C(t) = 0$, $H(t) = 1$, $a = 1$, $\sigma(t) = \sigma = \text{const}$. Hence, the system and the measurements are of the form

$$\dot{x}(t) = Bx(t-h), \qquad x(\tau) = 0, \quad -h \le \tau < 0, \qquad |x(0)| \le \sigma_-, \tag{2.47}$$

$$z(t) = x(t-\bar{h}) + \varrho(t), \qquad |\varrho(t)| \le \sigma, \qquad t \in [0, T], \qquad \sigma_-, \sigma > 0.$$

Now we are required to evaluate $x(T)$.

According to (2.23), we put $P_0 = \sigma_-^2$, $R = \beta_1 \sigma^2$ for some constant parameter $\beta_1 > 0$. In this case, the system of differential equations (2.27), (2.28) takes the form

$$\dot{P}(t) = -H(t+\bar{h})R^{-1}P^2(t) + 2P_1(t,0),$$

$$\left(\frac{\partial}{\partial t} + \frac{\partial}{\partial \tau}\right) P_1(t,\tau) = -H(t+\bar{h})R^{-1}P(t)P_1(t,\tau) + P_2(t,0,\tau),$$

$$\left(\frac{\partial}{\partial t} + \frac{\partial}{\partial \tau} + \frac{\partial}{\partial s}\right) P_2(t,\tau,s) = H(t+\bar{h})R^{-1}P_1(t,\tau)P_1(t,s),$$

$$0 \le t \le T, \qquad -h \le \tau, s \le 0$$

with the boundary conditions

$$P(0) = P_0, \qquad P_1(0,\tau) = P_2(0,\tau,s) = 0, \quad -h < \tau, s \le 0,$$

$$P_1(t,-h) = BP(t), \qquad P_2(t,-h,\tau) = P_2(t,\tau,-h) = BP_1(t,\tau),$$

$$0 \le t \le T, \qquad -h \le \tau \le 0.$$

It can easily be checked that this system has an explicit solution

$$P(t) = Y^2(t,0)L(t), \qquad L(t) = \left(R^{-1}\int_0^t Y^2(\tau,0)H(\tau+\bar{h})\,d\tau + P_0^{-1}\right)^{-1},$$

$$P_1(t,\tau) = BY(t,0)Y(t-\tau-h,0)L(t),$$

$$P_2(t,\tau,s) = B^2Y(t-\tau-h,0)Y(t-s-h,0)L(t),$$

where

$$H(t) = \begin{cases} 1, & 0 \le t \le T, \\ 0, & T < t \le T+\bar{h}, \end{cases}$$

and $Y(t,0)$ is defined by (2.9):

$$\dot{Y}(t,0) = BY(t-h,0), \qquad Y(0,0) = 1, \qquad Y(\tau,0) = 0, \quad -h \le \tau < 0.$$

It is not hard to prove that

$$Y(t,0) = 1 + \sum_{j=1}^{p} \frac{B^j(t-jh)^j}{j!}, \quad ph \le t < (p+1)h, \quad p = 0,1,\ldots. \tag{2.48}$$

The optimal estimator for the linear-quadratic problem is computed by formulas (2.24):

$$\varphi_- = \xi(0), \qquad \varphi(t) = \begin{cases} R^{-1}\eta(t-\bar{h}), & \bar{h} \le t \le T \\ 0, & 0 \le t < \bar{h}, \end{cases} \tag{2.49}$$

where $\xi(t)$, $\eta(t)$ satisfy the boundary value problem

$$\dot{\xi}(t) = -B\xi(t+h) + H(t+\bar{h})R^{-1}\eta(t),$$

$$\dot{\eta}(t) = B\eta(t-h), \qquad 0 \le t \le T,$$

with the boundary conditions

$$\xi(T) = 1, \qquad \xi(\tau) = 0, \qquad T < \tau \le T+h,$$

$$\eta(0) = P_0\xi(0), \qquad \eta(s) = 0, \qquad -h \le s < 0.$$

Furthermore, due to (2.26) and (2.30),

$$\eta(t) = P(t)\xi(t) + \int_{-h}^{0} P_1(t,\tau)\xi(t-\tau)\,d\tau, \qquad 0 \le t \le T, \tag{2.50}$$

and

$$\dot{\xi}(t) = -B\xi(t+h) + H(t+\bar{h})R^{-1}\left[P(t)\xi(t) + \int_{-h}^{0} P_1(t,\tau)\xi(t-\tau)\,d\tau\right] \tag{2.51}$$

or,

$$\dot{\xi}(t) = -B\xi(t+h) + H(t+\bar{h})\varphi(t+\bar{h}), \tag{2.52}$$

$$0 \le t \le T, \qquad \xi(T) = 1, \qquad \xi(\tau) = 0, \quad T < \tau \le T+h.$$

A regular way for finding φ is to solve equation (2.51) and then to use formulas (2.49), (2.50). Howerever, equation (2.51) has a quite complicated form to find its solution immediately. Therefore we shall use a trick.

For our case ($C(t) \equiv 0$), the computations can be greatly reduced. In fact, the measurements can be represented as

$$z(t) = Y(t-\bar{h},0)x(0) + \varrho(t),$$

and it is required to estimate the quantity $Y(T,0)x(0)$. Thus, another representation for our problem is obtained:

$$n = m = 1, \qquad A(t) = 0, \qquad B(t) = 0, \qquad C(t) = 0, \tag{2.53}$$

$$H(t) = Y(t-\bar{h},0), \qquad a = Y(T,0),$$

and delays are absent. Using (2.27), (2.28), and (2.53), we can easily show that (the bar is herein taken to mean that we use another representation)

$$\bar{P}(t) = \left(R^{-1}\int_0^t Y^2(\tau-\bar{h},0)\,d\tau + P_0^{-1}\right)^{-1}, \tag{2.54}$$

$$\bar{P}_1(t,\tau)\equiv 0, \qquad \bar{P}_2(t,\tau,s)\equiv 0,$$

$$\dot{\bar{\xi}}(t) = Y^2(t-\bar{h},0)R^{-1}\bar{P}(t)\bar{\xi}(t), \qquad \bar{\xi}(T) = Y(T,0).$$

Using (2.54), the last ordinary differential equation can be represented in the form

$$\frac{d}{dt}\left[\bar{\xi}(t)\left(R^{-1}\int_0^t Y^2(\tau-\bar{h},0)\,d\tau + P_0^{-1}\right)^{-1}\right] \equiv 0,$$

whence,

$$\bar{\xi}(t) = \omega R\left(R^{-1}\int_0^t Y^2(\tau-\bar{h},0)\,d\tau + P_0^{-1}\right), \qquad \bar{\eta}(t) = \bar{P}(t)\bar{\xi}(t),$$

$$\varphi(t) = R^{-1}Y(t-\bar{h},0)\bar{P}(t)\bar{\xi}(t) = \omega Y(t-\bar{h},0), \qquad \varphi_- = P_0^{-1}R\omega, \tag{2.55}$$

where

$$\omega = R^{-1}Y(T,0)\left(R^{-1}\int_0^T Y^2(t-\bar{h},0)\,dt + P_0^{-1}\right)^{-1} = \text{const}.$$

Now, using (2.55), we find from (2.52) and Corollary 1.2 that

$$\xi(t) = Y(T,t) - \omega\int_t^T Y(s,t)Y(s,0)H(s+\bar{h})\,ds, \qquad t\in[0,T], \tag{2.56}$$

where, in accordance with (2.9),

$$\frac{\partial Y(t,s)}{\partial s} = -BY(t,s+h),$$

$$Y(T,T)=1, \qquad Y(T,\tau)=0, \quad T<\tau\le T+h.$$

So, all elements of our variational problem are found explicitly.

As a useful exercise, let us prove by direct differentiation that $\xi(t)$ defined by (2.56) satisfies (2.51). For example, let us consider the case $t+h\le T$. In fact, from the one hand,

$$\dot{\xi}(t) + B\xi(t+h) = \omega H(t+\bar{h})Y(t,0).$$

From the other hand,

$$P(t)\xi(t) + \int_{-h}^{0} P_1(t,\tau)\xi(t-\tau)\,d\tau$$

$$= Y^2(t,0)L(t)\left(Y(T,t) - \omega\int_t^T Y(s,t)Y(s,0)H(s+\bar{h})\,ds\right)$$

$$+ \int_{-h}^{0} BY(t,0)Y(t-\tau-h,0)L(t)$$

$$\times\left(Y(T,t-\tau) - \omega\int_{t-\tau}^{T} Y(s,t-\tau)Y(s,0)H(s+\bar{h})\,ds\right)d\tau$$

$$= L(t)Y(t,0)\left\{\mathcal{Y}_1 - \omega\mathcal{Y}_2 - \omega B\mathcal{Y}_3\right\},$$

where

$$\mathcal{Y}_1 = Y(T,t)Y(t,0) + B\int_{-h}^{0} Y(T,t-\tau)Y(t-\tau-h,0)\,d\tau,$$

$$\mathcal{Y}_2 = Y(t,0)\int_t^T Y(s,t)Y(s,0)H(s+\bar{h})\,ds,$$

$$\mathcal{Y}_3 = \int_{-h}^{0} Y(t-\tau-h,0)\int_{t-\tau}^{T} Y(s,t-\tau)Y(s,0)H(s+\bar{h})\,dsd\tau,$$

or, by virtue of (1.6),

$$\mathcal{Y}_3 = \int_{-h}^{0} Y(t-\tau-h,0)\int_{0}^{T} Y(s,t-\tau)Y(s,0)H(s+\bar{h})\,dsd\tau.$$

In accordance with Lemma 1.1, for any $0 \le t \le s$,

$$Y(s,0) = Y(s,t)Y(t,0) + B\int_{t-h}^{t} Y(s,\tau+h)Y(\tau,0)\,d\tau \tag{2.57}$$

and, in particular,

$$Y(T,0) = Y(T,t)Y(t,0) + B\int_{t-h}^{t} Y(T,\tau+h)Y(\tau,0)\,d\tau$$

$$= Y(T,t)Y(t,0) + B\int_{-h}^{0} Y(T,t-\tau)Y(t-\tau-h,0)\,d\tau.$$

Hence, $\mathcal{Y}_1 = Y(T,0)$.

Furthermore, interchanging the integration and using (1.6), we have

$$\mathcal{Y}_3 = \int_0^T \int_{-h}^0 Y(t-\tau-h,0)Y(s,t-\tau)\,d\tau\, Y(s,0)H(s+\bar{h})\,ds$$

$$= \int_0^T \int_{t-h}^t Y(s,\tau+h)Y(\tau,0)\,d\tau\, Y(s,0)H(s+\bar{h})\,ds$$

$$= \int_t^T \int_{t-h}^t Y(s,\tau+h)Y(\tau,0)\,d\tau\, Y(s,0)H(s+\bar{h})\,ds.$$

By virtue of (2.57),

$$\mathcal{Y}_2 + B\mathcal{Y}_3 = \int_t^T Y^2(s,0)H(s+\bar{h})\,ds.$$

Since, obviously,

$$\int_0^T Y^2(\tau-\bar{h},0)\,d\tau = \int_0^T Y^2(s,0)H(s+\bar{h})\,ds,$$

we obtain finally that

$$H(t+\bar{h})R^{-1}\left(P(t)\xi(t) + \int_{-h}^0 P_1(t,\tau)\xi(t-\tau)\,d\tau\right)$$

$$= H(t+\bar{h})R^{-1}L(t)Y(t,0)\left(Y(T,0) - \omega\int_t^T Y^2(s,0)H(s+\bar{h})\,ds\right)$$

$$= H(t+\bar{h})R^{-1}L(t)Y(t,0)Y(T,0)L(T)L^{-1}(t) = \omega H(t+\bar{h})Y(t,0).$$

Thus, $\xi(t)$ defined by (2.56) satisfies (2.51) for $t+h \le T$. The case $t+h > T$ can be treated similarly.

REMARK 2.4. It directly follows from the proof of Theorem 2.2 that, for the scalar case (and with $C(t) = 0$), $\Delta_A = \Delta_A^0$, i.e., the estimate (2.31) is attainable.

Additionally, let us assume that there is no *a priori* information about $x(0)$:

$$\sigma_- = \varepsilon^{-1}\bar{\sigma}_-, \qquad \varepsilon \to 0.$$

TABLE 1. Values of $\Delta_A = \Delta_A^0$ $(h = \bar{h})$

	hT^{-1}					
BT	0.0	0.05	0.1	0.2	0.25	0.5
-1.0	1.46	1.39	1.32	1.21	1.15	1.00
-2.0	1.76	1.62	1.48	1.28	1.22	1.00
-5.0	1.98	1.55	1.29	1.36	1.22	1.00

Then, by virtue of Theorem 2.2 and formulas (2.55) (cf. Lemma 2.1[6]), the estimate Δ_A^0 takes the simplest form

$$\Delta_A = \Delta_A^0 = \frac{\int_{\bar{h}}^T |\varphi_\infty(t)|\,dt \ \max_{\bar{h}\le t\le T} |\varphi_\infty(t)|}{\int_{\bar{h}}^T \varphi_\infty^2(t)\,dt}, \tag{2.58}$$

$$\varphi_\infty(t) = \omega_1 Y(t-\bar{h},0), \qquad \omega_1 = Y(T,0)\left(\int_0^T Y^2(\tau-\bar{h},0)\,d\tau\right)^{-1} = \text{const}.$$

It follows from (2.58) that Δ_A^0 does not depend on β_1 and σ, but depends only on B, h, $\bar{h}$, and T. Hence, to find Δ_A^0, we must substitute (2.48) into (2.58). From (2.47) and (2.58) it follows that Δ_A^0 depends on BT, hT^{-1}, and $\bar{h}T^{-1}$.

Certain numerical results for $h = \bar{h}$ are presented in Table 1. The table shows that the use of the linear-quadratic estimator does not dramatically affect the estimation quality. In a number of cases, the accuracy of the linear-quadratic estimator is quite close to the accuracy of the optimal estimator. It should be noted that, in our example, the growth of delay aids in decreasing the nonoptimality level Δ_A^0.

REMARK 2.5. It is a simple matter to show that the optimal guaranteed estimator for our example is

$$\Phi^0(t) = Y(T,0)Y^{-1}(t^0-\bar{h},0)\delta(t-t^0), \qquad t^0 = \operatorname*{argmax}_{\bar{h}\le t\le T} |Y(t-\bar{h},0)|.$$

2.5. *Proof of Theorems 2.1 and 2.4*

Since constraints (2.18) do not depend on $\Phi(t)$, $0 \le t < \bar{h}$, the optimal estimator $\Phi(t)$ vanishes on $[0,\bar{h})$. Therefore, the first integral in (2.22) can be taken over $[\bar{h},T]$.

[6]The scalar system (2.47) satisfies condition (2.46).

Let us introduce a Hilbert space $\mathcal{H} = \mathbf{R}^n \times L_2^m[\bar{h}, T] \times L_2^r[0, T-\bar{h}]$ with elements $\Gamma = (\Phi_-, \Phi, w)$ and the inner product

$$\langle \Gamma_{(1)}, \Gamma_{(2)} \rangle_{\mathcal{H}} = \Phi'_{-(1)}\Phi_{-(2)} + \int_{\bar{h}}^{T} \Phi'_{(1)}(t)\Phi_{(2)}(t)\,dt + \int_0^{T-\bar{h}} w'_{(1)}(t)w_{(2)}(t)\,dt.$$

Then J in (2.22) maps $\mathcal{H}$ into $\mathbf{R}^1$.

The Hilbert space $\mathcal{H}$ is reflexive, the cost function (2.22) is convex, continuous, and coercive; and constraints (2.18) define a closed convex manifold. Hence an optimal solution (Φ_-^0, Φ^0, w^0) exists (see Section 1 of Chapter 2).

Let us also introduce the Lagrange multipliers $p \in L_2^r[0, T-\bar{h}]$, $\lambda \in \mathbf{R}^n$ for constraints (2.18). Then (by virtue of (2.15)) the Lagrange function for problem (2.21), (2.22), (2.18) is of the form

$$\mathcal{L}_{\mathcal{H}}(\Phi_-, \Phi, w; p, \lambda) \tag{2.59}$$

$$= \left\{ \Phi'_- P_0 \Phi_- + \int_{\bar{h}}^{T} \Phi'(t)R(t)\Phi(t)\,dt + \int_0^{T-\bar{h}} w'(t)Q(t)w(t)\,dt \right\}^{\frac{1}{2}}$$

$$+ \int_0^{T-\bar{h}} p'(t)\left\{ -w(t) - C'(t)\int_{t+\bar{h}}^{T} Y'(s-\bar{h}, t)H(s)\Phi(s)\,ds \right.$$

$$\left. + C'(t)Y'(T,t)a \right\} dt$$

$$+ \lambda'\left\{ Y'(T,0)a - \int_{\bar{h}}^{T} Y'(t-\bar{h}, 0)H(t)\Phi(t)\,dt - \Phi_- \right\}.$$

Using the Dirichlet formula (cf. Lemma 2.1 of Chapter 2)

$$\int_0^{T-\bar{h}} p'(t)C'(t)\int_{t+\bar{h}}^{T} Y'(s-\bar{h}, t)H(s)\Phi(s)\,dsdt$$

$$= \int_{\bar{h}}^{T} \Phi'(s)H'(s)\int_0^{s-\bar{h}} Y(s-\bar{h}, t)C(t)p(t)\,dtds,$$

from (2.59) we obtain

$$\mathcal{L}_{\mathcal{H}}(\Phi_-, \Phi, w; p, \lambda) = \left\| \left(P_0^{\frac{1}{2}}\Phi_-,\ R^{\frac{1}{2}}(t)\Phi(t),\ Q^{\frac{1}{2}}(t)w(t) \right) \right\|_{\mathcal{H}} \tag{2.60}$$

$$+ \left\langle \left(-P_0^{-\frac{1}{2}}\lambda,\ R^{-\frac{1}{2}}(t)H'(t)\tilde{\eta}(t-\bar{h}),\ -Q^{-\frac{1}{2}}(t)p(t) \right), \right.$$

$$\left. \left(P_0^{\frac{1}{2}}\Phi_-,\ R^{\frac{1}{2}}(t)\Phi(t),\ Q^{\frac{1}{2}}(t)w(t) \right) \right\rangle_{\mathcal{H}}$$

$$+\left[\int_0^{T-\bar{h}} p'(t)C'(t)Y'(T,t)\,dt + \lambda' Y'(T,0)\right] a,$$

where

$$\tilde{\eta}(t) = -\left[Y(t,0)\lambda + \int_0^t Y(t,s)C(s)p(s)\,ds\right], \qquad 0 \le t \le T-\bar{h}. \tag{2.61}$$

According to the duality theory (see Section 1 of Chapter 2), the dual problem has the form

$$J^0 = \sup_{p,\lambda}\left[\inf_{\Phi_-,\Phi,w} \mathcal{L}_{\mathcal{H}}(\Phi_-,\Phi,w;p,\lambda)\right]. \tag{2.62}$$

Moreover,

$$J^0 = J_0. \tag{2.63}$$

Since $\inf \mathcal{L}_{\mathcal{H}}$ is equal to

$$\begin{cases} a'\left[Y(T,0)\lambda + \int_0^{T-\bar{h}} Y(T,t)C(t)p(t)\,dt\right] \\ \qquad\qquad \text{if } \left\|\left(P_0^{-\frac{1}{2}}\lambda,\ R^{-\frac{1}{2}}(t)H'(t)\tilde{\eta}(t-\bar{h}),\ Q^{-\frac{1}{2}}(t)p(t)\right)\right\|_{\mathcal{H}} \le 1 \\ -\infty \qquad\qquad \text{otherwise}, \end{cases}$$

it follows from (2.62) that the dual problem is described by (2.37), (2.38). This completes the proof of Assertion 1 and the first equality of Asserrtion 2 of Theorem 2.4.

The Hilbert space $\mathbf{R}^n \times L_2^r[0, T-\bar{h}]$ is reflexive, the cost function (2.37) is continuous and convex (linear), and constraint (2.38) defines a convex bounded closed set. Hence (see Section 1 of Chapter 2), an optimal solution (p^0, λ^0) for (2.37), (2.38) exists.

By the Lagrange theorem (see Section 1 of Chapter 2 and Section 3 of Chapter 3) from (2.60) we find that

$$\Phi_-^0 = -kP_0^{-1}\lambda^0, \tag{2.64}$$

$$\Phi^0(t) = kR^{-1}(t)H'(t)\bar{\eta}(t-\bar{h}),$$

$$w^0(\tau) = -kQ^{-1}(\tau)p^0(\tau),$$

$$\bar{h} \le t \le T, \qquad 0 \le \tau \le T-\bar{h}, \qquad \bar{\eta}(t) = \tilde{\eta}(t)\Big|_{p=p^0,\ \lambda=\lambda^0}$$

for some $k \in \mathbf{R}^1$, which is determined from constraints (2.18).

Let us now find the optimal solution (p^0, λ^0) of dual problem (2.37), (2.38). Obviously, without loss of generality, the exponent $\frac{1}{2}$ in condition (2.38) can be omitted. The corresponding Lagrange function takes the form

$$\mathcal{L}^0(p,\lambda;\alpha) = -a'\left[Y(T,0)\lambda + \int_0^{T-\bar{h}} Y(T,t)C(t)p(t)\,dt\right] \tag{2.65}$$

$$+\alpha\left\{\lambda' P_0^{-1}\lambda + \int_{\bar{h}}^{T} \bar{\eta}'(t-\bar{h})H'(t)R^{-1}(t)H(t)\bar{\eta}(t-\bar{h})\,dt\right.$$

$$\left. + \int_0^{T-\bar{h}} p'(t)Q^{-1}(t)p(t)\,dt\right\}, \qquad \alpha \geq 0.$$

If

$$Y'(T,0)a = 0 \quad \text{and} \quad C'(t)Y'(T,t)a \equiv 0, \tag{2.66}$$

then it follows from (2.15) and (2.18) that $\Phi_- = 0$, $\Phi = 0$, $w = 0$ are admissible elements for the minimax problem. Consequently, the optimal estimator is zero and the minimax problem is resolved with $I_0 = 0$. In the sequel, we omit this trivial case. Then it follows from (2.65) that $\alpha > 0$, since the optimal solution exists. Indeed, if $\alpha = 0$ and (2.66) does not hold, then the necessary conditions of the Lagrange theorem is violated.

The function $\mathcal{L}^0$ is linear-quadratic (with positive definite quadratic form). Therefore its minimum is given by the stationary point

$$\frac{\partial \mathcal{L}^0}{\partial p} = 0, \qquad \frac{\partial \mathcal{L}^0}{\partial \lambda} = 0.$$

Direct calculations in these relations lead to the following equalities

$$-C'(t)Y'(T,t)a + 2\alpha Q^{-1}(t)p(t)$$

$$-2\alpha C'(t)\int_{t+\bar{h}}^{T} Y'(s-\bar{h},t)H(s)R^{-1}(s)H'(s)\bar{\eta}(s-\bar{h})\,ds = 0,$$

$$0 \leq t \leq T-\bar{h},$$

and

$$-Y'(T,0)a + 2\alpha P_0^{-1}\lambda$$

$$+2\alpha\int_{\bar{h}}^{T} Y'(t-\bar{h},0)H(t)R^{-1}(t)H'(t)Y(t-\bar{h},0)\,dt\,\lambda$$

$$+2\alpha\int_0^{T-\bar{h}}\int_{s+\bar{h}}^{T} Y'(t-\bar{h},0)H(t)R^{-1}(t)H'(t)Y(t-\bar{h},s)\,dt\,C(s)p(s)\,ds = 0.$$

Thus we get

$$p^0(t) = -Q(t)C'(t)\bar{\xi}(t), \qquad \lambda^0 = -P_0\bar{\xi}(0), \qquad 0 \le t \le T - \bar{h}, \tag{2.67}$$

where

$$\bar{\xi}(t) = -\int_{t+\bar{h}}^{T} Y'(s-\bar{h},t)H(s)R^{-1}(s)H'(s)\bar{\eta}(s-\bar{h})\,ds - \frac{Y'(T,t)a}{2\alpha},$$

$0 \le t \le T$.

From (2.61) and (2.67) we find that the functions $\bar{\xi}(t)$ and $\bar{\eta}(t)$ satisfy the boundary value problem

$$\dot{\bar{\xi}}(t) = -A'(t)\bar{\xi}(t) - B'(t+h)\xi(t+h) \tag{2.68}$$
$$+ H(t+\bar{h})R^{-1}(t+\bar{h})H'(t+\bar{h})\bar{\eta}(t),$$

$$\dot{\bar{\eta}}(t) = A(t)\bar{\eta}(t) + B(t)\bar{\eta}(t-h) + C(t)Q(t)C'(t)\bar{\xi}(t),$$

$$0 \le t \le T,$$

$$\bar{\xi}(T) = -\frac{a}{2\alpha}, \quad \bar{\xi}(s) = 0, \quad H(s)R^{-1}(s)H'(s) = 0, \quad T < s \le T + \max(h,\bar{h}),$$

$$\bar{\eta}(0) = P_0\bar{\xi}(0), \qquad \bar{\eta}(\tau) = 0, \quad -h \le \tau < 0.$$

In these formulas, the domain of definition of the function $\bar{\eta}(t)$ is the entire interval $[0,T]$.

Obviously,

$$\xi(t) = -2\alpha\bar{\xi}(t), \qquad \eta(t) = -2\alpha\bar{\eta}(t), \tag{2.69}$$

where $\xi(t)$ and $\eta(t)$ are defined by (2.25).

Problem (2.68) has a unique solution. In fact, in the contrary case there exist nonzero functions $\delta\xi(t)$ and $\delta\eta(t)$ which satisfy (2.68) for $a = 0$. Multiplying the first equation by $\delta\eta'(t)$, the second one by $\delta\xi'(t)$, and summing up, we obtain

$$\frac{d}{dt}(\delta\eta'(t)\delta\xi(t)) = -\delta\eta'(t)B'(t+h)\delta\xi(t+h) + \delta\xi'(t)B(t)\delta\eta(t-h) \tag{2.70}$$

$$+ \delta\eta'(t)H(t+\bar{h})R^{-1}(t+\bar{h})H'(t+\bar{h})\delta\eta(t) + \delta\xi'(t)C(t)Q(t)C'(t)\delta\xi(t),$$

$$\delta\xi(s) = 0, \quad T \le s \le T+h, \qquad \delta\eta(0) = P_0\delta\xi(0), \quad \delta\eta(\tau) = 0, \quad -h \le \tau < 0.$$

Integrating both sides of (2.70) with regard for boundary conditions, we obtain

$$0 = \delta\xi'(0)P_0\delta\xi(0) + \int_{\bar{h}}^{T} \delta\eta'(t-\bar{h})H(t)R^{-1}(t)H'(t)\delta\eta(t-\bar{h})\,dt$$

$$+\int_0^T \delta\xi'(t)C(t)Q(t)C'(t)\delta\xi(t)\,dt$$

(integration of first two terms on the right side of (2.70) yields zero). Consequently,

$$\delta\xi(0)=0, \qquad H'(t)\delta\eta(t-\bar{h})=0, \qquad C'(t)\delta\xi(t)=0,$$

$$\delta\xi(s)=0, \quad T\le s\le T+h, \qquad \delta\eta(\tau)=0, \quad -h\le\tau\le 0,$$

and, by virtue of (2.68), we find that $\delta\xi(t)=0$ and $\delta\eta(t)=0,\ 0\le t\le T$.

In choosing α from (2.65) and k from (2.64), we should note that (2.38), (2.67), and (2.69) imply

$$(4\alpha^2)^{-1}\left\{\xi'(0)P_0\xi(0)+\int_{\bar{h}}^{T}\eta'(t-\bar{h})H(t)R^{-1}(t)H'(t)\eta(t-\bar{h})\,dt\right. \tag{2.71}$$

$$\left.+\int_0^{T-\bar{h}}\xi'(t)C(t)Q(t)C'(t)\xi(t)\,dt\right\}=1.$$

Substituting (2.64) into (2.18), by virtue of (2.67), we obtain

$$\left(1+\frac{k}{2\alpha}\right)Y'(T,0)a=0, \tag{2.72}$$

$$\left(1+\frac{k}{2\alpha}\right)C'(t)Y'(T,t)a=0, \qquad 0\le t\le T-\bar{h}.$$

Since trivial case (2.66) is excluded, from (2.64), (2.67), (2.69), (2.72), and (2.25) it follows that

$$\varphi_-=\Phi_-^0=\xi(0), \quad \varphi(t)=\Phi^0(t)=R^{-1}(t)H'(t)\eta(t-\bar{h}), \tag{2.73}$$

$$w_\varphi(t)=C'(t)\xi(t), \qquad k=-2\alpha, \qquad \xi(t)=\xi_\varphi(t).$$

Substituting $(\varphi_-,\varphi,w_\varphi)$ from (2.73) into (2.22), we find that the term within braces in (2.71) is equal to $J_0^2={J^0}^2$ (cf. (2.63)). Therefore,

$$2\alpha=J_0=J^0. \tag{2.74}$$

Thus, we have demonstrated Assertion 1 and the first part of Assertion 3 of Theorem 2.1. By virtue of formulas (2.67) and (2.74), we obtain Assertion 3 of Theorem 2.4.

Let us now prove (a) Assertion 2 and the second part of Assertion 3 of Theorem 2.1, and (b) the second equality of Assertion 2 of Theorem 2.4.

Let us verify that the function $\eta(t)$ defined by (2.26) – (2.28) satisfies the second equation of (2.25). Substituting (2.26) into the second equation of (2.25), we obtain the following relations.

For $h \le t \le T$:

$$Z(t) \triangleq \dot{\eta}(t) - A(t)\eta(t) - B(t)\eta(t-h) - C(t)Q(t)C'(t)\xi(t)$$

$$= \dot{P}(t)\xi(t) + P(t)\Big\{-A'(t)\xi(t) - B'(t+h)\xi(t+h)$$

$$+ H(t+\bar{h})R^{-1}(t+\bar{h})H'(t+\bar{h})\Big[P(t)\xi(t) + \int_t^{t+h} P_1(t, t-\tau)\xi(\tau)\,d\tau\Big]\Big\}$$

$$+ P_1(t,-h)\xi(t+h) - P_1(t,0)\xi(t) + \int_t^{t+h} \frac{d}{dt}P_1(t,t-\tau)\,\xi(\tau)\,d\tau$$

$$- A(t)\Big[P(t)\xi(t) + \int_t^{t+h} P_1(t,t-\tau)\xi(\tau)\,d\tau\Big]$$

$$- B(t)\Big[P(t-h)\xi(t-h) + \int_{t-h}^{t} P_1(t-h, t-\tau-h)\xi(\tau)\,d\tau\Big]$$

$$- C(t)Q(t)C'(t)\xi(t)$$

$$= \mathcal{P}(t)\xi(t) + \int_t^{t+h} \mathcal{P}_1(t,t-\tau)\xi(\tau)\,d\tau + \Big[P_1(t,-h) - P(t)B'(t+h)\Big]\xi(t+h)$$

$$- B(t)\Big[P(t-h)\xi(t-h) + \int_{t-h}^{t} P_1(t-h, t-\tau-h)\xi(\tau)\,d\tau\Big]$$

$$+ \Big[P_1'(t,0) - P_1'(t,0)\Big]\xi(t) + \Big[P_1'(t-h,-h) - P_1'(t-h,-h)\Big]\xi(t-h),$$

where

$$\mathcal{P}(t) = \dot{P}(t) - A(t)P(t) - P(t)A'(t)$$

$$+ P(t)H(t+\bar{h})R^{-1}(t+\bar{h})H'(t+\bar{h})P(t) - P_1(t,0) - C(t)Q(t)C'(t),$$

$$\mathcal{P}_1(t,t-\tau) = \frac{d}{dt}P_1(t,t-\tau) - A(t)P_1(t,t-\tau)$$

$$+ P(t)H(t+\bar{h})R^{-1}(t+\bar{h})H'(t+\bar{h})P_1(t,t-\tau).$$

Continuing the calculations, we obtain

$$Z(t) = \Big[\mathcal{P}(t) - P_1'(t,0)\Big]\xi(t) + \int_t^{t+h} \mathcal{P}_1(t, t-\tau)\xi(\tau)\, d\tau$$

$$+ \Big[P_1(t,-h) - P(t)B'(t+h)\Big]\xi(t+h)$$

$$+ \Big[P_1'(t-h,-h) - B(t)P(t-h)\Big]\xi(t-h)$$

$$+ \int_{t-h}^{t} \left\{ \frac{d}{d\tau}\Big[P_1'(\tau, \tau - t)\xi(\tau)\Big] - B(t)P_1(t-h, t-\tau-h)\xi(\tau) \right\} d\tau$$

$$= \Big[\mathcal{P}(t) - P_1'(t,0)\Big]\xi(t) + \int_t^{t+h} \mathcal{P}_1(t, t-\tau)\xi(\tau)\, d\tau$$

$$+ \Big[P_1(t,-h) - P(t)B'(t+h)\Big]\xi(t+h)$$

$$+ \Big[P_1'(t-h,-h) - B(t)P(t-h)\Big]\xi(t-h)$$

$$+ \int_{t-h}^{t} \left\{ \frac{d}{d\tau}P_1'(\tau, \tau - t)\xi(\tau) + P_1'(\tau, \tau - t)\Big[-A'(\tau)\xi(\tau) - B'(\tau+h)\xi(\tau+h) \right.$$

$$+ H(\tau+\bar{h})R^{-1}(\tau+\bar{h})H'(\tau+\bar{h})\left(P(\tau)\xi(\tau) + \int_\tau^{\tau+h} P_1(\tau, \tau - s)\xi(s)\, ds\right)\Big]$$

$$\left. - B(t)P_1(t-h, t-\tau-h)\xi(\tau) \right\} d\tau$$

$$= \Big[\mathcal{P}(t) - P_1'(t,0)\Big]\xi(t)$$

$$+ \int_t^{t+h} \Big\{\mathcal{P}_1(t, t-\tau) - P_1'(\tau - h, \tau - t - h)B'(\tau)\Big\}\xi(\tau)\, d\tau$$

$$+ \Big[P_1(t,-h) - P(t)B'(t+h)\Big]\xi(t+h)$$

$$+ \Big[P_1'(t-h,-h) - B(t)P(t-h)\Big]\xi(t-h)$$

$$+ \int_{t-h}^{t} \Big\{\mathcal{P}_1'(\tau, \tau - t) - B(t)P_1(t-h, t-\tau-h)\Big\}\xi(\tau)\, d\tau$$

$$+\int_{t-h}^{t}\int_{\tau}^{\tau+h} P_1'(\tau,\tau-t)H(\tau+\bar{h})R^{-1}(\tau+\bar{h})H'(\tau+\bar{h})P_1(\tau,\tau-s)\xi(s)\,dsd\tau.$$

The last integral can be represented in another form using the formula

$$\int_{t-h}^{t}\int_{\tau}^{\tau+h} F(\tau,s)\xi(s)\,ds\,d\tau$$

$$=\int_{t-h}^{t}\int_{t-h}^{s} F(\tau,s)\,d\tau\,\xi(s)\,ds+\int_{t}^{t+h}\int_{s-h}^{t} F(\tau,s)\,d\tau\,\xi(s)\,ds$$

$$=\int_{t-h}^{t}\int_{t-h}^{\tau} F(s,\tau)\,ds\,\xi(\tau)\,d\tau+\int_{t}^{t+h}\int_{\tau-h}^{t} F(s,\tau)\,ds\,\xi(\tau)\,d\tau,$$

where

$$F(\tau,s)=P_1'(\tau,\tau-t)H(\tau+\bar{h})R^{-1}(\tau+\bar{h})H'(\tau+\bar{h})P_1(\tau,\tau-s).$$

Therefore,

$$Z(t)\Big|_{h\le t\le T}=\Big[\mathcal{P}(t)-P_1'(t,0)\Big]\xi(t)$$

$$+\int_{t}^{t+h}\Big\{\mathcal{P}_1(t,t-\tau)-P_1'(\tau-h,\tau-t-h)B'(\tau)$$

$$+\int_{\tau-h}^{t} P_1'(s,s-t)H(s+\bar{h})R^{-1}(s+\bar{h})H'(s+\bar{h})P_1(s,s-\tau)\,ds\Big\}\xi(\tau)\,d\tau$$

$$+\Big[P_1(t,-h)-P(t)B'(t+h)\Big]\xi(t+h)$$

$$+\Big[P_1'(t-h,-h)-B(t)P(t-h)\Big]\xi(t-h)$$

$$+\int_{t-h}^{t}\Big\{\mathcal{P}_1'(\tau,\tau-t)-B(t)P_1(t-h,t-\tau-h)$$

$$+\int_{t-h}^{\tau} P_1'(s,s-t)H(s+\bar{h})R^{-1}(s+\bar{h})H'(s+\bar{h})P_1(s,s-\tau)\,ds\Big\}\xi(\tau)\,d\tau.$$

Let us now consider the segment $0\le t<h$. In this case,

$$Z(t)\Big|_{0\le t<h}=\mathcal{P}(t)\xi(t)+\int_{t}^{t+h}\mathcal{P}_1(t,t-\tau)\xi(\tau)\,d\tau$$

$$+\Big[P_1(t,-h) - P(t)B'(t+h)\Big]\xi(t+h)$$

$$= \Big[\mathcal{P}(t) - P_1'(t,0)\Big]\xi(t) + \Big[P_1(t,-h) - P(t)B'(t+h)\Big]\xi(t+h)$$

$$+\int_0^t \frac{d}{ds}\Big[P_1'(s,s-t)\xi(s)\Big]\,ds + P_1'(0,-t)\xi(0) + \int_t^{t+h} \mathcal{P}_1(t,t-\tau)\xi(\tau)\,d\tau$$

$$= \Big[\mathcal{P}(t) - P_1'(t,0)\Big]\xi(t) + \Big[P_1(t,-h) - P(t)B'(t+h)\Big]\xi(t+h)$$

$$+\int_0^t \Big\{\frac{d}{ds}P_1'(s,s-t)\;\xi(s) + P_1'(s,s-t)\Big[-A'(s)\xi(s) - B'(s+h)\xi(s+h)$$

$$+H(s+\bar h)R^{-1}(s+\bar h)H'(s+\bar h)\Big(P(s)\xi(s) + \int_s^{s+h} P_1(s,s-\tau)\xi(\tau)\,d\tau\Big)\Big]\Big\}\,ds$$

$$+P_1'(0,-t)\xi(0) + \int_t^{t+h} \mathcal{P}_1(t,t-\tau)\xi(\tau)\,d\tau$$

$$= \Big[\mathcal{P}(t) - P_1'(t,0)\Big]\xi(t) + \Big[P_1(t,-h) - P(t)B'(t+h)\Big]\xi(t+h)$$

$$+P_1'(0,-t)\xi(0) + \int_0^t \mathcal{P}_1'(s,s-t)\xi(s)\,ds + \int_t^{t+h} \mathcal{P}_1(t,t-\tau)\xi(\tau)\,d\tau$$

$$+\int_0^t\int_s^{s+h} P_1'(s,s-t)H(s+\bar h)R^{-1}(s+\bar h)H'(s+\bar h)P_1(s,s-\tau)\xi(\tau)\,d\tau ds$$

$$-\int_h^{t+h} P_1'(\tau-h,\tau-t-h)B'(\tau)\xi(\tau)\,d\tau.$$

The double integral in this formula can be represented as

$$\int_0^t\int_s^{s+h}(\cdots)\,d\tau ds = \int_0^t\int_0^\tau(\cdots)\,dsd\tau$$

$$+\int_t^h\int_0^t(\cdots)\,dsd\tau + \int_h^{t+h}\int_{\tau-h}^t(\cdots)\,dsd\tau.$$

Therefore,

$$Z(t)\Big|_{0\le t<h} = \Big[\mathcal{P}(t) - P_1'(t,0)\Big]\xi(t) + \Big[P_1(t,-h) - P(t)B'(t+h)\Big]\xi(t+h)$$

$$+ P_1'(0,-t)\xi(0) + \int_0^t \Big\{\mathcal{P}_1'(\tau,\tau-t)$$

$$+ \int_0^\tau P_1'(s,s-t)H(s+\bar{h})R^{-1}(s+\bar{h})H'(s+\bar{h})P_1(s,s-\tau)\,ds\Big\}\,\xi(\tau)\,d\tau$$

$$+ \int_h^{t+h} \Big\{\mathcal{P}_1(t,t-\tau)$$

$$+ \int_{\tau-h}^t P_1'(s,s-t)H(s+\bar{h})R^{-1}(s+\bar{h})H'(s+\bar{h})P_1(s,s-\tau)\,ds$$

$$- P_1'(\tau-h,\tau-t-h)B'(\tau)\Big\}\,\xi(\tau)\,d\tau + \int_t^h \Big\{\mathcal{P}_1(t,t-\tau)$$

$$+ \int_0^t P_1'(s,s-t)H(s+\bar{h})R^{-1}(s+\bar{h})H'(s+\bar{h})P_1(s,s-\tau)\,ds\Big\}\,\xi(\tau)\,d\tau.$$

Let us introduce the function $P_2(t,t-\tau,t-s)$ such that

$$\frac{d}{dt}P_2(t,t-\tau,t-s) = -P_1'(t,t-\tau)H(t+\bar{h})R^{-1}(t+\bar{h})H'(t+\bar{h})P_1(t,t-s).$$

Using this function, we finally obtain

$$Z(t)\Big|_{0\le t<h} = \Big[\mathcal{P}(t) - P_1'(t,0)\Big]\xi(t) + \Big[P_1(t,-h) - P(t)B'(t+h)\Big]\xi(t+h)$$

$$+ P_1'(0,-t)\xi(0) + \int_0^t \Big\{\mathcal{P}_1'(\tau,\tau-t) - P_2(\tau,\tau-t,0) + P_2(0,-t,-\tau)\Big\}\xi(\tau)\,d\tau$$

$$+ \int_h^{t+h} \Big\{\mathcal{P}_1(t,t-\tau) - P_2(t,0,t-\tau)$$

$$+ \Big[P_2(\tau-h,\tau-t-h,-h) - P_1'(\tau-h,\tau-t-h)B'(\tau)\Big]\Big\}\,\xi(\tau)\,d\tau$$

$$+ \int_t^h \Big\{\mathcal{P}_1(t,t-\tau) - P_2(t,0,t-\tau) + P_2(0,-t,-\tau)\Big\}\,\xi(\tau)\,d\tau$$

and

$$Z(t)\Big|_{h\le t\le T} = \Big[\mathcal{P}(t) - P_1'(t,0)\Big]\xi(t) + \Big[P_1(t,-h) - P(t)B'(t+h)\Big]\xi(t+h)$$

$$+ \Big[P_1'(t-h,-h) - B(t)P(t-h)\Big]\xi(t-h)$$

$$+ \int_{t-h}^{t} \Big\{P_1'(\tau,\tau-t) - P_2(\tau,\tau-t,0) + P_2(t-h,-h,t-\tau-h)$$

$$- B(t)P_1(t-h,t-\tau-h)\Big\}\xi(\tau)\,d\tau$$

$$+ \int_{t}^{t+h} \Big\{P_1(t,t-\tau) - P_2(t,0,t-\tau) + P_2(\tau-h,\tau-t-h,-h)$$

$$- P_1'(\tau-h,\tau-t-h)B'(\tau)\Big\}\xi(\tau)\,d\tau.$$

It follows directly from the last equalities for $Z(t)$ that if $P(t)$, $P_1(t,\tau)$, $P_2(t,\tau,s)$ obey the system of functional differential equations

$$\frac{d}{dt}P(t) = A(t)P(t) + P(t)A'(t) \tag{2.75}$$

$$- P(t)H(t+\bar{h})R^{-1}(t+\bar{h})H'(t+\bar{h})P(t) + C(t)Q(t)C'(t)$$

$$+ P_1(t,0) + P_1'(t,0),$$

$$\frac{d}{dt}P_1(t,t-\tau_1) = A(t)P_1(t,t-\tau_1)$$

$$- P(t)H(t+\bar{h})R^{-1}(t+\bar{h})H'(t+\bar{h})P_1(t,t-\tau_1) + P_2(t,0,t-\tau_1),$$

$$\frac{d}{dt}P_2(t,t-\tau_1,t-s_1)$$

$$= - P_1'(t,t-\tau_1)H(t+\bar{h})R^{-1}(t+\bar{h})H'(t+\bar{h})P_1(t,t-s_1),$$

$$0 \le t \le T, \qquad -h \le t-\tau_1, t-s_1 \le 0,$$

with the boundary conditions

$$P(0) = P_0, \qquad P_1(0,\tau) = P_2(0,\tau,s) = 0, \quad -h < \tau, s \le 0, \tag{2.76}$$

$$P_1'(t,-h) = B(t+h)P(t),$$

$$P_2(t,-h,\tau) = P_2'(t,\tau,-h) = B(t+h)P_1(t,\tau),$$

$$0 \le t \le T, \qquad -h \le \tau \le 0,$$

then $Z(t) = 0,\ \ 0 \le t \le T$.

It is proved in [96], [91], [97] that the system (2.75), (2.76) has a unique solution in the class of integrable functions $P(t)$, $P_1(t,\tau)$, $P_2(t,\tau,s)$ such that $P(t)$, $P_1(t, t-\tau)$, $P_2(t, t-\tau, t-s)$ are absolutely continuous in t. If the corresponding partial derivatives exist, then

$$\frac{d}{dt}P_1(t, t-\tau_1) = \left(\frac{\partial}{\partial t} + \frac{\partial}{\partial \tau}\right) P_1(t,\tau)\bigg|_{t,t-\tau_1},$$

$$\frac{d}{dt}P_2(t, t-\tau_1, t-s_1) = \left(\frac{\partial}{\partial t} + \frac{\partial}{\partial \tau} + \frac{\partial}{\partial s}\right) P_2(t,\tau,s)\bigg|_{t,t-\tau_1,t-s_1}$$

and equations (2.75), (2.76) turn into equations (2.27), (2.28).

REMARK 2.6. The necessary additional conditions for the existence of the derivatives in t, τ, and s were studied in [96]. Though the equations (2.75), (2.76), generally speaking, are more rigorous than (2.27), (2.28), we conventionally use the more traditional representation (2.27), (2.28), treating the symbols $\left(\frac{\partial}{\partial t} + \frac{\partial}{\partial \tau}\right) P_1(t,\tau)$ and $\left(\frac{\partial}{\partial t} + \frac{\partial}{\partial \tau} + \frac{\partial}{\partial s}\right) P_2(t,\tau,s)$ as the derivatives in the corresponding directions along the characteristics of (2.27), (2.28).

Thus the second function of the boundary value problem $\eta(t)$ is defined by (2.26) and Assertion 2 of Theorem 2.1 holds.

Substituting the optimal solution (2.40) into functional (2.37) and using (2.26) and (2.63), we obtain (2.39) (the second equality of Assertion 2 of Theorem 2.4). Formula (2.30) (the second part of Assertion 3 of Theorem 2.1) follows immediately from (2.24) and (2.26).

Strictly speaking, we have proved Theorem 2.1 only for the case that P_0 and $Q(t)$ are nonsingular. The proof of Theorem 2.1 for the case $P_0 \ge 0$ and $Q(t) \ge 0$ is identical with the proof of Theorem 2.1 of Chapter 3 for singular P_0 and $Q(t)$. □

2.6. *Proof of Theorem 2.3*

Let us introduce a normed space $\mathcal{X} = \mathbf{R}^n \times \mathcal{F}^m \times L_1^r[0, T-\bar{h}]$ with elements (Φ_-, Φ, w) and the norm $\|\cdot\|_{\mathcal{X}} = \max(\|\cdot\|_{\mathbf{R}^n}, \|\cdot\|_{\mathcal{F}^m}, \|\cdot\|_{L_1^r})$.

The Lagrange function for problem (2.13), (2.19), (2.18) has the form (see Section 1 of Chapter 2)

$$\mathcal{L}(\Phi_-, \Phi, w; p, \lambda) \tag{2.77}$$

$$= \sum_{d=1}^{n} \sigma_{-d}|\Phi_{-d}| + \sum_{l=1}^{m} \int_{\bar{h}}^{T} \sigma_l(t)|\Phi_l(t)|\,dt + \sum_{j=1}^{r} \int_0^{T-\bar{h}} \gamma_j(t)|w_j(t)|\,dt$$

$$+\int_0^{T-\bar{h}} p'(t)\Big\{-w(t) - C'(t)\int_{t+\bar{h}}^{T} Y'(s-\bar{h},t)H(s)\Phi(s)\,ds + C'(t)Y'(T,t)a\Big\}\,dt$$

$$+\lambda'\Big\{Y'(T,0)a - \int_{\bar{h}}^{T} Y'(t-\bar{h},0)H(t)\Phi(t)\,dt - \Phi_-\Big\},$$

where $p \in L_\infty^r$, $\lambda \in \mathbf{R}^n$.

Using the Dirichlet formula, from (2.77) we obtain

$$\mathcal{L}(\Phi_-, \Phi, w; p, \lambda) \tag{2.78}$$

$$= \Big[\sum_{d=1}^{n} \sigma_{-d}|\Phi_{-d}| - \lambda'\Phi_-\Big] + \int_{\bar{h}}^{T}\Big\{\sum_{l=1}^{m} \sigma_l(t)|\Phi_l(t)|$$

$$- \Phi'(t)H'(t)\Big[\int_0^{t-\bar{h}} Y(t-\bar{h},s)C(s)p(s)\,ds + Y(t-\bar{h},0)\lambda\Big]\Big\}\,dt$$

$$+\int_0^{T-\bar{h}}\Big\{\sum_{j=1}^{r} \gamma_j(t)|w_j(t)| - p'(t)w(t)\Big\}\,dt$$

$$+a'\Big[\int_0^{T-\bar{h}} Y(T,t)C(t)p(t)\,dt + Y(T,0)\lambda\Big].$$

According to the duality theory (see Section 1 of Chapter 2) the dual problem is

$$I^0 = \sup_{p,\lambda}\Big[\inf_{\Phi_-,\Phi,w} \mathcal{L}(\Phi_-, \Phi, w; p, \lambda)\Big]. \tag{2.79}$$

Moreover,

$$I^0 = I_0. \tag{2.80}$$

From (2.78) – (2.80) it directly follows that the dual problem is of the form (2.34), (2.35) and equality (2.36) holds (see the proof of Theorem 5.2 of Chapter 2 for technical details). □

So, an efficient method for analyzing guaranteed estimation problems in dynamic systems with delay is developed. The initial complex problem is replaced by a much simpler problem that admits constructive numerical solution. Using the duality theory of convex variational problems, a nonoptimality level of the simplified algorithm is constructed without solving the initial minimax filtering problem. Thus, a relatively simple algorithm is constructed that yields the "key" solution for minimax estimation problems with an evaluation of the influence of neglected factors.

3. Level of Nonoptimality in Stochastic Problem

Similar to Chapter 4, the optimal guaranteed estimation problem for stochastic time-delay dynamic system with uncertain statistics can be considered. The main result has the same form.

3.1. *Stochastic Guaranteed Estimation Problem*

Let us consider a linear stochastic dynamic system with time delay

$$\dot{x}(t) = A(t)x(t) + B(t)x(t-h) + C(t)u(t), \qquad t \in [0,T],$$

$$x(\tau) = 0, \qquad -h \le \tau < 0,$$

and the measurements (cf. (2.1), (2.3))

$$z(t) = H'(t)x(t) + \varrho(t).$$

In contrast to Section 2, we assume for simplicity that there is no delay in the measurements.

Let us suppose that the following stochastic hypotheses hold:
(a) The initial state $x(0)$ is a random variable, the perturbations $u(t)$, $\varrho(t)$ are white noise processes:

$$\mathbf{E}(x(0) - \mathbf{E}x(0))(x(0) - \mathbf{E}x(0))' = \operatorname{diag}(c_1^x, \ldots, c_n^x),$$

$$\mathbf{E}(u(t) - \mathbf{E}u(t))(u(s) - \mathbf{E}u(s))' = \operatorname{diag}(q_1^u(t), \ldots, q_r^u(t))\,\delta(t-s),$$

$$\mathbf{E}(\varrho(t) - \mathbf{E}\varrho(t))(\varrho(s) - \mathbf{E}\varrho(s))' = \operatorname{diag}(r_1^\varrho(t), \ldots, r_m^\varrho(t))\,\delta(t-s),$$

where $\delta(t)$ is the Dirac delta-function;
(b) The first moments and the intensities of the perturbations are unknown but bounded:

$$|\mathbf{E}x_d(0)| \le \sigma_{-d}, \qquad |\mathbf{E}u_j(t)| \le \gamma_j(t), \qquad |\mathbf{E}\varrho_l(t)| \le \sigma_l(t),$$

$$c_d^x \le c_d, \qquad q_j^u(t) \le q_j(t), \qquad r_l^\varrho(t) \le r_l(t),$$

$$d = 1, \ldots, n, \qquad j = 1, \ldots, r, \qquad l = 1, \ldots, m;$$

(c) The initial state $x(0)$ and the perturbations $u(t)$, $\varrho(t)$ are mutually independent.

Here σ_{-d}, c_d are specified positive numbers, $\gamma_j(t)$, $q_j(t)$, $\sigma_l(t)$, $r_l(t)$ are specified positive functions ($\gamma_j(t) \geq \gamma^0 > 0$, $q_j(t) \geq q^0 > 0$, $\sigma_l(t) \geq \sigma^0 > 0$, $r_l(t) \geq r^0 > 0$). The entries of the matrices $A(t)$, $B(t)$, $H(t)$, $C(t)$, the quantities $\mathbf{E}u_j(t)$, $\mathbf{E}\varrho_l(t)$, $\gamma_j(t)$, $q_j^u(t)$, $q_j(t)$, $\sigma_l(t)$, $r_l^\rho(t)$, $r_l(t)$ are assumed to be piecewise continuos functions on $[0,T]$. As in Section 1 of Chapter 4, if the elements $x(0)$, $u(t)$, and $\varrho(t)$ satisfy hypotheses (a) – (c), we shall write

$$(x(0), u(t), \varrho(t)) \in \mathcal{S}.$$

The problem is to evaluate a scalar quantity $l_* = a'x(T)$, where $a \in \mathbf{R}^n$ is a given vector. We use the linear estimators of the form

$$\hat{l} = \int_0^T \Phi'(t)z(t)\,dt, \qquad \Phi(t) = (\Phi_1(t), \ldots, \Phi_m(t))' \in \mathbf{R}^m, \qquad \Phi \in L_2^m.$$

Let us define the quality of estimation by the guaranteed value of the second moment of the estimation error:

$$\mathcal{D}_{\text{guar}}(\Phi) = \sup_{(x(0),u,\varrho)\in\mathcal{S}} \mathbf{E}\left(\hat{l} - l_*\right)^2.$$

The optimal guaranteed estimation problem is to find an optimal estimator Φ^0 that minimizes the guaranteed value of the second moment

$$\mathcal{D}_{\text{guar}}(\Phi^0) = \mathcal{J}_0 \triangleq \inf_{\Phi\in L_2^m} \mathcal{D}_{\text{guar}}(\Phi).$$

Let us introduce the following auxiliary quantities (they are identical to (2.12), (2.13) for $\bar{h} = 0$):

(1) the function $\xi_\Phi(t)$

$$\dot{\xi}_\Phi(t) = -A'(t)\xi_\Phi(t) - B'(t+h)\xi_\Phi(t+h) + H(t)\Phi(t), \qquad 0 \leq t \leq T,$$

$$\xi_\Phi(T) = a, \qquad \xi_\Phi(\tau) = 0, \qquad B(\tau) = 0, \qquad T < \tau \leq T+h;$$

(2) the functional $I(\Phi_-, \Phi, w)$: $\mathbf{R}^n \times L_2^m \times L_2^r \to \mathbf{R}^1$

$$I(\Phi_-, \Phi, w) = \sum_{d=1}^n \sigma_{-d}|\Phi_{-d}| + \sum_{l=1}^m \int_0^T \sigma_l(t)|\Phi_l(t)|\,dt + \sum_{j=1}^r \int_0^T \gamma_j(t)|w_j(t)|\,dt.$$

Using the stochastic hypotheses and the auxiliary quantities it is not hard to show that

$$\mathcal{D}_{\text{guar}}(\Phi) = \mathcal{J}(\Phi_-(\Phi), \Phi, w(\Phi)), \tag{3.1}$$

where

$$\mathcal{J}(\Phi_-, \Phi, w) = I^2(\Phi_-, \Phi, w)$$
$$+ \Phi'_- c\, \Phi_- + \int_0^T \Phi'(t) r(t) \Phi(t)\, dt + \int_0^T w'(t) q(t) w(t)\, dt$$

and

$$\Phi_- = \xi_\Phi(0), \qquad w(t) = C'(t)\xi_\Phi(t).$$

Thus the optimal stochastic guaranteed estimation problem reduces to the following nonsmooth variational problem:

$$\mathcal{J}_0 = \inf_{\Phi_- \in \mathbf{R}^n,\, \Phi \in L_2^m,\, w \in L_2^r} \mathcal{J}(\Phi_-, \Phi, w) \tag{3.2}$$

under the constraints

$$\Phi_- - \xi_\Phi(0) = 0, \qquad w(t) - C'(t)\xi_\Phi(t) = 0. \tag{3.3}$$

The finding of the optimal estimator is a very difficult problem. In order to obtain an approximate solution of (3.2), (3.3), we replace the initial problem by a certain approximating one. Let us denote by φ the corresponding approximate estimator.

We define the quality of the approximate estimator by the ratio

$$\Delta_{AS} = \left(\frac{\mathcal{D}_{\text{guar}}(\varphi)}{\mathcal{D}_{\text{guar}}(\Phi^0)} \right)^{1/2} = \left(\frac{\mathcal{D}_{\text{guar}}(\varphi)}{\mathcal{J}_0} \right)^{1/2}, \qquad \Delta_{AS} \geq 1.$$

The numerator $\mathcal{D}_{\text{guar}}(\varphi)$ is known as φ is known. Nevertheless, the denominator $\mathcal{J}_0$ is unknown and the value of Δ_{AS} is also unknown. We shall construct an upper bound Δ_{AS}^0 for Δ_{AS} that can be computed without exact solving the initial problem (3.2), (3.3). If the value of Δ_{AS}^0 is not large, then the use of the approximate estimator φ will be justified.

3.2. *Approximating Problem for Stochastic Case*

Let us simplify the problem (3.2), (3.3) replacing the first summand in the formula for $\mathcal{J}(\Phi_-, \Phi, w)$ by an appropriate quadratic term

$$\beta_0 \Phi'_- \tilde{\sigma}_- \Phi_- + \beta_1 \int_0^T \Phi'(t)\tilde{\sigma}(t)\Phi(t)\, dt + \beta_2 \int_0^T w'(t)\tilde{\gamma}(t) w(t)\, dt.$$

Here β_0, β_1, and β_2 are chosen positive parameters and

$$\tilde{\sigma}_- = \operatorname{diag}(\sigma_{-1}^2, \ldots, \sigma_{-n}^2),$$

$$\tilde{\sigma}(t) = \operatorname{diag}(\sigma_1^2(t), \dots, \sigma_m^2(t)), \qquad \tilde{\gamma}(t) = \operatorname{diag}(\gamma_1^2(t), \dots, \gamma_r^2(t)).$$

Thus the approximating variational problem takes the form

$$\mathcal{J}_{a0} = \inf_{\Phi_- \in \mathbf{R}^n,\, \Phi \in L_2^m,\, w \in L_2^r} \mathcal{J}_a(\Phi_-, \Phi, w), \tag{3.4}$$

$$\mathcal{J}_a(\Phi_-, \Phi, w) = \Phi'_- P_0 \Phi_- + \int_0^T \Phi'(t) R(t) \Phi(t)\, dt + \int_0^T w'(t) Q(t) w(t)\, dt$$

under constraints (3.3), where

$$P_0 = \operatorname{diag}(P_{01}, \dots, P_{0n}) = \beta_0 \tilde{\sigma}_- + c,$$

$$R(t) = \operatorname{diag}(R_1(t), \dots, R_m(t)) = \beta_1 \tilde{\sigma}(t) + r(t),$$

$$Q(t) = \operatorname{diag}(Q_1(t), \dots, Q_r(t)) = \beta_2 \tilde{\gamma}(t) + q(t).$$

As it was already mentioned in Section 2, there exist constructive numerical algorithms to solve the problem (3.4), (3.3).

3.3. *Bound for Level of Nonoptimality*

Let us introduce the notation

$$s_d = \frac{\sigma_{-d}^2}{c_d}, \qquad S_d = \frac{P_{0d}\varphi_{-d}}{\sigma_{-d}}, \qquad d = 1, \dots, n,$$

$$f_l(t) = \frac{\sigma_l^2(t)}{r_l(t)}, \qquad F_l(t) = \frac{R_l(t)\varphi_l(t)}{\sigma_l(t)}, \qquad l = 1, \dots, m,$$

$$g_j(t) = \frac{\gamma_j^2(t)}{q_j(t)}, \qquad G_j(t) = \frac{Q_j(t) w_{\varphi j}(t)}{\gamma_j(t)}, \qquad j = 1, \dots, r,$$

$$\varphi_- = (\varphi_{-1}, \dots, \varphi_{-n})', \qquad \varphi(t) = (\varphi_1(t), \dots, \varphi_m(t))',$$

$$w_\varphi(t) = (w_{\varphi 1}(t), \dots, w_{\varphi r}(t))'.$$

$$\zeta_{\max} = \max\left\{ \max_{d=1,\dots,n} |S_d|, \sup_{l=1,\dots,m,\; t\in[0,T]} |F_l(t)|, \sup_{j=1,\dots,r,\; t\in[0,T]} |G_j(t)| \right\}.$$

These quantities are uniquely determined by the solution of (3.4), (3.3).

Let the function $\mathcal{E}(\zeta)\colon [0, \zeta_{\max}] \to \mathbf{R}^1$ be defined by the formula

$$\mathcal{E}(\zeta) = \sum_{|S_d| > \zeta} s_d \Big(|S_d| - \zeta \Big) + \sum_{l=1}^{m} \int_{|F_l(t)| > \zeta} f_l(t) \Big(|F_l(t)| - \zeta \Big)\, dt$$

$$+\sum_{j=1}^{r} \int\limits_{|G_j(t)|>\zeta} g_j(t)\Big(|G_j(t)| - \zeta\Big)\, dt - \zeta.$$

LEMMA 3.1. *The function $\mathcal{E}(\zeta)$ is a strictly monotonically decreasing function on the interval $[0, \zeta_{\max}]$. The equation*

$$\mathcal{E}(\zeta) = 0 \tag{3.5}$$

has a unique root on $[0, \zeta_{\max}]$.

The proof of Lemma 3.1 is completely analogous to the proof of Lemma 2.1 from Section 2 of Chapter 4. By virtue of Lemma 3.1 the root of (3.5) can easily be determined by sequential halving the interval $[0, \zeta_{\max}]$.

THEOREM 3.1. *Let $\{\varphi_-, \varphi, w_\varphi\}$ be the solution of the linear-quadratic problem (3.4), (3.3). Then the level of nonoptimality is bounded with the quantity Δ^0_{AS}:*

$$1 \le \Delta_{AS} \le \Delta^0_{AS} = \frac{\sqrt{\mathcal{D}_{\text{guar}}(\varphi)\ \Omega(\varphi_-, \varphi, w_\varphi)}}{\mathcal{J}_{a0}},$$

where

$$\Omega(\varphi_-, \varphi, w_\varphi) = \sum_{|S_d|>\tilde{\zeta}} s_d\Big(|S_d| - \tilde{\zeta}\Big)^2$$

$$+\sum_{l=1}^{m} \int\limits_{|F_l(t)|>\tilde{\zeta}} f_l(t)\Big(|F_l(t)| - \tilde{\zeta}\Big)^2 dt$$

$$+\sum_{j=1}^{r} \int\limits_{|G_j(t)|>\tilde{\zeta}} g_j(t)\Big(|G_j(t)| - \tilde{\zeta}\Big)^2 dt + \tilde{\zeta}^2,$$

$\mathcal{D}_{\text{guar}}(\Phi)$ is defined by (3.1), $\mathcal{J}_{a0}$ is defined by (3.4), and $\tilde{\zeta}$ is the root of equation (3.5).

The proof of Theorem 3.1 is analogous to the proof of Theorem 2.2 from Section 2 of Chapter 4 with obvious modifications like in the proofs of Theorems 2.2 – 2.4.

4. Simplified Algorithms for Mean-Square Filtering Problem

In Sections 2 and 3 we approximated the minimax filtering problems by an important basic problem: the optimal mean-square filtering problem, which has a constructive numerical solution. However, the numerical algorithm for this problem contains a block of solving the system of functional partial differential equations. Hence, the solution of the mean-square filtering problem contains nontrivial numerical procedures. Let us discuss possible ways for the simplification of the solution of the filtering problem.

The structure of the optimal algorithm allows us to hope that we can significantly simplify the estimation algorithm (having excluded the procedure for solving the system of functional partial differential equations) without considerable loss of optimality. At this point, it is important to evaluate the nonoptimality level of the simplified algorithm without solving the initial optimal filtering problem. We shall show that, using our general approach, we can evaluate this level. Moreover, a new state-space representation of the optimal filter will be obtained.

4.1. *Mean-Square Filtering Problem*

Let us formulate the optimal mean-square filtering problem, which we have already touched on in Sections 2 and 3. Consider a linear dynamic system with delay

$$\dot{x}(t) = A(t)x(t) + B(t)x(t-h) + C(t)u(t), \qquad t \in [0,T], \tag{4.1}$$

$$x(\tau) = 0, \qquad -h \le \tau < 0, \qquad h = \text{const}, \qquad h < T,$$

where $x(t) = (x_1(t), \ldots, x_n(t))' \in \mathbf{R}^n$ is the system state vector, $A(t), B(t) \in \mathbf{R}^{n\times n}$, $C(t) \in \mathbf{R}^{n\times r}$ are specified matrices, $u(t) = (u_1(t), \ldots, u_r(t))' \in \mathbf{R}^r$ is a vector of disturbances.

Let the measurements of the state vector be made on $[0,T]$:

$$z(t) = H'(t)x(t-\bar{h}) + \varrho(t), \qquad z(t), \varrho(t) \in \mathbf{R}^m, \tag{4.2}$$

$$H(t) \in \mathbf{R}^{n\times m}, \qquad \bar{h} = \text{const}, \quad \bar{h} < T.$$

We assume that $x(0)$ is a Gaussian random variable and $u(t)$, $\varrho(t)$ are Gaussian white noise processes:

$$\mathbf{E}x(0) = 0, \qquad \mathbf{E}x(0)x'(0) = P_0, \tag{4.3}$$

$$\mathbf{E}u(t) = 0, \qquad \mathbf{E}u(t)u'(s) = Q(t)\delta(t-s),$$

$$\mathbf{E}\varrho(t) = 0, \qquad \mathbf{E}\varrho(t)\varrho'(s) = R(t)\delta(t-s), \qquad R(t) \ge R_0 > 0.$$

Moreover, $x(0)$, $u(t)$, and $\varrho(t)$ are independent.

We find an estimate for the scalar quantity $l_* = a'x(T)$, where $a \in \mathbf{R}^n$ is a given vector. Consider the class of linear estimates for l_*

$$\hat{l}(\Phi) = \int_0^T \Phi'(t)z(t)\,dt, \qquad \Phi \in L_2^m. \tag{4.4}$$

The optimal mean-square filtering problem (see, e.g., [97], [105], [108], [106]) is to find an estimator $\Phi^0 \in L_2^m$ that minimizes the mean-square estimation error:

$$\mathcal{D}(\Phi^0) \leq \mathcal{D}(\Phi), \qquad \mathcal{D}(\Phi) = \mathbf{E}\Big(\hat{l}(\Phi) - l_*\Big)^2, \qquad \Phi \in L_2^m.$$

This problem is an extension of the classical optimal filtering problem (studied in Section 6 of Chapter 4) to dynamic systems with delay.

REMARK 4.1. Since all the processes are Gaussian, the use of nonlinear estimates does not improve the estimation quality [10], [108].

Using the pseudo-fundamental matrix $Y(t,s)$ defined by (2.9) and the auxiliary function $\xi_\Phi(t)$ defined by (2.12), we can easily show by straightforward computations that

$$\hat{l}(\Phi) - l_* = -\xi_\Phi'(0)x(0) + \int_0^T \Phi'(t)\varrho(t)\,dt \tag{4.5}$$

$$- \int_0^{T-\bar{h}} \xi_\Phi'(\tau)C(\tau)u(\tau)\,d\tau - \int_{T-\bar{h}}^T \xi_0'(\tau)C(\tau)u(\tau)\,d\tau.$$

Then it follows immediately from (4.3) and (4.5) that

$$\mathcal{D}(\Phi) = J^2(\Phi_-(\Phi), \Phi, w(\Phi)) + \kappa_Q, \tag{4.6}$$

where the functional J is specified by (2.22), the quantities $\Phi_-(\Phi)$ and $w(\Phi)$ are defined by the relations

$$0 = \Phi_- - \xi_\Phi(0), \qquad 0 = w(t) - C'(t)\xi_\Phi(t), \qquad 0 \leq t \leq T - \bar{h}, \tag{4.7}$$

and

$$\kappa_Q = \int_{T-\bar{h}}^T \xi_0'(t)C(t)Q(t)C'(t)\xi_0(t)\,dt. \tag{4.8}$$

Thus the optimal mean-square filtering problem reduces to the linear-quadratic variational problem

$$J_0 = \inf_{\Phi_- \in \mathbf{R}^n,\, \Phi \in L_2^m,\, w \in L_2^r} J(\Phi_-, \Phi, w) \tag{4.9}$$

under constraints (4.7).

The solution of the optimal mean-square filtering problem is given by Theorem 2.1. Formulas (2.24) – (2.30) determine a numerical algorithm for computing the optimal estimator $\varphi(t)$ for the filtering problem. The most complicated part of this algorithm is the solving of the system of functional partial differential equations (2.27), (2.28).

There are exist constructive numerical procedures for solving (2.27), (2.28) (see Section 2). Nevertheless, these procedures are not so easy for implementation and it is desirable to construct a simplified algorithm that does not require the resolving of (2.27), (2.28).

Denote by φ^a the estimator that corresponds to a simplified algorithm. We define the level of the nonoptimality of φ^a by the ratio

$$\Delta_E = \left(\frac{\mathcal{D}(\varphi^a)}{\mathcal{D}(\varphi)}\right)^{\frac{1}{2}}. \tag{4.10}$$

Obviously, $\Delta_E \geq 1$. Since we prefer not to solve the original full problem, it is important to evaluate the nonoptimality level Δ_E without finding the optimal estimator φ. Further the upper bound Δ_E^0 for Δ_E will be constructed. This upper bound can be calculated *without solving the original optimal mean-square filtering problem.* If Δ_E^0 is not large, then the use of the simplified algorithm φ^a will be justified.

4.2. *Simplification of Optimal Estimation Algorithm*

Let us obtain a state-space representation for the optimal estimator φ. This representation will allow us to simplify the optimal algorithm in a quite natural way.

THEOREM 4.1. *The optimal estimate $\hat{l}(\varphi)$ determined by (4.4) and (2.24) can be represented in the state-space form*

$$\hat{l}(\varphi) = \hat{l}_S, \qquad \hat{l}_S = a'\hat{x}(T), \tag{4.11}$$

$$\dot{\hat{x}}(t) = A(t)\hat{x}(t) + B(t)\hat{x}(t-h) + K(t)\big(z(t+\bar{h}) - H'(t+\bar{h})\hat{x}(t)\big)$$

$$+ \int_{\max(0,t-h)}^{t} K_1(s,t)\big(z(s+\bar{h}) - H'(s+\bar{h})\hat{x}(s)\big)\,ds,$$

$$\hat{x}(\tau) = 0, \qquad -h \leq \tau \leq 0, \qquad\qquad 0 \leq t \leq T,$$

where the gains are defined by the formulas

$$K(t) = P(t)H(t+\bar{h})R^{-1}(t+\bar{h}),$$

$$K_1(s,t) = P_1'(s,s-t)H(s+\bar{h})R^{-1}(s+\bar{h})$$

and

$$H(s) = 0, \quad T < s \le T + \bar{h}. \tag{4.12}$$

PROOF. In accordance with Theorem 2.1 the optimal estimate is determined by the relation

$$\begin{aligned} \hat{l}(\varphi) &= \int_{\bar{h}}^{T} \varphi'(t)z(t)\,dt \qquad (4.13)\\ &= \int_{\bar{h}}^{T} \left\{ \int_{-h}^{0} \xi'(t-\bar{h}-\tau)P_1'(t-\bar{h},\tau)\,d\tau + \xi'(t-\bar{h})P(t-\bar{h}) \right\} \\ &\qquad \times H(t)R^{-1}(t)z(t)\,dt \\ &= \int_{\bar{h}}^{T} \left\{ \int_{-h}^{0} \xi'(t-\bar{h}-\tau)K_1(t-\bar{h},t-\bar{h}-\tau)\,d\tau + \xi'(t-\bar{h})K(t-\bar{h}) \right\} z(t)\,dt \\ &= \int_{0}^{T-\bar{h}} \xi'(t)K(t)z(t+\bar{h})\,dt + \int_{0}^{T-\bar{h}} \int_{-h}^{0} \xi'(t-\tau)K_1(t,t-\tau)\,d\tau\, z(t+\bar{h})\,dt \\ &= \int_{0}^{T} \xi'(t)K(t)z(t+\bar{h})\,dt + \int_{0}^{T} \int_{t}^{t+h} \xi'(s)K_1(t,s)z(t+\bar{h})\,dsdt, \end{aligned}$$

where $\xi(t)$ is determined by (2.30). The last equality from (4.13) is valid due to (4.12).

Let us introduce the auxiliary matrix $Y_K(t,s)$ that satisfies the the direct and conjugate equations for $(t,s) \in [0,T] \times [0,T]$ (cf. Section 1)

$$\begin{aligned} \frac{\partial Y_K(t,s)}{\partial t} &= \big(A(t) - K(t)H'(t+\bar{h})\big)Y_K(t,s) \qquad (4.14)\\ &+ B(t)Y_K(t-h,s) - \int_{-h}^{0} K_1(t+\theta,t)H'(t+\theta+\bar{h})Y_K(t+\theta,s)\,d\theta, \end{aligned}$$

$t \ge s$,

$$\begin{aligned} \frac{\partial Y_K(t,s)}{\partial s} &= -Y_K(t,s)\big(A(s) - K(s)H'(s+\bar{h})\big) \qquad (4.15)\\ &- Y_K(t,s+h)B(s+h) + \int_{0}^{h} Y_K(t,s+\theta)K_1(s,s+\theta)H'(s+\bar{h})\,d\theta, \end{aligned}$$

$s \le t$,

with the boundary conditions

$$Y_K(s,s) = E_n, \qquad Y_K(t,s) = 0, \qquad s > t. \tag{4.16}$$

In accordance with (4.16), the values of $K_1(\lambda,\theta)$ for $\lambda < 0$ have no influence on the integrals in (4.11), (4.13), (4.14). Hence, up to the end of the proof we shall put for convenience that

$$K_1(\lambda,\theta) = 0 \quad \text{for} \quad \lambda < 0. \tag{4.17}$$

By virtue of (4.17), the integral in (4.11) can be represented in the form

$$\int_{\max(0,t-h)}^{t} K_1(s,t)\big(z(s+\bar{h}) - H'(s+\bar{h})\hat{x}(s)\big)\,ds \tag{4.18}$$

$$= \int_{t-h}^{t} K_1(s,t)\big(z(s+\bar{h}) - H'(s+\bar{h})\hat{x}(s)\big)\,ds.$$

It follows from (4.14), (4.17), and (4.18) that the state-space estimate from (4.11) can be written in the form

$$\hat{l}_S = \int_0^T a'Y_K(T,s)K(s)z(s+\bar{h})\,ds + \int_0^T \int_{s-h}^{s} F(t,s)\,dtds, \tag{4.19}$$

$$F(t,s) = a'Y_K(T,s)K_1(t,s)z(t+\bar{h}).$$

Interchanging the integrations, we get

$$\int_0^T \int_t^{t+h} F(t,s)\,dsdt = \int_T^{T+h} \int_{s-h}^{T} F(t,s)\,dtds \tag{4.20}$$

$$+ \int_0^T \int_{s-h}^{s} F(t,s)\,dtds - \int_0^h \int_{s-h}^{0} F(t,s)\,dtds.$$

It is easy to verify that the first and the third integrals in the right side of (4.20) are equal to zero (as $Y_K(T,s) = 0$ for $s > T$ and (4.17) holds). Furthermore, it follows from (4.15) that

$$\xi(s) = Y_K'(T,s)a.$$

Hence, by virtue of (4.13), (4.19), and (4.20), $\hat{l}(\varphi) = \hat{l}_S$. □

COROLLARY 4.1. *Let an estimate $\hat{l}(\Phi)$ be defined by the equalities*

$$\hat{l}(\Phi) = \int_0^T \Phi'(t)z(t)\,dt,$$

$$\Phi(t) = \begin{cases} K'(t-\bar{h})\xi(t-\bar{h}) + \int_{-h}^{0} K_1'(t-\bar{h}, t-\bar{h}-\tau)\xi(t-\bar{h}-\tau)\,d\tau, & \\ & \text{for } \bar{h} \le t \le T \\ 0 & \text{for } 0 \le t < \bar{h}, \end{cases}$$

where $K(t)$ and $K_1(s,t)$ are certain coefficients (not necessarily optimal) and $\xi(t) = \xi_\Phi(t)$ is defined by the equation

$$\dot{\xi}(t) = -A'(t)\xi(t) - B'(t+h)\xi(t+h)$$

$$+ H(t+\bar{h}) \times \left[K'(t)\xi(t) + \int_{-h}^{0} K_1'(t, t-\tau)\xi(t-\tau)\,d\tau \right],$$

$$\xi(T) = a, \qquad \xi(\tau) = 0, \qquad T < \tau \le T + h.$$

Then $\hat{l}(\Phi)$ has the state-space representation (4.11).

PROOF. All calculations in the proof of Theorem 4.1 are valid for arbitrary $K(t)$ and $K_1(s,t)$. □

COROLLARY 4.2. *Each component of $\hat{x}(t)$, $t \in [0,T]$, is the optimal mean-square estimate of the corresponding component of $x(t)$ based on the measurements $\{z(s),\ 0 \le s \le \min(t+\bar{h}, T)\}$.*

PROOF. Let us introduce modified measurements without delay

$$\tilde{z}(v) = \tilde{H}'(v)x(v) + \tilde{\varrho}(v), \qquad 0 \le v \le t^*, \qquad \tilde{\varrho}(v) = \varrho(v+\bar{h}),$$

$$\tilde{H}(v) = H(v+\bar{h}), \qquad \tilde{z}(v) = z(v+\bar{h}). \tag{4.21}$$

Denote by $\hat{l}(t)[\,z(v),\ v \in V\,]$ the optimal mean-square estimate for $a'x(t)$ by the measurements $\{z(v),\ v \in V\}$. Using convention (4.12) and taking into account that the values of $z(v)$ for $0 \le v < \bar{h}$ have no influence on the optimal estimate, it is not hard to prove that, for any $t^* \le T$,

$$\hat{l}(t^*)[\,z(v),\ 0 \le v \le \min(t^*+\bar{h}, T)\,] = \hat{l}(t^*)[\,z(v),\ 0 \le v \le t^*+\bar{h}\,] \tag{4.22}$$

$$= \hat{l}(t^*)[\,z(v),\ \bar{h} \le v \le t^*+\bar{h}\,] = \hat{l}(t^*)[\,\tilde{z}(v),\ 0 \le v \le t^*\,].$$

By virtue of Theorem 4.1, the estimate $\hat{l}(t^*)[\,\tilde{z}(v),\ 0 \le v \le t^*\,]$, which is determined by relations (4.4), (2.24) – (2.30) with substitutions (4.21), coincides with the dynamic estimate $a'\hat{x}(t^*)$ defined by (4.11) with substitutions (4.21). Then, using again (4.21) and (4.22), we obtain that

$$a'\hat{x}(t^*) = \hat{l}(t^*)[\,z(v),\ 0 \le v \le t^*+\bar{h}\,].$$

The solution of the system of partial differential equations (2.27), (2.28) in the domain $\{0 \le t \le T, -h \le \tau, s \le 0\}$ satisfies the system (2.27), (2.28) in the domain $\{0 \le t \le t^*, -h \le \tau, s \le 0\}$ for any $t^* \le T$. Since the solution of (2.27), (2.28) is unique, the functions $P(t)$, $P_1(t,\tau)$, and $P_2(t,\tau,s)$ are the same for any instant t^*. Therefore,

$$a'\hat{x}(t) = \hat{l}(t)[\,z(v),\, 0 \le v \le t + \bar{h}\,].$$

Since the system (2.27), (2.28) does not depend on a, our arguments are valid for each component of $x(t)$. □

Estimate (4.11) has the Kalman structure: it duplicates the original system perturbed by residuals. The form (4.11) allows us to get natural simplifications of the optimal algorithm by modifying the gains $K(t)$ and $K_1(s,t)$. These modifications retain the Kalman structure and, hence, possess the following important property. If $\mathbf{E}x(0) = m_0 \ne 0$ and $\hat{x}(0) = m_0$, then the estimate is unbiased: $\mathbf{E}\,(\hat{x}(t) - x(t)) = 0, \quad t \in [0,T]$.

It is most desirable to avoid the exact solving of the system of functional partial differential equations (2.27), (2.28). Let us discuss the corresponding simplifications.

Quasi-Stationary Gains

The equation for $P_1(t,\tau)$ can be rewritten in the form

$$\frac{d}{dt}P_1(t,t-v) = A(t)P_1(t,t-v) \tag{4.23}$$

$$- P(t)H(t+\bar{h})R^{-1}(t+\bar{h})H'(t+\bar{h})P_1(t,t-v) - P_2(t,0,t-v)$$

$$0 \le t \le T, \qquad -h \le t - v \le 0.$$

The domain of variables (t,τ) is denoted by $\mathcal{V}$ and is depicted in Figure 2. It follows from (4.23) that, given $P(t)$ and $P_2(t,0,t-\tau)$, the boundary conditions propagate along the directions determined by the vector $e^0 = (1,1)'$. Therewith, the boundary condition on the left segment $\{t = 0, -h < \tau \le 0\}$ determines the values of $P_1(t,\tau)$ in the triangle formed by the coordinate axes and the line $\tau - t = -h$; the boundary condition on the lower segment $\{0 \le t \le T, \tau = -h\}$ determines the values of $P_1(t,\tau)$ in the rest part of $\mathcal{V}$.

The main idea is to regard the strip $\mathcal{V}$ as "quite narrow" and to neglect the dynamics along the directions indicated, i.e., to replace $P_1(t,\tau)$ by $P_1^{(1)}(t,\tau)$, where

$$\frac{d}{dt}P_1^{(1)}(t,t-\tau) = 0 \qquad \text{or} \qquad \left(\frac{\partial}{\partial t} + \frac{\partial}{\partial \tau}\right) P_1^{(1)}(t,\tau) = 0. \tag{4.24}$$

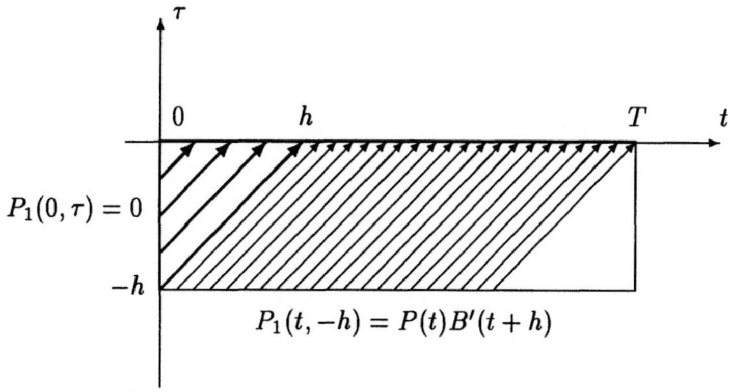

Fig. 2. Domain $\mathcal{V}$

In accordance with the boundary conditions (2.28), the equation (4.24) has the solution

$$P_1^{(1)}(t,\tau)=\begin{cases}0 & \text{for } 0\le t<\tau+h,\\ P(t-\tau-h)B'(t-\tau) & \text{for } \tau+h\le t\le T.\end{cases} \tag{4.25}$$

Following Theorem 2.1, we can obtain the corresponding approximate elements

$$\varphi_-^{(1)}=\xi^{(1)}(0), \tag{4.26}$$

$$\varphi^{(1)}(t)=\begin{cases}R^{-1}(t)H'(t)\eta^{(1)}(t-\bar{h}) & \text{for } \bar{h}\le t\le T\\ 0 & \text{for } 0\le t<\bar{h},\end{cases}$$

$$w_\varphi^{(1)}(t)=C'(t)\xi^{(1)}(t), \qquad 0\le t\le T-\bar{h},$$

and

$$\eta^{(1)}(t)=P^{(1)}(t)\xi^{(1)}(t)+\int_{-h}^{0}P_1^{(1)}(t,\tau)\xi^{(1)}(t-\tau)\,d\tau,$$

$$\begin{aligned}\dot{\xi}^{(1)}(t)&=-A'(t)\xi^{(1)}(t)-B'(t+h)\xi^{(1)}(t+h)\\ &\quad+H(t+\bar{h})R^{-1}(t+\bar{h})H'(t+\bar{h})\\ &\quad\times\left[P^{(1)}(t)\xi^{(1)}(t)+\int_{-h}^{0}P^{(1)}(t-\tau-h)B'(t-\tau)\xi^{(1)}(t-\tau)\,d\tau\right],\end{aligned} \tag{4.27}$$

$$\xi^{(1)}(T) = a, \qquad \xi^{(1)}(\tau) = 0, \qquad T < \tau \le T + h.$$

Obviously, $\xi^{(1)}(t) = \xi_{\varphi^{(1)}}(t)$, where $\xi_{\varphi^{(1)}}(t)$ is determined by (2.12), just like for the exact optimal filter. By virtue of (4.25), the matrix $P^{(1)}(t)$ is determined by the equation

$$\begin{aligned}\dot{P}^{(1)}(t) &= A(t)P^{(1)}(t) + P^{(1)}(t)A'(t) \\ &\quad + B(t)P^{(1)}(t-h) + P^{(1)}(t-h)B'(t) \\ &\quad - P^{(1)}(t)H(t+\bar{h})R^{-1}(t+\bar{h})H'(t+\bar{h})P^{(1)}(t) + C(t)Q(t)C'(t), \\ P^{(1)}(0) &= P_0, \qquad P^{(1)}(\tau) = 0, \quad -h \le \tau < 0, \qquad 0 \le t \le T.\end{aligned} \tag{4.28}$$

It is interesting to note that equation (4.28) can be represented in the Riccati type form

$$\begin{aligned}\dot{P}^{(1)}(t) &= \mathcal{A}[P^{(1)}(t)] + \mathcal{A}[P^{(1)}(t)]' \\ &\quad - P^{(1)}(t)H(t+\bar{h})R^{-1}(t+\bar{h})H'(t+\bar{h})P^{(1)}(t) + C(t)Q(t)C'(t),\end{aligned}$$

where the operator $\mathcal{A}$ is defined by the equality

$$\mathcal{A}[P^{(1)}(t)] = A(t)P^{(1)}(t) + B(t)P^{(1)}(t-h).$$

By virtue of Corollary 4.1, the equivalent state-space representation takes the form

$$\begin{aligned}\dot{\hat{x}}^{(1)}(t) &= A(t)\hat{x}^{(1)}(t) + B(t)\hat{x}^{(1)}(t-h) \\ &\quad + P^{(1)}(t)H(t+\bar{h})R^{-1}(t+\bar{h})\big(z(t+\bar{h}) - H'(t+\bar{h})\hat{x}^{(1)}(t)\big) \\ &\quad + B(t)P^{(1)}(t-h)\int_{\max(0,t-h)}^{t} H(s+\bar{h})R^{-1}(s+\bar{h}) \\ &\qquad\qquad \times \big(z(s+\bar{h}) - H'(s+\bar{h})\hat{x}^{(1)}(s)\big)\,ds, \\ \hat{x}^{(1)}(\tau) &= 0, \quad -h \le \tau \le 0, \qquad 0 \le t \le T.\end{aligned} \tag{4.29}$$

Further simplifications can be obtained by neglecting the integral terms in (4.27), (4.29), and by putting $h = 0$ in (4.28)[7]. Then, we obtain

$$\varphi_-^{(2)} = \xi^{(2)}(0), \tag{4.30}$$

[7] We simplify the gains only.

$$\varphi^{(2)}(t) = \begin{cases} R^{-1}(t)H'(t)\eta^{(2)}(t-\bar{h}) & \text{for } \bar{h} \le t \le T \\ 0 & \text{for } 0 \le t < \bar{h}, \end{cases}$$

$$w_\varphi^{(2)}(t) = C'(t)\xi^{(2)}(t), \qquad 0 \le t \le T - \bar{h},$$

and

$$\eta^{(2)}(t) = P^{(2)}(t)\xi^{(2)}(t),$$

$$\dot{\xi}^{(2)}(t) = -A'(t)\xi^{(2)}(t) - B'(t+h)\xi^{(2)}(t+h) \tag{4.31}$$

$$+ H(t+\bar{h})R^{-1}(t+\bar{h})H'(t+\bar{h})P^{(2)}(t)\xi^{(2)}(t),$$

$$\xi^{(2)}(T) = a, \qquad \xi^{(2)}(\tau) = 0, \qquad T < \tau \le T + h,$$

where $P^{(2)}(t)$ is subjected to the Riccati equation

$$\dot{P}^{(2)}(t) = \big(A(t) + B(t)\big)P^{(2)}(t) + P^{(2)}(t)\big(A'(t) + B'(t)\big)$$

$$- P^{(2)}(t)H(t+\bar{h})R^{-1}(t+\bar{h})H'(t+\bar{h})P^{(2)}(t) + C(t)Q(t)C'(t),$$

$$P^{(2)}(0) = P_0, \qquad 0 \le t \le T.$$

Stationary Gains

Let the system (4.1) – (4.3) be stationary: $A(t) = A$, $B(t) = B$, $C(t) = C$, $H(t) = H$, $R(t) = R$, $Q(t) = Q$, $t \in [0, T]$. By our assumption, $B(t) = 0$, $H(t)R^{-1}(t)H'(t) = 0$ for $t > T$. We put

$$P(t) = P = \text{const}, \qquad P_1(t, \tau) = P_1 = \text{const}.$$

Thus,

$$K(t) = \begin{cases} K = PHR^{-1} & \text{for } 0 \le t \le T - \bar{h} \\ 0 & \text{for } T - \bar{h} < t \le T, \end{cases}$$

$$K_1(s,t) = \begin{cases} K_1 = P_1'HR^{-1} & \text{for } 0 \le s \le T - \bar{h} \\ 0 & \text{for } T - \bar{h} < s \le T. \end{cases}$$

In this case, the approximating solution has the form

$$\varphi_-^{(3)} = \xi^{(3)}(0), \tag{4.32}$$

$$\varphi^{(3)}(t) = \begin{cases} K'\xi^{(3)}(t-\bar{h}) + K_1' \int_{-h}^{0} \xi^{(3)}(t-\tau-\bar{h})\, d\tau & \text{for } \bar{h} \le t \le T \\ 0 & \text{for } 0 \le t < \bar{h}, \end{cases}$$

$$w_\varphi^{(3)}(t) = C'(t)\xi^{(3)}(t), \qquad 0 \le t \le T - \bar{h},$$

and

$$\dot{\xi}^{(3)}(t) = -A'\xi^{(3)}(t) - B'\xi^{(3)}(t+h) \tag{4.33}$$
$$+ HK'\xi^{(3)}(t) + HK_1' \int_{-h}^{0} \xi^{(3)}(t-\tau)\, d\tau, \qquad 0 \le t \le T - \bar{h},$$

$$\dot{\xi}^{(3)}(t) = -A'\xi^{(3)}(t) - B'\xi^{(3)}(t+h), \qquad T - \bar{h} < t \le T,$$

$$\xi^{(3)}(T) = a, \qquad \xi^{(3)}(\tau) = 0, \quad T < \tau \le T + h.$$

The corresponding state-space representation takes the form

$$\dot{\hat{x}}(t) = A\hat{x}(t) + B\hat{x}(t-h) + K\big(z(t+\bar{h}) - H'\hat{x}(t)\big)$$
$$+ K_1 \int_{\max(0,t-h)}^{t} \big(z(s+\bar{h}) - H'\hat{x}(s)\big)\, ds, \qquad 0 \le t \le T - \bar{h},$$

$$\dot{\hat{x}}(t) = A\hat{x}(t) + B\hat{x}(t-h) + K_1 \int_{\max(0,t-h)}^{T-\bar{h}} \big(z(s+\bar{h}) - H'\hat{x}(s)\big)\, ds,$$
$$T - \bar{h} < t \le T + \min(h - \bar{h}, 0),$$

$$\dot{\hat{x}}(t) = A\hat{x}(t) + B\hat{x}(t-h), \qquad T + \min(h - \bar{h}, 0) < t \le T,$$

$$\hat{x}(\tau) = 0, \qquad -h \le \tau \le 0.$$

Thus the functional optimization problem is reduced to a parametric optimization problem over $K,\ K_1 \in \mathbf{R}^{n\times m}$. Such approach can be successful for systems with small dimensions of n and m.

The equation for estimation error $e(t) = \hat{x}(t) - x(t)$ for $0 \le t \le T - \bar{h}$ takes the form

$$\dot{e}(t) = (A - KH')e(t) + Be(t-h) - K_1 H' \int_{\max(0,t-h)}^{t} e(\tau)\, d\tau$$
$$+ K\varrho(t+\bar{h}) + K_1 \int_{\max(0,t-h)}^{t} \varrho(\tau+\bar{h})\, d\tau - C(t)u(t),$$

with the initial condition

$$e(0) = -x(0), \qquad e(\tau) = 0, \quad -h \le \tau < 0.$$

It is natural to require the following property:[8] if $u(t) = 0$ and $\varrho(t) = 0$, then $e(t) \to 0$ as $t \to \infty$. So, the trivial solution of the corresponding homogeneous equation should be asymptotically stable. The homogeneous equation (for $t \geq h$) can be represented in the form

$$\dot{\tilde{e}}(t) = \int_{-\infty}^{0} dS(\theta)\,\tilde{e}(t+\theta), \tag{4.34}$$

$$S(\theta) = (A - KH')\big(\chi(\theta) - 1\big) + B\big(\chi(\theta+h) - 1\big)$$

$$+ K_1 H'\big((\theta+h)\chi(\theta+h) - h\big), \qquad -\infty < \theta \leq 0,$$

where $\chi(\theta)$ is the Heaviside function: $\chi(\theta) = 0,\ \ \theta < 0$, and $\chi(\theta) = 1,\ \ \theta \geq 0$.

General conditions for asymptotic stability of (4.34) are contained, in particular, in [107], [105], [106]. They reduce to the investigation of the characteristic function of (4.34).

As in the quasi-stationary case, further simplification is to set $K_1 = 0$. Then the corresponding approximations $\varphi_-^{(4)}$, $\varphi^{(4)}(t)$, $w_\varphi^{(4)}(t)$, $\xi^{(4)}(t)$ can be obtained.

Many other methods of simplification can be proposed. For example, as an approximation for $P(t)$ and $P_1(t,\tau)$, we can take $P^{(k)}(t)$ and $P_1^{(k)}(t,\tau)$ from a k-th step of the iteration procedure for solving the system (2.27), (2.28) discussed in Section 2.

4.3. *Estimate of Nonoptimality*

Several approximate filtering algorithms were proposed above. Each of them generates certain approximating elements $\xi^a(t)$, φ_-^a, $\varphi^a(t)$, $w_\varphi^a(t)$. Let us construct the guaranteed levels of nonoptimality for these approximate algorithms.

Let us introduce the following auxiliary quantities:

(1) the process $\tilde{x}(t;p,\lambda)$ (given $p \in L_2^r, \lambda \in \mathbf{R}^n$)

$$\dot{\tilde{x}}(t;p,\lambda) = A(t)\tilde{x}(t;p,\lambda) + B(t)\tilde{x}(t-h;p,\lambda) + C(t)p(t), \tag{4.35}$$

$$t \in [0,T], \qquad \tilde{x}(0;p,\lambda) = \lambda, \qquad \tilde{x}(\tau;p,\lambda) = 0, \quad -h \leq \tau < 0;$$

$$p(t) = 0, \qquad T - \bar{h} < t \leq T;$$

(2) the pair (p^a, λ^a)

$$p^a(t) = Q(t)C'(t)\xi^a(t), \quad 0 \leq t \leq T - \bar{h}, \qquad \lambda^a = P_0\xi^a(0); \tag{4.36}$$

[8] We assume that $T - \bar{h}$ is sufficiently large.

(3) the value

$$\nu_0^2 = \left(\xi^{a\prime}(0) P_0 \xi^a(0) \right. \tag{4.37}$$

$$+ \int_{\bar{h}}^{T} \tilde{x}'(t-\bar{h}; p^a, \lambda^a) H(t) R^{-1}(t) H'(t) \tilde{x}(t-\bar{h}; p^a, \lambda^a)\, dt$$

$$\left. + \int_0^{T-\bar{h}} \xi^{a\prime}(t) C(t) Q(t) C'(t) \xi^a(t)\, dt \right)^{-1}$$

(provided the expression in the parentheses is not equal to zero).
Note that the expression in parentheses in (4.37) is equal to zero if and only if

$$P_0 \xi^a(0) = 0, \quad R^{-1}(t) H'(t) \tilde{x}(t-\bar{h}; p^a, \lambda^a) = 0, \quad Q(t) C'(t) \xi^a(t) = 0.$$

Obviously, this is an extremely degenerate case.

THEOREM 4.2. *Let $\xi^a(t)$, φ_-^a, $\varphi^a(t)$, $w_\varphi^a(t)$ be the corresponding approximating elements. Then*

$$1 \le \Delta_E \le \Delta_E^0, \qquad \Delta_E^0 = \left(\frac{J^2(\varphi_-^a, \varphi^a, w_\varphi^a) + \kappa_Q}{\nu_0^2 \{a'\tilde{x}(T; p^a, \lambda^a\}^2 + \kappa_Q} \right)^{\frac{1}{2}}. \tag{4.38}$$

PROOF. For the sake of simplicity, let us first consider the case that P_0 and $Q(t)$ are nonsingular and $Q^{-1}(t)$ is bounded.

In accordance with (4.6), (4.10), the denominator in the formula for Δ_E is equal to

$$\mathcal{D}^{\frac{1}{2}}(\varphi) = \left(J^2(\varphi_-, \varphi, w_\varphi) + \kappa_Q\right)^{\frac{1}{2}} = \left(J_0^2 + \kappa_Q\right)^{\frac{1}{2}}.$$

Hence, if we construct a lower bound for J_0, then we obtain an upper bound for Δ_E.

THEOREM 4.3. 1^0. *The problem* (4.9), (4.7) *has the dual representation*

$$J_0 = \sup_{p \in L_2^r,\, \lambda \in \mathbf{R}^n} a'\tilde{x}(T; p, \lambda) \tag{4.39}$$

under the constraint

$$\left\{ \tilde{x}'(0; p, \lambda) P_0^{-1} \tilde{x}(0; p, \lambda) \right. \tag{4.40}$$

$$+\int_{\bar h}^{T} \tilde x'(t-\bar h;p,\lambda)H(t)R^{-1}(t)H'(t)\tilde x(t-\bar h;p,\lambda)\,dt$$

$$+\int_0^{T-\bar h} p'(t)Q^{-1}(t)p(t)\,dt\Bigg\}^{\frac{1}{2}} \le 1,$$

where $\tilde x(t;p,\lambda)$ *is given by* (4.35).
2^0. *The solution of* (4.39), (4.40) *is given by the formulas*

$$p^0(t) = J_0^{-1}Q(t)C'(t)\xi(t), \qquad 0 \le t \le T-\bar h, \tag{4.41}$$

$$\lambda^0 = J_0^{-1}\eta(0) = J_0^{-1}P_0\xi(0),$$

where $\xi(t)$, $\eta(t)$ *are determined by the boundary value problem* (2.25).

PROOF. Let us represent $\tilde x(t;p,\lambda)$ in the form

$$\tilde x(t;p,\lambda) = Y(t,0)\lambda + \int_0^t Y(t,s)C(s)p(s)\,ds,$$

where $Y(t,s)$ is defined by (2.9). Then Theorem 4.3 follows directly from Theorem 2.4. □

If we choose an admissible pair (p,λ) for the problem (4.39), (4.40), then we obtain the estimate from below for J_0:

$$J_0^2 \ge \left(a'\tilde x(T;p,\lambda)\right)^2. \tag{4.42}$$

Every approximate estimator generates an approximation $\xi^a(t)$ for $\xi(t)$. Then, following formulas (4.41), we shall search an appropriate pair on the one-dimensional manifold

$$p(t;\nu) = \nu p^a(t), \qquad \lambda(\nu) = \nu\lambda^a, \qquad \nu \in \mathbf{R}^1, \tag{4.43}$$

where (p^a,λ^a) is given by (4.36).

The pair that yeilds the maximal value for the lower bound for J_0 on the manifold (4.43) is the pair with $\nu=\nu_0$ that turns inequality (4.40) into strict equality. Hence, ν_0 satisfies the condition

$$\nu_0^2 = \Bigg(\lambda^{a\prime}P_0^{-1}\lambda^a \tag{4.44}$$

$$+\int_{\bar h}^{T} \tilde x'(t-\bar h;p^a,\lambda^a)H(t)R^{-1}(t)H'(t)\tilde x(t-\bar h;p^a,\lambda^a)\,dt$$

$$+\int_0^{T-\bar h} p^{a\prime}(t)Q^{-1}(t)p^a(t)\,dt\Bigg)^{-1}.$$

Obviously, the two solutions of (4.44) give the same value for the lower bound. Note that the expression in the parentheses is equal to zero only if $p^a(t) \equiv 0$ and $\lambda^a = 0$.

By virtue of (4.36), the guaranteed level of the nonoptimality takes the form (4.38). Certainly, we also assume that $|a'\tilde{x}(T; p^a, \lambda^a)| + \kappa_Q \neq 0$; obviously this is a very weak restriction.

Now, let P_0 or $Q(t)$ be singular. Then consider the perturbed nonsingular matrices

$$P_0(\varepsilon) = P_0 + \varepsilon E_n, \qquad Q(t,\varepsilon) = Q(t) + \varepsilon E_r, \qquad \varepsilon > 0, \tag{4.45}$$

and the corresponding perturbed mean-square filtering problem (4.9), (4.7), (2.22), (4.45). Let us denote by $\xi(t,\varepsilon)$, $\varphi(t,\varepsilon)$, $\nu_0(\varepsilon)$, $p^a(t,\varepsilon)$, $\lambda^a(\varepsilon)$, $J_0(\varepsilon)$ the corresponding elements of the perturbed problem.

LEMMA 4.1. If $\varepsilon \to 0^+$, then

$$\lim_{\varepsilon\to 0+} \|\varphi(t,\varepsilon) - \varphi(t)\|_{L_2^m} = 0, \qquad \lim_{\varepsilon\to 0+} \max_{t\in[0,T]} \|\xi(t,\varepsilon) - \xi(t)\|_{\mathbf{R}^n} = 0.$$

PROOF. The assertions of Lemma 4.1 can be proved similar to the corresponding assertions in the proof of Theorem 2.1 of Chapter 3 for the case of singular P_0 and $Q(t)$. □

In accordance with (4.42) we obtain that

$$J_0^2(\varepsilon) \geq \nu_0^2(\varepsilon)\left\{a'\tilde{x}(T; p^a(\varepsilon), \lambda^a(\varepsilon))\right\}^2, \qquad \varepsilon \to 0^+. \tag{4.46}$$

By virtue of Lemma 4.1, all elements in (4.46) have the corresponding limits. Consequently, the estimate (4.38) remains valid for singular P_0, $Q(t)$ as well. □

Let us discuss the algorithm for computing the level of nonoptimality. For definiteness, we choose the first approximation scheme. Then the algorithm for computing the estimate (4.38) consists of the following steps.

Step 1. Solution of the differential equation with delay (4.28) to determine the function $P^{(1)}(t)$.

Step 2. Solution of the differential equation with aftereffect (4.27) to determine the function $\xi^{(1)}(t)$.

Step 3. Computation of the approximating elements $\varphi_-^{(1)}$, $\varphi^{(1)}(t)$, $w_\varphi^{(1)}(t)$, $p^a(t)$, and λ^a by formulas (4.26) and (4.36) ($\xi^a(t) = \xi^{(1)}(t)$).

Step 4. Computation of the auxiliary process $\tilde{x}(t; p^a, \lambda^a)$ by (4.35) and the quadratures $J(\varphi_-^{(1)}, \varphi^{(1)}(t), w_\varphi^{(1)}(t))$, κ_Q, and ν_0^2 by (2.22), (4.8), and (4.37) ($\xi^a(t) = \xi^{(1)}(t)$).

TABLE 2. Values of Δ_E^0

		$h = \bar{h}$			
B	Type	0.2	0.3	0.4	0.5
-1.0	FS	1.01	1.19	breakdown	
	SS	1.01	1.01	1.03	1.05
-0.5	FS	1.01	1.01	1.03	1.16
	SS	1.01	1.02	1.05	1.10
1.0	FS	1.05	1.14	1.24	1.33
	SS	1.01	1.02	1.05	1.07

Step 5. Computation of the value Δ_E^0 defined by (4.38).

The steps 1–5 are quite simple for numerical implementation.

4.4. *Example of the Mean-Square Filtering Problem*

EXAMPLE 4.1. Let us suppose that $n = m = r = 1$, $A(t) = 0$, $B(t) = B =$ const, $C(t) = 1$, $H(t) = 1$, $a = 1$, $h = \bar{h}$. In other words, the system and the measurements have the simplest form

$$\dot{x}(t) = Bx(t-h) + u(t), \qquad x(\tau) = 0, \quad -h \leq \tau < 0,$$

$$z(t) = x(t-h) + \varrho(t), \qquad l_* = x(T).$$

We put $T = 3.0$, $P_0 = 25.0$, $R = 1.0$, $Q = 1.0$ and use two approximating schemes: the first scheme (FS), which is described by (4.26) – (4.29), and the second scheme (SS), which is described by (4.30), (4.31). The results are presented in Table 2.

Table 2 shows that the use of the simplified estimators do not influence substantially the estimation accuracy. Moreover, for some values of $h = \bar{h}$ and B, the simplified estimators virtually do not break the optimality. We see from Table 2 that, for our parameters, a more sophisticated first scheme can be worse than a more primitive second scheme. The breakdown of the first scheme for $B = -1.0$ and $h = 0.4$ or $h = 0.5$ occurs because of the nonextensibility of the solution of (4.28) for the parameters chosen. Our guaranteed level of nonoptimality allows us to choose the best version for the specific problem with the specific parameters.

REMARK 4.2. For more complex than in Example 4.1 systems, the simple approximation schemes proposed above can fail. The main feature of our approach is that, for any approximate scheme, we can efficiently construct the guaranteed level of nonoptimality. Having analyzed a current scheme, we can easily propose a more accurate scheme, if necessary. Thus, early or late, we can fit the successful algorithm with the desired level of nonoptimality.

5. Control Algorithms for Systems with Aftereffect

In Sections 2 and 4 we have studied the case of lumped delays. Of course, all results of these sections can be extended to the case of distributed delays. It is well known that, in linear systems, the estimation problem is equivalent to a control problem. Since the control problem itself is of great theoretical and practical importance, we shall study the case of lumped and distributed delays for control problems. In this section, we shall investigate two optimal control problems: the linear-quadratic problem and the L_1-cost problem.

The linear-quadratic control problem has a well-known solution (see, e.g., [96], [105], [106]). In order to compute this solution, a system of functional partial differential equations must be resolved. Though there exists an iterative algorithm for solving this partial differential system, it is a difficult numerical task. Therefore, it is highly desirable to avoid the direct solution of this system, even if the optimality is somewhat lost. The structure of the linear-quadratic optimal algorithm suggests natural ways of its simplification. The corresponding simplified algorithms contain only ordinary differential equations with aftereffect. The character of simplifications allows us to hope that the loss in optimality is not substantial. Here, it is important to evaluate the nonoptimality level for the approximate solution to the optimal control problem without solving the original optimal problem exactly. Such an estimate can be obtained with the help of the duality theory for convex variational problems, much as it was done above for estimation problems.

The L_1-cost problem is a very difficult nonsmooth control problem. In contrast to the linear-quadratic problem, the solving algorithms to this problem are unknown. As a constructive step for approximate solving the L_1-cost problem we can use the optimal control law from an appropriate linear-quadratic control problem. A simple estimate for the nonoptimality level of this suboptimal control law will be constructed without solving the L_1-cost problem.

Since the implementation of the linear-quadratic law is not so easy (in view of the necessity of solving the functional partial differential equations), the corresponding simplified control law for the linear-quadratic problem can be used. The estimate of the nonoptimality level of this easily implemented simplified control law (similar to the linear-quadratic case) will be also obtained.

5.1. *Linear-Quadratic Control Problem*

Consider a linear dynamic control system with lumped and distributed delays

$$\dot{x}(t) = A(t)x(t) + B(t)x(t-h) + \int_{-h}^{0} G(t,s)x(t+s)\,ds + C(t)u(t), \quad (5.1)$$

$$x(\tau) = \varphi(\tau), \qquad -h \le \tau \le 0, \qquad h = \text{const}, \qquad h < T, \qquad t \in [0,T],$$

where $x(t) = (x_1(t), \dots, x_n(t))' \in \mathbf{R}^n$ is the state vector, $A(t), B(t), G(t,s) \in \mathbf{R}^{n\times n}$, $C(t) \in \mathbf{R}^{n\times r}$ are given matrices, $u(t) = (u_1(t), \dots, u_r(t))' \in \mathbf{R}^r$ is the control vector, $\varphi(\tau)$ is a given function.

The classical linear-quadratic control problem is to find $u(t) \in L_2^r$ that minimizes the quadratic cost functional

$$\mathcal{I}(u) = \left\{ x'(T)N_0x(T) + \int_0^T \big(u'(t)N_1(t)u(t) + x'(t)N_2(t)x(t)\big)\,dt \right\}^{\frac{1}{2}}. \quad (5.2)$$

The elements of the matrices $A(t)$, $B(t)$, $G(t,s)$, $C(t)$, $N_1(t)$, $N_2(t)$ are assumed to be piecewise continuous (in each variable for $G(t,s)$). In the discontinuity points in t these functions are continuous from the left, possibly, except for the point $t = 0$, at which they are continuous from the right. The matrices N_0 and $N_2(t)$ are positive semidefinite, $N_1(t)$ is a positive definite matrix.

For simplicity we also assume that the discontinuity points of $G(t,\tau)$ in t may lie only on the segments $\{t = t_{p},\ {\scriptstyle p=1,\dots,p^0};\ -h \le \tau \le 0\}$ and the discontinuity points of $G(t,\tau)$ in τ may lie only on the segments $\{0 \le t \le T;\ \tau = \tau_{q},\ {\scriptstyle q=1,\dots,q^0}\}$. Moreover, we assume that the points (t_p, τ_q) partition the domain $[0,T] \times [-h,0]$ into a finite number of rectangles $\Pi_\alpha,\ {\scriptstyle \alpha=1,\dots,\alpha^0}$, such that (a) $G(t,\tau)$ is continuous (in both variables) in the interior of Π_α, ${\scriptstyle \alpha=1,\dots,\alpha^0}$; (b) at each boundary point of $\Pi_\alpha,\ {\scriptstyle \alpha=1,\dots,\alpha^0}$, $G(t,\tau)$ has a finite limit (in both variables) from inside of Π_α.

The solution of control problem (5.1), (5.2) is given by the following theorem.

THEOREM 5.1. ([96], [91], [105], [106], [108]) *The optimal control $u^0(t)$ has the form*

$$u^0(t) = -N_1^{-1}(t)C'(t)\left[P(t)x^0(t) + \int_{-h}^{0} Q(t,\tau)x^0(t+\tau)\,d\tau\right], \quad (5.3)$$

where $x^0(t)$ is the optimal process

$$\dot{x}^0(t) = \big(A(t) - C(t)N_1^{-1}(t)C'(t)P(t)\big)x^0(t) + B(t)x^0(t-h) \quad (5.4)$$

$$+ \int_{-h}^{0} \big(G(t,\tau) - C(t)N_1^{-1}(t)C'(t)Q(t,\tau)\big)\, x^0(t+\tau)\,d\tau,$$

$$x^0(\tau) = \varphi(\tau), \qquad -h \leq \tau \leq 0.$$

The matrices $P(t)$, $Q(t,\tau)$, $R(t,\tau,s)$ are described by the system of functional partial differential equations:

$$\dot{P}(t) + A'(t)P(t) + P(t)A(t) - P(t)C(t)N_1^{-1}(t)C'(t)P(t) \tag{5.5}$$

$$+ N_2(t) + Q(t,0) + Q'(t,0) = 0,$$

$$\left(\frac{\partial}{\partial t} - \frac{\partial}{\partial \tau}\right) Q(t,\tau) + A'(t)Q(t,\tau) + P(t)G(t,\tau) \tag{5.6}$$

$$- P(t)C(t)N_1^{-1}(t)C'(t)Q(t,\tau) + R(t,0,\tau) = 0,$$

$$\left(\frac{\partial}{\partial t} - \frac{\partial}{\partial \tau} - \frac{\partial}{\partial s}\right) R(t,\tau,s) + G'(t,\tau)Q(t,s) + Q'(t,\tau)G(t,s) \tag{5.7}$$

$$- Q'(t,\tau)C(t)N_1^{-1}(t)C'(t)Q(t,s) = 0,$$

$$0 \leq t \leq T, \qquad -h \leq \tau, s \leq 0,$$

with the boundary conditions

$$P(T) = N_0, \qquad Q(T,\tau) = R(T,\tau,s) = 0, \quad -h < \tau, s \leq 0, \tag{5.8}$$

$$Q'(t,-h) = B'(t)P(t), \qquad R(t,-h,\tau) = R'(t,\tau,-h) = B'(t)Q(t,\tau),$$

$$0 \leq t \leq T, \qquad -h \leq \tau \leq 0.$$

The proof of this theorem is given in Subsection 5.4. Theorem 5.1 is the analog of Theorem 2.1. The most laborious part of the optimal control computation is the solving of the system of functional partial differential equations (5.5) – (5.8). In order to solve (5.5) – (5.8) an iterative functional algorithm was proposed in [96], [91]. This algorithm is based on the Bellman method of successive approximations. Another algorithm is based on the direct fiberwise in t solution of (5.5) – (5.8). Both methods were briefly discussed in Section 2 for the equivalent estimation problem.

Nevertheless, an implementation of this algorithm is a nontrivial numerical procedure. So, it is highly desirable to use simpler equations for an approximate calculation of $P(t)$ and $Q(t,\tau)$ in (5.3), (5.4).

Let us describe some simple ways of reduction in (5.5)–(5.8).

First Scheme

Let us set an explicit approximate formulas for $Q(t,\tau)$. Then, in particular, (5.5) is separated from (5.6). The matrix $Q(t,\tau)$ is defined in the rectangle $\mathcal{V} = \{(t,\tau) \mid 0 \le t \le T; -h < \tau \le 0\}$. Usually, $h \ll T$, and we simplify the description of $Q(t,\tau)$ dynamics in (5.6) in the "narrow" strip $\mathcal{V}$. Namely, we neglect the quadratic term $P(t)C(t)N_1^{-1}(t)C'(t)Q(t,\tau)$ and the term $R(t,0,\tau)$. Then (5.6) can be represented in the form of ordinary differential equation with a parameter s:

$$\frac{d}{dt}Q(t,s-t) + A'(t)Q(t,s-t) + P(t)G(t,s-t) = 0 \tag{5.9}$$

with boundary conditions (5.8).

The solution of (5.9) can be obtained in explicit form. If, in addition, the term $A'(t)Q(t,s-t)$ is also neglected, then (5.9) has a particular simple solution

$$Q(t,s-t) = \begin{cases} \int_t^T P(v)G(v,s-v)\,dv & \text{for } T-h < s \le T, \quad s \le t \le T \\ \int_t^{s+h} P(v)G(v,s-v)\,dv + P(s+h)B(s+h) & \\ & \text{for } 0 \le s \le T-h, \quad s \le t \le s+h \\ & \text{or } \; -h \le s < 0, \quad 0 \le t \le s+h. \end{cases}$$

Thus, the approximate simplified equations for $P(t)$, $Q(t,\tau)$ are defined by the formulas

$$\dot{P}^a(t) + A'(t)P^a(t) + B'(t+h)P^a(t+h) + \int_0^h G'(s+t,-s)P^a(t+s)\,ds$$
$$+ P^a(t)A(t) + P^a(t+h)B(t+h) + \int_0^h P^a(t+s)G(t+s,-s)\,ds$$
$$- P^a(t)C(t)N_1^{-1}(t)C'(t)P^a(t) + N_2(t) = 0,$$

$$P^a(T) = N_0, \qquad P^a(\tau) = 0, \qquad T < \tau \le T+h,$$

$$Q^a(t,\tau) = \int_0^{h+\tau} P^a(t+s)G(s+t,\tau-s)\,ds + P^a(t+\tau+h)B(t+\tau+h).$$

Therefore, the approximate control law $u^a(t)$ and the corresponding process $x^a(t)$ are defined by (5.3), (5.4), where $P(t)$ and $Q(t,\tau)$ are replaced by $P^a(t)$ and $Q^a(t,\tau)$:

$$u^a(t) = -N_1^{-1}(t)C'(t)\left[P^a(t)x^a(t) + \int_{-h}^0 Q^a(t,\tau)x^a(t+\tau)\,d\tau\right], \tag{5.10}$$

$$\dot{x}^a(t) = \big(A(t) - C(t)N_1^{-1}(t)C'(t)P^a(t)\big)\, x^a(t) + B(t)x^a(t-h) \qquad (5.11)$$

$$+ \int_{-h}^{0} \big(G(t,\tau) - C(t)N_1^{-1}(t)C'(t)Q^a(t,\tau)\big)\, x^a(t+\tau)\, d\tau.$$

Second Scheme

Let us disregard the delays when calculating $P(t)$, $Q(t,\tau)$, $u^0(t)$. In this case, (cf. (5.8))

$$Q(t,0) = Q(t,-h) = B'(t)P(t).$$

Then (5.5) implies that the simplified equation for the coefficient $P(t)$ takes the form

$$\dot{P}^a(t) + \big(A'(t) + B'(t)\big)P^a(t) + P^a(t)\big(A(t) + B(t)\big)$$

$$- P^a(t)C(t)N_1^{-1}(t)C'(t)P^a(t) + N_2(t) = 0,$$

$$P^a(T) = N_0.$$

We define the approximate control law $u^a(t)$ by the relation

$$u^a(t) = -N_1^{-1}(t)C'(t)P^a(t)x^a(t). \qquad (5.12)$$

Then the corresponding process $x^a(t)$ is defined by the equation

$$\dot{x}^a(t) = \big(A(t) - C(t)N_1^{-1}(t)C'(t)P^a(t)\big)x^a(t) \qquad (5.13)$$

$$+ B(t)x^a(t-h) + \int_{-h}^{0} G(t,\tau)x^a(t+\tau)\, d\tau.$$

Note that, for an approximation for $P(t)$ and $Q(t,\tau)$, we can also take $P^{(k)}(t)$ and $Q^{(k)}(t,\tau)$ from the k-th step of the iterative solving algorithm for (5.5) – (5.8) [96], [91], [108].

The obtained approximate control algorithms are considerably simpler than the optimal algorithm, and, that is very important, they can be efficiently implemented by simple numerical means. Now, the main problem is to evaluate the quality of such approximations. The nonoptimality level is defined by the ratio

$$\Delta_C = \frac{\mathcal{I}(u^a)}{\mathcal{I}(u^0)}, \qquad (5.14)$$

where $u^a(t)$ and $u^0(t)$ are the approximate and the optimal control laws. As far as $u^0(t)$ is the optimal control, $\Delta_C \geq 1$. Since the optimal control $u^0(t)$ is unknown, Δ_C is also unknown. Then it is reasonable to construct an upper bound for Δ_C that can be actually calculated. The efforts needed for this calculations must be substantially lesser than the calculation of the optimal law.

THEOREM 5.2. *Let the approximate control $u^a(t)$ be determined by* (5.10) (*or* (5.12). *Then the approximation quality satisfies the following inequalities*

$$1 \leq \Delta_C \leq \Delta_C^0, \qquad \Delta_C^0 = \frac{\mathcal{I}(u^a) \cdot \mu_C}{|\mathcal{Q}^a|}, \tag{5.15}$$

where

$$\mathcal{Q}^a = \varphi'(0)\eta^a(0) + \int_{-h}^{0} \varphi'(s)B'(s+h)\eta^a(s+h)\,ds + \int_0^h \int_{-\theta}^{0} \varphi'(s)G'(s+\theta,-\theta)\eta^a(s+\theta)\,ds d\theta, \tag{5.16}$$

$$\dot{\eta}^a(t) = -A'(t)\eta^a(t) - B'(t+h)\eta^a(t+h) - \int_0^h G'(t+\theta,-\theta)\eta^a(t+\theta)\,d\theta - p^a(t), \qquad t \in [0,T], \tag{5.17}$$

$$\eta^a(T) = \lambda^a, \qquad \eta^a(\tau) = 0, \quad T < \tau \leq T+h,$$

$$\lambda^a = N_0 x^a(T), \qquad p^a(t) = N_2(t)x^a(t), \tag{5.18}$$

$$\mu_C = \left\{ x^{a\prime}(T)N_0x^a(T) + \int_0^T \eta^{a\prime}(t)C(t)N_1^{-1}(t)C'(t)\eta^a(t)\,dt + \int_0^T x^{a\prime}(t)N_2(t)x^a(t)\,dt \right\}^{\frac{1}{2}}.$$

The proof of the theorem is given in Subsection 5.5. Theorem 5.2 is the analog of Theorem 4.2.

Theorem 5.2 provides the following algorithm for Δ_C^0 calculation. Basing on the process $x^a(t)$ (defined by (5.11) or (5.13)), the boundary value λ^a and the function $p^a(t)$ (defined by (5.18)) are computed. Then $\eta^a(t)$ is obtained from the ordinary differential equation with aftereffect (defined by (5.17)). Finally, the values $\mathcal{I}(u^a)$, μ_C, and Q^a are determined by simple quadratures.

So, the simplified control algorithms are presented. They are quite easy for implementation. Moreover, the estimate of the nonoptimality level (5.15) can be calculated by relatively simple numerical means.

5.2. *L_1-cost Control Problem*

Consider the control system (5.1). The L_1-cost control problem is to minimize the nonsmooth functional

$$\mathcal{J}(u) = \sum_{d=1}^{n} \sigma_{-d}|x_d(T)| + \int_0^T \left(\sum_{i=1}^{r} \sigma_i(t)|u_i(t)| + \sum_{d=1}^{n} \gamma_d(t)|x_d(t)| \right) dt, \quad (5.19)$$

where $\sigma_{-d} \geq 0$, $\gamma_d(t) \geq 0$, $d = 1, \dots, n$, $\sigma_i(t) \geq \sigma^0 > 0$, $i = 1, \dots, r$, $t \in [0, T]$, are given coefficients, $u \in L_1^r$.

The L_1-cost control problem (5.1), (5.19) is a very difficult variational problem. In order to obtain its approximate solution let us replace the nonsmooth functional (5.19) by the quadratic one (5.2) with

$$N_0 = \mathrm{diag}(\sigma_{-1}^2, \dots, \sigma_{-n}^2), \qquad N_1(t) = \beta_1 \mathrm{diag}(\sigma_1^2(t), \dots, \sigma_r^2(t)), \quad (5.20)$$

$$N_2(t) = \beta_2 \mathrm{diag}(\gamma_1^2(t), \dots, \gamma_n^2(t)),$$

where β_1, β_2 are specified positive scalars. The solution of the auxiliary linear-quadratic problem (5.1), (5.2), (5.20) is used as an approximate solution to the nonsmooth problem (5.1), (5.19). Similar to (5.14), the approximation quality is defined by the ratio

$$\Delta_{L_1} = \frac{\mathcal{J}(u^0)}{\mathcal{J}_0}, \qquad \mathcal{J}_0 = \inf_u \mathcal{J}(u), \quad (5.21)$$

where $u^0(t)$ is the optimal control for (5.1), (5.2), (5.20), and $\mathcal{J}_0$ is the optimal cost value for (5.1), (5.19).

THEOREM 5.3. *Let $u^0(t)$ and $x^0(t)$ be the optimal control and the optimal process for the problem* (5.1), (5.2), (5.20). *Then*

$$1 \leq \Delta_{L_1} \leq \Delta_{L_1}^0 = \frac{\mathcal{M}_1 \cdot \mathcal{M}_\infty}{\mathcal{M}_2}, \quad (5.22)$$

where

$$\mathcal{M}_1 = \sum_{d=1}^{n} |\varphi_d| + \sum_{i=1}^{r} \beta_1^{-1} \int_0^T |\Psi_i(t)|\, dt + \sum_{d=1}^{n} \beta_2^{-1} \int_0^T |\chi_d(t)|\, dt,$$

$$\mathcal{M}_2 = \sum_{d=1}^{n} \varphi_d^2 + \sum_{i=1}^{r} \beta_1^{-1} \int_0^T \Psi_i^2(t)\, dt + \sum_{d=1}^{n} \beta_2^{-1} \int_0^T \chi_d^2(t)\, dt,$$

$$\mathcal{M}_\infty = \max\left\{ \max_{1 \leq d \leq n} |\varphi_d|, \sup_{1 \leq i \leq r,\ t \in [0,T]} |\Psi_i(t)|, \sup_{1 \leq d \leq n,\ t \in [0,T]} |\chi_d(t)| \right\},$$

$$\varphi_d = \sigma_{-d} x_d^0(T), \qquad \Psi_i(t) = \beta_1 \sigma_i(t) u_i^0(t), \qquad \chi_d(t) = \beta_2 \gamma_d(t) x_d^0(t),$$

$$d = 1, \ldots, n, \qquad i = 1, \ldots, r.$$

The proof is contained in Subsection 5.6. Note that Theorem 5.3 is the analog of Theorem 2.2.

So, the approximate solution to the difficult nonsmooth problem is obtained with the help of specially fitted linear-quadratic problem. The latter problem has a numerical solving algorithm. Though our approximate solution can be rough, it has a guaranteed approximation quality determined by (5.22).

It was emphasized above that, though the calculation of $u^0(t)$, $x^0(t)$ is possible, this is a nontrivial procedure. The reduced approximate solution (5.10), (5.11) or (5.12), (5.13) (with (5.20)) described above can be used for approximate solving the nonsmooth problem.

Let us introduce the ratio

$$\Delta_{CL_1} = \frac{\mathcal{J}(u^a)}{\mathcal{J}_0}, \qquad \mathcal{J}_0 = \inf_u \mathcal{J}(u). \tag{5.23}$$

Then another estimate, simpler for numerical implementation than (5.22), can be obtained.

THEOREM 5.4. *Let $u^a(t)$ and $x^a(t)$ from* (5.10), (5.11) (*or* (5.12), (5.13)) *be the approximate control and the corresponding process for the problem* (5.1), (5.2), (5.20). *Then*

$$1 \leq \Delta_{CL_1} \leq \Delta_{CL_1}^0 = \frac{\mathcal{M}_1^a \cdot \mu_L}{|Q^a|},$$

where

$$\mathcal{M}_1^a = \sum_{d=1}^n \sigma_{-d} |x_d^a(T)| + \int_0^T \left(\sum_{i=1}^r \sigma_i(t) |u_i^a(t)| + \sum_{d=1}^n \gamma_d(t) |x_d^a(t)| \right) dt,$$

$$\mu_L = \max \Bigg\{ \max_{1 \leq d \leq n} |\sigma_{-d} x_d^a(T)|, \quad \sup_{1 \leq i \leq r,\ t \in [0,T]} |\sigma_i^{-1}(t) C_i'(t) \eta^a(t)|, \quad \sup_{1 \leq d \leq n,\ t \in [0,T]} |\beta_2 \gamma_d(t) x_d^a(t)| \Bigg\},$$

and Q^a, $\eta^a(t)$ are determined by (5.16), (5.18), (5.17); $C_i(t)$ is the i-th column of $C(t)$.

The idea of the proof is described in Subsection 5.7.

TABLE 3. LQ-problem; $h = 0.5$, $N_0 = N_1 = 1$, $G = N_2 = 0$, $T = 5.0$

	First scheme				Second scheme			
B	$\mathcal{I}(u^a)$	μ_C	Q^a	Δ_C^0	$\mathcal{I}(u^a)$	μ_C	Q^a	Δ_C^0
1.0	1.77	0.28	0.47	1.05	1.69	1.36	2.29	1.01
2.0	3.06	0.007	0.018	1.15	2.67	2.81	7.44	1.01
5.0	7.75	55.0	-280.0	1.52	5.20	27.6	141.0	1.02

TABLE 4. LQ-problem; $h = 0.5$, $B = 0$, $N_0 = 1.2$, $N_1 = 1.3$, $N_2 = 1.1$, $T = 5.0$

	First scheme				Second scheme			
G	$\mathcal{I}(u^a)$	μ_C	Q^a	Δ_C^0	$\mathcal{I}(u^a)$	μ_C	Q^a	Δ_C^0
1.0	1.50	1.13	1.67	1.01	1.70	6.20	8.50	1.24
2.0	1.96	0.84	1.49	1.10	4.22	115.0	205.0	2.37
3.0	2.43	0.55	0.95	1.41	16.8	2285.0	5166.0	7.43

Thus, the approximate solution to the difficult nonsmooth control problem is also obtained with the help of an approximate solution for specially fitted linear-quadratic problem. The estimate of the approximation quality can be calculated as well.

5.3. *Examples of Control Systems*

EXAMPLE 5.1. Consider a simplest one-dimenisonal control problem (5.1), (5.2) with stationary coefficients ($n = r = 1$, $A(t) = 0$, $B(t) = \text{const}$, $G(t,\tau) = \text{const}$, $C(t) = 1$):

$$\dot{x}(t) = Bx(t-h) + G\int_{-h}^{0} x(t+s)\,ds + u(t),$$

$$x(\tau) = 1, \qquad \tau \in [-h, 0],$$

$$\mathcal{I}(u) = N_0 x^2(T) + \int_0^T \Big(N_1 u^2(t) + N_2 x^2(t)\Big)\,dt \ \to \min.$$

For solving this linear-quadratic problem (LQ-problem) let us apply Theorem 5.2. Some numerical results are presented in Tables 3 and 4. Tables 3 and 4 show that we have obtained simple control algorithms with near-optimal performance. In the case $G = N_2 = 0$, both schemes are valid. But the primitive second scheme proves to be more successful. It follows from the proof of Theorem 5.1 that if $N_2 = 0$ and $n = 1$, then the corresponding dual problem is one-dimensional. Hence, every nonzero admissible element of the dual problem generates the exact value for $\mathcal{I}(u^0)$ and $\Delta_C = \Delta_C^0$. Too small and too large numbers in the third and fourth row in Table 3 indicate that, for $B = 2.0$ and $B = 5.0$, the first scheme gives a poor approximation for the solution of the dual problem. However, the first scheme provides acceptable control algorithms: $1.05 \le \Delta_C = \Delta_C^0 \le 1.52$.

In the case $B = 0$, the nontrivial first scheme is effective, but the second scheme can be poor. It results from the fact that, for $B = 0$, the first scheme has a better approximation order. Nevertheless, for $G = 1.0$, the second scheme is also acceptable. So, our method allows us to choose the most adequate approximate control algorithm.

EXAMPLE 5.2. Now, let us investigate the L_1-cost problem:

$$\mathcal{J}(u) = \sigma_-|x(T)| + \int_0^T \Big(\sigma|u(t)| + \gamma|x(t)|\Big)\, dt \;\to \inf .$$

In accordance with (5.20) we construct the auxiliary linear-quadratic control problem, in which we put

$$N_0 = \sigma_-^2, \qquad N_1 = \beta_1\sigma^2, \qquad N_2 = \beta_2\gamma^2;$$

β_1 and β_2 are positive free parameters, which can be adjusted.

Let us consider the case

$$B = 0, \qquad G = 1.0, \qquad h = 0.5, \qquad T = 5.0,$$
$$\sigma_- = 1.0, \qquad \sigma = 2.0, \qquad \gamma = 1.0, \qquad \beta_1 = 1.0, \qquad \beta_2 = 2.0.$$

We prefer to refuse from exact solving the linear-quadratic problem. Since $B = 0$, we use the first approximation scheme. In accordance with Theorem 5.4 we obtain

$$\mathcal{J}(u^a) = 4.39, \qquad \Delta_{CL_1}^0 = 1.98. \tag{5.24}$$

Let us study another variant of system parameters

$$B = 0, \qquad G = 1.0, \qquad T = 1.0,$$
$$\sigma_- = 5.0, \qquad \sigma = 1.0, \qquad \gamma = 1.0, \qquad \beta_1 = 1.0, \qquad \beta_2 = 1.0.$$

The corresponding results are presented in Table 5.

It follows from (5.24) and Table 5 that, in our example, the linear-quadratic control law in the L_1-cost problem yields the same order of L_1-performance as the unknown optimal control.

Mathematical description of real control processes is not quite accurate. Therefore, for practical purposes there is no necessity to resolve optimal problem with high accuracy. Design decisions are often based on performance

TABLE 5. L_1-cost problem; variation of h

h	0.1	0.2	0.3	0.4	0.5
$\Delta^0_{CL_1}$	1.65	1.67	1.67	1.66	1.63

orders rather than on exact characteristics of inaccurate models. From this point of view, an approximate synthesis of control systems without critical loss in optimality, perhaps, is more desirable. Thus, our estimates are useful for a designer.

We recall that Remark 4.2 is true for the control problems as well as for estimation problems.

5.4. *Proof of Theorem 5.1*

Let us represent the optimal problem (5.1), (5.2) in the form of the following variational problem: minimize the convex functional

$$\mathcal{I}(u^0) = \mathcal{I}_0 = \inf_{x(T)\in\mathbf{R}^n,\, u\in L_2^r,\, x\in L_2^n} I(x(T), u, x) \tag{5.25}$$

under affine constraints on $(x(T), u, x)$:

$$V(x(T), u) = 0, \qquad W(u, x) = 0, \tag{5.26}$$

where

$$I(x(T), u, x) = \Big\{ x'(T)N_0x(T) + \int_0^T \Big(u'(t)N_1(t)u(t) + x'(t)N_2(t)x(t)\Big)\, dt \Big\}^{\frac{1}{2}}, \tag{5.27}$$

$$V(x(T), u) = -x(T) + Y(T,0)\varphi(0) + \int_{-h}^0 Y(T, s+h)B(s+h)\varphi(s)\, ds \tag{5.28}$$

$$+ \int_0^h \int_{-\theta}^0 Y(T, s+\theta)G(s+\theta, -\theta)\varphi(s)\, ds d\theta + \int_0^T Y(T,s)C(s)u(s)\, ds$$

and

$$W(u, x) = -x(t) + Y(t,0)\varphi(0) + \int_{-h}^0 Y(t, s+h)B(s+h)\varphi(s)\, ds \tag{5.29}$$

$$+ \int_0^h \int_{-\theta}^0 Y(t, s+\theta)G(s+\theta, -\theta)\varphi(s)\, ds d\theta + \int_0^t Y(t,s)C(s)u(s)\, ds,$$

$t \in [0, T]$.

The constraints (5.26), (5.28), (5.29) are the well-known explicit formulas for the solution of linear dynamic system with aftereffect (5.1). These formulas follow immediately from Lemma 1.1. They contain a pseudo-fundamental matrix $Y(t,s)$ defined by (1.4) – (1.6).

For simplicity, we assume further that $N_0 > 0$, $N_1(t) \geq N_1 > 0$, and $N_2(t) \geq N_2 > 0$. The final results remain valid without this technical restriction (cf. the proof of Theorem 2.1 of Chapter 3 for singular matrices).

Introduce the Hilbert space $\mathcal{K} = \mathbf{R}^n \times L_2^r \times L_2^n$ with elements $\omega = (x(T), u, x)$ and the inner product

$$\langle \omega^{(1)}, \omega^{(2)} \rangle_{\mathcal{K}} = \langle x^{(1)}(T), x^{(2)}(T) \rangle_{\mathbf{R}^n} + \langle u^{(1)}, u^{(2)} \rangle_{L_2^r} + \langle x^{(1)}, x^{(2)} \rangle_{L_2^n},$$

$$x^{(1)}(T), x^{(2)}(T) \in \mathbf{R}^n, \qquad u^{(1)}, u^{(2)} \in L_2^r, \qquad x^{(1)}, x^{(2)} \in L_2^n.$$

Then $I(x(T), u, x) : \mathcal{K} \to \mathbf{R}^1$. The space $\mathcal{K}$ is reflexive, I is continuous, convex, and coercive; moreover, the affine constraints define a closed convex manifold. Hence, by virtue of Lemma 1.2 of Chapter 2 the solution of (5.25), (5.26) exists.

LEMMA 5.1. 1^0. *The problem dual to* (5.25) – (5.29) *has the form*

$$\mathcal{I}^0 = \sup_{p \in L_2^n, \, \lambda \in \mathbf{R}^n} \mathcal{A}(p, \lambda) \tag{5.30}$$

under the constraint

$$S(p, \lambda) \leq 1, \tag{5.31}$$

where

$$\mathcal{A}(p, \lambda) = \varphi'(0) F(0) + \int_{-h}^{0} \varphi'(s) B'(s+h) F(s+h) \, ds \tag{5.32}$$

$$+ \int_0^h \int_{-\theta}^{0} \varphi'(s) G'(s+\theta, -\theta) F(s+\theta) \, ds d\theta,$$

$$S(p, \lambda) = \left\{ \lambda' N_0^{-1} \lambda + \int_0^T F'(t) C(t) N_1^{-1}(t) C'(t) F(t) \, dt \right. \tag{5.33}$$

$$\left. + \int_0^T p'(t) N_2^{-1}(t) p(t) \, dt \right\},$$

$$F(s) = Y'(T, s) \lambda + \int_s^T Y'(t, s) p(t) \, dt. \tag{5.34}$$

2^0. *The following equality holds*

$$\mathcal{I}_0 = \mathcal{I}^0. \tag{5.35}$$

Lemma 5.1 is the analog of Assertions 1 and 2 of Theorem 2.4.

PROOF. The Lagrange function for (5.25), (5.26) has the form

$$\mathcal{L}_{\mathcal{K}}(x(T), u, x; p, \lambda) = I(x(T), u, x) + \langle p, W(u, x)\rangle_{L_2^n} + \langle \lambda, V(x(T), u)\rangle_{\mathbf{R}^n},$$

where $p \in L_2^n$, $\lambda \in \mathbf{R}^n$ are Lagrange multipliers. By changing the integrations and grouping the terms, it can be easily shown that

$$\mathcal{L}_{\mathcal{K}}(x(T), u, x; p, \lambda) = \Big[\langle N_0^{\frac{1}{2}} x(T), N_0^{\frac{1}{2}} x(T)\rangle_{\mathbf{R}^n} \tag{5.36}$$

$$+ \langle N_1^{\frac{1}{2}} u, N_1^{\frac{1}{2}} u\rangle_{L_2^r} + \langle N_2^{\frac{1}{2}} x, N_2^{\frac{1}{2}} x\rangle_{L_2^n}\Big]^{\frac{1}{2}} + \mathcal{A}(p, \lambda)$$

$$- \langle N_0^{-\frac{1}{2}} \lambda, N_0^{\frac{1}{2}} x(T)\rangle_{\mathbf{R}^n} + \langle N_1^{-\frac{1}{2}} C'F, N_1^{\frac{1}{2}} u\rangle_{L_2^r} - \langle N_2^{-\frac{1}{2}} p, N_2^{\frac{1}{2}} x\rangle_{L_2^n}$$

$$= \mathcal{A}(p, \lambda) + \|(N_0^{\frac{1}{2}} x(T),\ N_1^{\frac{1}{2}} u,\ N_2^{\frac{1}{2}} x)\|_{\mathcal{K}}$$

$$+ \langle(-N_0^{-\frac{1}{2}} \lambda,\ N_1^{-\frac{1}{2}} C'F,\ -N_2^{-\frac{1}{2}} p),\ (N_0^{\frac{1}{2}} x(T),\ N_1^{\frac{1}{2}} u,\ N_2^{\frac{1}{2}} x)\rangle_{\mathcal{K}}.$$

In accordance with the duality theory of convex variational problems (see Theorems 1.7 – 1.9 of Chapter 2), the dual problem has the form

$$\mathcal{I}^0 = \sup_{p,\lambda} \mathcal{G}(p, \lambda), \qquad \mathcal{G}(p, \lambda) = \inf_{x(T),u,x} \mathcal{L}_{\mathcal{K}}(x(T), u, x; p, \lambda).$$

Moreover,

$$\mathcal{I}_0 = \mathcal{I}^0.$$

It follows from (5.36) that

$$\mathcal{G}(p, \lambda) = \begin{cases} \mathcal{A}(p, \lambda) & \text{if } \|(-N_0^{-\frac{1}{2}} \lambda,\ N_1^{-\frac{1}{2}} C'F,\ -N_2^{-\frac{1}{2}} p)\|_{\mathcal{K}} \le 1 \\ -\infty & \text{otherwise.} \end{cases}$$

Thus, the dual problem can be written in the form (5.30) – (5.34), and (5.35) holds. □

The cost functional of the dual problem $\mathcal{A}(p, \lambda)$ is linear and continuos, and constraint (5.31) defines a bounded closed convex set. Therefore, by virtue of Lemma 1.2 of Chapter 2, the dual problem has a solution (p^0, λ^0).

LEMMA 5.2. 1^0. *The solution of the dual problem* (5.30), (5.31) *is determined by the formulas*

$$p^0(t) = (2\alpha)^{-1}N_2(t)\xi(t), \qquad \lambda^0 = (2\alpha)^{-1}N_0\xi(T), \tag{5.37}$$

where $\xi(t)$ *satisfies the boundary value problem*

$$\dot{\xi}(t) = A(t)\xi(t) + B(t)\xi(t-h) + \int_{-h}^{0} G(t,s)\xi(t+s)\,ds - C(t)N_1^{-1}(t)C'(t)\eta(t), \tag{5.38}$$

$$\dot{\eta}(t) = -A'(t)\eta(t) - B'(t+h)\eta(t+h) - \int_0^h G'(t+\theta,-\theta)\eta(t+\theta)\,d\theta - N_2(t)\xi(t),$$

$$\xi(\tau) = \varphi(\tau), \qquad -h \le \tau \le 0,$$

$$\eta(T) = N_0\xi(T), \qquad \eta(\tau) = 0, \quad T < \tau \le T+h,$$

and α *is a positive scalar.*
2^0. *The following equality holds:*

$$\eta(t) = P(t)\xi(t) + \int_{-h}^{0} Q(t,\tau)\xi(t+\tau)\,d\tau, \tag{5.39}$$

where $P(t)$ *and* $Q(t,\tau)$ *are subjected to* (5.5) – (5.8).

Lemma 5.2 is analogous to Assertion 2 of Theorem 2.1 and Assertion 3 of Theorem 2.4.

PROOF. The problem (5.30), (5.31) is a smooth linear-quadratic problem with a scalar constraint. The corresponding Lagrange function has the form (see Section 1 of Chapter 2)

$$\mathcal{L}^0(\lambda, p; \alpha) = -\mathcal{A}(\lambda, p) + \alpha S(\lambda, p), \qquad \alpha > 0.$$

(It can easily be shown that if $\alpha = 0$, then $\mathcal{I}_0 = \mathcal{I}(u^0) = 0$.)

In accordance with the Lagrange theorem, the optimal pair (p^0, λ^0) is determined by the stationary point: $\partial\mathcal{L}^0/\partial p = 0$, $\partial\mathcal{L}^0/\partial\lambda = 0$. These formulas are equivalent to the functional equations:

$$-Y(t,0)\varphi(0) - \int_{-h}^{0} Y(t,s+h)B(s+h)\varphi(s)\,ds$$

$$-\int_0^h \int_{-\theta}^{0} Y(t,s+\theta)G(s+\theta,-\theta)\varphi(s)\,ds d\theta$$

$$+ 2\alpha N_2^{-1}(t)p(t) + 2\alpha \int_0^t Y(t,s)C(s)N_1^{-1}(s)C'(s)F(s)\,ds = 0$$

and

$$-Y(T,0)\varphi(0) - \int_{-h}^0 Y(T,s+h)B(s+h)\varphi(s)\,ds$$

$$-\int_0^h \int_{-\theta}^0 Y(T,s+\theta)G(s+\theta,-\theta)\varphi(s)\,dsd\theta$$

$$+ 2\alpha N_0^{-1}\lambda + 2\alpha \int_0^T Y(T,s)C(s)N_1^{-1}(s)C'(s)F(s)\,ds = 0.$$

It follows directly from (1.5) and (5.34) that

$$\dot{F}(t) = -A'(t)F(t) - B'(t+h)F(t+h) \qquad (5.40)$$

$$- \int_0^h G'(t+\theta,-\theta)F(t+\theta)\,d\theta - p(t),$$

$$F(T) = \lambda, \qquad F(\tau) = 0, \quad T < \tau \le T+h.$$

So, the following implicit (with respect to p and λ) relations are valid:

$$p(t) = (2\alpha)^{-1}N_2(t)\Big\{Y(t,0)\varphi(0) + \int_{-h}^0 Y(t,s+h)B(s+h)\varphi(s)\,ds \qquad (5.41)$$

$$+ \int_0^h \int_{-\theta}^0 Y(t,s+\theta)G(s+\theta,-\theta)\varphi(s)\,dsd\theta$$

$$- 2\alpha \int_0^t Y(t,s)C(s)N_1^{-1}(s)C'(s)F(s)\,ds\Big\},$$

$$\lambda = (2\alpha)^{-1}N_0\Big\{Y(T,0)\varphi(0) + \int_{-h}^0 Y(T,s+h)B(s+h)\varphi(s)\,ds$$

$$+ \int_0^h \int_{-\theta}^0 Y(T,s+\theta)G(s+\theta,-\theta)\varphi(s)\,dsd\theta$$

$$- 2\alpha \int_0^T Y(T,s)C(s)N_1^{-1}(s)C'(s)F(s)\,ds\Big\}.$$

In accordance with Lemma 1.1, equations (5.41) imply that

$$p(t) = (2\alpha)^{-1}N_2(t)\xi(t), \qquad \lambda = (2\alpha)^{-1}N_0\xi(T), \qquad (5.42)$$

where

$$\dot{\xi}(t) = A(t)\xi(t) + B(t)\xi(t-h) + \int_{-h}^{0} G(t,s)\xi(t+s)\,ds - C(t)N_1^{-1}(t)C'(t)2\alpha F(t),$$

$$\xi(\tau) = \varphi(\tau), \qquad -h \le \tau \le 0.$$

Hence, if we introduce the function $\eta(t) = 2\alpha F(t)$, then we see from (5.40) that the functions $\xi(t)$ and $\eta(t)$ satisfy the boundary value problem (5.38).

By straightforward but laborious calculations (cf. the proof of Theorem 2.1) it can be shown that if $\xi(t)$, $\eta(t)$ satisfy (5.38) and $P(t)$, $Q(t,\tau)$ satisfy (5.5) – (5.8), then (5.39) holds. The solution of (5.5) – (5.8) exists and is unique [96], [105], [108], [106]. Lemma 5.2 is proved. □

It follows from (5.38) and (5.39) that

$$\dot{\xi}(t) = \left(A(t) - C(t)N_1^{-1}(t)C'(t)P(t)\right)\xi(t) + B(t)\xi(t-h) + \int_{-h}^{0} \left(G(t,s) - C(t)N_1^{-1}(t)C'(t)Q(t,s)\right)\xi(t+s)\,ds, \tag{5.43}$$

$$\xi(\tau) = \varphi(\tau), \qquad -h \le \tau \le 0.$$

By virtue of Theorem 1.8 of Chapter 2, if (p^0, λ^0) is the solution of the dual problem, then

$$\mathcal{L}_{\mathcal{K}}(x^0(T), u^0, x^0; p^0, \lambda^0) \le \mathcal{L}_{\mathcal{K}}(x(T), u, x; p^0, \lambda^0) \tag{5.44}$$

for all $x(T) \in \mathbf{R}^n$, $u \in L_2^r$, $x \in L_2^n$. Then it follows from (5.36) and (5.44) that, for a $k \in \mathbf{R}^1$, $k \le 0$,

$$x^0(T) = -\,kN_0^{-1}\lambda^0, \tag{5.45}$$

$$u^0(t) = kN_1^{-1}(t)C'(t)F(t)\big|_{p=p^0,\,\lambda=\lambda^0},$$

$$x^0(t) = -\,kN_2^{-1}(t)p^0(t).$$

Substituting $p^0(t)$ and λ^0 from (5.42) into (5.45), we get

$$x^0(T) = -\,(2\alpha)^{-1}k\xi(T), \tag{5.46}$$

$$u^0(t) = (2\alpha)^{-1}kN_1^{-1}(t)C'(t)\eta(t),$$

$$x^0(t) = -\,(2\alpha)^{-1}k\xi(t).$$

Substituting the variables (5.46) into the constraints of the initial variational problem (5.26), it can easily be shown that $k = -2\alpha$. Thus,

$$x^0(t) = \xi(t), \qquad u^0(t) = -N_1^{-1}(t)C'(t)\eta(t), \qquad t \in [0, T], \tag{5.47}$$

and

$$p^0(t) = (2\alpha)^{-1}N_2(t)x^0(t), \qquad \lambda^0 = (2\alpha)^{-1}N_0x^0(T). \tag{5.48}$$

By substituting (5.48) into (5.31), it can be also shown that $(2\alpha)^2\mathcal{I}_0^2 = 1$. Since $\alpha > 0$ and $\mathcal{I}_0 > 0$, we obtain that

$$2\alpha = \mathcal{I}_0. \tag{5.49}$$

Formulas (5.43), (5.39), and (5.47) imply the assertions of Theorem 5.1. The formulas (5.48), (5.49) strengthen the first assertion of Lemma 5.2. □

5.5. *Proof of Theorem 5.2*

By virtue of the duality relation (5.35),

$$\mathcal{I}(u^0) = \mathcal{I}_0 = \mathcal{I}^0 = \mathcal{A}(p^0, \lambda^0) \geq \mathcal{A}(p, \lambda) \tag{5.50}$$

for any admissible pair (p, λ) (restricted by (5.31)).

The numerator in (5.14) can be easily computed. Let us construct a lower bound for the denominator. Then the upper bound for Δ_C will be obtained.

Since $x^0(t)$ is unknown, the Lagrange multipliers (p^0, λ^0) can not be calculated from (5.48), (5.49). Nevertheless, we can construct an approximate control $u^a(t)$ and a process $x^a(t)$ in accordance with (5.10) and (5.11) (or (5.12) and (5.13)). Then, in complete analogy with the exact formulas (5.48), the straight line

$$p(\nu) = p(t; \nu) = \nu N_2(t)x^a(t), \qquad \lambda(\nu) = \nu N_0 x^a(T), \qquad \nu \in \mathbf{R}^1, \tag{5.51}$$

provides an appropriate set for approximate maximization in (5.50). Then the value ν, that gives the maximum to $\mathcal{A}$ on the straight line (5.51), yields the best estimate for $\mathcal{I}_0$. Obviously, this is the value such that $S(p(\nu), \lambda(\nu)) = 1$. Hence,

$$\mathcal{I}_0 \geq \frac{|\mathcal{A}(p^a, \lambda^a)|}{\mu_C}. \tag{5.52}$$

Certainly, we assume that $\mathcal{A}(p^a, \lambda^a) \neq 0$ and $\mu_C \neq 0$. Obviously, this condition holds almost without exception. Then the estimate (5.15) follows from (5.52). □

5.6. *Proof of Theorem 5.3*

Let us represent the optimal problem (5.1), (5.19) in the form of the following variational problem: minimize the convex functional

$$\mathcal{J}_0 = \inf_{x(T)\in\mathbf{R}^n,\, u\in L_1^r,\, x\in L_1^n} J(x(T), u, x), \tag{5.53}$$

where

$$J(x(T), u, x) = \sum_{d=1}^{n} \sigma_d |x_d(T)| + \int_0^T \left(\sum_{i=1}^{r} \sigma_i(t)|u_i(t)| + \sum_{d=1}^{n} \gamma_d(t)|x_d(t)| \right) dt$$

under affine constraints (5.26), (5.28), (5.29).

LEMMA 5.3. 1^0. *The problem dual to* (5.53), (5.26), (5.28), (5.29) *has the form*

$$\mathcal{J}^0 = \sup_{p\in L_\infty^n,\, \lambda\in\mathbf{R}^n} \mathcal{A}(p,\lambda) \tag{5.54}$$

under the restrictions

$$|C_i'(t)F(t)| \le \sigma_i(t), \qquad |p_d(t)| \le \gamma_d(t), \qquad |\lambda_d| \le \sigma_d, \tag{5.55}$$

$$i = \overline{1,r}, \qquad d = \overline{1,n}, \qquad t \in [0,T],$$

where $\mathcal{A}(p,\lambda)$ *and* F *are defined by* (5.32), (5.34).
2^0. *The following equality holds*

$$\mathcal{J}_0 = \mathcal{J}^0. \tag{5.56}$$

The proof of Lemma 5.3 is similar to the proof of Lemma 5.1. Lemma 5.3 is the analog of Theorem 2.3.

If the linear-quadratic control problem is resolved, the numerator in (5.21) becomes known. The lower bound for $\mathcal{J}^0$ can be obtained by the argument similar to that in the proof of Theorem 5.2. Namely,

$$\mathcal{J}_0 = \mathcal{J}^0 \ge \mathcal{A}(p,\lambda) \tag{5.57}$$

for any admissible pair (p,λ) (i.e., the pair subjected to (5.55)).

For an appropriate set for evaluating the functional $\mathcal{J}^0$ we choose the straight line

$$\bar{p}(\nu) = \bar{p}(t;\nu) = \nu p^0(t), \qquad \bar{\lambda}(\nu) = \nu\lambda^0, \qquad \nu \in \mathbf{R}^1,$$

where (p^0, λ^0) are the optimal Lagrange multipliers (5.37), (5.49) for the linear-quadratic problem (5.1), (5.2), (5.20). Then it follows from (5.54), (5.56), (5.57), and Lemma 5.1 that

$$\mathcal{J}_0 \geq \nu \mathcal{A}(p^0, \lambda^0) = \nu \mathcal{I}_0, \tag{5.58}$$

where $\mathcal{I}_0$ is the optimal value of the functional for the linear-quadratic problem.

It follows from (5.37), (5.38), (5.40), and (5.49) that

$$F(t)\big|_{p^0,\lambda^0} = \mathcal{I}_0^{-1}\eta(t). \tag{5.59}$$

Then, by virtue of (5.20), (5.37), and (5.59), the pair $(\bar{p}(\nu), \bar{\lambda}(\nu))$ is admissible for the dual problem if and only if

$$\nu \mathcal{I}_0^{-1}|C_i'(t)\eta(t)| \leq \sigma_i(t),$$

$$\nu \mathcal{I}_0^{-1}|\beta_2\gamma_d^2(t)\xi_d(t)| \leq \gamma_d(t),$$

$$\nu \mathcal{I}_0^{-1}|\sigma_{-d}^2\xi_d(T)| \leq \sigma_{-d},$$

$$i = \overline{1,r}, \qquad d = \overline{1,n}, \qquad t \in [0,T].$$

Taking into account formulas (5.47), (5.20), the latter inequalities can be rewritten in the form

$$\nu \mathcal{I}_0^{-1} \max \left\{ \max_{1 \leq d \leq n} |\sigma_{-d}x_d^0(T)|, \sup_{1\leq i \leq r,\ t\in[0,T]} |\beta_1\sigma_i(t)u_i^0(t)|, \right. \tag{5.60}$$

$$\left. \sup_{1\leq d\leq n,\ t\in[0,T]} |\beta_2\gamma_d(t)x_d^0(t)| \right\} \leq 1.$$

Then the desired estimate (5.22) follows from (5.21), (5.58), and (5.60). □

5.7. *Proof of Theorem 5.4*

The proof of Theorem 5.4 is similar to the proof of Theorem 5.2. The only distinction is that Lemma 5.3 should be used instead of Lemma 5.1. □

REMARK 5.1. In the proofs of Theorems 5.2 – 5.4 we use the same approximate algorithm to compute the numerator in the level of nonoptimality and to obtain an estimate for the denominator in the level of nonoptimality. As it was already remarked in Chapter 3, we can use different algorithms for these purposes and to decouple the procedure. Then our estimate of nonoptimality can be improved.

6. Reduced Algorithms for Systems with Weakly Connected Blocks

The classical stochastic mean-square filtering problem was studied in Sections 2 and 4. As discussed above, there exists a constructive algorithm for finding the solution of this problem. Nevertheless, this algorithm contains a block of solving a system of functional partial differential equations. Hence the optimal filtering algorithm, particulary for large-scale systems, is hard for direct implementation.

It is not unfrequent that a researcher needs to estimate only a part of components of a system, and this part is weakly connected with the other system components (see, e.g., [191]). Then it is reasonable (and it is a usual engineering technique) to ignore "unimportant" components. So, the filter is designed for the corresponding reduced system. This reduced filter has a lesser dimension and, hence, the reduced filter requires considerably lesser efforts for implementation. It is important to evaluate the reduction quality without solving the initial full-dimensional complicated filtering problem.

6.1. *Weakly Connected Blocks in Dynamic Systems with Delay*

Consider a linear stochastic dynamic system with delay

$$\dot{x}(t) = A(t)x(t) + B(t)x(t-h) + C(t)u(t), \qquad t \in [0,T],$$

$$x(0) = x_0, \qquad x(\tau) = 0, \quad -h \le \tau < 0,$$

and the measurements

$$z(t) = H'(t)x(t-\bar{h}) + \varrho(t), \qquad z(t), \varrho(t) \in \mathbf{R}^m,$$

where $x(t) = (x_1'(t), x_2'(t))' \in \mathbf{R}^n$, $x_1(t) \in \mathbf{R}^k$, $x_2(t) \in \mathbf{R}^{n-k}$, $u(t) \in \mathbf{R}^r$,

$$A(t) = \begin{pmatrix} A_{11}(t) & \varepsilon A_{12}(t) \\ A_{21}(t) & A_{22}(t) \end{pmatrix}, \qquad B(t) = \begin{pmatrix} B_{11}(t) & \varepsilon B_{12}(t) \\ B_{21}(t) & B_{22}(t) \end{pmatrix},$$

$$H(t) = \begin{pmatrix} H_1(t) \\ \varepsilon H_2(t) \end{pmatrix}, \qquad \varepsilon \ll 1.$$

$$A_{11}(t),\ B_{11}(t) \in \mathbf{R}^{k\times k}, \qquad A_{22}(t),\ B_{22}(t) \in \mathbf{R}^{(n-k)\times(n-k)},$$

$$A_{12}(t),\ B_{12}(t),\ A_{21}'(t),\ B_{21}'(t) \in \mathbf{R}^{k\times(n-k)}, \qquad C(t) \in \mathbf{R}^{n\times r},$$

$$H_1(t) \in \mathbf{R}^{k\times m}, \qquad H_2(t) \in \mathbf{R}^{(n-k)\times m}.$$

Clearly, the subsystem for $x_1(t) \in \mathbf{R}^k$ is weakly connected with the whole system.

We assume that $x(0)$ is a Gaussian zero-mean random variable with covariance matrix P_0 and $u(t)$, $\varrho(t)$ are Gaussian zero-mean white noise processes with intesities $Q(t)$ and $R(t)$, respectively. The vector $x(0)$ and the processes $u(t)$, $\varrho(t)$ are mutually independent.

The filtering problem is to find an estimate for a scalar quantity $l_* = a'x(T)$, where $a = (a_1', 0_{n-k}')'$ is a given vector with $a_1 \in \mathbf{R}^k$, $0_{n-k} \in \mathbf{R}^{n-k}$.

Let us consider a class of linear estimates $\hat{l}$ for l_* of the form

$$\hat{l}(\Phi) = \int_0^T \Phi'(t)z(t)\,dt, \qquad \Phi \in L_2^m, \tag{6.1}$$

where $\Phi(t)$ is an estimator. The optimal mean-square filtering problem is to find an estimator $\Phi^0(t)$ that minimizes the mean-square estimation error:

$$\mathbf{E}(\hat{l}(\Phi^0) - l_*)^2 \le \mathbf{E}(\hat{l}(\Phi) - l_*)^2, \qquad \forall \Phi \in L_2^m. \tag{6.2}$$

It was shown in Section 4 that the optimal estimate can be represented in the state-space form (similar to the well-known Kalman filter)

$$\hat{l}(\Phi^0) = a'\hat{x}(T), \tag{6.3}$$

$$\dot{\hat{x}}(t) = A(t)\hat{x}(t) + B(t)\hat{x}(t-h) + K(t)\big(z(t+\bar{h}) - H'(t+\bar{h})\hat{x}(t)\big)$$

$$+ \int_{\max(0,t-h)}^{t} K_1(s,t)\big(z(s+\bar{h}) - H'(s+\bar{h})\hat{x}(s)\big)\,ds,$$

$$0 \le t \le T, \qquad \hat{x}(\tau) = 0, \quad -h \le \tau \le 0,$$

$$H(s) = 0, \quad T < s \le T + \bar{h}.$$

The gains in (6.3) are specified by the formulas

$$K(t) = P(t)H(t+\bar{h})R^{-1}(t+\bar{h}),$$

$$K_1(s,t) = P_1'(s, s-t)H(s+\bar{h})R^{-1}(s+\bar{h}),$$

where $P(t)$ and $P_1(t,\tau)$ are determined by the system of functional partial differential equations with respect to three $n \times n$ matrices $P(t)$, $P_1(t,\tau)$, $P_2(t,\tau,s)$ defined by (2.27), (2.28).

It is most desirable to simplify the estimation algorithm by reducing the dimension of the system of functional partial differential equations. A natural way for the simplification of the optimal filtering algorithm is to put formally $\varepsilon = 0$ in (6.3), (2.27), and (2.28). Then the simplified estimate takes the form

$$\hat{l}(\Phi^R) = a_1'\hat{x}_1(T), \tag{6.4}$$

$$\dot{\hat{x}}_1(t) = A_{11}(t)\hat{x}_1(t) + B_{11}(t)\hat{x}_1(t-h)$$
$$+ \hat{K}(t)\big(z(t+\bar{h}) - H_1'(t+\bar{h})\hat{x}_1(t)\big)$$
$$+ \int_{\max(0,t-h)}^{t} \hat{K}_1(s,t)\big(z(s+\bar{h}) - H_1'(s+\bar{h})\hat{x}_1(s)\big)\, ds,$$
$$0 \le t \le T, \qquad \hat{x}_1(\tau) = 0, \quad -h \le \tau \le 0.$$

The gains in (6.4) are determined by the formulas

$$\hat{K}(t) = \hat{P}(t)H_1(t+\bar{h})R^{-1}(t+\bar{h}), \tag{6.5}$$
$$\hat{K}_1(s,t) = \hat{P}_1'(s, s-t)H_1(s+\bar{h})R^{-1}(s+\bar{h}),$$

where $\hat{P}(t)$ and $\hat{P}_1(t,\tau)$ are defined by the system of functional partial differential equations with respect to three $k \times k$ matrices $\hat{P}(t)$, $\hat{P}_1(t,\tau)$, $\hat{P}_2(t,\tau,s)$, which are completely analogous to (2.27), (2.28):

$$\dot{\hat{P}}(t) = A_{11}(t)\hat{P}(t) + \hat{P}(t)A_{11}'(t) \tag{6.6}$$
$$- \hat{P}(t)H_1(t+\bar{h})R^{-1}(t+\bar{h})H_1'(t+\bar{h})\hat{P}(t)$$
$$+ C_1(t)Q(t)C_1'(t) + \hat{P}_1(t,0) + \hat{P}_1'(t,0),$$

$$\left(\frac{\partial}{\partial t} + \frac{\partial}{\partial \tau}\right)\hat{P}_1(t,\tau) = A_{11}(t)\hat{P}_1(t,\tau)$$
$$- \hat{P}(t)H_1(t+\bar{h})R^{-1}(t+\bar{h})H_1'(t+\bar{h})\hat{P}_1(t,\tau) + \hat{P}_2(t,0,\tau),$$

$$\left(\frac{\partial}{\partial t} + \frac{\partial}{\partial \tau} + \frac{\partial}{\partial s}\right)\hat{P}_2(t,\tau,s)$$
$$= -\hat{P}_1'(t,\tau)H_1(t+\bar{h})R^{-1}(t+\bar{h})H_1'(t+\bar{h})\hat{P}_1(t,s),$$
$$C(t) = (C_1'(t), C_2'(t))', \qquad C_1(t) \in \mathbf{R}^{k\times r}, \qquad C_2(t) \in \mathbf{R}^{(n-k)\times r},$$
$$0 \le t \le T, \qquad -h \le \tau, s \le 0.$$

The boundary conditions for (6.6) have the form

$$\hat{P}(0) = (P_0)_{11}, \qquad \hat{P}_1(0,\tau) = \hat{P}_2(0,\tau,s) = 0, \quad -h < \tau, s \le 0, \tag{6.7}$$
$$\hat{P}_1'(t,-h) = B_{11}(t+h)\hat{P}(t),$$
$$\hat{P}_2(t,-h,\tau) = \hat{P}_2'(t,\tau,-h) = B_{11}(t+h)\hat{P}_1(t,\tau),$$

$$0 \le t \le T, \qquad -h \le \tau \le 0,$$

$$P_0 = \begin{pmatrix} (P_0)_{11} & (P_0)_{12} \\ (P_0)_{21} & (P_0)_{22} \end{pmatrix},$$

$$(P_0)_{11} \in \mathbf{R}^{k\times k}, \quad (P_0)_{12} = (P_0)'_{21} \in \mathbf{R}^{k\times(n-k)}, \quad (P_0)_{22} \in \mathbf{R}^{(n-k)\times(n-k)}.$$

This system has the same type as the full-dimensional one but it has a lesser dimension. In some cases, the reduced algorithm (6.4) – (6.7) is considerably simpler for numerical implementation. Such is the case if, e.g., $k \ll n$ or $C_1(t) = 0$, i.e., the reduced dynamic system does not contain perturbations. In the latter case, matrix system (6.6), (6.7) has an explicit analytical solution [97], [95], [105], [108], [106].

The reduced filter (6.4), (6.5) generates a simplified estimate for l_* that can be represented by (6.1) with a reduced estimator $\Phi^R(t)$. By virtue of Theorems 2.1 and 4.1, for $\bar{h} \le t \le T$,

$$\Phi^R(t) = \hat{K}'(t-\bar{h})\hat{\xi}(t-\bar{h}) + \int_h^0 \hat{K}_1'(t-\bar{h}, t-\bar{h}-\tau)\hat{\xi}(t-\bar{h}-\tau)\,d\tau \tag{6.8}$$

and $\Phi^R(t) = 0$ for $0 \le t < \bar{h}$. Here,

$$\dot{\hat{\xi}}(t) = -A_{11}'(t)\hat{\xi}(t) - B_{11}'(t)\hat{\xi}(t+h) + H_1'(t+\bar{h}) \tag{6.9}$$

$$\times \left[\hat{K}'(t)\hat{\xi}(t) + \int_h^0 \hat{K}_1'(t, t-\tau)\hat{\xi}(t-\tau)\,d\tau\right],$$

$$\hat{\xi}(T) = a_1, \qquad \hat{\xi}(\tau) = 0, \quad T < \tau \le T+h, \qquad \hat{\xi}(t) \in \mathbf{R}^k.$$

We define the reduction quality by the ratio

$$\Delta_R = \left\{ \frac{\mathbf{E}\big(\hat{l}(\Phi^R) - l_*\big)^2}{\mathbf{E}\big(\hat{l}(\Phi^0) - l_*\big)^2} \right\}^{\frac{1}{2}}, \tag{6.10}$$

where $\hat{l}(\Phi^R)$ and $\hat{l}(\Phi^0)$ are provided by the reduced filter and the optimal filter, respectively. Obviously, $\Delta_R \ge 1$. Since we prefer not to solve the initial full-dimensional problem, Δ_R is unknown.

In order to evaluate Δ_R, we shall construct an upper bound Δ_R^0 for Δ_R that can be easily computed without solving the initial full-dimensional filtering problem. If this guaranteed nonoptimality level is not large, then the use of the reduced algorithm will not cause the substantial loss of optimality.

The idea for the construction of Δ_R^0 is just the same as in Section 4. The numerator in (6.10) can be easily computed. The nontrivial problem is to construct the lower bound for the denominator. It was already shown in

Section 4 that the optimal problem (6.1), (6.2) is equivalent to a certain convex variational problem in minimum. In accordance with the duality theory the corresponding dual problem in maximum has the same optimal value of the cost functional as the primal one. Basing on the structure of the optimal filter, a "good" admissible element for the dual problem can be constructed and, then, the desired lower bound can be obtained. This admissible element depends on the reduced matrices $\hat{P}(t)$ and $\hat{P}_1(t,\tau)$ only. Together with the actual estimation error $\mathbf{E}(\hat{l}(\Phi^R)-l_*)^2$, which can be easily found, Δ_R^0 describes the guaranteed quality of reduction.

6.2. *Algorithm for Calculating* Δ_R^0

Let us define the approximating function $\xi^a(t) = \xi_{\Phi^R}(t)$ for (2.29) by the equation

$$\dot{\xi}^a(t) = -A'(t)\xi^a(t) - B'(t+h)\xi^a(t+h) + H(t+\bar{h})\Phi^R(t+\bar{h}), \tag{6.11}$$

$$\xi^a(T) = a, \qquad \xi^a(\tau) = 0, \qquad T < \tau \le T + h,$$

where $\Phi^R(t)$ is specified by (6.8).

Let us construct the approximating triple $\{\xi^a(0), \Phi^R, w^a\}$ for $\{\varphi_-, \varphi, w_\varphi\}$, where

$$w^a(t) = C'(t)\xi^a(t), \qquad 0 \le t \le T - \bar{h}, \tag{6.12}$$

and the auxiliary quantities $\tilde{x}(t;p,\lambda)$, $(p^a(t),\lambda^a)$, and ν_0 defined by (4.35), (4.36), and (4.37), respectively.

THEOREM 6.1. *The guaranteed level of nonoptimality for the reduced filter is given by the inequalities:*

$$1 \le \Delta_R \le \Delta_R^0, \qquad \Delta_R^0 = \left(\frac{J^2(\xi^a(0), \Phi^R, w^a) + \kappa_Q}{\nu_0^2\{a'\tilde{x}(T;\lambda^a,p^a\}^2 + \kappa_Q} \right)^{\frac{1}{2}}, \tag{6.13}$$

where $J(\Phi_-, \Phi, w)$ *and* κ_Q *are defined by* (2.22) *and* (4.8).

PROOF. The proof is completely analogous to the proof of Theorem 4.2.

The algorithm for computing the upper bound (6.13) consists of the following steps.

Step 1. Solution of the reduced system of functional partial differential equations (6.6) with boundary conditions (6.7).

Step 2. Solution of differential equation (6.9) to find the estimator $\Phi^R(t)$ from (6.8).

Step 3. Solution of differential equation (6.11) to determine the vector $\xi^a(0)$ and the functions $\xi^a(t)$, $w^a(t)$ (using (6.12)).

Step 4. Computation of auxiliary function (4.35) and quadratures (2.22), (4.8), (4.37).

Step 5. Computation of Δ_R^0 defined by (6.13).

The implementation of all these steps is considerably simpler than the solving of the full-dimensional filtering problem.

REMARK 6.1. With $h = \bar{h} = 0$ the optimal estimation algorithm turns into the Kalman filter, and the system of functional partial differential equations turns into the Riccati equation. For large-scale systems ($n \gg 1$), the resolving of the Riccati equation, generally, is a nontrivial procedure. So, our result is useful for the verification of reduction in ordinary dynamic systems without delay as well.

6.3. *Examples for Systems with Small Parameter*

EXAMPLE 6.1. Suppose the system and the measurements have the form

$$\begin{aligned} \dot{x}_1(t) &= B_1 x_1(t-h) + \varepsilon x_2(t), \qquad \varepsilon \ll 1, \\ \dot{x}_2(t) &= B_2 x_2(t-h) + u(t), \qquad x_{1,2}(\tau) = 0, \quad -h \le \tau < 0, \end{aligned} \tag{6.14}$$

$$z(t) = x_1(t) + \varrho(t), \qquad t \in [0, T],$$

$$\mathbf{E}x(0) = 0, \qquad \mathbf{E}u(t) = \mathbf{E}\varrho(t) = 0,$$

$$\mathbf{E}x(0)x'(0) = P_0, \qquad \mathbf{E}u(t)u(s) = Q\delta(t-s), \qquad \mathbf{E}\varrho(t)\varrho(s) = R\delta(t-s).$$

The problem is to estimate $x_1(T)$ with a minimal variance.

In general notation, $n = 2$, $r = 1$, $m = 1$, $\bar{h} = 0$,

$$A = \begin{pmatrix} 0 & \varepsilon \\ 0 & 0 \end{pmatrix}, \qquad B = \begin{pmatrix} B_1 & 0 \\ 0 & B_2 \end{pmatrix}, \qquad C = \begin{pmatrix} 0 \\ 1 \end{pmatrix},$$

$$H = \begin{pmatrix} 1 \\ 0 \end{pmatrix}, \qquad a = \begin{pmatrix} 1 \\ 0 \end{pmatrix}.$$

The reduced system of partial differential equations (6.6), (6.7) takes the form

$$\dot{\hat{P}}(t) = -\hat{P}^2(t)R^{-1} + 2\hat{P}_1(t,0), \tag{6.15}$$

$$\left(\frac{\partial}{\partial t} + \frac{\partial}{\partial \tau}\right)\hat{P}(t,\tau) = -\hat{P}(t)\hat{P}_1(t,\tau)R^{-1} + \hat{P}_2(t,0,\tau),$$

$$\left(\frac{\partial}{\partial t}+\frac{\partial}{\partial \tau}+\frac{\partial}{\partial s}\right)\hat{P}_2(t,\tau,s)=-\hat{P}_1(t,\tau)\hat{P}_1(t,s)R^{-1},$$

$$0\le t\le T, \qquad -h\le \tau,s\le 0,$$

with the boundary conditions

$$\hat{P}(0)=(P_0)_{11}, \qquad \hat{P}_1(0,\tau)=\hat{P}_2(0,\tau,s)=0, \quad -h<\tau,s\le 0, \tag{6.16}$$

$$\hat{P}_1(t,-h)=B_1\hat{P}(t), \qquad \hat{P}_2(t,-h,\tau)=\hat{P}_2(t,\tau,-h)=B_1\hat{P}_1(t,\tau),$$

$$0\le t\le T, \qquad -h\le \tau\le 0.$$

This system has the explicit solution (cf. Example 2.1)

$$\hat{P}(t)=Y^2(t,0)L(t), \tag{6.17}$$

$$\hat{P}_1(t,\tau)=B_1Y(t,0)Y(t-\tau-h,0)L(t),$$

$$\hat{P}_2(t,\tau,s)=B_1^2Y(t-\tau-h,0)Y(t-s-h,0)L(t),$$

where

$$L(t)=\left(R^{-1}\int_0^t Y^2(\tau,0)\,d\tau+(P_0)_{11}^{-1}\right)^{-1}, \qquad 0\le t\le T,$$

and

$$\dot{Y}(t,0)=B_1Y(t-h,0), \tag{6.18}$$

$$Y(0,0)=1, \qquad Y(\tau,0)=0, \quad -h\le\tau<0.$$

It can be easily checked that

$$Y(t,0)=1+\sum_{j=1}^{p}\frac{(B_1)^j(t-jh)^j}{j!}, \qquad ph\le t<(p+1)h, \quad p=0,1,\dots\,.$$

Therefore, the function $\hat{\xi}(t)$ from (6.9), the reduced estimator $\Phi^R(t)$ from (6.8), the function $\xi^a(t)$ from (6.11) and all other elements from Theorem 6.1 can be easily calculated.

REMARK 6.2. In the case under consideration, the one-dimensional reduced system (with $\varepsilon=0$) can be represented in another static form

$$\dot{q}=0, \qquad z(t)=Y(t,0)q+\varrho(t), \qquad q=x_1(0),$$

TABLE 6. The values of Δ_R^0 $(n = 2)$

	h	0.1	0.2	0.3	0.4	0.5
$\varepsilon = 0.01$	Δ_R^0	1.01	1.02	1.03	1.09	1.16
$\varepsilon = 0.1$	Δ_R^0	1.02	1.08	1.12	1.18	1.25

where $Y(t,0)$ is given by (6.18). With this notation, the value $l_* = x_1(T) = Y(T,0)q$ should be estimated. Using this representation, the reduced estimator can be found by a simple formula (see Example 2.1)

$$\Phi^R(t) = \frac{R^{-1}Y(t,0)Y(T,0)}{(P_0)_{11}^{-1} + R^{-1}\int_0^T Y^2(t,0)\,dt}.$$

Then the function (6.11) and other elements can be found in a particularly simple way.

Let us set the following values for the parameters of the process:

$$B_1 = -0.5, \qquad B_2 = -1.0, \qquad \varepsilon = 0.01 \text{ or } 0.1,$$

$$P_0 = \operatorname{diag}(25.0,\ 25.0), \qquad Q = 1.0, \qquad R = 1.0,$$

$$T = 3.0, \qquad \bar{h} = 0, \qquad h = 0.1 \div 0.5.$$

The values of Δ_R^0 defined by (6.13) are presented in Table 6.

EXAMPLE 6.2. Consider a ten-dimensional dynamic system that is an extension of (6.14). Namely we put $n = 10,\ r = 1,\ m = 1,$

$$A = \begin{pmatrix} 0 & \varepsilon & 0 & \dots & 0 \\ 0 & 0 & \varepsilon & \dots & 0 \\ & & & \dots & \\ 0 & 0 & 0 & \dots & \varepsilon \\ 0 & 0 & 0 & \dots & 0 \end{pmatrix}, \qquad B = \begin{pmatrix} B_1 & 0 & \dots & 0 \\ 0 & B_2 & \dots & 0 \\ & & \dots & \\ 0 & 0 & \dots & B_{10} \end{pmatrix},$$

where $B_1 = -0.5,\ B_2 = \ldots = B_{10} = -1.0,$

$$C = \begin{pmatrix} 0 \\ \vdots \\ 0 \\ 1 \end{pmatrix}, \qquad H = \begin{pmatrix} 1 \\ 0 \\ \vdots \\ 0 \end{pmatrix}, \qquad a = \begin{pmatrix} 1 \\ 0 \\ \vdots \\ 0 \end{pmatrix}.$$

The problem is to estimate $x_1(T)$.

TABLE 7. The values of Δ_R^0 $(n = 10)$

	h	0.1	0.2	0.3	0.4	0.5
$\varepsilon = 0.01$	Δ_R^0	1.01	1.05	1.06	1.09	1.13
$\varepsilon = 0.1$	Δ_R^0	1.05	1.12	1.17	1.20	1.25

The reduced partial differential system and its solution again have the form (6.15), (6.16) and (6.17), respectively.

Let us put

$$P_0 = \mathrm{diag}(25.0, \ldots, 25.0), \qquad Q = 1.0, \qquad R = 1.0,$$

$$T = 3.0, \qquad \varepsilon = 0.1, \qquad \bar{h} = 0, \qquad h = 0.1 \div 0.5.$$

The values of Δ_R^0 are given in Table 7. It follows from Tables 6 and 7 that the quality of the simplified filters is no more than by twenty five percent worse than the quality of the optimal filters (even for the ten-dimensional system). Moreover, in number of cases, the loss of optimality does not exceed several percent. Thus, the estimate of nonoptimality is useful for the estimation problems in dynamic systems with weakly connected blocks.

7. Comments to Chapter 5

7.1. *Section 1*

A great many of works is devoted to the theory of dynamic systems with aftereffect. See, e.g., [212], [213], [205], [171], [111], [110], [179], [112], [27], [53], [114], [115], [116], [121], [89], [90], [176], [122], [52], [58], [96], [91], [123], [97], [95], [107], [145], [4], [68], [178], [56], [57], [146], [69], [105], [10], [70], [119], [92], [48], [99], [100], [147], [16], [93], [101], [102], [94], [103], [104], [108], [17], [148], [50], [106], [43a]. Certainly, this list of references is not complete. The same is true for other references in this book as well.

The existence and uniqueness for a solution of (1.1) with piecewise continuous initial function (1.2) can be established by the same methods as in [105], [70], [106]. When deriving Lemma 1.1 we use the technique described in [27].

7.2. *Section 2*

When posing the guaranteed estimation problem for systems with aftereffect we follow the problem statement for the mean-square filtering problem given in [97], [105], [106]. Theorem 2.1 was formulated and proved in a different form in [96], [91], [97], [105], [10], [108], [106]. Our modification, however, is more suitable for the theory we outline.

The results of this section was first presented in [101]. The idea of the proof of the main result for systems with delay is the same as for systems without delay (see Chapter 3). However, the implementation of our approach for systems with aftereffect is much more tedious. So, we presented the necessary nontrivial calculations. Example 2.1 was computed by A. Bakhaev.

7.3. *Section 3*

Using the technique developed in Section 2, we obtained the result, which is analogous to the main result of Chapter 4 for systems without delay [104].

7.4. *Section 4*

It follows from Theorem 4.1 that the optimal filter for systems with delay is an immediate extension of the classical Kalman filter in its form as well. It is surprising that such representation did not appear early in the literature. The construction of simplified algorithms for systems with delay is necessary even for the mean-square filtering problem. Clearly, the corresponding estimate of nonoptimality for simplified algorithms is very important [98]. Example 4.1 was studied by S. Chernatskii.

7.5. *Section 5*

The control problem is equivalent to the estimation problem considered in Section 2. Here we consider the case of distributed delay. The early results (without proofs) are contained in [102], [103]. Numerical experiments were performed by K. Kivi. Since the justification of the applicability of the duality theory to our problem is absolutely the same as for systems without aftereffect (see Chapter 3, where this question is studied in detail), we pay the main attention to calculations. The estimation and control algorithms for systems with aftereffect are extremely difficult for derivation and implementation. Hence, the use of our approach to systems with aftereffect can be even more efficient than for systems without delay.

7.6. *Section 6*

In essence, this section can be considered as a useful modification of Section 4. The computation for Example 6.1 were carried out by S. Chernatskii.

8. Exercises to Chapter 5

1. Prove Lemma 1.1 by direct differentiation. For simplicity put $G(t,\tau) = 0$.
2. Give a detailed proof of Corollary 1.1.
3. Give an example of a pseudo-fundamental matrix $Y(t,s)$ that has no the inverse one.
4. Consider the system (2.1), (2.3). Let the initial function $x^0(\tau)$ be an unknown stochastic process with zero mean and the covariance matrix $\mathbf{E}x^0(\tau)x^{0\prime}(s) = K(\tau,s)$, $-h \le \tau, s \le 0$ (in contrast to the second condition of (2.1)). What is the form of the corresponding variational problem for the optimal estimator $\Phi(t)$?
5. Consider the estimation problem (2.1), (2.3). Let the estimated parameter be $l_* = a'x(T) + \int_{T-h}^{T} \alpha'(s)x(s)\,ds$, where $a, \alpha(s) \in \mathbf{R}^n$ are specified quantities. What is the form of the corresponding variational problem for the optimal estimator $\Phi(t)$? How does the corresponding boundary value problem change? For simplicity put $\bar{h} = 0$.
6. Prove by direct differentiation that the explicit formulas for the solution of the system of functional differential equations from Example 2.1 are valid.
7. Consider the stochastic guaranteed estimation problem with combined noises from Section 4 with $\bar{h} \neq 0$, i.e.,
   $z(t) = H'(t-\bar{h})x(t) + \varrho(t)$.
   Construct the variational problem for the optimal estimator.
8. Consider the optimal mean-square filtering problem from Section 4. Suppose $h \ll 1$. Obtain the equations for the null and the first approximation for the system of functional differential equations (2.27) – (2.28).
9. Prove equalities (4.22) using formulas (2.10) – (2.13).
10. Consider the linear dynamic system without delay (see Section 6):
    $\dot{x}_1 = \varepsilon x_2, \qquad \varepsilon \ll 1,$
    $\dot{x}_2 = 0, \qquad x = (x_1, x_2)' \in \mathbf{R}^2,$
    and the measurements
    $z(t) = x_1 + \varrho(t), \qquad t \in [0,T],$
    where
    $\mathbf{E}x(0)x'(0) = P_0, \qquad \mathbf{E}\varrho(t)\varrho(s) = R\delta(t-s), \qquad l_* = x_1(T).$
    What is the simplified algorithm for this problem? Compute the level of nonoptimality Δ_R^0 for some values of parameters.

Bibliography

1. Afanasyev, V.N., Kolmanovskii, V.B., and Nosov, V.R., *Mathematical Theory of Control Systems Design,* Kluwer Academic Publishers, Dordrecht–Boston–London, (1995).
2. Aleksandrov, V.M., Minimax Approach for Solving Data Processing Problem, *Tekhnicheskaya Kibernetika,* no. 5 (1966), pp. 124–136; English transl. in *Engineering Cybernetics,* no. 5 (1966).
3. Aleksandrov, V.V., Linear Accuracy Analysis of Inertial Navigation Systems, *Proceedings of the Institute of Mechanics,* **33** (1974), pp. 32–37 (in Russian).
4. Aleksandrov, V.V., Gomez Fernandez, A., and Castro Figueroa, A., *Mathematical Modelling for Dynamic Control Systems,* vol. 1, The Ministry of Higher Education of the Republica Cuba, Habana, (1989) (in Spanish).
5. Alekseev, V.M., Tikhomirov, V.M., and Fomin, S.V., *Optimal Control,* Nauka, Moscow (1979) (in Russian); French transl.: *Commande Optimale,* Mir, Moscou, (1982).
6. Anan'ev, B.I., Minimax Mean-Square Estimates in Statistically Uncertain Systems, *Differentsialnye Uravneniya,* **20**, no. 8 (1984), pp. 1291-1297; English transl. in *Differential Equations,* **20**, no. 8 (1984).
7. Anan'ev, B.I., On Minimax State Estimates for Multistage Statistically Uncertain Systems, *Problems of Control and Information Theory,* **18**, no. 1 (1989), pp. 27–41.
8. Anan'ev, B.I., Minimax Linear Filtering of Multistage Processes with Uncertain Distributions of Random Disturbances, *Avtomatika i Telemekhanika,* no. 1 (1993), pp. 131–139; English transl. in *Automation and Remote Control,* no. 1 (1993).
9. Anan'ev, B.I., Minimax Estimation of Statistically Uncertain Systems under the Choice of a Feedback Parameter, *Journal of Mathematical Systems, Estimation, and Control,* **5**, no. 2 (1995), pp. 263–266.
10. Andreyeva, E.A., Kolmanovskii, V.B., and Shaikhet, L.E., *Control for Hereditary Systems,* Nauka, Moscow, (1992) (in Russian).
11. D'Appolito, J.A. and Hutchinson, C.E., A Minimax Approach to the Design of Low Sensitivity State Estimator, *Automatica,* **18**, no. 5 (1972), pp. 599–608.
12. Arkhangelskii, V.A. and Belousov, L.Yu., Minimax Accuracy Estimate for Spacecraft Orbit Determination under Unmodeled Accelerations, *Kosmicheskie Issledovaniya,* **17**, no. 3 (1979), pp. 345–353; English transl. in *Cosmic Research,* **17**, no. 3 (1979).
13. Ashmanov S.A., *Linear Programming,* Nauka, Moscow, (1981) (in Russian).
14. Astrovskii A.I., *A priori* Method for the Construction of Guaranteed Operations in Initial State Estimation Problems, *Doklady AN BSSR,* **31**, no. 9 (1987), pp. 795–803 (in Russian).
15. Babenko, K.I., *Fundamentals of Numerical Analysis,* Nauka, Moscow, (1986) (in Russian).
16. Baker, C.T.H., Kolmanovskii, V.B., Stability of Chemostat Equations for Two Species Competing for Renewal Resource, *Dynamic Systems and Applications,* **4**, no. 4 (1995), pp. 601–611.
17. Baker, C.T.H., Tang, A., Stability of Grazing Systems with Infinite Delay, *Mathematics and Computers in Simulation,* no. 4 (1997), pp. 18–26.
18. Bakhshiyan, B.Ts., Guaranteed Accuracy Characteristics in Linear Estimation: Properties and Applications, *Preprint No. 1332,* Space Research Institute of the Russian Academy of Sciences, Moscow, (1987) (in Russian).
19. Bakhshiyan, B.Ts., Simplex-Type Algorithm for Solving the Optimal Guaranteed Estimation Problem with Unmodeled Acceleration, *Kosmicheskie Issledovaniya,* **26**, no. 1 (1988), pp. 127–143; English transl. in *Cosmic Research,* **26**, no. 1 (1988).
20. Bakhshiyan, B.Ts., Nazirov, R.R., and Elyasberg, P.E., *Determination and Correction of Motion,* Nauka, Moscow, (1980) (in Russian).
21. Bakhshiyan, B.Ts. and Soloviov, V.N., Application of Duality Theorem to the Optimal Guaranteed Estimation Problem, *Kosmicheskie Issledovaniya,* **28**, no. 2 (1990),

pp. 163–169; English transl. in *Cosmic Research,* **28**, no. 2 (1990).

22. Bakhshiyan B.Ts., Sukhanov A.A., and Elyasberg P.E., *A priori* Accuracy for Prediction of the Location of Halley's Comet by Ground and On-Board Measurements, *Kosmicheskie Issledovaniya,* **23**, no. 6 (1985), pp. 876–884; English transl. in *Cosmic Research,* bf 23, no. 6 (1985).
23. Bakhvalov, N.S., On the optimality of linear methods for the approximation of operators on convex classes of functions, *Zhurnal Vychislitelnoi Matematiki i Matematicheskoi Fiziki,* **11**, no. 4 (1971), pp. 1014–1019; English transl. in *USSR Journal of Computational Mathematics and Mathematical Physics,* **11**, no. 4 (1971).
24. Balakrishnan, A.V., *Introduction to Optimization Theory in a Hilbert Space,* Springer, Berlin–Heidelberg–New York, (1971).
25. Balakrishnan, A.V., *Applied Functional Analysis,* Springer, New York–Heidelberg–Berlin, (1976).
26. Basar, T., Optimum Performance Levels for Minimax Filters, Predictors, and Smoothers, *System and Control Letters,* **16** (1991), pp. 309–317.
27. Bellman, R. and Cooke, K.L., *Differential - Difference Equations,* Academic Press, New York–London, (1963).
28. Belousov, L.Yu., Determination of Optimal Measurement Instants, *Kosmicheskie Issledovaniya,* **7**, no. 1 (1969), pp. 28–34; English transl. in *Cosmic Research,* **7**, no. 1 (1969).
29. Belousov, L.Yu., On Multiple Objective Planning with Arbitrary Correlation Measurement Model, *Kosmicheskie Issledovaniya,* **18**, no. 5 (1980), pp. 790–799; English transl. in *Cosmic Research,* **18**, no. 5 (1980).
30. Belousov, L.Yu. and Krupen, V.Ya., On Asymptotic Estimates for Initial Parameters on Ranging, *Kosmicheskie Issledovaniya,* **12**, no. 2 (1974), pp. 191–196; English transl. in *Cosmic Research,* **12**, no. 2 (1974).
31. Belousov, L.Yu. and Kuzmin, E.A., Duality Theory Application to Evaluation of Accuracy of Parameters of Spacecraft Orbits, *Kosmicheskie Issledovaniya,* **23**, no. 6 (1985), pp. 849–859; English transl. in *Cosmic Research,* **23**, no. 6 (1985).
32. Birkuhn, G.I. and Matasov, A.I., Least-Squares Method in the Guaranteed Estimation Problem, *Mathematical Modelling of the Dynamics of Control Systems, Machines, and Mechanisms,* No. 217, Moscow Energy Institute Press, Moscow, (1989), pp. 47–57 (in Russian).
33. Blagoveschenskii, Yu.N. and Legostaeva, I.L., Parametric Bounds for non-Parametric Trend, *Doklady Akademii Nauk SSSR,* **264**, no. 4 (1982), pp. 791–795; English transl. in *Soviet Mathematics Doklady* (1982).
34. Bobrick, G.I., Guaranteed Parameter Estimation Problems in Navigation Systems, *Ph.D. Thesis,* Faculty of Mechanics and Mathematics, M.V. Lomonosov Moscow State University, (1991) (in Russian).
35. Bobrick, G.I., Golovan, A.A., and Matasov, A.I., The Kalman–Bucy Filter in the Guaranteed Estimaion Problem, *Technical Report No. 4024,* Institute of Mechanics, M.V. Lomonosov Moscow State University (1990) (in Russian).
36. Bobrick, G.I., Golovan, A.A., and Matasov, A.I., The Kalman–Bucy Filter Accuracy in the Guaranteed Parameter Estimation Problem, *Proceedings of the 30th IEEE Conference on Decision and Control,* Brighton, United Kingdom, **3** (1991), pp. 3066–3071.
37. Bobrick, G.I., Golovan, A.A., and Matasov, A.I., Application of Guaranteed Approach to the Solution of Topographical Referencing Problem, *Avtomatika i Telemekhanika,* no. 10 (1997), pp. 34–47; English transl. in *Automation and Remote Control,* no. 10 (1997).
38. Bobrick, G.I. and Matasov, A.I., Optimal Guaranteed Parameter Estimation for the Unit of Accelerometers, *Izvestiya RAN, Mekhanika Tverdogo Tela,* no. 5 (1993), pp. 8–14; English transl. in *Mechanics of Solids,* no. 5 (1993).
39. Boguslavskii, I.A., On Linear System Phase Coordinates Estimation in Statistically Uncertain Situations, *Avtomatika i Telemekhanika,* no. 1 (1971), pp. 31–40; English

transl. in *Automation and Remote Control,* no. 1 (1971).

40. Bryson, A.E. and Ho, Y.C., *Applied Optimal Control Revised: Optimization, Estimation, and Control,* Hemisphere Publishing Corporation, Washington D.C., (1975).
41. Bucy, R.S. and Joseph, P.D., *Filtering for Stochastic Processes with Application to Guidance,* Wiley Interscience, New York, (1968).
42. Bulgakov, B.V., *Applied Theory of Gyroscopes,* Gostekhizdat, Moscow–Leningrad, (1939) (in Russian).
43. Chernousko, F.L., *Estimation of Phase State in Dynamic Systems,* Nauka, Moscow, (1988) (in Russian).
    (a) Chernousko, F.L. and Kolamnovskii, V.B., *Optimal Control under Stochastic Perturbations,* Nauka, Moscow, (1968) (in Russian).
44. Clark, F.H., *Optimization and Nonsmooth Analysis,* John Wiley and Sons, New York-Chichester-Brisbane-Toronto-Singapore, (1983).
45. Christensen, G.S. and Soliman, S.A., Optimal Filtering of Linear Discrete Dynamic Systems Based on Least Absolute Value Approximations, *Automatica,* **26** (1990), pp. 389–395.
46. Dantzig, G.B., *Linear Programming and Extensions,* Princeton University Press, Princeton, New Jersey, (1963).
47. Davis, M.H.A., *Linear Estimation and Stochastic Control,* Chapman and Hall, London, (1977).
48. Drozdov, A.D. and Kolmanovskii, V.B. *Stability in Viscoelasticity,* North–Holland, Elsevier, (1994).
49. Dzyadyk V.K., *Introduction to the Theory of Uniform Approximation of Functions by Polinomials,* Nauka, Moscow, (1977) (in Russian).
50. Dugard, L. and Virrist, E.I. (Eds.), *Stability and Control of Time-Delay Systems,* Springer, Berlin–Heidelberg–New York–Tokyo, (1997).
51. Ekeland, I. and Temam, R., *Convex Analysis and Variational Problems,* North-Holland, Amsterdam, Elsevier, New York, (1976).
52. Eller, D.H., Aggarwal, A.G., and Banks, H.T., Optimal Control of Linear Time Delay Systems, *IEEE Transactions on Automatic Control,* **AC-14**, no. 2 (1969), pp. 678–685.
53. Elsgolts, L.E., *Introduction to the Theory of Differential Equations with Deviating Argument,* Nauka, Moscow, (1964); English transl.: *Introduction to the Theory of Differential Equations with Deviating Argument,* Holden-Day Inc., New York, (1966).
54. Elyasberg, P.E., *Measurement Data: How much it is required? How to process it?* Nauka, Moscow, (1983) (in Russian).
55. Elyasberg, P.E. and Bakhshiyan, B.Ts., Determination of the Spacecraft Trajectory under Uncertainty in the Probability Distribution of Measurement Errors, *Kosmicheskie Issledovaniya,* **7**, no. 1 (1969); English transl. in *Cosmic Research,* **7**, no. 1 (1969).
56. Fiagbedzi, Y.A. and Pearson, A.E., A State Observer for Systems Described by Functional Differential Equations, *Automatica,* **26**, no. 2 (1990), pp. 321–331.
57. Fiagbedzi, Y.A. and Pearson, A.E., Output Feedback Stabilization of Delay Systems via Generalization of the Transformation Method, *International Journal of Control,* **51**, no. 4 (1990), pp. 801–822.
58. Gabasov, R. and Kirillova, F.M., *A Qualitative Theory of Optimal Processes,* Nauka, Moscow, (1971) (in Russian).
59. Gantmaher, F.R., *Theory of Matrices* (forth edition), Nauka, Moscow, (1988) (in Russian); English transl. of the first edition: *Applications of the Theory of Matrices,* Interscience Publications, New York–London, (1959).
60. Gass, S.L., *Linear Programming,* McGraw-Hill, New York, (1969).
61. Genkin, M.D. and Ryaboi, V.M., *Elastic-Mass Vibroisolation Systems. Limiting Potentialities, Optimal Structures,* Nauka, Moscow, (1988) (in Russian).
62. Geurkov, V.L., Parametric Diagnostics of Failures by Residual Method in Linear Static Systems with Unknown Input, *Avtomatika i Telemekhanika,* no. 6 (1990),

pp. 135–138; English transl. in *Automation and Remote Control,* no. 6 (1990).
63. Golovan, A.A., Goritskii, A.Yu., Parusnikov, N.A., and Tikhomirov, V.V., Algorithms for the Correcting Inertial Navigation Systems that Solve the Topographical Referencing Problem (N.A. Parusnikov, Ed.), *Preprint no. 2,* Faculty of Mechanics and Mathematics, M.V. Lomonosov Moscow State University, (1994) (in Russian).
64. Golovan, A.A., Martirosyan, S.R., and Matasov, A.I., Numerical Comparison of the Optimal Guaranteed Algorithm with the Least-Squares Method, *Kosmicheskie Issledovaniya,* **26**, no. 2 (1988), pp. 319–322; English transl. in *Cosmic Research,* **26**, no. 2 (1988).
65. Golovan, A.A. and Matasov, A.I., The Computational Formulas for the Estimate of the Kalman–Bucy Filter Nonoptimality Degree in the Guaranteed Estimation Problem with Combined Disturbances, *Technical Report No. 4108,* Institute of Mechanics, M.V. Lomonosov Moscow State University (1991) (in Russian).
66. Golovan, A.A. and Matasov, A.I., The Kalman–Bucy Filter in the Guaranteed Estimation Problem, *IEEE Transactions on Automatic Control,* **AC-39**, no. 6 (1994), pp. 1282–1286.
67. Golovan, A.A. and Matasov, A.I., Efficiency of the Kalman–Bucy Filter in the Presence of Uncertain Disturbances, *Proceedings of the 10th IFAC Symposium on System Identification,* Copenhagen, Denmark, **3** (1994), pp. 481–486.
68. Górecki, H., Fuksa, S., Grabowski, P., and Korytowski, A., *Analysis and Synthesis of Time Delay Systems,* Polish Scientific Publishers, Warszawa; John Wiley and Sons, Chichester–New York, (1989).
69. Gopalsamy, K., *Equations of Mathematical Ecology,* Kluwer Academic Publishers, Dordrecht–Boston–London, (1992).
70. Hale, J.K. and Lunel, S.M.V., *Introduction to Functional Differential Equations,* Springer, New York–Berlin, (1993).
71. Huber, P.J., Robust Estimation of a Location Parameter, *Annals of Mathematical Statistics,* **35** (1964), pp. 73–101.
72. Huber, P.J., *Robustness in Statistics,* John Wiley and Sons, (1981).
73. Ibragimov, I.A. and Has'minskii, R.Z., On the Nonparametric Estimation of a Value of a Linear Functional in the Gaussian White Noise, *Teoriya Veroyatnostei i eyo Primeneniya,* **29**, no. 1 (1984), pp. 19–32; English transl. in *The Theory of Probability and its Applications,* **29** (1984).
74. Ibragimov, I.A. and Has'minskii, R.Z., On the Estimation of Linear Functionals in Gaussian Noise, *Teoriya Veroyatnostei i eyo Primeneniya,* **32**, no. 1 (1987), pp. 35–44; English transl. in *The Theory of Probability and its Applications,* **32** (1987).
75. Ioffe, A.D. and Tikhomirov, V.M., *Theory of Extremal Problems,* Nauka, Moscow, (1974) (in Russian); English transl.: *Theory of Extremal Problems,* North-Holland, Amsterdam–New York–Oxford, (1979).
76. Irger, D.S., On Optimal Filtering by Minimax Criterion, *Tekhnicheskaya Kibernetika,* no. 5 (1966), pp. 137–144; English transl. in *Engineering Cybernetics,* no. 5 (1966).
77. Isaacs, R., *Differential Games,* John Wiley and Sons, New York–London–Sydney, (1965).
78. Jualin, L. and Walter, E., Set Inverison via Interval Analysis for Nonlinear Bounded Error Estimation, *Automatica,* **29**, no. 4 (1993), pp. 1053–1064.
79. Jazwinski, A.H., *Stochastic Processes and Filtering Theory,* Academic Press, New York, (1970).
80. Kailath, T., An Innovation Approach to Least-Squares Estimation. Part I: Linear Filtering in Additive White Noise, *IEEE Transactions on Automatic Control* , **AC-13**, no. 6 (1968), pp. 646–655.
81. Kailath, T. and Frost, P., An Innovation Approach to Least-Squares Estimation. Part II: Linear Smoothing in Additive White Noise, *IEEE Transactions on Automatic Control* , **AC-13**, no. 6 (1968), pp. 655–660.
82. Kailath, T., *Lectures on Wiener and Kalman Filtering,* Springer-Verlag, New York, (1988).

83. Kallianpur, G., *Stochastic Filtering Theory,* Springer-Verlag, New York, (1980).
84. Kalman, R.E., A New Approach to Linear Filtering and Prediction Problems, *Transactions of the ASME, Journal of Basic Engineering,* **82D** (1960), pp. 35–45.
85. Kalman, R.E. and Bucy, R.S., New Results in Linear Filtering and Prediction Theory, *Transactions of the ASME, Journal of Basic Engineering,* **83D** (1961), pp. 95–108.
86. Kantorovich, L.V., *Mathematical Methods for Organization and Planning of Production,* Leningrad University Press, Leningrad, (1939) (in Russian).
87. Karmanov, V.G., *Mathematical Programming,* Nauka, Moscow, (1975) (in Russian).
88. Katz, I.Ya. and Kurzhanskii, A.B., Minimax Estimation in Multi-Step Systems, *Doklady Akademii Nauk SSSR,* **221**, no. 3 (1975), pp. 535–538; English transl. in *Soviet Mathematics Doklady* (1975).
89. Kharatishvili, G.L., *Optimal Processes with Delay,* Metsniereba, Tbilisi, (1966) (in Russian).
90. Kharatishvili, G.L., A Maximum Principle in Extremal Problems with Delay, *Mathematical Theory of Control* (Eds. Balakrishnan and Neustadt), p. 26, Academic Press, New York, (1967).
91. Kolmanovskii, V.B., On the Approximation of Linear Control Systems with Time Lag, *Problems of Control and Information Theory,* **3**, no. 1 (1974), pp. 63–76.
92. Kolmanovskii, V.B., Stability of Some Systems with Arbitrary Aftereffect, *Doklady Rossiiskoi Akademii Nauk,* **331**, no. 4 (1993), pp. 421–425; English transl. in *Physics – Doklady,* (1994).
93. Kolmanovskii, V.B., Stability of Some Nonlinear Functional Differential Equations, *Nonlinear Differential Equations,* no. 2 (1995), pp. 185–198.
94. Kolmanovskii, V.B., Stability of Discrete Volterra Equations, *Doklady Rossiiskoi Akademii Nauk,* **349**, no. 5 (1996), pp. 610–614; English transl. in *Doklady Mathematics,* (1997).
95. Kolmanovskii, V.B. and Frolov, A.L., Optimal State Estimates for Systems with Delay by Continuous and Discrete Measurements, *Prikladnaya Matematika i Mekhanika,* **50**, no. 4 (1986), pp. 544–550; English transl. in *Applied Mathematics and Mechanics,* **50**, no. 4 (1986).
96. Kolmanovskii, V.B. and Maisenberg, T.L., Optimal Control for Stochastic Systems with Delay, *Avtomatika i Telemekhanika,* no. 1 (1973), pp. 47–61; English transl. in *Automation and Remote Control,* no. 1 (1973).
97. Kolmanovskii, V.B. and Maisenberg, T.L., Optimal Estimates of System State and Certain Control Problems with Aftereffect, *Prikladnaya Matematika i Mekhanika,* **41**, no. 3 (1977), pp. 446–456; English transl. in *Applied Mathematics and Mechanics,* **41**, no. 3 (1977).
98. Kolmanovskii, V.B., Mao, X., and Matasov, A.I., On Approximate Solving the Mean-Square Filtering Problem in Hereditary Systems, *Dynamic Systems and Applications,* (1998) (to be published).
99. Kolmanovskii, V.B. and Matasov, A.I., A New Approach to Approximate Solving Minimax Filtering Problems in Dynamic Systems with Delay, *Proceedings of the 1st Asian Control Conference,* Tokyo, Japan, **3** (1994), pp. 499–502.
100. Kolmanovskii, V.B. and Matasov, A.I., Evaluation of the Accuracy of the Approximation Method for Solving Minimax Filtering Problem in Systems with Delay, *Doklady Rossiiskoi Akademii Nauk,* **339**, no. 1 (1994), pp. 37–39; English transl. in *Doklady Physics,* **39**, no. 11 (1994), pp. 758–760.
101. Kolmanovskii, V.B. and Matasov, A.I., Minimax Filtering Problems in Hereditary Systems. A New Approach to Approximate Solution, *Avtomatika i Telemekhanika,* no. 6 (1996), pp. 125–147; English transl. in *Automation and Remote Control,* no. 6 (1996).
102. Kolmanovskii, V.B. and Matasov, A.I., Efficient Control Algorithms for Hereditary Dynamic Systems, *Proceedings of the 13th IFAC World Congress,* San-Francisco, California, USA, **C** (1996), pp. 403–408.
103. Kolmanovskii, V.B. and Matasov, A.I., An Approximate Method for Solving Optimal

Control Problem in Hereditary Systems, *Doklady Rossiiskoi Akademii Nauk,* **354**, no. 4 (1997), pp. 465–468; English transl. in *Doklady Mathematics,* **55**, no. 3 (1997), pp. 475–478.

104. Kolmanovskii, V.B., Matasov, A.I., and Kopylova, N.K., On Approximate Method for Solving Stochastic Guaranteed Estimation Problem in Hereditary Systems, *Proceedings of the 15th IMACS World Congress on Scientific Computation, Modelling, and Applied Mathematics* Berlin, Germany, **1** (1997), pp. 251–256.
105. Kolmanovskii, V.B. and Myshkis, A.D., *Applied Theory of Functional Differential Equations,* Kluwer Academic Publishers, Dordrecht–Boston–London, (1992).
106. Kolmanovskii, V.B. and Myshkis, A.D., *Introduction to the Theory and Applications of Functional Differential Equations,* Kluwer Academic Publishers, Dordrecht–Boston–London, (1998).
107. Kolmanovskii, V.B. and Nosov, V.R., *Stability of Functional Differential Equations,* Academic Press, New York–London, (1986).
108. Kolmanovskii, V.B. and Shaikhet, L.E., *Control of Systems with Aftereffect,* American Mathematical Society, Providence RI, (1997).
109. Kolmogorov, A.N. and Fomin, S.V., *Elements of Function Theory and Functional Analysis,* Nauka, Moscow, (1972) (in Russian).
110. Kramer, J., On Control of Linear Systems with Time Lags, *Information and Control,* **3**, no. 4 (1960).
111. Krasovskii, N.N., *Stability of Motion,* Fizmatgiz, Moscow, (1959) (in Russian); English transl.: *Stability of Motion,* Stanford University Press, Stanford, California, (1963).
112. Krasovskii, N.N., On the Analytical Design of an Optimal Controller in Systems with Time Lags, *Prikladnaya Matematika i Mekhanika,* **26**, no. 1 (1962), pp. 39–51; English transl. in *Applied Mathematics and Mechanics,* **26**, no. 1 (1962).
113. Krasovskii, N.N., To the Theory of Controllability and Observability of Linear Dynamic Systems, *Prikladnaya Matematika i Mekhanika,* **28**, no. 1 (1964), pp. 3–14; English transl. in *Applied Mathematics and Mechanics,* **28**, no. 1 (1964).
114. Krasovskii, N.N., Problem on Observing a Linear Dynamic System and Equations with Argument Lags, *Differentsialnye Uravneniya,* **1**, no. 12 (1965), pp. 1551–1556; English transl. in *Differential Equations,* **1**, no. 12 (1965).
115. Krasovskii, N.N., On an Approximation of a Problem for Optimal Control in Systems with Time Lag, *Doklady Akademii Nauk SSSR,* **167**, no. 3 (1966), pp. 540–542; English transl. in *Soviet Mathematics Doklady* (1966).
116. Krasovskii, N.N., On the Control of an Object with Aftereffect, *Prikladnaya Matematika i Mekhanika,* **30**, no. 5 (1966), pp. 938–946; English transl. in *Applied Mathematics and Mechanics,* **30**, no. 5 (1966).
117. Krasovskii, N.N., *Theory of Motion Control,* Nauka, Moscow, (1968) (in Russian).
118. Krein, M.G. and Nudelman, A.A., *Markov Moment Problem and Extremal Problems,* Nauka, Moscow, (1973) (in Russian).
119. Kuang, Y., *Delay Differential Equations with Applications to Population Dynamics,* Academic Press, London, (1993).
120. Kurkin, O.M., Korobochkin, Yu.B., and Shatalov, S.A., *Minimax Processing of Information,* Energoatomizdat, Moscow, (1990) (in Russian).
121. Kurzhanskii, A.B., On the Control Problem for a System of Differential Equations with Delay, *Prikladnaya Matematika i Mekhanika,* **30**, no. 6 (1966), pp. 1121–1124; English transl. in *Applied Mathematics and Mechanics,* **30**, no. 5 (1966).
122. Kurzhanskii, A.B., On an Approximation of Linear Differential Equations with Delay, *Differentsialnye Uravneniya,* **3**, no. 12 (1967), pp. 2094–2107; English transl. in *Differential Equations,* **3**, no. 12 (1967).
123. Kurzhanskii, A.B., *Control and Observation under Uncertainty Conditions,* Nauka, Moscow, (1977) (in Russian).
124. Kurzhanskii, A.B., On Stochastic Filtering Approximations of Estimation Problems for Systems with Uncertainty, *Stochastics,* **23** (1988), pp. 109–130.

125. Kurzhanskii, A.B. and Valyi, I., *Ellipsoidal Calculas for Estimation and Control,* Laxenburg, Austria IIASA, Birkhauser, Boston, (1997).
126. Kwakernaak, H. and Sivan, R., *Linear Optimal Control Systems,* Wiley Interscience, New York, (1972).
127. Lankaster, P., *Theory of Matrices,* Academic Press, New York–London, (1969).
128. Lasdon, L.S., *Optimization Theory for Large Systems,* Macmillan Series in Operations Research, London, (1970).
129. Laurent, P.-J., *Approximation et Optimisation,* Hermann, Paris, (1972).
130. Lidov, M.L., On an *a priori* Accuracy of Parameter Estimation by Least-Squares Method, *Kosmicheskie Issledovaniya,* **2**, no. 5 (1964), pp. 713–718; English transl. in *Cosmic Research,* **2**, no. 5 (1964).
131. Lidov, M.L., Mathematical Analogy between Optimal Problems for Trajectory Correction and Problems for Choice of Measurements, *Kosmicheskie Issledovaniya,* **9**, no. 5 (1971), pp. 687–694; English transl. in *Cosmic Research,* **9**, no. 5 (1971).
132. Lidov, M.L., Minimax Continuous Parameter Estimation Problem, *Kosmicheskie Issledovaniya,* **22**, no. 4 (1984), pp. 483–498; English transl. in *Cosmic Research,* **22**, no. 4 (1984).
133. Lidov, M.L., Game Theory Estimation Problem with Unmodeled Accelerations and the Solving Algorithm, *Kosmicheskie Issledovaniya,* **24**, no. 2 (1986), pp. 246–276; English transl. in *Cosmic Research,* **24**, no. 2 (1986).
134. Lidov, M.L., On a Duration of Measurements under Weak Influence of White Noise, *Kosmicheskie Issledovaniya,* **26**, no. 2 (1988), pp. 179–183; English transl. in *Cosmic Research,* **26**, no. 2 (1988).
135. Lidov, M.L., Parameter Estimation Algorithm in the Problem with Unmodeled Accelerations, *Doklady Akademii Nauk SSSR,* **300**, no. 1 (1988), pp. 51–54; English transl. in *Soviet Physics Doklady* (1988).
136. Lidov, M.L., On a Modification of Simplex Method in the Degeneracy Case, *Kosmicheskie Issledovaniya,* **29**, no. 4 (1991), pp. 499–508; English transl. in *Cosmic Research,* **29**, no. 4 (1991).
137. Lidov, M.L., On Guaranteed Estimation Problem, *Kosmicheskie Issledovaniya,* **29**, no. 6 (1991), pp. 803–814; English transl. in *Cosmic Research,* **29**, no. 6 (1991).
138. Lidov, M.L., Bakhshiyan, B.Ts., and Matasov, A.I., On a Trend in Guaranteed Estimation Problem, *Kosmicheskie Issledovaniya,* **29**, no. 5 (1991), pp. 659–684; English transl. in *Cosmic Research,* **29**, no. 5 (1991).
139. Lidov, M.L. and Bakuma, L.M., Application of Optimal Correction Algorithm for Solving the Estimation Problem with Unmodeled Accelerations, *Kosmicheskie Issledovaniya,* **26**, no. 3 (1988), pp. 339–352; English transl. in *Cosmic Research,* **26**, no. 3 (1988).
140. Lidov, M.L. and Bakuma, L.M., On the Experimental Testing of a New Algorithm for the Estimation Problem with Unmodeled Perturbations, *Kosmicheskie Issledovaniya,* **29**, no. 1 (1991), pp. 115–126; English transl. in *Cosmic Research,* **29**, no. 1 (1991).
141. Lidov, M.L. and Matasov, A.I., A Generalization of the "Worst Correlation Problem", *Kosmicheskie Issledovaniya,* **27**, no. 3 (1989), pp. 454–456; English transl. in *Cosmic Research,* **27**, no. 3 (1989).
142. Lidov M.L., Teslenko N.M., *A priori* Estimates of the Accuracy in Determining the Value of Flattening of the Sun, *Kosmicheskie Issledovaniya,* **22**, no. 1 (1984), pp. 131–132; English transl. in *Cosmic Research,* bf 22, no. 1 (1984).
143. Liptser, R.S. and Shiryayev, A.N., *Statistics of Random Processes,* Nauka, Moscow, (1974) (in Russian); English transl.: *Statistics of Random Processes,* Springer, New York–Heidelberg–Berlin, (1978).
144. Loève, M., *Probability Theory,* Van Nostrand, Princeton, (1963).
145. Malek-Zavarei, M. and Jamshidi, M., *Time-Delay Systems. Analysis, Optimization, and Applications,* North-Holland, Amsterdam–New York–Oxford–Tokyo, (1987).
146. Mao, X., *Stability of Stochastic Differential Equations with Respect to Semimartingales,* Longman Scientific and Technical, London, (1991).

147. Mao, X., *Exponential Stability of Stochastic Differential Equations,* Marcel Dekker, New York, (1994).
148. Mao, X., *Stochastic Differential Equations and Applications,* Horwood, Chichester, (1997).
149. Marchuk, A.G. and Osipenko, K.Yu., Best Approximations for Functions Specified with an Error at a Finite Number of Points, *Matematicheskie Zametki,* **17**, no. 3 (1975), pp. 359–368; English transl. in *Mathematical Notes of Russian Academy of Sciences,* **17**, no. 3 (1975).
150. Martin, C.J. and Mintz, M., Robust Filtering and Prediction for Linear Systems with Uncertain Dynamic: A Game-Theoretic Approach, *IEEE Transactions on Automatic Control,* **AC-28** (1983), pp. 888–896.
151. Matasov, A.I., On the Problem of the Guaranteed Parameter Estimation of Inertial Navigation System, *Trends in Navigation and Control,* M.V. Lomonosov Moscow State University Press, Moscow, (1983) (in Russian).
152. Matasov, A.I., On the *a priori* Accuracy of the Kalman-Bucy Filter in the Guaranteed Estimation Problem, *Technical Report No. 3740,* Institute of Mechanics, M.V. Lomonosov Moscow State University (1988) (in Russian).
153. Matasov, A.I., On the Optimality of Linear Guaranteed Estimation Algorithms, Part I, *Kosmicheskie Issledovaniya,* **26**, no. 5 (1988), pp. 643–653; English transl. in *Cosmic Research,* **26**, no. 5 (1988).
154. Matasov, A.I., On the Optimality of Linear Guaranteed Estimation Algorithms, Part II, *Kosmicheskie Issledovaniya,* **26**, no. 6 (1988), pp. 807–812; English transl. in *Cosmic Research,* **26**, no. 6 (1988).
155. Matasov, A.I., An Optimality of Linear Algorithms in the "Worst Correlation Problem", *Vestnik Moskovskogo Gosudarctvennogo Universiteta,* Seriya 1, Matematika – Mekhanika, no. 1 (1989), pp. 61–64; English transl. in *Moscow University Mechanics Bulletin,* no. 1 (1989).
156. Matasov, A.I., On the Necessary and Sufficient Conditions for Optimality in the Estimation Problem with Unmodeled Disturbances, *Kosmicheskie Issledovaniya,* **27**, no. 5 (1989), pp. 652–659; English transl. in *Cosmic Research,* **27**, no. 5 (1989).
157. Matasov, A.I., On the *a priori* Accuracy of the Least-Squares Method in the Guaranteed Estimation Problems, Part I, *Kosmicheskie Issledovaniya,* **28**, no. 1 (1990), pp. 11–16; English transl. in *Cosmic Research,* **28**, no. 1 (1990).
158. Matasov, A.I., On the *a priori* Accuracy of the Least-Squares Method in the Guaranteed Estimation Problems, Part II, *Kosmicheskie Issledovaniya,* **28**, no. 2 (1990), pp. 170–185; English transl. in *Cosmic Research,* **28**, no. 2 (1990).
159. Matasov, A.I., On the Sensitivity of the Kalman–Bucy Filter to the *a priori* Covariance Matrices Values, *Avtomatika i Telemekhanika,* no. 1 (1991), pp. 78–87; English transl. in *Automation and Remote Control,* no. 1 (1991).
160. Matasov, A.I., On the Guaranteed Parameter Estimation with Outliers in the Measurements, *Kosmicheskie Issledovaniya,* **29**, no. 3 (1991), pp. 323–327; English transl. in *Cosmic Research,* **29**, no. 3 (1991).
161. Matasov, A.I., On the *a priori* Kalman–Bucy Filter Accuracy in the Guaranteed Parameter Estimation Problem with Combined Noises, *Technical Report No. 4059,* Institute of Mechanics, M.V. Lomonosov Moscow State University (1991) (in Russian).
162. Matasov, A.I., On the Phase Constraints in the Estimation Problem with Unmodeled Disturbances, *Kosmicheskie Issledovaniya,* **30**, no. 1 (1992), pp. 3–9; English transl. in *Cosmic Research,* **30**, no. 1 (1992).
163. Matasov, A.I., The Kalman–Bucy Filter Accuracy in the Guaranteed Parameter Estimation Problem with Unknown Statistics, *IEEE Transactions on Automatic Control,* **AC-39**, no. 3 (1994), pp. 635–639.
164. Matasov, A.I., Guaranteed Sensitivity of the Kalman Filter to Uncertainty in Noise Intensities, *Proceedings of the 3rd European Control Conference,* Rome, Italy, **2** (1995), pp. 2144–2148.
165. Matasov, A.I. and Birkuhn, G.I., The *a priori* Kalman–Bucy Filter Accuracy in

the Guaranteed Estimation Problems. Formulas for Computation, *Technical Report No. 3859,* Institute of Mechanics, M.V. Lomonosov Moscow State University (1989) (in Russian).

166. Matasov, A.I. and Martirosyan, S.R., Minimax Algorithms for the Position Correction of Inertial Navigation Systems, *Izvestiya AN SSSR, Mekhanika Tverdogo Tela,* no. 2 (1988), pp. 4–14; English transl. in *Mechanics of Solids,* no. 2 (1988).
167. Matasov, A.I. and Samokhvalov, V.G., Guaranteeing Parameter Estimation with Anomalous Measurement Errors, *Automatica,* **32**, no. 9 (1996), pp. 1317–1322.
168. Milanese, M. and Tempo, R., Optimal Algorithms Theory for Robust Estimation and Prediction, *IEEE Transactions on Automatic Control,* **AC-30**, no. 8 (1985), pp. 730–743.
169. Milanese, M. and Vicino, A., Optimal Estimation Theory for Dynamic Systems with Set Membership Uncertainty: an Overview, *Automatica,* **27**, no. 6 (1991), pp. 997–1009.
170. Mintz, M., A Kalman Filter as a Minimax Estimator, *Journal of Optimization Theory and Applications,* **9**, no. 2 (1972), pp. 99–111.
171. Myshkis, A.D., *Linear Differential Equations with Retarded Argument,* Gostekhizdat, Moscow, (1951) (in Russian).
172. Nagpal, K.M. and Khargonekar, P.P., Filtering and Smoothing in an H^∞-Setting, *IEEE Transactions on Automatic Control,* **AC-36**, no. 2 (1991), pp. 152–166.
173. Neustadt, L.W., Optimization, a Moment Problem and Nonlinear Programming, *Journal of SIAM,* **A2**, no. 1 (1964).
174. Norton, J.P., Identification and Application of Bounded-Parameter Models, *Automatica,* **23** (1987), pp. 497–507.
175. Norton, J.P. and Mo, S.H., Parameter Bounding for Time-Varying Systems, *Mathematics and Computers in Simulation,* **32** (1990), pp. 527–534.
176. Oguztoreli, M.N., *Time-Lag Control Systems,* Academic Press, New York, (1966).
177. Parusnikov, N.A., Morozov, V.M., and Borzov, V.I., *Correction Problem in Inertial Navigation,* Moscow University Press, Moscow, (1982) (in Russian).
178. Pearson, A.E. and Fiagbedzi, Y.A., An Observer for Time Lag Systems, *IEEE Transactions on Automatic Control,* **AC-34**, no. 7 (1989), pp. 775–777.
179. Pontryagin, L.S., Boltyanskii, V.G., Gamkrelidze, R.V., and Mishchenko, E.F., *The Mathematical Theory of Optimal Processes,* Fizmatgiz, Moscow, (1961) (in Russian); English transl.: *The Mathematical Theory of Optimal Processes,* Interscience Publishers, John Wiley and Sons, New York–London, (1965).
180. Poor, H.V. and Looze, D.P., Minimax State Estimation for Stochastic Systems with Noise Uncertainty, *IEEE Transactions on Automatic Control,* **AC-26**, no. 4 (1981), pp. 902–906.
181. Pronzato, L. and Walter, E., Minimal Volume Ellipsoids, *International Journal of Adaptive Control and Signal Processing,* **8** (1994), pp. 15–30.
182. Pshenichnyi, B.N., *Convex Analysis and Extremal Problems,* Nauka, Moscow, (1980) (in Russian).
183. Pshenichnyi, B.N., *Necessary Conditions for Extremum,* Nauka, Moscow, (1982) (in Russian).
184. Pshenichnyi B.N., Pokotilo V.G., Minimax Approach to Estimation of Parameters of Linear Regression, *Tekhnicheskaya Kibernetika,* no. 2 (1983), pp. 94–99; English transl. in *Engineering Cybernetics,* no. 2 (1983).
185. Rockafellar, R.T., *Convex Analysis,* Princeton University Press, Princeton, New Jersey, (1970).
186. Rockafellar R.T., *Conjugate Duality and Optimization,* CBMS-NSF Regional Conference Series in Applied Mathematics, No. 16, SIAM, Philadelphia, Pennsylvania, (1989) (Second printing).
187. Roytenberg, Ya.N., *Automatic Control,* Nauka, Moscow, (1978) (in Russian).
188. Sacks, J. and Ylvisaker, D., Linear Estimation for Approximately Linear Models, *Annals of Mathematical Statistics,* **6**, no. 5 (1978), pp. 1122–1137.

189. Schweppe, F.C., Recursive State Estimation: Unknown but Bounded Error and System Inputs, *IEEE Transactions on Automatic Control,* **AC-13** (1968), pp. 22–28.
190. Schweppe, F.C., *Uncertain Dynamic Systems,* Prentice Hall, Englewood Cliffs, (1973).
191. Siljak, D.D., *Decentralized Control of Complex Systems,* Academic Press, New York–London, (1991).
192. Smolyak, S.A., On the Optimal Recovering of Functions and Functionals, *Candidate of Sciences Thesis,* M.V. Lomonosov Moscow State University (1965).
193. Soloviov, V.N., Solving Algorithms for Quadratic Programming Problems and Optimal Guaranteed Estimation Problems, *Avtomatika i Telemekhanika,* no. 9 (1990), pp. 67–73; English transl. in *Automation and Remote Control,* no. 9 (1990).
194. Soloviov, V.N., Dual Algorithms for Minimax Estimation Problem in Continuous Case, *Kosmicheskie Issledovaniya,* **29**, no. 1 (1991), pp. 127–132; English transl.in *Cosmic Research,* **29**, no. 1 (1991).
195. Soloviov, V.N., Dual Algorithms for Guaranteed Estimation, *Kosmicheskie Issledovaniya,* **30**, no. 1 (1992), pp. 10–20; English transl. in *Cosmic Research,* **30**, no. 1 (1992).
196. Soloviov, V.N., Minimax Estimation and Least-Squares Method, *Stochastics and Stochastic Reports,* **42** (1993), pp. 209–223.
197. Soloviov, V.N., On the Optimality of Linear Algorithms for Guaranteed Estimation under Random Measurement Errors, *Kosmicheskie Issledovaniya,* **32**, no. 2 (1994), pp. 122–124; English transl. in *Cosmic Research,* **32**, no. 2 (1994).
198. Soloviov, V.N., Dual Algorithms for Optimal Guaranteed Estimation and Trimmed Least-Squares Method, *Kosmicheskie Issledovaniya,* **33**, no. 1 (1995), pp. 3–11; English transl. in *Cosmic Research,* **33**, no. 1 (1995).
199. Soloviov, V.N., Minimax-Bayesian Estimation on the Classes of Disturbances with Bounded Second Moments, *Uspekhi Matematicheskikh Nauk,* **50**, no. 1 (1995), pp. 171-172; English transl. in *Russian Mathematical Surveys,* **50**, no. 1 (1995).
200. Soloviov, V.N., Dual Extremal Problems and Their Application to Minimax Estimation Problems, *Uspekhi Matematicheskikh Nauk,* **52**, no. 4 (1997), pp. 49-86; English transl. in *Russian Mathematical Surveys,* **52**, no. 4 (1997).
201. Sorenson, H.W., Least-Squares Estimation: from Gauss to Kalman, *IEEE Spectrum,* **7** (1970), pp. 63–68.
202. Swerling, P., First Order Error Propagation in a Stagewise Smoothing Procedure for Satellite Observations, *Journal of Astronautical Sciences,* **6** (1959), pp. 46–52.
203. Tikhomirov V.M., Convex Analysis. In the book: *Advances in Science and Technology. Modern Problems of Mathematics. Analysis-2,* VINITI, Moscow, (1987), pp. 5–101 (in Russian).
204. Tikhomirov V.M., Approximation Theory. In the book: *Advances in Science and Technology. Modern Problems of Mathematics. Analysis-2,* VINITI, Moscow, (1987), pp. 103–260 (in Russian).
205. Tsypkin, Ya.Z., Stability of Systems with Delayed Feedback, *Avtomatika i Telemekhanika,* no. 2 (1946), pp. 107–116; English transl. in *Frequency-Response Methods in Control Systems,* Ed. A.G.J. MacFarlane, IEEE Press, (1975), pp. 45–56.
206. Tsypkin, Ya.Z., *Fundamentals of Informational Identification Theory,* Nauka, Moscow, (1984) (in Russian).
207. Vandelinde, V.D., Robust Properties of Solutions to Linear-Quadratic Estimation and Control Problems, *IEEE Transactions on Automatic Control,* **AC-22**, no. 1 (1977), pp. 138–139.
208. Vastola, K.S. and Poor, H.V., Robust Wiener-Kolmogorov Theory, *IEEE Transactions on Information Theory,* **IT-30** (1984), pp. 316–327.
209. Verdu, S. and Poor, H.V., On Minimax Robustness: a General Approach and Application, *IEEE Transactions on Information Theory,* **IT-30**, no. 2 (1984), pp. 328–340.
210. Verdu, S. and Poor, H.V., Minimax Linear Observers and Regulators for Stochastic Systems with Uncertain Second-Order Statistics, *IEEE Transactions on Automatic*

Control, **AC-29**, no. 6 (1984), pp. 499–511.

211. Veres, S. and Norton, J.P., Structure Selection for Bounded Parameter Models: Consistency Conditions and Selection Criterion, *IEEE Transactions on Automatic Control,* **36**, no. 4 (1991), pp. 474–481.
212. Volterra, V., Sur la Théorie Mathematique des Phenomenes Hereditaires, *J. Math. Pures Appl.,* **7** (1928).
213. Volterra, V., *Théorie Mathematique de la Lutte pour la Vie,* Gauthier-Villars, Paris, (1931).
214. Voyskovskii, M.I. and Merinov, I.E., Simplex-Like Algorithm for Solving the Minimax Estimation Problem, *Preprint No. 1697,* Space Research Institute of Russian Academy of Sciences (1990).
215. Witsenhausen, H.S., Sets of Possible States of Linear Systems Given Perturbed Observaitons, *IEEE Transactions on Automatic Control,* **AC-13** (1968), pp. 556–558.

Index

Other *Mathematics and Its Applications* titles of interest:

A.G. Butkovskiy: *Phase Portraits of Control Dynamical Systems*. 1991, 180 pp. ISBN 0-7923-1057-8

A.A. Zhigljavsky: *Theory of the Global Random Search*. 1991, 360 pp. ISBN 0-7923-1122-1

G. Ruhe: *Algorithmic Aspects of Flows in Networks*. 1991, 220 pp. ISBN 0-7923-1151-5

S. Walukuwiecz: *Integer Programming*. 1991, 196 pp. ISBN 0-7923-0726-7

M. Kisielewicz: *Differential Inclusions and Optimal Control*. 1991, 320 pp. ISBN 0-7923-0675-9

J. Klamka: *Controllability of Dynamical Systems*. 1991, 260 pp. ISBN 0-7923-0822-0

V.N. Fomin: *Discrete Linear Control Systems*. 1991, 302 pp. ISBN 0-7923-1248-1

L. Xiao-Xin: *Absolute Stability of Nonlinear Control Systems*. 1992, 180 pp. ISBN 0-7923-1988-5

A. Halanay and V. Rasvan: *Applications of Liapunov Methods in Stability*. 1993, 238 pp. ISBN 0-7923-2120-0

D. den Hertog: *Interior Point Approach to Linear, Quadratic and Convex Programming*. 1994, 208 pp. ISBN 0-7923-2734-9

V.S. Tanaev, V.S. Gordon and Y.M. Shafranksy: *Scheduling Theory*. Single-Stage Systems. 1994, 380 pp. ISBN 0-7923-2853-1

V.S. Tanaev, Y.N. Sotskov and V.A. Strusevich: *Scheduling Theory*. Multi-Stage Systems. 1994, 412 pp. ISBN 0-7923-2854-X

L.D. Akulenko: *Problems and Methods of Optimal Control*. 1994, 356 pp. ISBN 0-7923-2855-8

C. Udrişte: *Convex Functions and Optimization Methods on Riemannian Manifolds*. 1994, 348 pp. ISBN 0-7923-3002-1

B.S. Jensen: *The Dynamic Systems of Basic Economic Growth Models*. 1994, 355 pp. ISBN 0-7923-3091-9

V. Barbu: *Mathematical Methods in Optimization of Differential Systems*. 1994, 259 pp. ISBN 0-7923-3176-1

A. Prékopa: *Stochastic Programming*. 1995, 556 pp. ISBN 0-7923-3482-5

R. Lucchetti and J. Revalski (eds.): *Recent Developments in Well-Posed Variational Problems*. 1995, 266 pp. ISBN 0-7923-3576-7

A. Cheremensky and V. Fomin: *Operator Approach to Linear Control Systems*. 1995, 410 pp. ISBN 0-7923-3765-4

Other *Mathematics and Its Applications* titles of interest:

V.N. Afanas'ev, V.B. Kolmanovskii and V.R. Nosov: *Mathematical Theory of Control Systems Design*. 1996, 681 pp. ISBN 0-7923-3724-7

D. Pallaschke and S. Rolewicz: *Foundations of Mathematical Optimization*. Convex Analysis without Linearity. 1997, 594 pp. ISBN 0-7923-4424-3

A. Chikrii: *Conflict-Controlled Processes*. 1997, 424 pp. ISBN 0-7923-4522-3

I.M. Stancu-Minasian: *Fractional Programming*. Theory, Methods and Applications. 1997, 426 pp. ISBN 0-7923-4580-0

A.B. Piunovskiy: *Optimal Control of Random Sequences in Problems with Constraints*. 1997, 360 pp. ISBN 0-7923-4571-1

A.P. Abramov: *Connectedness and Necessary Conditions for an Extremum*. 1998, 212 pp. ISBN 0-7923-4910-5

A.I. Matasov: *Estimators for Uncertain Dynamic Systems*. 1998, 432 pp. ISBN 0-7923-5278-5

MIX
Papier aus verantwortungsvollen Quellen
Paper from responsible sources
FSC® C105338

If you have any concerns about our products,
you can contact us on
ProductSafety@springernature.com

In case Publisher is established outside the EU,
the EU authorized representative is:
Springer Nature Customer Service Center GmbH
Europaplatz 3, 69115 Heidelberg, Germany

Printed by Libri Plureos GmbH
in Hamburg, Germany